TEXTILE MECHANICS AND CALCULATIONS

TEXTILE MECHANICS AND CALCULATIONS

J. Hayavadana

WOODHEAD PUBLISHING INDIA PVT LTD

New Delhi

Published by Woodhead Publishing India Pvt. Ltd.
Woodhead Publishing India Pvt. Ltd.,
303, Vardaan House, 7/28, Ansari Road,
Daryaganj, New Delhi - 110002, India
www.woodheadpublishingindia.com

First published 2018, Woodhead Publishing India Pvt. Ltd.
© Woodhead Publishing India Pvt. Ltd., 2018
Reprint 2020

Woodhead Publishing India Pvt. Ltd. ISBN: 978-93-85059-05-6
Woodhead Publishing India Pvt. Ltd. e-ISBN: 978-93-85059-86-5

Typeset by Allen Smalley, Chennai
Printed and bound in India by Replika Press Pvt. Ltd.

Contents

The book is specially dedicated to

My God
My Parents
My Wife Dr. M. Vanitha (MD)
&
My Son J. Arvind

Foreword

The entire textile production uses a gamut of machineries varying in different principle from Mechanical to Electronics. Every textile machine has its own mechanics as far as performance and working is concerned. There is a saying that "Weaving is an art and spinning is technology". It is true because art has many forms. It involves human skill, as opposed to nature. Therefore the art of textile production involves the use of several machines with a unique principle and designed for a specific operation. Textile mechanics is a foundation for any production process in textile industry.

The idea of giving this brief was to highlight the importance of various mechanisms of textile machineries employed in textile production. In this book the author has confined to discuss aspects of laws of motion, kinetics of shedding, picking, Beat up, and other spinning processes with simple numerical examples.

The contents of this book is so beautifully drafted that even the beginner can follow the steps to understand the concepts of Textile Mechanics. This book is a very useful guide not only to the textile students but also to the textile industry.

The rich experience of the author in the area of academic, research, and industrial work is clearly reflected in the book. The approach to the contents and the flow of information is reader friendly and industry oriented. In anticipation of many more such contributions from the author to the textile freternity, I wish him good luck.

Dr. H. V. S. Murthy
Professor and Head of the Department (Retired), Textile Manufacture, Veermata Jijabai Technological Institute (VJTI), Mumbai, India.
Chairman- Professional Awards Committee
The Textile Association of India, Central office, Mumbai, India.

Preface

I am happy to release my fourth book titled "Textile Mechanics and Calculations" to esteemed Textile fraternity and feel elated in continuing my service to the academic world. I am dedicating this book to my family. Indeed, every textile machine is not free from mechanics and mechanical and with this aim I have prepared a book with a number of examples solved and also equal number of constructions in respective titles. 'Textile Mechanics and Calculations', will be very interesting for any reader as it gives a total picture about the performances of the machines. From the point of view of any textile technological student, textile machines is always a very tough subject as it involves a very high number of derivations and numerical examples. But in reality, the subject is very simple, interesting and stimulating.

After teaching the subject for the past three decades, a challenge was always steadfast in my mind in writing a book on Textile Mechanics and Calculations with a simple approach and finally I decided to dedicate a book to my textile fraternity.

This book has made an attempt to discuss the Laws of Motion, kinetics of spinning and weaving machine with simple examples, construction of cone drums for scutcher, cams and tappets etc.

Any suggestions in improving the quality and content of the book from any corner is always welcomed and can be intimated to me so as to improve the readability and accessibility further. Lastly I feel that the book shall fulfill the requirements of reader and will satisfy his/her demands. I thank Woodhead Publishing India in bringing the first edition of Textile Mechanics and Calculations.

Lastly, I thank Mrs. G.M. Sridevi, Freelance designer for her continued patronage support in getting the text typed without any mistakes.

Prof. Dr. J. Hayavadana

Introduction to mechanics of textile machineries

1.1   Introduction

Textile production includes the use of wide range of machineries differing in principle of operation and working like non-automatic, semi-automatic, automatic and fully automatic etc., for converting the feed raw material or semi-finished material to final output. The selection machine will however depend on a number of factors like:

1. Type of material processed
2. Nature of raw material
3. Specifications of the product
4. Capacity of the machines
5. Type of the machines used and the automation required there of
6. Type of working conditions like type of floor, power required, amount of lighting, number of labour required etc., and so on.

But the final product do depend on all these factors.

Thus, it is of paramount significance to know the necessity of chemistry of the process and machines involved. In other words, one need to know the mechanics of the machinery and the book has tried to help the reader to understand about the mechanics of textile machines used in textile production.

Need for study of mechanics of textile machinery
Each machine in textile production starting from fibre production, yarn formation, fabric formation, wet processing, garment production, etc., will involve a gamut of pieces of power transmission units like belt drives, gears, chain drives, etc. The product produced will certainly depend on the functioning of the machine and in turn on the functioning of various machinery parts.

Scope of mechanics of textile machinery
Following are the aspects that need to be understood by a reader to acquire the knowledge about the mechanics of textile production.

1. Mechanics of simple linear and circular motion.
2. Power transmission pieces like gears, belts, ropes and chain.

3. Balancing of revolving and reciprocating masses.
4. Role of accessories like loose and fast pulley, jockey pulley, idler pulley and grooved pulley.
5. Role of Sun and Planet gears in spinning and weaving machines.
6. Role of stepped pulleys and their method of designing.
7. Role of breaks and clutches in sizing and warping along with other textile machines.
8. Feed regulation motion in conventional scutcher.
9. Role of displacement, velocity and acceleration diagrams in understanding the working of machine parts.
10. Kinetics of winding, warping, sizing and looming operations.
11. Kinetics of shedding, picking, Beat-up, Take-up and Let-off motions.
12. Cams and tappets and their role.

1.2   Equations of motion

1.3   Introduction

Textile machines are of different types with respect to material receiving, working, transporting, cleaning, clearing, winding, depending on the type of machine, type of action, type of material etc. During this transition, the material or member of the machine is observed to follow some of the principles of laws of motion. This motion of the member or material or a conveyer may be linear or circular. In this context it is very essential to know the movement of the material or member or other auxiliary part of the machine in understanding the process and this is possible when we analyse the process in depth. Fundamental concepts of motion are found to be more useful in controlling the process or process parameter or product quality or characteristic. Thus the main thrust of this section is to explore the possible ways of using the fundamental physics concept of motion.

Type of motions normally found are:

1. Linear motion
2. Circular motion
3. Non-circular motion
4. Simple harmonic motion
5. Cycloidal motion, etc.

1.3.1   Linear motion

The following are the equations governed by the linear movement:

$$v = u + at$$
$$s = ut + \tfrac{1}{2} at^2$$
$$v^2 = u^2 - 2as$$

Following are the few examples to illustrate the use of the above equations:

Fibre production

1. Movement of monomers to polymerisation.
2. Motion of spinning dope through spinning duct (vertical or horizontal).
3. Motion of the staple fibres after cutting and crimping.
4. Baling of the staple fibres after spin finish application.
5. Motion of the filaments on to the package in continuous filament production.

Cotton mixing

1. Motion of the fibres in bale blender or plucker.
2. Motion of the fibres in mixing equipment (up or down or horizontal).
3. Motion of bales of cotton from storage to mixing area.

Cotton through blow room

1. Cotton movement on horizontal feed lattice, inclined spiked lattice, bottom lattice.
2. Fibre movement through bye-pass.
3. Fibre flow through ERM cleaner.
4. Movement of belt on Scutcher cone drums.
5. Fibre flow through feed roller and cages.
6. Chute feed of fibres to card.

Cotton through carding

1. Cotton movement from blow room to licker-in via feed roller (movement till feed roller only will be linear).
2. Movement of falt in carding.
3. Web movement from doffer to table calendar and table calendar to coiler calendar roller.

Material through draw frame, comber, simplex, ring frame and TFO

1. Passage of card slivers in draw frame.
2. Passage of slivers in sliver lap machine and motion of the lap in ribbon lap machine.
3. Passage of lap through the nipper in comber.
4. Motion of the combed sliver through table calendar and final drafting in comber.
5. Passage of drawn sliver through simplex.
6. Movement of horse head drive or bobbin rail in simplex.
7. Movement of belt on cone drums in simplex.
8. Passage of roving through ring frame.
9. Passage of twisted yarn through Two – for One Twister.

Yarn through weaving preparatory and loom

1. Passage of yarn through winding machine.
2. Passage of yarn through warping machine.
3. Passage of yarn through sizing machine.
4. Passage of yarn through drawing, denting and (old fully closed drop conventional) drop wire.
5. Passage of yarn through shuttle and shuttles loom.

Linear movement in loom

1. Movement of shuttle in loom.
2. Movement of picker and picking stick.
3. Movement of vertical or horizontal shaft in cone under or over picking motion.
4. Movement of weft fork cam in weft stop motion.
5. Movement of drop wires in warp stop motion.
6. Movement of oscillating back rest in warp easing motion.
7. Movement of healds by tappet shedding motion.
8. Movement of Dobby crank, T-lever, knives & jacks in Dobby.
9. Movement of hooks, needles, harness, lingoues in Jacquard machine.
10. Movement of rapier, projectile, jet of water and air, multiple projectiles in multiphase weaving, rigid and flexible rapier.

Passage of fabric through textile wet processing (open width)

1. Passage of fabric through singing machine.
2. Passage of fabric through scouring.
3. Passage of fabric through desizing machine.
4. Passage of fabric through mercerising machine.
5. Passage of fabric through bleaching machine.
6. Passage of fabric through Jigger dyeing machine.
7. Fabric movement in long jet dyeing machine.
8. Passage of fabric through rotary screen printing machine.
9. Passage of fabric through thermo-fixation machine.
10. Passage of fabric through stenter machine.
11. Passage of fabric through fabric folding machine.

Passage of fabric in apparel production

1. Passage of fabric in fabric spreader machine.
2. Movement of marker planner and straight cutting machine.
3. Movement of fabric parts in sewing machine.
4. Movement of stitched garment through fusing and pressing machine.
5. Movement of Jeans/Denim trouser through hot air size setting machine.
6. Movement of sand over Jeans/Denim in sand blasting machine.
7. Movement of garment through printing machine.

Linear movement of material in other machines

1. Movement of webs from carding machine towards web laying machine.
2. Movement of webs in cross lapper.
3. Movement of pre-laid web in needle punching machine.
4. Movement of polymer web from spinnerets in direct polymer spin process.
5. Movement of pre-laid web through spraying machine in chemical bonding process.
6. Movement of fibres in Rando-Webber and DOA web preparation machines.

7. Movement of fibres in Malivlies, Malimo, Malipol, Voltex, Multiknit stitch bonding system.
8. Passage of fibre drift through paper making machine.
9. Movement of Heddles in vertical shedding of carpet production.

1.3.2   Circular motion

Examples from spinning or yarn formation field

1. Motion of cotton through mono cylinder, ERM cleaner (near grid bar section), beaters in blow room, winding of lap through calendar roller, unwinding of lap in carding.
2. Motion of the fibres around the cylinder, lincker-in and Doffer in carding.
3. Motion of the sliver in nearly circular style in coiler of carding and draw frame.
4. Motion of fibres in lap formation in comber preparatory machines.
5. Motion of the fibres through cylinder of comber.
6. Motion of the fibres through condenser card in long staple spinning.
7. Movement of tape on tin roller in ring frame.
8. Movement of traveller in ring frame around the ring.

Examples from weaving field

1. Movement of yarn on to package through winding drums in drum winding machine.
2. Movement of wet sized warp yarn onto drying cylinders of sizing machine.
3. Movement of sized warp on to weavers beam.
4. Movement of yarn on double flanged bobbin and unwinding of yarn from hanks in swift of precision winders.
5. Winding of warp in beam and sectional warping machines.
6. Unwinding of warp from wrapper beams in creel zone of sizing machine.
7. Winding of sized warp yarn on to weavers beam.
8. Circular motion of size paste in pressure cooker.
9. Movement of warp from weavers beam in loom and winding of fabric on to cloth roller.
10. Movement of weft carriers in circular weaving machine.

Examples from wet processing field

1. Preparation of Jumbo rolls of grey cloth from small cloth rollers in singing.
2. Movement of mercerised fabric onto the beam.
3. Movement of dyed fabric in Jigger dyeing machine.
4. Movement of finished fabric on Jumbo for printing machine.

Numerical examples on equations of motion

(*Note*: Reader is directed to convert the units into SI for other than SI units)
1. At the time of entering the shed, the shuttle is 11 m/s, the shuttle is being started at 3.65 m/s^2. If it travels a distance of 1.5 m through the shed, what time will it take to pass through and what will be its velocity when leaving?

Solution :

$$t = ?; \quad v = ?; \quad u = 11 \text{ m/s}; \quad a = 3.65 \text{ m/s}; \quad s = 1.5 \text{ m}$$
$$v^2 = u^2 - 2as$$
$$= 11^2 - 2\,(1.5)(3.65)$$
$$= 110.05$$
$$v = 10.49 \text{ m/s}$$
$$v = u - at$$
$$at = 11 - 10.49$$
$$t = 0.51/3.65$$
$$t = 0.1397 \text{ s}.$$

2. The yarn guide of a winding frame makes 48 double traverses/min, each traverse is being 25 cm. Calculate the velocity of the yarn guide.

Solution :

Number of double traverse/min = 48

Time for one double traverse = 1/48 min

$$t = \frac{\frac{1}{48}}{60} \text{ or } \frac{60}{48} s$$

$$s = 25 \times 2 = 50 \text{ cm}.$$

$$v = \frac{D}{t} = \frac{50}{\frac{60}{48}} = 40 \quad cm/s.$$

3. When entering the shed, a shuttle has a speed of 16 m/s. If the retardation during its passage through the shed is 0.9 m/s² and the time for the passage is 0.08 s, determine the speed of the shuttle as it leaves the shed.

Solution :

$$u = 16 \text{ m/s} \qquad a = 0.9 \text{ m/s}^2 \qquad t = 0.08 \text{ s}$$
$$v = u - at$$
$$v = 16 - 0.9 \times 0.08 = 15.928 \text{ m/s.}$$

4. A yarn of linear density 50 tex is wound onto an empty package continuously for 20 min. At the end of this time, the mass of yarn on the package is found to be 47.1 g. Calculate the winding speed if the delivery rollers used in winding are running at 600 rev/min and also find their diameter.

Solution :

Yarn linear density = 50 tex, i.e., 50 g of yarn have a length of 1000 m, and thus 1 g of yarn occupies 20 m and 47.1 g of yarn occupy 942 m, i.e., 942 m of yarn are wound in 20 min. Hence winding speed = 942 m/20 min = 0.785 m/s and surface speed of delivery roller must also be 0.785 m/s.
Now,

$$\text{Speed of rotation} = 600 \text{ rev/min}$$
$$= 10 \text{ rev/s,}$$

i.e., in one revolution, yarn delivered = 0.785 m.
Hence:

Roller circumference = 7.85 cm

And

$$\text{Roller diameter} = \frac{7.85}{3.14}$$
$$= 2.5 \text{ cm.}$$

5. A body is moving at 60 ft/s in a tunnel and it moves 18 ft before it is thrown into the tunnel. During its passage it is subjected to retardation which may be considered as uniform at 36 ft/s². Find the time taken to reach the other end of the tunnel and its velocity when leaving.

Solution :

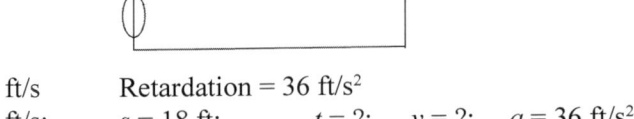

$$u = 60 \text{ ft/s} \qquad \text{Retardation} = 36 \text{ ft/s}^2$$
$$u = 60 \text{ ft/s;} \qquad s = 18 \text{ ft;} \qquad t = ?; \qquad v = ?; \qquad a = 36 \text{ ft/s}^2$$

$$v^2 = u^2 - 2as$$
$$= (60)^2 - 2\ (36)\ (18) = 2304$$
$$v = 48 \text{ ft/s}$$
$$v = u - at$$
$$48 = 60 - 36\ (t)$$
$$= 36t = 12$$
$$t = 0.33 \text{ s.}$$

6. A shuttle is moving at 40 ft/s when it enters a warp shed and it moves 5 ft before it is through the shed. During its passage it is subjected to a retardation of 30 ft/s². Find the time taken to reverse the shed and velocity when it is leaving.

Solution :

$$u = 40 \text{ ft/s}; \qquad s = 5 \text{ ft}; \qquad \text{set} = 30 \text{ ft/s}^2; \qquad t = ?; \qquad v = ?$$
$$v^2 = u^2 - 2as \qquad v = u - at$$
$$= 40^2 - 2(30)(5) \qquad 36.055 = 40 - 30t$$
$$= 1600 - 300 \qquad \therefore 30t = 40 - 36.055$$
$$= 1300 \qquad\qquad t = 0.13 \text{ s.}$$
$$\therefore v = 36.055 \text{ ft/s.}$$

7. A loom shuttle is required to pass through the warp shed, a distance of 4 ft in 1/12th of a second, what is the average speed during the passage, if it is subjected to a retardation of 32 ft/s². During the passage what are its speeds at the beginning and at the end?

Solution :

$$s = 4t; \qquad\qquad t = 1/12 \text{ lb/s}; \qquad\qquad u = 32 \text{ ft/s}^2$$
$$u = ?; \quad v = ? \quad \text{average speed} = \frac{u + v}{2} \quad \text{or average speed} = s/t$$
$$s = ut - \tfrac{1}{2}\,at^2$$
$$4 = u\ (1/12) - \tfrac{1}{2}\ (32)\ (1/12)^2$$
$$4 + \tfrac{1}{2}\ (32)\ (1/12)^2 = u\ /\ 12$$
$$u = 49.55 \text{ ft/s}^2$$
$$v = u - at$$
$$= 46.85 - 32(1/12)$$
$$= 46.6 \text{ ft/s.}$$

8a. A heavy shuttle loom enters the warp shed at a speed of 36 ft/s and is retarded at a rate of 12 ft/s². What time will it take to pass through a shed of about 5 ft. Find also the final speed at which it leaves the shed.

Solution :

$u = 36$ ft/s; $a = 12$ ft/s²; $t = ?$; $s = 5t$; $v = ?$

$v^2 = u^2 - 2as$

$= (36)^2 - 2 (12)(5)$

$= 1296 - 120 = 1176$

$= 34.29$ ft/s

$v = u - at$

$at = u - v$

$12t = 36 - 34.29$

$t = 0.142$ s.

average speed $= (u + v) / 2$

$$= \frac{34.29 + 36}{2} = 35.145 \text{ft/s}$$

or average speed $= s / t$

$= 5 / 0.1425$

$= 35.087$ ft/s.

8b. In auto pirn-winder the fully automatic pirn replenishment chute is 30 cm long. Compute the drop time of each empty pirn. (*Note:* The pirn is held by holder before being dropped into the chute.)

Solution :

$u = 0$, $t = ?$ $d = 30$ cm, (0.3 m) $a = 9.81$ m/s²

$d = ut + \frac{1}{2} at^2$

$0.3 = 0 + \frac{1}{2} \times 9.81 \times t^2$

$t = 0.25$ s.

Referring to the above problem, workout the speed of the new pirn holder which receives the pirn.

Solution :

$v^2 = u^2 + 2as = 0$

$= 0 + 2 \times 9.81 \times 0.3$

$v = 2.43$ m/s.

9. A carding machine is started from rest to full speed of 180 rpm with a uniform acceleration of 36 rad/s². Find the time taken to reach the full speed and the angle moved through by the driving shaft during the acceleration.

Solution :

$$a = 36 \text{ rad/s}^2; \qquad n = 180 \text{ rpm}; \qquad u = 0; \qquad t = ?$$

$n = 180$ rpm

$n = 180/60 = 3$ rps

$\omega = 3 (2\pi)$ rad

$v = u + at$	$s = ut + \frac{1}{2} at^2$
$6\pi = 0 + 36 (t)$	$= 0 + \frac{1}{2} (36) (0.524)^2$
$6\pi/36 = t$	$= 4.95$ rad
$\therefore t = \pi/6$	$= 4.95 \times 57.3$
$t = 0.524$ s.	$= 284.2°.$

10a. A beam warping machine with a normal warping speed of 480 yards/min is stopped with uniform retardation in 2 s. What length of yarn will be run on the beam during the stoppage? If the beam diameter is 1 ft, what will be the angular retardation of the beam in rad/s².

Solution :

$u = 480$ yards/min; $\qquad t = 2$ s; $\qquad a = ? \qquad s = ?$

$= 480 \times 3$ ft/min; $\qquad v = 0$

$= 480 \times 3$ ft/s;

$u = 24$ ft/s

$v = u - at$

$0 = 24 - a (2)$

$\therefore a = 12 \text{ ft/s}^2$

Angular retardation = Linear retardation / Radius of beam

$$= 12/0.5 = 24 \text{ rad/s}^2$$

$v^2 = u^2 - 2as$

$0 = 24^2 - 2 (12)s$

$24s = 24^2$

$s = 24$ ft.

10b. A beam warping machine with a warping speed of 8 m/s is brought to rest exactly 2 s after the drive is stopped. What is the retardation, if it is assumed to be uniform?

Solution :

We have

Initial speed, $u = 8$ m/s; Final speed, $v = 0$ m/s,; and Time, $t = 2$s.

Substituting these values in

$$v = u + ft$$
$$0 = 8 + f2$$
$$f = -4 \text{ m/s}^2.$$

11. A beam warping machine is operated at 435 m/min and is stopped with a uniform retardation in 4 s. What length of yarn has been wound on the wrappers beam before it actually stops? If the beam diameter is 30 cm, what is the acceleration in rad/s².

Solution :

$u = 435$ m/min

$u = 7.25$ m/s

$v = u - at$

$0 = 7.25 - a(4)$

$a = 7.25/4 = 1.8125$ m/s².

Angular retardation = Linear retardation / Radius of beam = 1.8125/0.15

$= 12.05 \text{ rad/s}^2$

$v^2 = u^2 - 2as$

$0 = (7.25)^2 - 2(1.8125)s$

$3.62s = (7.25)^2$

$s = 14.5$ m.

12. A shuttle has a mass of 0.5 kg and is given a speed of 15 m/s in 0.05 s. How much work is done in accelerating it? If the work done is 200 times per minute, what is the rate of power dissipation?

Solution :

$m = 0.5$ kg; $t = 0.05$ s; $u = 0$ m/s; $v = 15$ m/s;

speed = 200 tones in one minute.

$F = m \times a = 0.5 \times 300$

$F = 150$ kg/s²;

$v = u + at$

$15 = 0 + a(0.05)$

$a = 300$ m/s.

Work done = $F \times$ displacement (s)

Now $v^2 = u^2 + 2(a)$

$(15)^2 = 0 + 2(300)s$

$s = 0.375$ m

Work done $= F \times s = 150 \times 0.375$

Work done $= 56.25$ J

    This is work done per one time.

    $\therefore$ Work done $= 56.25 \times 200$

    $= 11{,}250$ J

Or power $= 11{,}250/60 =$ work done$/60$

    $= 187.5$ watts.

13. The swell exerts a force of 65 N on the back wall of shuttle. If the friction between the contacting surfaces is 0.25, find the force exerted by the picker on the shuttle in overcoming the friction.

Solution :

We know that normally when a shuttle enters the box, firstly the swell exerts a force on the shuttle to spoil the momentum of it and equally the shuttle front wall will exert an equal force on the box front plate. Therefore taking this fact and in the same way the friction between the swell and shuttle and shuttle and box front we can write

    $65 \, (0.25 + 0.25) = 32.5$ N

14. The mass of the shuttle is 0.5 kg and it is uniformly accelerated from rest to a speed of 12.5 m/s over a distance of 0.2 m. Find the force.

Solution :

We know that $v^2 = u^2 + 2as$ and it is mentioned that initial velocity is zero therefore,

$v^2 = 2as$ and we also know that $F = ma$, then we have,

$$F = \frac{mv^2}{2s}$$

where v is the final velocity in m/s, a is the uniform acceleration in m/s^2
F is the force in N, m is the mass of the shuttle in kg, and s is the distance over which acceleration occurs in m.

$$F = \frac{0.5 \times 12.5^2}{2 \times 0.2} = 195 \, N.$$

15. In a loom the shuttle speed is 13.3 m/s and shuttle weights 510 g with length as 165 cm. Find the accelerating force acting on it.

Solution :

$$F = \frac{0.51 \times 13.3^2}{2 \times 0.165} = 273.5\,N.$$

16. In a loom the mass of the shuttle is 320 g with the distance moved by the picker is 20 cm. The shuttle speed is 12.2 m/s and the picking shaft reversing spring has stiffness equals to 3.65 kN/m. What force is acting on the shuttle?

Solution :

$$F = \frac{0.32 \times 12.2^2}{2 \times 0.2} = 119\,N.$$

17a. The mass of the shuttle is 510 g when it carries a full pirn and it weighs 480 g when the pirn is nearly empty. The shuttle to have an impact speed of not less than 4.5 m/s at any time while the loom is functioning correctly, when it strikes the swell is 13.75 m/s and is uniformly retarded over a distance of 20 cm. Work out the necessary impact velocity of the picker when the pirn is full and empty. What is the pressure that the swell must exert on the shuttle?

Solution :

Now, we know that $v^2 - u^2 = 2as$,

The impact velocity is least when the pirn is nearly empty, i.e., when:
$$13.75^2 - 4.5^2 = 2\,a \times 0.20, \text{ and thus } a = 422 \text{ m/s}^2$$

As $F = ma$, the retardation is inversely proportional to the mass being retarded, so that, when the pirn is full, we have,

$$a = 422 \times \frac{0.48}{0.51} = 397\,\text{m/s}^2$$

$$F = 0.51 \times 397 = 202.5 \text{ N},$$

Pressure exerted by swell on shuttle $= \dfrac{202.5}{2 \times 0.25} = 405\,N.$

The initial velocity is
$$13.75^2 - u^2 = 2 \times 397 \times 0.20.$$
$$u = 5.5 \text{ m/s}.$$

Thus the impact velocity on the picker is 22% greater when the pirn is full than when it is empty.

17b. Fibres are given a draft of 8 between two pairs of rollers, each roller having a diameter of 3 cm, in a roller-drafting system. A short fibre moves between the two with uniform acceleration and takes exactly 1 s to move from the first nip to the second. If the delivery rollers are rotating at 120 revolutions/min, calculate the ratch of the system.

Solution :

We have,

Circumference of each roller $= \pi d$
$$= 3.14 \times 3 \text{ cm}$$
$$= 9.42 \text{ cm, or } 0.0942 \text{ m,}$$

and

angular speed of delivery roller $= 120$ rev /min
$$= 2 \text{ rev/s,}$$

i.e., surface speed of fibres after drafting $= 0.1884$ m/s.

Thus, surface speed of fibres before drafting $= 0.02355$ m/s, and we have
$$u = 0.02355 \text{ m/s,}$$
$$v = 0.1884 \text{ m/s,}$$

and
$$t = 1 \text{ s.}$$

Since,
$$v = u + ft,$$
$$0.1884 = 0.02355 + f$$

from which,
$$f = 0.1648 \text{ m /s}^2.$$

We also have,
$$s = ut + \tfrac{1}{2} ft^2$$
$$= 0.02355 + \tfrac{1}{2}(0.1648)$$
$$= 0.1059 \text{ m,}$$

i.e.
$$\text{ratch} = 10.6 \text{ cm.}$$

18. A loom has reed space of 115 cm and the average shuttle speed is 13.75 m/s. When crank turns through 135° for the passage of the shuttle. The effective length of the shuttle is 0.30 m, find the loom speed.

Solution :

Speed = distance travelled/time taken

$$p = \frac{13.75 \times 135}{6(1.15+0.30)} = 213 \text{ picks /min.}$$

19a. A loom is running at 216 picks/min and the shuttle moves 0.075 m between contacting the swell and displacing it fully with an average velocity of 12.25 m/s. Find the time taken by the shuttle.

Solution :

$$\frac{216 \times 360 \times 0.075}{60 \times 12.25} = 7.9° \text{ of crank shaft rotation.}$$

19b. A Jacquard loom produces a fabric with 20 picks/cm at a rate of 2.4 m/h. If the pattern repeat in the fabric occurs at 36 cm intervals and the width of a card is 8 cm, calculate the length of the pattern chain and the average speed of its motion.

Solution :

We have,

Number of picks/cm = 20,

and thus;

Number of picks in 36 cm = 20 × 36
= 720.

i.e.,   there are 720 cards in the pattern,

We also have,

Card width = 8 cm.

So that,

Length of pattern chain = 720 × 8 cm
= 57.6 m.

Fabric is produced at 2.4 m/h = 240 cm/h, and thus,

$$\text{Time to produce one repeat of 36 cm} = \frac{36 \times 60}{240} \text{ min.}$$

$$= 9 \text{ min.}$$

Hence

$$\text{Average speed of motion of pattern chain} = \frac{57.6}{9}$$

$$= 6.4 \text{ m/min.}$$

20. The vertical distance from the bottom end of the central hook in the Jacquard to the comber board is 150 cm, and the machine is 180 cm in width, find the length of the harness cord controlling this end operated by same hook. The Jacquard is timed such that the shed depth is 7.5 cm. Workout the percentage loss in the lift of the end by the hook in the sides.

Solution :

$$x = \sqrt{150^2 + 90^2} = 174.9\,\text{cm.}$$

$$y = \sqrt{157.5^2 + 90^2} = 181.4\,\text{cm.}$$

Thus, the lift at the side would be $181.4 - 174.9 = 6.5$ cm or a loss of $13\dfrac{1}{3}\%$ in shed depth.

21. A shuttle has a mass of 0.5 kg with weft package loaded. Find the force required to accelerate the shuttle at speed of 13 m/s assume that the picking shaft reversing spring is attached to vertical shaft at a point two fifths of the distance from its fulcrum and that the spring contracts 50 mm while the shuttle is being accelerated. Workout necessary statistics.

Solution :

The stroke of the picker will therefore be $5 / 2 \times 50 = 125$ mm neglecting the mass of picking elements like picking stick, picker, and other related parts we can write

$$v^2 = 2as.$$

Where v is the final shuttle velocity in m/s, a is the average acceleration in m/s^2 and s is the distance (meters) over which the shuttle accelerates

$$a = \frac{v^2}{2s} = \frac{13 \times 13 \times 1000}{2 \times 125} = 676 \text{ m/s}^2$$

Let P be the average force acting on the picker and F is the average tension in the spring, then,

$$P = ma = 0.5 \times 676 = 338 \text{ N.}$$

And $\quad F = 338 \times \dfrac{5}{2} = 845\,N$

We also have, $\quad s = \dfrac{1}{2}at^2$

Where t in seconds is the accelerating time in seconds

$$t^2 = \frac{2s}{a} = \frac{2 \times 125}{1000 \times 676}$$

$$t = 0.019 \text{ s.}$$

22. The swell of a loom reduces the speed of the shuttle from 15 to 10 m/s. While the shuttle moves 10 cm. If a shuttle has a mass of 0.2 kg, and the retardation is uniform, find the time taken and the value of the checking force applied by the swell.

Solution :

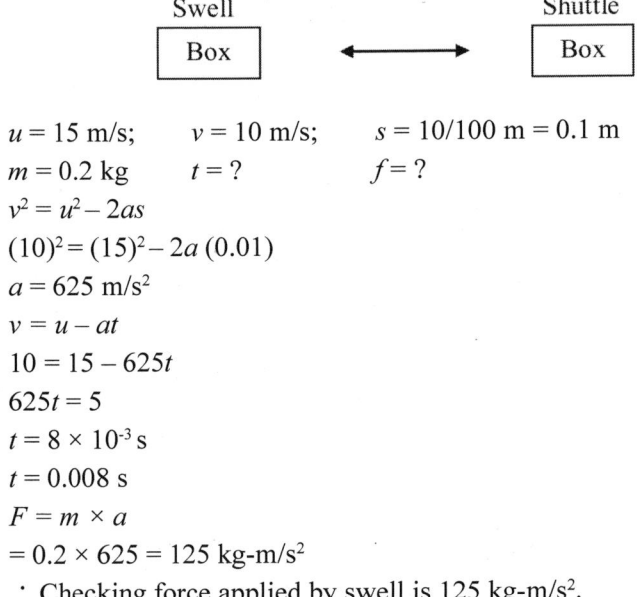

$u = 15$ m/s; $v = 10$ m/s; $s = 10/100$ m $= 0.1$ m

$m = 0.2$ kg $t = ?$ $f = ?$

$v^2 = u^2 - 2as$

$(10)^2 = (15)^2 - 2a\,(0.01)$

$a = 625$ m/s^2

$v = u - at$

$10 = 15 - 625t$

$625t = 5$

$t = 8 \times 10^{-3}$ s

$t = 0.008$ s

$F = m \times a$

$= 0.2 \times 625 = 125$ kg-m/s^2

$\therefore$ Checking force applied by swell is 125 kg-m/s^2.

23. A loom shuttle offers an average resistance of 135 N (Newton) during picking over a district of 20 cm. How much work is done/min in picking the shuttle if the loom speed is 180 pick per minute (ppm).

Solution :

Work done $= F \times$ distance

$= 135 \times 0.2$

$= 27$ J

Work done per 180 picks $= 27 \times 180 = 4860$ J

24. Cloth is run forward by the take up motion of a loom against a tension of 1400 N. If there are 24 picks/cm in the cloth and the loom speed is 180 picks/min, what is the power consumption of take up motion?

Solution :

Distance moved $=$ cloth taken up $= 180/24 = 7.5$ cm $= 0.075$ m.

Work done $= F \times$ distance

$= 1400 \times 0.095$

$$= 105 \text{ J}$$
Power $= 105/60 = $ work done $/60$
$$= 1.75 \text{ watts.}$$

25. The resistance offered to stretching by a viscose thread was found to increase uniformly with a stretch to a load of 1 N. When the stretch was 5 mm, calculate the work done in stretching.

Solution :

Uniformly increase in force $= 1$ N
$\therefore$ Average force $= (0 + 1)/2 = 0.5$ N
Work done $=$ force $\times$ distance moved
$$= 0.5 \times 5 \times 10^{-3}$$
$$= 2.5 \times 10^{-3} \text{J}$$
$$= 0.0025 \text{ J.}$$

26. The driving belt from a motor to a line shaft is removed from the motor pulley. With the machine belt on the loose pulley, it is found that a force of 90 N is needed to turn the shaft and the machine belt. Assuming that the resistance remains constant, find the power needed to run the shaft at 200 rpm if the driving pulley in the two shafts is l m in diameter.

Solution :

$F = 90$ N   distance moved $= \pi$ DN
$F = m \times a$
Distance moved $= \pi DN = $ s. s. of pulley
$$= \pi \times 1 \times 200 = 628.32 \text{ m}$$
Work done $= F \times$ distance moved $= 90 \times 628.32$
Work done $= 56548.668$ J
Power $= 942.48$ watts.

27. A shuttle of mass 0.22 kg is given a speed of 16 m/s in 0.03 s. What is the gain in momentum and what average force has acted on it?

Solution :

Momentum $=$ mass $\times$ velocity
$$= 0.22 \times 16 = 3.52 \text{ kg-m/s}$$
$F = m \times a$ $v = u + at$
$F = 0.22 \times 533.33$ $16 = 0 + a \,(0.03)$
$= 117.33$ NJ $a = 533.33$ m/s^2

28. A shuttle of a heavy loom weighs 3.2 pounds and is moving at 40 ft/s when it has passed through the valve. It must be stopped before the cram shaft running at 150 rpm has turned 60°. What is the least average force required to slope the shuttle?

Solution :

$$m = 3.21 \text{ lbs} \qquad\qquad t = ?$$
$$u = 40 \text{ ft/s} \qquad\qquad a = ?$$
$$v = 0 \qquad\qquad\qquad F = ?$$
$$N = 150 \text{ rpm}$$

Number of degree moved by crank shaft/min $= 150 \times 360 = 54{,}000°$

$$T == 1/15 \text{ s}$$

$$v = u - at \qquad\qquad \frac{60 \times 60}{150 \times 360} \qquad F = m \times a$$

$$0 = 40 - a\,(1/15) \qquad\qquad = 3.2 \times 600$$
$$a = 600 \text{ ft /s}^2 \qquad\qquad = 1920 \text{ lb.}$$

29. A shuttle of a loom weighing 0.75 lbs is given a velocity of 50 ft/s in the time the loom crank shaft revolving at 200 rpm turns 50. What is the average force acting on the shuttle during that time to change its velocity?

Solution :

$$m = 0.75 \text{ lbs}; \qquad F = ?; \qquad v = 50 \text{ ft/s}$$
$$u = 0 \quad N = 200 \text{ rpm}$$
$$t = \frac{60 \times 50}{200 \times 360} = \frac{1}{24} \text{ s}$$
$$v = u + at$$
$$50 = 0 + a\,(1/24)$$
$$a = 1200 \text{ ft/s}^2$$
$$F = m \times a$$
$$= 0.75 \times 1200$$
$$F = 900 \text{ lbs.}$$

30. A loom sley weighs 96 lbs concentrated on the loom sword pen. At the beat of position it is stationary. After ¼ rev of the crank shaft, the sley word pin has a speed is equal to that of the crank pin which is revolving in a 5″ diameter circle at 180 rpm. What average force has acted on the sley during this time to change the velocity?

Solution :

$u = 0$; $N = 180$ rpm; $m = 96$ lbs; $D = 5''$;   angular velocity $= 2 \pi N$
$= 2 \pi \times 180$

$= 1131$ rad/min

$= 1131/60$ rad/s

Velocity $=$ angular velocity $\times$ radius of pulley

$= 18.85 \times 2.5$

$= 47.12'' / s = 3.93$ ft/s

$v = u + at$

$3.93 = 0 + a \, (1/12)$

$a = 47.12$ ft/s^2

$F = m \times a$ weight $/32 =$ mass

$= 96 \times 47.12 = 4523.89/32$ lbs   mass $= 96/32$

$F = 141.37$ lbs.

31. A loom shuttle offers a resistance during picking of 30 lbs over a distance of 8''. How much work is done per min in picking the shuttle if the loom speed is 180 rpm?

Solution :

Work done per pick $= F \times$ distance moved

$= 30 \times (8/12)$

$= 20$ ft lbs/min

Work done per min $= 20 \times 180 = 3 \times 600$ ft lbs/min.

Work done in hp (Horse Power) $= 3,600/33,000 = 0.1091$ hp

32. A cloth is drawn forward by the take up motion of a loom against a cloth tension of 300 lbs. If there are 60 picks/inch in the cloth and the loom speed is 180 picks/min. What is the work done per minute in taking up the cloth?

Solution :

Work done $= F \times$ distance

$= 300 \times (180/60)$

$= 900''$

$= 900 / 12 = 75$ ft lbs.

Work done in hp $= 75/33,000$

$= 2.27 \times 10^{-3}$ hp.

33. The inertia resistance of the shuttle and the picker of an under pick loom increases uniformly with displacement from 0 to 54 lbs during a shuttle displacement of 8″. How much work is done in overcoming the resistance? In addition to the inertia displacement. The loom speed is 165 rpm. What horse power is extended in the shuttle?

Solution :

Average force $= (0 + 54) / 2 = 27$ lbs

Displacement $= 8″ = 8/12$ ft $= 2/3$ ft

Work done $= 27 \times 2/3$ ft lbs

Work done $= 18$ ft lbs

Frictional resistance $= 3$ lbs

Displacement $= 8/12 = 2/3$ ft

Work done $= F \times$ distance

$= 3 \times (2/3) = 2$ ft lbs $= 2 \times 165$

Work done $= 330$ ft lbs

Horse Power $= 330 / 33,000 = 0.01.$

34. A rapier loom having a reed space of 180 cm and running at 225 picks/min will not have excessive velocities if the rapier can be made to enter the shed at 60° and leave at 300°. These figures are typical for a range of looms, and find the average velocity.

Solution :

$$\text{average } v = \frac{\dfrac{180}{100}}{\dfrac{60}{225} \times \left(\dfrac{300 - 60}{360}\right)} = \frac{180}{100} \times \frac{225}{60} \times \frac{360}{240} = 10.125 \, \text{m/s}$$

The maximum velocity will thus be $10.125 \times \pi / 2 = 15.91$ m/s.

35. A shuttle of mass 0.5 kg, moving at a velocity of 10 m/s. strikes a picker of mass 0.15 kg connected by means of a check strap of mass 0.4 kg to the identical picker at the other end of the sley. Assuming shuttle and pickers to move together after the impact, and ignoring the effect of swell friction or check-strap extensibility, find the velocity with which shuttle and picker strike the box end. How much energy is lost in the impact of shuttle and picker?

Solution :

Momentum before impact $= 0.5 \times 10.$

Total mass after impact $= 0.5 + 0.4 + 2 \times 0.15$ kg $= 1.2$ kg

and
$$\text{Momentum after impact} = 1.2\,v$$
Where v is the common velocity of shuttle and picker.
By the principle of conservation of momentum
$$5 = 1.2\,v$$
i.e., $\quad v = 4.17$ m/s
Thus velocity at which box end is struck is 4.17 m/s.
Kinetic energy before impact is given by
$$K_1 = \tfrac{1}{2} \times 0.5 \times 10^2 = 25 \text{ J}$$

Kinetic energy after impact is given by
$$K_2 = \tfrac{1}{2} \times 1.2 \times (4.17)^2 = 10.43 \text{ J}$$

i.e., loss of kinetic energy on impact is 14.57 J.

36. In an experimental set up to study the impulse forces, a shuttle is dropped from a height of 5 m and allowed to strike a block of material similar to that used for making pickers. If the shuttle rebounds 15 cm. what is the coefficient of restitution?

Solution :

For initial fall,
$$u = 0,$$
$$s = 5 \text{ m, and}$$
$$f = 9.81 \text{ m/s}^2$$
thus: $\quad v^2 = u^2 + 2fs$
$$= 0 + 2 \times 9.81 \times 5$$
i.e., $\quad v = 9.90$ m/s

after rebound,
$$s = 15 \text{ cm} = 0.15 \text{ m}$$
$$f = -9.81 \text{ m/s}^2,$$
$$v = 0.$$

Substituting in
$$v^2 = u^2 + 2fs$$
gives, $\quad 0 = u^2 - 2 \times 9.81 \times 0.15,$
i.e., $\quad u = 1.72$ m/s
but

$$e = \frac{\text{Velocity after rebound}}{\text{Velocity before rebound}}$$
$$= \frac{1.72}{9.90} = 0.174$$

Hence coefficient of restitution = 0.17.

37. The weighting lever for a loom let-off is 50 cm in length, pivoted at one end and supported by a chain placed 10 cm from that end. If the mass of the lever is 2 kg and masses of 8 kg and 12 kg are placed 40 cm and 45 cm, respectively from the fulcrum find the tension in the chain.

Solution :
The mass of the bar may be considered being located at its midpoint i.e., at 25 cm from the fulcrum.

There are thus four forces maintaining equilibrium in the bar, these being:

a)   The tension, T, in the chain,

b)   The force due to the lever's mass, (2×9.81) N at a distance of 0.25 m from the fulcrum,

c)   A force of (8×9.81) N at 0.40 m from the fulcrum, and

d)   A force of (12×9.81) N at 0.45 from the fulcrum.

The tension acts anti-clockwise and the other three forces act clockwise around. Taking moments about the fulcrum.

Anti-clockwise moment = $T \times 0.1$

Clockwise moment = $(2 \times 9.81 \times 0.25) + (8 \times 9.81 \times 0.45) +$
$$(12 \times 9.81 \times 0.45)$$
$$= 4.905 + 31.392 + 52.974$$
$$= 89.271$$

Then, by principle of moments we can write
$$0.1\, T = 89.271$$
i.e.,
$$T = 892.71$$
i.e., tension in chain is 892.7 N.

38. A material handling device is used to lift a fibre bale of mass 800 kg from a delivery lorry. If the two wheels in the machine have radii of 15 cm and 12 cm, respectively, and the efficiency is 44%, find how many men (each capable of exerting a force of 500 N) would have to work to unload the bale.

Solution :
    We have
$$\text{Velocity ratio (VR)} = \text{effort / load}$$
$$\text{VR} \quad = 2\,N / (N - n)$$

$$= 2 \times 15/ (15 - 12)$$
$$= 10$$

and mechanical advantage (MA) = load / effort

$$MA = \text{efficiency} \times VR$$
$$= 0.44 \times 10$$
$$= 4.4$$

then, since

Force exerted by load = 800×9.81 N
$$= 7848 \text{ N.}$$

Force exerted by effort = $7848/4.4$ N
$$= 1783.6 \text{ N.}$$

i.e., four men are needed to operate the lifting device.

39. A motor-driven handler is used to lift a fibre bale of mass 560 kg over a ramp with a height of 1 m and a slope length of 5 m. The bale moves up the ramp in 20 s and the 440-V motor draws a constant current of 2.1 A throughout this time. If the coefficient of friction, μ between bale and ramp is 0.22, calculate the efficiency of the ramp and the efficiency of the motor and winch combined.

Solution :

Angle of slope of plane $\alpha = 0.201$ rad, since sin $\alpha = 0.200$
Hence

$$\cos \alpha = 0.98$$

Since force exerted vertically down wards by bale = 560×9.81 N
$$= 5493.6 \text{ N}$$

The reaction normal to the plane is given by

$$R = 5493.6 \cos \alpha$$
$$= 5493.6 \times 0.98$$
$$= 5383.7 \text{ N}$$

and, if μ is the coefficient of friction:

$$\text{frictional force} = \mu R$$
$$= 0.22 \times 5383.7$$
$$= 1184.4 \text{ N}$$

we also have,

Component of load down slope = $5493.6 \sin \alpha$
$$= 1098.7 \text{ N.}$$

Thus,

Total force required to move load = $(1098.7 + 1184.4)$ N = 2283.1 N,
now

$$VR = \frac{1}{\sin \alpha}$$

Since this force is applied up the slope i.e.,

$$VR = 5$$

and, by definition,

$$
\begin{aligned}
MA &= \text{load/effort} \\
&= 5493.6/2283.1 \\
&= 2.41.
\end{aligned}
$$

Thus,

$$
\begin{aligned}
\text{efficiency} &= MA\ /VR \\
&= 2.41/5.
\end{aligned}
$$

i.e., efficiency of inclined plane = 48%.

Now

work done in moving bale up slope = force × distance moved

$$
\begin{aligned}
&= 2283.1 \times 5\ J \\
&= 11415.5\ J.
\end{aligned}
$$

This work is done in 20 s, i.e.,

$$\text{rate of doing work} = 11415.5\ /20\ J\ /s.$$

thus,

$$\text{power expended} = 570.8\ W.$$

but

power used by 440 W motor in drawing 2.1 A = 440 × 2.1 W

$$= 924.0\ W.$$

and thus,

$$
\begin{aligned}
\text{efficiency} &= \text{useful power / power put in} \\
&= 570.8 / 924.0.
\end{aligned}
$$

i.e., efficiency of motor and winch = 62%.

40. Calculate the average shuttle velocity from the following particulars. Width of warp in reed = 1.2 m, Effective length of shuttle = 30 cm, Loom speed = 240 rpm

Degree of crank shaft rotation available for the passage of shuttle = 140°

Solution :

From the given data

$$R = 1.2\ m, \qquad L = 0.3\ m, \qquad P = 240\ \text{picks/min}, \qquad \theta = 140°$$

$$\text{Shuttle velocity} = v \frac{6\,P\,(R+L)}{\theta} = \frac{6 \times 240 \times (1.2 + 0.3)}{140} = 15.4\ m\,/\,s$$

So, average shuttle velocity is 15.4 m/s.

41. Calculate the loom speed and weft insertion rate of a loom if the average velocity of shuttle 50 km/h, degree of crank shaft rotation available for the passage of shuttle across the shed 135°, shuttle length 40 cm, width of warp in reed 2 m.

Solution :

From the given data

$$R = 2 \text{ m}, \qquad L = 0.4 \text{ m}, \qquad \theta = 140°, \qquad v = \frac{50 \times 1000}{3600} = 13.9 \, m/s.$$

Loom speed $P = \dfrac{v\,\theta}{6(R+L)} = \dfrac{13.9 \times 135}{6 \times (2+0.4)} = 130$ picks/min.

Weft insertion rate (WIR) $= P(R+L) = 130 \times (2+0.4) = 312$ m/min.
So, loom speed and weft insertion rate are 130 picks/min and 312 m/min.

42. Calculate the power required from the following particulars.

Width of warp in reed = 1.5 m, Effective length of shuttle = 32 cm, Loom speed = 224 rpm, Shuttle mass = 500 g, Shuttle enters in the shed at 110°, Shuttle leaves the shed at 245°.

Solution :

From the given data
$$R = 1.5 \text{ m}, \qquad L = 0.32 \text{ m}, \qquad m = 0.5 \text{ kg}, \qquad P = 224 \text{ picks/min}$$
$$\theta = (245° - 110°) = 135°$$
Power required for picking =

$$\frac{3\,mP^3(R+L)^2}{\theta^2} \times 10^{-4} = \frac{3 \times 0.5 \times 224^3 \times (1.5+0.32)^2}{135^2} \times 10^{-4} = 0.306 \, kW$$

Power required for picking is 0.306 kW.

43. Calculate the power requirement for picking and work done per pick from the following parameters.

Shuttle mass, 0.5 kg; Reed width, 1.5 m; Shuttle length, 0.25 m; Loom speed, 200 PPM; Degree of crank shaft rotation for passage of shuttle, 120.

Solution :

$$R + L = (1.5 + 0.25) \text{ m} = 1.75 \text{ m}$$

$$P = \frac{3\,mp^3(R+L)^2}{\theta^2} \times 10^{-4} \, kW$$

$$m = 500 \text{ g} = 0.5 \text{ kg}.$$

So, $P = \dfrac{3 \times 0.5 \times 200^3 \times (1.75)^2}{120^2} \times 10^{-4} = 0.255 \, kW$

Work done /pick =

$$\frac{18\,mP^2(R+L)^2}{\theta^2} \quad J = \frac{18 \times 0.5 \times 200^2 \times (1.75)^2}{120^2} = 76.56 \, J$$

Therefore power requirement is 0.255 kW work done per pick is 76.56 J.

44. A leather picking strap, 40 cm long and with a cross-section of 30 mm, wide by 6 mm thick is subjected to a tensile load of 50 kg. How much will it stretch, if its Young's modulus is $1.75 \times 10^8 \, \text{N/m}^2$?

Solution :

We have

$$\text{Stress} = \frac{50 \times 9.81}{0.030 \times 0.006} \, \text{N/m}^2$$

$$= 2,725,000 \, \text{N/m}^2$$

Since

$$Y = \frac{\text{Stress}}{\text{Strain}}$$

$$\text{Strain} = \frac{2,725,000}{1.75 \times 10^8} \qquad = 1.56 \times 10^{-2}$$

But

$$\text{Strain} = \frac{\Delta l}{l^2}$$

Hence

Change in length $= 1.56 \times 10^{-2} \times 0.4$ m.

Thus change in length is 6.2 mm.

45. A carding engine cylinder has a mass of 560 kg and is carried by two bearings each 6.4 cm in diameter. What is the shear stress on the shaft at the bearings as a result of the cylinder mass?

Solution :

We have,

Total force exerted by cylinder $= 560 \times 9.81$ N

$= 5494$ N

Thus,

Force on each bearing $= 2747$ N

Now,

Area carrying load $= \pi r^2$

$= \pi \times (0.032)^2 \, \text{m}^2$

$= 3.22 \times 10^{-3} \, \text{m}^2$

Hence,

$$\text{Stress} = \frac{2747 \times 10^3}{3.22} \, \text{N} / \text{m}^2$$

Thus, shear stress is 8.5×10^5 N /m².

46. A spiral spring has a stiffness of 500 N/m. How much work is it capable of doing when stretched 15 cm?

Solution :

We have,

$$\text{Straining load} = 500 \times 0.15 \text{ N}$$
$$= 75 \text{ N}$$

Thus,

$$\text{Energy stored} = \frac{75}{2} \times \text{stretch}$$

$$= \frac{75}{2} \times 0.15$$

$$= 5.63 \text{ J}$$

Hence work available is 5.6 J.

47. A loom sley is brought to rest, during a fast reed stoppage, by causing it to deflect two flat springs. If the energy of the sley, which must be absorbed by the springs is 50 J and the spring's deflection must not exceed 5 mm, what stiffness of spring is needed?

Solution :

Work done on springs = ½ Pl where P is the total load and l is the distance moved.

This must be equal to the energy absorbed 50 J i.e.,

$$P = 2 \times 50 \times \frac{1}{0.005}$$

Or $P = 20$ kN.

Maximum load applied to each spring is 10 kN.

Thus, stiffness (which is force required to cause a deflection of 1 m) is given by

$$\text{Stiffness} = 10 \times 10^3 \times \frac{1}{0.005} N / m.$$

Hence stiffness required = 2 MN/m.

48. On a shuttle-less loom weaving a cloth having a reed width of 225 cm, the weft projectile travels a total distance of 245 cm at an average velocity of 27 m/s. If the period during which it is in motion occupies 0.4 of the loom cycle, calculate: (a) the maximum loom speed in picks per minute; and

(b) the rate of weft insertion in metres per minute if the over-all running efficiency of the loom is 90%.

Solution :

a) The total distance travelled by the weft carrier is 245 cm. Hence
The time taken at 27 m/s = 2.45 / 27 = 0.09074 s.
This time is 0.4 of the loom cycle time, so one cycle takes,

$$\frac{0.09074}{0.4} = 0.2268 \text{ s.}$$

From this, we have

The number of loom cycles / min = $\dfrac{60}{0.2268}$ = 264.6 say 265 picks / min.

b) At 265 picks/min, the rate of weft insertion is given by
reed width × picks/min × efficiency
= 2.25 × 265 × 0.90 ≅ 537 m/min.

49. If, on a water-jet loom, the consumption of water is proportional to the loom width, how many litres per hour would be consumed on a loom 165 cm wide, running at 500 picks/min, if a loom 125 cm wide consumes 0.5 ml of water per pick?

Solution :

We have

Hourly consumption in litres = $\dfrac{165}{125} \times \dfrac{0.5}{1000} \times 500 \times 60 = 19.8$ l/h

(it is assumed that the loom speed as 500 picks/min).

50. The yarn width in the reed is 100 cm and the shuttle length, with the curved ends neglected, is 30 cm. The loom speed is 200 picks/min, and 100° of crank shaft rotation are available for shuttle traverse through the shed. If the retardation is 950 cm/s², determine (a) the mean velocity, (b) the maximum velocity, and (c) the minimum velocity i.e., that before the shuttle enters the box.

Solution :

a) At 200 picks/min, the time available for the shuttle passage is:

$$\frac{60}{200} \times \frac{100}{360} = \frac{1}{12} \text{ s}$$

The distance travelled in this time is the yarn width in the reed plus the nominal shuttle length i.e., 100 + 30 = 130 cm. Hence,

$$\text{Mean velocity} = \text{distance / time} = \frac{130}{1/12} \text{ cm/s}$$
$$= 15.6 \text{ m/s.}$$

b) The equation of motion s = ut + ½ at2 is used here. The distance s is 130 cm, the time is 1/12 s, and the initial velocity, u, is the maximum velocity we require. The acceleration a in this case is 950 cm/s2, since there is a retardation. Substituting, we obtain

$$130 = \frac{u}{12} - \frac{950}{2 \times 12^2}$$
$$= \frac{24u - 950}{2 \times 144}$$

From which
$$130 \times 288 = 24\,u - 950$$
And $u = \dfrac{38,390}{24} = 1{,}600$ cm/s (very nearly).

Hence the maximum velocity is 16 m/s.

c) The minimum velocity can be calculated by using the equation of motion
$$v = u + at$$

Substituting, we obtain
$$v = 1600 - \frac{950}{12}$$
$$= 1600 - 79$$
$$= 1521 \text{ cm/s.}$$

Hence the minimum velocity is 15.2 m/s.

51. A shuttle of 500 g mass is uniformly accelerated by the picker of a cone under-pick system. The length of the picking stroke is 15 cm. The final velocity of the shuttle is 15 m/s. If AB:BC = 1:1 (Figure 1.1), then calculate the force acting on the picking strap (lug strap).

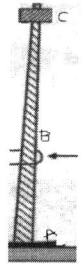

Figure 1.1   Cone under-pick mechanism

Solution :

For uniform acceleration, $v^2 = 2fs$

Where v is final velocity, f is acceleration and s is distance travelled by the shuttle.

$$f = \frac{v^2}{2s} = \frac{15^2}{2 \times 0.15}$$

$$= 750 \text{ m/s}^2$$

Force applied by the picker on shuttle $= P = mf = \frac{1}{2} \times 750 = 375 \, N$.

If F is the force acting on the lug-strap then,

$P \times AC = F \times AB$

Or, $P \times 2 = F \times 1$

So, $F = 750$ N

So, force acting on the lug-strap $= 750$ N.

52. You are given reed space of loom 45"; width of warp in reed 41"; length of shuttle 13"; weight of shuttle 10 oz; loom speed 180 picks per minute; time for the passage of the shuttle 96° of the crank shaft revolution; time for development of the pick 45° of the tappet shaft revolution; effective length of picking arm 21"; angle moved by arm and cone during acceleration 36°. Work out necessary statistics.

Solution :

The distance moved by the shuttle during its passage will be approximately

$$= (41" + 13" - 6") = 48" = s;$$

the time for this movement $= \dfrac{96}{360} \times \dfrac{60}{180} = 0.089 \, s = t$.

$\therefore$ Average shuttle speed during passage $= \dfrac{s}{t} = \dfrac{4}{0.089} = 45$ ft/s.

The retardation of the shuttle at this speed of loom will be about $f = 24$ ft/s/s, corresponding to a retarding force of

$$\frac{24}{32} W = 0.75W = 0.75 \times 10 = 7.5 \text{ ounces.}$$

The maximum shuttle velocity

$$V = \frac{s + \frac{1}{2} f t^2}{t} = \frac{4 + \left(\frac{1}{2} \times 24 \times 0.089^2\right)}{0.089} = \frac{4.095}{0.089} = 46 \, ft/s$$

Or the maximum velocity may be found as follows:

Kinetic energy in shuttle at average speed = $\dfrac{Wv^2}{2g}$, where v =average velocity = 45 ft/s.

$$= \frac{10}{16} \times \frac{45^2}{2 \times 32.2} = 19.8 \text{ ft lb.}$$

This may be assumed to be the energy in the shuttle when halfway through its passage, though that is not quite correct. The energy lost by the shuttle in moving from its maximum speed position to its average speed position is equal to the work done in overcoming the resistance to the shuttle through that distance, viz.

Energy lost = $\dfrac{7.5}{16} \times 2$ ft = 0.94 foot – lb approximately.

∴ Kinetic energy in shuttle at maximum speed

$$= 19.8 + 0.94 = 20.74 \text{ ft lbs} = \frac{WV^2}{2g}$$

and maximum velocity—

$$v = \sqrt{\frac{2g\,KE}{W}} = \sqrt{\frac{2 \times 32.2 \times 20.74 \times 16}{10}} = 46 \quad \text{ft/s approximately.}$$

The time for developing the pick is 45° of the tappet shaft or 90° of the crank shaft, the time in seconds.

$$t_1 = \frac{90}{360} \times \frac{60}{180} = \frac{1}{12}\text{s} = 0.0835\,\text{s}$$

The time occupied by the arm in moving 36° is, as found above, 0.0835 s, and in that time the arm end moves.

$$\frac{36}{360} \times 2\pi \times 21'' = 13.2''.$$

Hence the average arm end velocity must be $\dfrac{13.2}{12 \times 0.0835} = 13.2$ ft.

53. The details of an under pick motion is as follows:

Picking disc dia 8″, diameter of follower 3″, length of picking stick 28″ inches, dDistance between the toggle point and picking stick 4.75″, distance between shoe and fulcrum 16″, total length of side lever 29″. Reed space is 36″; cloth width 32″; shuttle length 13″; speed 150 picks per minute; time for passage of shuttle 90° of crank shaft revolution; time for development of pick 20° of crank shaft or 10° of bottom shaft revolution; movement of picker during acceleration 4″. Centre of picking bowl vertically below centre of bottom shaft when maximum shuttle speed is reached.

Solution :

Distance moved by shuttle during passage
$$s = 32 + 13 - 6 = 39'' = 3.25 \text{ ft.}$$

Time for shuttle passage $= \dfrac{90}{360} \times \dfrac{60}{150} = 0.1 \text{ s}$

Average shuttle velocity during passage $= \dfrac{3.25}{0.1} = 32.5 \text{ ft/s}$

Taking the retardation of the shuttle $f = 20$ ft/s/s, then,
Maximum shuttle velocity

$$V = \frac{S + \dfrac{1}{2}ft^2}{t} = \frac{3.25 + \left(\dfrac{1}{2} \times 20 \times 0.1^2\right)}{0.1} = 33.5 \text{ ft/s}$$

Time for development of pick $= \dfrac{20}{360} \times \dfrac{60}{150} = \dfrac{1}{45} = 0.0222 \text{ s.}$

Average velocity of picker during acceleration
$$= \frac{4''}{12 \times 0.0222} = 15 \text{ ft/s.}$$

54. Calculate the work done in lifting a bale of fibres of mass 120 kg from the ground to a delivery bay 3 m above ground level. If the bale accidentally slips from the bay, what power is expended in its fall?

Solution :

The mass of 120 kg is moved against gravity i.e.,
$$\text{Force exerted} = 120 \text{ kg} \times 9.81 \text{ m/s}^2$$
$$= 1177.2 \text{ N.}$$

We also have
$$\text{Work done} = \text{force} \times \text{distance moved}$$
$$= 1177.2 \times 3 \text{ J,}$$
i.e.,work done $= 3531.6$ J.
In the bale's fall from the bay, the relevant details are:
$$s = 3 \text{ m,}$$
$$u = 0,$$
and $f = 9.81$ m/s²

Substituting these values in $s = ut + \frac{1}{2}ft^2$
gives,

$$3 = 0 + \frac{1}{2}(9.81)\, t^2,$$

From which $t = 0.782\ s,$

The potential energy stored during lifting, 3531.6 J, is expended during the fall. Hence

$$\text{Rate of doing work} = 3531.6\ J\, /\, 0.782\ s,$$

and thus,   Power expended = 4516 W.

55. The two springs for reversing the motion of a heald shaft each have to be stretched 15 cm to put them in position with the heald shaft down. If the stiffness of each spring is 1.5 N/cm, find the work done in putting the springs in position.

Solution :

A spring offers a resistance that varies uniformly with the stretch, and the term "'stiffness"' denotes the force required to stretch it big unit length. We have,

Total force required to stretch each spring = 1.5 × 15 N

$$= 22.5\ N,$$

i.e.,   Total force = 2 × 22.5 N

$$= 45\ N.$$

We then have

Work done = force × distance moved

$$= 45 \times 15\ N\ cm.$$
$$= 675\ N\ cm.$$
$$= 6.75\ N\ cm.$$

Hence work done in stretching springs   = 6.75 J.

56. A leather brake block is held against the rim of a 40 cm, diameter pulley by a 15 kg load. If the coefficient of kinetic friction is 0.45, calculate the power used in braking if the pulley speed is maintained at 150 rev/min.

Solution :

We have

Mass of braking load = 15 kg.

and thus,

force applied by braking load= 15 × 9.81 N,

i.e.,

normal reaction = 147.2 N.

The frictional force at the rim is given by,

$$F = \mu N$$
$$= 0.45 \times 147.2 \text{ N},$$

i.e.,

frictional resistance = 66.2 N.

In one revolution, a point on the circumference moves 40π cm, or 1.26

i.e.,

work done during each revolution = 1.26×66.2 J
$$= 83.4 \text{ J}.$$

Pulley rotates at 150 rev/min of 2.5 rev/s thus,

Work done in 1 s = 83.4×2.5 J,

i.e., Power used in braking = 208.5 W.

57. A leather brake block presses against the rim of a pulley with a normal force of 120 N. If $\mu = 0.5$, what power is dissipated in stopping the machine in 10 s if the diameter of the pulley is 28 cm and it completes 18 revolutions in the stopping process?

Solution :

We have

Normal force = 120 N.

Hence

$$F = \mu N$$
$$= (0.5)(120),$$

i.e.,

Frictional resistance of brake = 60 N,

since

Pulley diameter = 28 cm,
Circumference = 0.28π m
$$= 0.88 \text{ m}.$$

Hence

Work done in 18 revolutions = 18 (0.88) (60) J
$$= 950.4 \text{ J}.$$

This work is done in 10 s, and thus power dissipated = 95.0 W.

58. A bale of fibre of mass 120 kg rests on a ramp of length 4 m, one end of which is gradually raised by means of a hydraulic system. The bale slips when the raised end reaches a point 1.2 m higher than the lower end. Calculate the frictional force, the normal reaction, and the coefficient of friction for the contact between the bale and the ramp when slipping occurs.

Solution :

The conditions at the angle of repose and using geometry

$$F = W \sin \theta,$$
$$N = W \cos \theta,$$

and

$$\mu = \tan \theta.$$

The force exerted by the weight of the bale is given by

$$W = 120 \times 9.81 \text{ N}$$
$$= 1177.2 \text{ N}.$$

We also have,

$$\sin \theta \quad = 1.2/4$$
$$= 0.300$$

i.e.,

$$\theta = 0.305 \text{ rad},$$

from which

$$\cos \theta = 0.954$$
$$\tan \theta = 0.314.$$

Hence

$$F = (1177.2)(0.300) \text{ N},$$
$$N = (1177.2)(0.954) \text{ N},$$

and

$$\mu = \tan \theta = 0.314.$$

Thus, the frictional force is 353.2 N, the normal reaction 1123.0 N, and the coefficient of friction 0.31.

59. A loom is running at 240 PPM. The shuttle could enter and remain within the shed when the sley displacement is at least 50% of the maximum. If the loom width is 1.75 m and shuttle length is 25 cm then determine the minimum average velocity of the shuttle (m/s) during its flight. (assume $e = 1/3$).

Solution :

$$\text{Sley displacement} = s = r\left(1 - \cos\theta + \frac{r \sin^2 \theta}{2l}\right)$$

Where "r" is the crank radius and "l" is the connecting arm length

Here, $s = 2r \times \dfrac{1}{2} = r$ and $e = \dfrac{r}{l} = \dfrac{1}{3}$

So, $r = r\left(1 - \cos\theta + \dfrac{\sin^2 \theta}{6}\right)$

Or, $\sin^2 \theta - 6 \cos \theta = 0$

Or, $\cos^2 \theta + 6 \cos \theta - 1 = 0$

Hence, $\cos \theta = \dfrac{-6 \pm \sqrt{36 + 4}}{2} = 0.162$

Or, $\theta = 80.66°$ is approximately equal to 80° and 280°

So, the maximum degree of crank shaft movement for shuttle flight
$= 280° - 80° = 200°$

$$\text{Maximum available time for shuttle flight} = \frac{60}{240} \times \frac{200}{360} s = 0.139 \, s$$

Distance travelled by shuttle $= 2$ m.

$$\text{Therefore the minimum average velocity of shuttle} = \frac{2}{0.139} = 14.4 \, m/s.$$

Circular movement

1. Cloth emerging from a senter is being collected on a take-up roller and at a time when the diameter of the cloth on the roller is 60 cm, the period of rotation of the roller is found to be exactly 10 s. Calculate the speed at which cloth is moving through the stenter.

Solution :

We have

Diameter of cloth on roller $= 60$ cm.

i.e.,

radius of rotation $= 0.30$ m.

and

period of rotation $= 10$ s.

The surface speed of rotation, v, is given by

$$v = \frac{2\pi r}{T}$$

$$= \frac{2\pi \times 0.30}{10}$$

$$= 0.1884.$$

Hence cloth speed through tenter is 0.19 m/s.

The other approach for the problem is as follows

Period of rotation $= 10$ s.

i.e.,

$$\text{angular velocity} = \frac{2\pi}{t}$$

$$= 0.628 \text{ rad/s.}$$

Thus, surface speed,

$$v = rw$$
$$= 0.628 \times 0.30$$
$$= 0.1884$$

As before, cloth speed through tenter is 0.19 m/s.

2. The traveller on a ring spinning frame is rotating at 8400 rev/min. Calculate its angular speed in rad/s. If the speed is reduced to 2000 rad/s, and the ring rail has a diameter of 10 cm, calculate the surface speed of the traveller.

Solution :

Speed of rotation of traveller = 8400 rev/min.

$$= 700 \text{ rev/s.}$$

1 revolution of $360° = 2 \pi$ rad,

Speed of rotation $= 700 \times 2 \pi$ rad/s

$$= 4396 \text{ rad/s.}$$

Since it is given that diameter = 10 cm,

Then, radius = 5 cm, $v = r w$

And distance along arc for 1 radian of rotation = 5 cm, i.e., distance travelled per second = 5 × 2000 cm.

$$= 10,000 \text{ cm} \qquad = 100 \text{ m}$$

Surface speed $= 100$ m/s.

3. A carding engine is started from rest to its full speed of 180 rev/min with a uniform acceleration of 36 rad/s². Find the time taken to reach full speed and the angle through which the driving shaft moves during this time.

Solution :

The final angular velocity is:

$$\omega = 180 \text{ rev/min}$$
$$= 3 \text{ rev/s}$$
$$= 18.85 \text{ rad /s}$$

Thus,

$$\omega = 0,$$
$$\omega = 18.85 \text{ rad/s}$$

and $\alpha = 36$ rad/s

Substituting these values in,

$$\omega = \omega_0 + \alpha t$$

gives,

$$18.85 = 0 + 36t$$

i.e.,

$$t = 18.85 / 36$$

Hence time to reach running speed = 0.524 s

Furthermore,

$$\omega_0 = 0$$
$$\alpha = 36 \text{ rad /s}^2$$

and

$$t = 0.524 \text{ s.}$$

$$\theta = \omega_0 t + \tfrac{1}{2}\, a r^2$$
$$= 0 + \tfrac{1}{2}.36(0.524)^2$$

Hence angle moved by shaft = 4.94 rad.

4. A machine is driven by a belt from an overhead line shaft rotating at 140 rev/min. The line-shaft and machine pulleys have diameters of 40 and 30 cm, respectively. Calculate the angular speed of the machine pulley in rad/s.

Solution :

We have,

Line-shaft speed = 140 rev/min

$$= \frac{140}{60} \times 2\pi \text{ rad/s}$$

$$= 14.66 \text{ rad/s.}$$

Diameter of line-shaft pulley = 40 cm = 0.40 m;
Diameter of machine pulley = 30 cm = 0.30 m.

Since

$$d_1 n_1 = d_2 n_2,$$
$$0.40 \times 14.66 = 0.30 \times n_2$$

from which;

$$n_2 = \frac{0.4 \times 14.66}{0.3} = 19.55.$$

5. A loom is driven by a V-rope drive from a motor running at 960 rev/min. The effective pulley diameters when the rope is new are 8 cm and 40 cm for motor and loom, respectively. After wear, however, the rope sinks in by 1.5 mm, and slip (which was not present initially) becomes 5%. Calculate the percentage change in running speed of the loom-drive shaft.

Solution :

We have

Angular speed of motor pulley = 960 rev/min
$$= 16 \text{ rev/s}$$
$$= 100.5 \text{ rad/s.}$$

Hence

Initial driven speed of loom shaft = $100.5 \times \dfrac{0.08}{0.40} = 20.1$ rad/s.

After wear, the rope sinks in by 1.5 mm at each side, i.e., the pulley diameter decreases by 3 mm in each case.

Thus, new diameters are 0.077 m and 0.397 m, respectively.

In addition only 95% of the power is transmitted sine 5% is lost in slip.

Thus,

$$\text{Loom speed} = 100.5 \times \frac{0.077}{0.397} \times \frac{95}{100}$$

$$= 18.5 \text{ rad/s}.$$

Hence,

Loss in running speed = 1.6 rad/s.

i.e.,

$$\text{Change in speed} = \frac{1.6}{20.1} \times 100\%.$$

Hence running speed decreases by 8.0%.

6. A circular-knitting machine, of diameter 40 cm, performs a revolution in 1.2 s. What is the acceleration, if uniform angular velocity may be assumed?

Solution :

We have,

$$\text{Circumference} = \pi\, d$$

$$= 3.14 \times 0.40 \text{ m}$$

$$= 1.256 \text{ m}.$$

i.e.,

$$\text{Speed of rotation} = 1.256\, /1.2 \text{ m/s}$$

$$= 1.05 \text{ m/s}.$$

Now

$$\text{Radius} = 20 \text{ cm}$$

$$= 0.2 \text{ m}.$$

i.e.,

$$\text{Acceleration} = \frac{v^2}{r} = \frac{(1.05)^2}{0.2} = \text{m/s}^2$$

$$= 5.51 \text{ m/s}^2.$$

Hence acceleration = 5.51 m/s² towards the centre.

7. A card cylinder is disconnected from its drive and from the remaining parts of the machine so that it can rotate freely in its bearings for a test of bearing friction. In this test, a rope is coiled round a pulley on the cylinder shaft, passed round a frictionless pulley system suspended from the roof, and connected to a 20-kg weight. This weight is allowed to fall from a height of 3 m to the ground; it takes 9 s in doing so and rotates the cylinder through 1.5 revolutions. At the instant that it strikes the ground, the rope disengages itself from the pulley and the cylinder comes to rest after further 7 revolutions. Calculate the kinetic energy in the weight when it reaches the floor, in the cylinder at the same instant, and in the cylinder running at its normal speed of 3 revolutions on the assumption that frictional resistance at the bearings remains unchanged throughout the various arrangements.

Solution :

Consider the motion of the falling weight,

$$u = 0$$
$$s = 3 \text{ m,}$$

and

$$t = 9 \text{ s.}$$

Substituting these values in

$$s = ut + \tfrac{1}{2} ft^2$$

gives

$$3 = 0 + \tfrac{1}{2} f(81),$$

from which

$$f = \frac{6}{81} \text{ m/s}^2.$$

Substituting this value with the above values in

$$v = u + ft$$
$$v = 0 + \frac{6}{81} \times 9$$
$$= 0.67 \text{ m/s.}$$

Now the kinetic energy of weight on reaching floor is

$$\text{KE} = \tfrac{1}{2} mv^2,$$
$$= \tfrac{1}{2} (20) (0.67^2) \text{ J}$$
$$= 4.49 \text{ J.}$$

Now, the energy transferred from the weight to the cylinder is the original potential energy in the weight minus the kinetic energy remaining in the weight at the end of its fall. The initial potential energy is given by:

$$P = mgh$$
$$= 20 \text{ kg} \times 9.81 \text{ m/s}^2 \times 3 \text{ m}$$
$$= 588.6 \text{ J,}$$

and thus

$$\text{energy transferred to cylinder} = (588.6 - 4.49) \text{ J}$$
$$= 584.1 \text{ J.}$$

During the fall, some of this energy is absorbed by friction at the bearings, and this is given by,

average energy absorbed by bearings per revolution

$$= \frac{\text{Total energy supplied}}{\text{total revolutions of cylinder}}$$

$$= \frac{584.1}{(1.5 + 7)} \text{ J}$$

$$= 68.7 \text{ J}$$

i.e.,

Energy absorbed during all $= (68.7 \times 1.5)$ J

$$= 103.0 \text{ J}.$$

Now

Kinetic energy in cylinder as weight touches floor = energy transferred – energy absorbed

$$= (584.1 - 103.0) \text{ J}.$$

Consider the motion of the cylinder during the fall of the weight:

$$u = 0$$
$$t = 9 \text{ s},$$

and

$$s = 1.5 \text{ rev}.$$

Substituting these values in

$$s = ut + \tfrac{1}{2} ft^2,$$

gives

$$1.5 = 0 + \tfrac{1}{2} f (81),$$

from which

$$f = 1/27 \text{ rev/s}^2 \quad .$$

Substituting this value with the above values in the velocity – distance relation now gives

$$v^2 = u^2 + 2 fs$$
$$= 0 + 2 \,(1/27)\,(1.5)$$
$$= 1/9,$$

i.e., velocity of cylinder at instant of release = 0.33 rev/s.

Since kinetic energy $= \tfrac{1}{2} mv^2$, the kinetic energies at this slow speed and at the normal running speed, K_{run} and K_{test}, must be in the same ratio as the square of the velocities, i.e.,

$$\frac{K_{run}}{K_{test}} = \frac{\left(v_{run}\right)^2}{\left(v_{test}\right)^2} = \frac{3^2}{0.33^2}$$

from which;

$$K_{run} = 81 \, K_{test},$$
$$= 81 \times 481.1 \text{ J}.$$

Hence kinetic energy when cylinder is running normally is 38969.1 J, or approximately 39 kJ.

8. A carding engine cylinder has mass of 500 kg and its radius of gyration is 60 cm. What is its kinetic energy at 180 rev/min? If the moment of inertia of the cylinder is 90% of that of the entire machine, what torque and maximum power are required to accelerate the engine from rest to this operating speed in 30 s?

Solution :

We have

angular speed = 180 rev/min = 6 π rad/s.

Hence

$$KE \quad = \frac{1}{2}\,m\,\omega^2 k^2$$
$$= \frac{1}{2} \times 500 \times 36\,\pi^2 \times (0.6)^2 \, J$$

i.e., kinetic energy = 31.978 J.

Angular acceleration during starting is the change of angular velocity per unit time, and is given by

$$\alpha \;=\; \frac{6\pi}{30} \; rad/s^2.$$

Now

moment of inertia of cylinder $= mk^2$
$$= 500 \times 0.6^2$$
$$= 180 \; m^2\,kg.$$

Hence

$$\text{moment of inertia of machine} = 180 \times \frac{100}{90}$$
$$= 200 \; m^2\,kg.$$

Now

$$\text{torque} = 1\,\alpha$$
$$= 200 \times \pi\,/\,5 \; J.$$

i.e.,

$$\text{torque} = 125.7 \; J.$$

Since

Work done = torque × angular movement.

The maximum value of work done occurs when the final angular speed of 6 π rad /s has just been reached and thus:

Maximum work = 125.7 × 6 π J
$$= 2369.4 \; J.$$

Power dissipated is equal to work done per second.

Thus maximum power = 2.4 kW.

9. In a detailed practical study of the conventional carding machine a student records the data as follows:

Final speed of cylinder = 160 rpm
Weight of the cylinder = 1120 pounds
Radii of gyration = 2″
Effective driving force = 200 pounds
Diameter of ring pulley = 15″

Horse power required to overcome the bearing friction resistance to keep the cylinder 160 rpm = 0.8 hp

Find out following detail

1. Energy (spent in the cylinder).
2. Total work done/revolution during starting.
3. Work spent in overcoming friction /revolution.
4. Power available for acceleration /revolution.
5. Number of revolution required to stop.
6. Average speed.
7. Time taken to start.
8. Number of revolutions required to bring the machine to stop.
9. Time taken to stop.

Solution :

1. Heavy starting load is due to inertia of heavy cylinder.

Kinetic energy stored at 160 rpm =

$$= \frac{1}{2}mv^2$$

$$= \frac{1}{2} \times \frac{1120}{32} \times v^2$$

$$= \frac{1}{2} \times \frac{1120 \times (WK)}{32}$$

$$= \frac{1}{2} \times \frac{1120 \times (WK)}{32}$$

$$= \frac{1}{2} \times \frac{1120 \times (6.28)}{32}$$

$$= 19{,}500 \text{ ft lb}$$
$$= 20{,}000 \text{ ft lb.}$$

$V^2 = WK^2$

$W = 2\pi n$

$$= (WK)^2 = \left[2 \times \pi \times \frac{160}{60}\right]^2 \times 2^2$$

$V^2 = (WK)^2$

This energy is stored up during starting and is used up in overcoming bearing.

2. Total work done / revolution

(Effective driving = 200 pounds) $= \pi D \times 200$

$$= 200 \times \pi \times 15/12 = 785 \text{ ft lb.}$$

3. To overcome frictional resistance card required 0.8 HP to keep it running steadily at 160 rpm.

$\therefore$ work spend in overcoming friction/revolution $= \dfrac{0.8hp \times 33,000}{160}$

$$= 165 \text{ ft lb.}$$

4. Work spend in accelerating /revolution $= 785 - 165 = 620$ ft /rev

5. Number of revolutions required to start =

$$= \frac{\text{Kinetic energy stored}}{\text{Work spend in accelerating /revolution}}$$

$$= \frac{20,000}{620}$$

$$= 32 \text{ rev.}$$

6. Average speed =

$$= \frac{0 + 160}{2}$$

$$= 80 \text{ rpm.}$$

7. Time taken to start =

$$= \frac{\text{Number of revolutions taken to start}}{\text{Average speed}}$$

$$= \frac{32}{80} \times 60 = 24 \ s$$

8. Number of revolutions required to stop the machine =

$$= \frac{\text{Kinetic energy stirred}}{\text{Work spind in friction/revolution}}$$

$$= \frac{20,000}{165}$$

$$= 120 \text{ rev.}$$

9. Time taken to stop =

$$= \frac{\text{Number of revolutions required to stop}}{\text{Average speed}}$$

$$= \frac{120}{80}$$

$$= 1.5 \text{ m} = 90 \text{ s.}$$

10. The carding engine weighs 1200 lbs and is supported by bearings of 3″ diameter when the cylinder is disconnected from the other part of machine it required 0.25 hp running 165 rpm. What is μ coefficient of friction for the bearings?

Solution :

$$\text{hp} = \frac{\mu R \pi D N}{33,000}$$

$$0.25 = \frac{\mu(1200)3.142 \times \dfrac{3}{12} \times 165}{33,000}$$

$$\mu = 0.058 \ (1 \text{ ft} \rightarrow 12″)$$

11. The carding engine require 0.35 hp to drive it at 160 rpm overcoming the friction at the bearings which has 15,000 ft lbs kinetic energy stored in it. Calculate energy required to start the cylinder from rest to full speed in 20 rev. The belt pulley is 20″ diameter. Calculate also time required for cylinder to come to rest from the moment the power supply stops.

Solution :

Work done due to friction/revolution =

$$= \frac{0.35 \times 33,000}{160} = 72 \text{ ft lb}$$

Work done due to acceleration / revolution =

$$= \frac{\text{Kinetic energy}}{\text{Number of revolutions}}$$

$$= \frac{15,000}{20} = 750 \text{ ft lb}$$

∴ Total work done/revolution = 750 +72 = 822 ft lb.

Number of revolutions in which cylinder comes to rest =

$$= \frac{\text{Kinetic energy}}{\text{work done /revolution in friction}}$$

$$= \frac{15,000}{72} = 208 \text{ revolutions}$$

∴ Time taken to stop the cylinder =

$$\frac{\text{Number of revolutions}}{\text{average speed}} = \frac{208}{80} = 2.6 \text{ m} = 156 \text{ s}.$$

12. A carding engine requires 0.35 hp to keep it running at 165 rpm against the bearing friction. Kinetic energy stored in cylinder is 15,000 ft lb. Find approximately the effort required to start the cylinder from rest to full speed in 20 revolutions with ordinary bearings and also with ball bearings which reduces the friction by 90%.

Solution :

Given data:

Power = 0.35 hp, Kinetic energy = 15,000 lb, Speed = 165 rpm.

Number of revolutions required to start = 20

Work done in overcoming friction/revolution = 0.35 × 33,000/165 = 70 ft lb.

Work done in accelerating / revolutions = Kinetic energy stored / Number of revolutions to reach full speed.

= 15,000/20 = 750 ft lb

Total work done /revolution = 750 + 70 = 820 ft lb

With ball bearings work done in overcoming friction = 70 × 0.1 = 7 ft lb.

Total work done /revolution= 750 + 7 = 757 ft lb.

i) Kinetic energy stored in cylinder at 160 rpm = ½ mv^2

$$= \frac{1}{2} \times \frac{510}{\text{Acceleration due to gravity}} \times v^2$$

$$= \frac{1}{2} \times \frac{\pi D}{9.8} \times (WK)^2$$

Where K = radius of gyration

W = Angular velocity in radians /s

$$= \frac{1}{2} \times \frac{510}{9.8} \left(2\pi \times \frac{160}{60} \right)^2 \times 2^2$$

$$= (26.02 \times 4) \left(\pi^2 \times \frac{256}{9} \right)$$

$$= 104.08 \times \left(\pi^2 \times \frac{256}{9} \right)$$

$$= 13{,}557.89 \text{ N.}$$

ii) Total work done / rev = Effective driving force $\times \pi D$
$$= 98 \times \pi \times 15/12$$
(Assure diameter of pulse as 5″) = - 384.65 N.

iii) Work spent in overcoming friction = 0.8 × hp

$$= \frac{0.8 \times 4500}{160} = 22.5 \text{ N.}$$

Work spent in accelerating for revolution = 384.65 - 22.5
(power available)   = 362.15 N.

13. A high speed warping machine has a warping speed of 360 m/min. The beam, frictionally driven by surface contact between the yarn and the motor-driven driving roller has a mass of 300 kg a diameter of 75 cm and a radius of gyration of 25 cm when nearly full. The roller has a mass of 225 kg, a diameter of 60 cm and a radius of gyration of 30 cm. The machine requires 2 KW for normal operation.

a)   What torque is required at the driving roller for normal operation?

b)   If the machine must stop with a beam surface movement of not more than 90 cm. What retarding torque must the brake apply to the driving roller, assuming all machine inertia other than that of the beam and the roller to be insignificant?

c)   If the machine must accelerate from rest to full speed in 5 s, what maximum drive motor output power is needed, assuming uniform motion?

d)   If the coefficient friction between yarn surface and driving roller is 0.25, what minimum pressure is needed between the two to prevent slippage during starting and stopping?

Solution :

We have

Surface speed of beam and roller is equal to warping speed = 6 m/s.
So that

angular velocity of driving roller $= \dfrac{6}{0.30} = 20$ rad/s
and

angular velocity of beam $= \dfrac{6}{0.375} = 16$ rad/s.

a) Since power dissipated = work done/s
$$= \text{torque} \times \omega,$$
then torque at driving roller $= \dfrac{2000}{20} = 100$ J.

b) During uniform deceleration of beam
Initial angular velocity, $\omega_0 = 16$ rad/s
Final angular velocity, $\omega = 0$
Displacement $= 0.9$ m $= \dfrac{0.9}{0.375}$ or 2.4 rad.
Substituting in $\omega^2 = \omega_0^2 + 2\,\alpha\theta$ gives
$$0 = 16^2 + 2\alpha \times 2.4.$$
Hence
$$\alpha = \text{-}53.3 \text{ rad/s}^2.$$
Now substituting $\omega = \omega_0 + at$ gives
$$0 = 16 - 53.3\ t.$$
Thus time required to stop $= 0.30$ s.
For the driving roller
$$\omega_0 = 20 \text{ rad/s.}$$
$$\omega = 0$$
$$t = 0.30 \text{ s.}$$
Substituting in $\omega = \omega_0 + at$ gives
$$0 = 20 + \alpha \times 0.30.$$
i.e.,
$$\alpha = -67.7 \text{ rad/s}^2.$$

Now
$$\begin{aligned}
\text{MI of roller} \quad &= mk^2 \\
&= 225 \times 0.3^2 \\
&= 20.25 \text{ m}^2\,\text{kg.}
\end{aligned}$$
and
$$\begin{aligned}
\text{MI of beam} \quad &= 300 \times 0.25^2 \\
&= 18.75 \text{ m}^2 \text{ kg.}
\end{aligned}$$
Referred to the roller, the MI of the beam is

$$I_r = I_0 \dfrac{\omega_0^2}{\omega_r^2}$$

$$= 18.75 \times \dfrac{16^2}{20^2} \quad = 12 \text{ m}^2\text{kg.}$$

Thus,

Total MI referred to roller = 32.25 m² kg.

and

Torque required to retard roller = –I α

$$= - 32.25 \times (- 67.7)$$
$$= 2183.3 \text{ J.}$$

In normal running, the torque required is 100 J, which must be used in overcoming friction. This friction, during retardation, is still present, so net retarding torque required from brake is given by subtracting running torque from total torque required.

Thus braking torque needed is 2083.3 J.

c) During start-up of roller

$$\omega_0 = 0$$

$$\omega = 20 \text{ rad/s.}$$

$$t = 5 \text{ s}$$

Substituting in $\omega = \omega_0 + \alpha t$ gives:

$$20 = 0 + \alpha \times 5$$

i.e.,

$$\alpha = 4 \text{ rad/s}^2$$

Thus, torque required for acceleration = Iα

$$= 32.25 \times 4$$
$$= 129 \text{ J.}$$

In addition, the frictional resistance is present and requires a torque of 100 J to overcome it. Thus total torque required = 229 J.

Maximum power is needed at the end of the acceleration period when full speed is just reached. At this point

Power = torque × ω.

$$= 229 \times 20.$$

Thus maximum power required during start = 4.6 kW.

d) Slip between roller and beam is most likely to occur during stopping, since the deceleration then (-67.7 rad /s² at the roller) is much greater than the acceleration (4 rad/s²) during starting.

Beam retardation during stop = -53.3 rad /s².

Thus, retarding torque required by beam = -Iα

$$= -18.75 \times (- 53.3)$$
$$= 999.4 \text{ J.}$$

Frictional force required at beam surface to produce this torque is found by dividing by the radius. Thus,

Force needed = 999.4 / 0.375

$$= 2665 \text{ J.}$$

From the equation $F = \mu N$.

$$2665 = 0.25 N.$$

i.e., minimum normal pressure between yarn and roller = 10.7 kN.

14. A flyer frame requires 1.5 kW to drive it at 320 rev/min. When the belt is thrown-off, the rotational movement of the driving shaft is found to decrease by 0.1 rad in successive time intervals of 0.1 s after the first. Calculate the moment of inertia of the machine, referred to the driving shaft, and the kinetic energy which it possesses at full speed. What effective pull must the belt apply to the pulley, of diameter 30 cm and what maximum power is needed to start the machine if it must reach full speed, from rest, in the same time that it takes to stop when the belt is thrown-off?

At full speed, angular speed of driving shaft $= \dfrac{320 \times 2\pi}{60} = \text{rad/s.}$

$$= 33.5 \text{ rad/s.}$$

Its displacement in 0.1 is thus 3.35 rad.

When the belt is thrown-off, displacement in first 0.1s is 3.3 rad, so average speed is 33 rad/s.

In the second time interval, displacement is 3.2 rad and average speed is 32 rad/s.

In the third interval, the two values are 3.1 rad and 31 rad/s, respectively, and so on.

Thus angular speed is changing by 1 rad/s in each 0.1s so that angular retardation (which is uniform) is 10 rad/s².

(It should be noted incidentally, that the displacement decreases only half as much in the first time interval as in subsequent ones, despite the fact that retardation begins immediately the belt is thrown off. This results because the change in speed 33.5–33.0 rad/s differs only by 0.5 rad/s from the zero–time point to the mid-point at the first time interval, whereas successive changes are measured over an entire 0.1 s interval, as becomes clear by reference to the deceleration curve.

Torque required during normal running is T_t where

$$T_t = \frac{P}{\omega}$$

$$= \frac{1500}{33.5}$$

$$= 44.78 \text{ J.}$$

So that

$$I = \frac{T_1}{\alpha}$$

$$= \frac{44.78}{10}.$$

Thus moment of inertia of machine = 4.48 m²kg referred to driving shaft.
Now

$$\text{KE} \quad = \frac{1}{2} \, l\omega^2$$
$$= \frac{1}{2} \times 4.48 \times (33.5)^2$$

Thus kinetic energy at full speed = 2513.8 J.

If the machine is accelerated from rest to full speed in the same time as is required for it to stop, as above, then acceleration and retardation must have the same numerical value.

Thus starting acceleration is 10 rad/s².

Now

Torque required = Iα

. So that torque needed for acceleration = 4.48 × 10 = 44.8 J.

This is equal to the torque required in overcoming friction, since acceleration and deceleration are equal. Thus

Total torque required during acceleration = 2 × 44.8
$$= 89.6 \text{ J.}$$

Since

$$\text{Force in driving belt} = \frac{\text{torque}}{\text{radium}}$$

$$= \frac{89.6}{0.15} \text{ J.}$$

Then effective pull in belt = 597.3 J.

Now

Power consumed = torque × ω

And maximum power is required as full speed of 33.5 rad/s is reached. i.e.,

Maximum power = 89.6 × 33.5 W
$$= 3001.6 \text{ W.}$$

Thus maximum power needed = 3 kW.

15. A flexible rapier loom has simple harmonic motion (SHM), and the two rapier's y each move through a distance of 100 cm per pick. Calculate the maximum permissible loom speed in picks / min if the motion of the rapiers may occupy up to three fifths of a pick and if their velocity must not exceed 14 m /s.

Solution :

The equation for SHM is $v = \omega r_s$

Where v is the velocity, ω the angular velocity in rad/s, and r the radius of the circle. The velocity v is the tangential velocity of a point moving round the circle of radius r. As it is mentioned that the motion is SHM it can be mentioned that total movement is $2\,r$. Thus, in the present problem $2\,r = 100$ cm.

The velocity at any instant is given by $\omega r \sin \theta$.

The maximum velocity occurs when $\sin \theta$ is a maximum i.e., when $\sin \theta = 1$ and

$\theta = \pi/2$ rad (90°).

It is given that maximum permitted velocity, 14 m/s, or 1400 cm/s, and r is $100/2 = 50$ cm, so we can now calculate the value of ω, the angular velocity,

$1400 = \omega \times 50$.

Hence $\omega = 1400 / 50 = 28$ rad/s.

One cycle of SHM is equivalent to $2\,\pi$ rad, and the time taken for the rapiers to move to the centre of the loom, transfer the weft, and move back to their starting points will therefore be

$2\,\pi / 28$ s $= 0.2244$ s.

Since this time is equal to three-fifths of a pick, the time per pick is

$0.2244 \times 5/3 = 0.374$ s.

The maximum value of picks /min $= 60/0.374 = 160$ picks/min.

Time may then be expressed as the number of degrees past front centre (or beat-up).

16. In a loom running at 220 picks / min (i.e., 220 rev/min.), the shuttle begins to move at 100° past beat-up and comes to rest at 225° past beat-up, how long is the shuttle moving?

Solution :

In this case, the time for 360° of crank shaft rotation is 0.2728 s (calculated in example 8.10). The shuttle movement occupies 2250°-100° = 125°. The time taken is therefore,

$$0.2728 \times \frac{125}{360} = 0.0947 \text{ s}.$$

If, in this example, the distance moved by the shuttle is 130 cm, its mean velocity may be calculated.

Thus,

$$\text{Mean velocity} = \frac{\text{distance}}{\text{time}} = \frac{130}{0.0947} \text{ cm/s}$$

$$= 1090 \text{ cm/s}$$
$$= 10.9 \text{ m/min}$$
$$\cong 66 \text{ km/h}.$$

17. A loom of reed width 114 cm runs at 200 picks/min. At what speed would a wider loom, of reed width 228 cm, run?

Solution :

We have

$$\text{Loom speed of 228-cm loom} = \text{speed of 114 cm loom} \times \sqrt{\frac{114}{228}}$$

$$= 200 \times \sqrt{0.5} \quad = 141 \text{ picks/min.}$$

The weft-insertion rate of the narrower loom is

$$200 \times \text{reed width} = 200 \times 114 \text{ cm /min} = 228 \text{ m/min.}$$

For the wider loom, the weft-insertion rate is

$$141 \times 228 \text{ cm/min} = 321 \text{ m/min.}$$

The wider loom, although running at 59 picks/ min, i.e., more slowly than the other, has a 41% higher rate of weft insertion.

18. Ring frame traveller moving round a ring of 1.5″ diameter at 8000 rpm presses against the ring with a normal force of 2000 grains. How much horse power is spent in overcoming this friction if $\mu = 0.1$ and there are 400 travellers in the frame. Assume that the frictional force acts at 0.75″ from the centre of ring.

Solution :

Given 7000 grain = 1 lb

$$\text{HP} = \frac{\mu R \pi D N}{33,000} \times 400$$

$$= \frac{(0.1)\left(\dfrac{2000}{7000}\right) \times 3.142 \times 8000 \times \dfrac{1.5}{1.2} \times 400}{33,000}$$

$$= 1.09 \text{ hp.}$$

19. A ring frame traveller, moving in a circle of 5-cm diameter at 9000 rev/min, offers a resistance to movement of 0.15 N. If the frame has 240 spindles, calculate the power expended in moving the travellers.

Solution :

We have

$$\text{Distance moved by traveller in one revolution} = 5 \pi \text{ cm}$$

$$= 15.70 \text{ cm.}$$

$$\text{Distance moved per second} = \frac{15.70 \times 9000}{60} \text{ cm,}$$

$$= 23.55 \text{ m.}$$

Hence

Work done per second on each traveller $= 23.55 \times 0.15$ Nm

$$= 3.53 \text{ J.}$$

And

Total work done $= 3.53$ J on each of 240 spindles per second.

So that,

Power expended $= 847.2$ W.

20. A spindle of a ring-spinning frame is driven at 9000 rev/min by a tape that passes with a right-angled lap around the 2.5 cm diameter spindle wharve. The coefficient of friction is 0.22 and the frictional force 1.5 N. Calculate the maximum power required to drive the spindle and the tensions in the tape.

Solution :

The maximum power is needed only if the tape is on the point of slipping. Under these conditions

$$\frac{T_e}{T_i} = e^{\mu\theta}$$

$$= e^{0.11\pi}$$
$$= 1.412.$$

We also have

frictional force $= 1.5$ N,

i.e.,

$$T_e - T_i = 1.5,$$

i.e.,

$$0.412 \, T_i = 1.5,$$

and thus,

$$T_i = 3.64 \text{ N.}$$

And

$$T_e = 5.14 \text{ N.}$$

Hence the tensions in the tape are 5.1 N and 3.6 N on the exist and incoming sides, respectively.

We also have

angular velocity $= 9000$ rev/min.

$$= 150 \text{ rev/s}$$
$$= 300 \, \pi \text{ rad/s,}$$

and

diameter of wharve $= 2.5$ cm

$$= 0.025 \text{ m}$$

i.e.,

$$\text{power transmitted} = \frac{1}{2} \, (T_e - T_i) \, d\omega$$
$$= \frac{1}{2} \, (1.5)(0.025) \, (300\pi) \text{ W},$$

and thus,

$$\text{maximum power needed} = 17.7 \text{ W}.$$

2

Transmission of motion by wheel gearing

Motion or power can be transmitted by various means like belt, rope, and gear as a driving means and we find these in maximum usage for all the stages of textile production. Especially the spinning machineries are designed with full-fledged gear drive for moving various parts of a machine. Thus in this chapter the reader is given an information about the gear drives. For example consider the following gearing diagram found in non-woven machine which include chain drive, belt drive and also gear drive (Figure 2.1a–c).

Figure 2.1a   Showing the different means of driving the machine

Figure 2.1b   Woolen carding machine showing all power transmission

Figure 2.1c   Woolen condenser carding machine

2.1   What is a gear drive?

Gears are compact, positive-engagement, power transmission elements that determine the speed, torque, and direction of rotation of driven machine elements. Gear types may be grouped into five main categories – Spur, Helical, Bevel, Hypoid, and Worm. Typically, shaft orientation, efficiency, and speed determine which of these types should be used for a particular application.

Every machine is run with the help of different pieces of power transmission and gear trains are one of the tools. In each textile machine the use of specific gear wheels is in line with the requirement of the process.

Following example will help the reader to understand how one has chosen different types of gears to suit the specific end use.

Find RPM of bottom shaft, crank shaft and PPM from data if the crank wheel is 48 T counter shaft speed is 75 rpm and if four picks are to be inserted per repeat.

Solution :

Figure 2.2 showing the typical arrangement in a power-loom with crank shaft deriving drive from motor and transferring to bottom shaft with 1:2 gearing. From here the drive will be transferred to counter shaft for producing the weave which have more than 2 picks per repeat. Generally the speed ration with crank and counter shaft will be 1:4. In other words, for every one revelation of crank shaft, the counter or auxiliary shaft make one fourth of the revolution we can write it as,

Counter or auxiliary shaft speed = A / B where A is loom speed in picks per minute and B is number of picks per repeat or number of tappets 75 × 4 = 300 rpm and the speed of the bottom shaft is 150 rpm, bottom shaft gear size is 96 teeth.

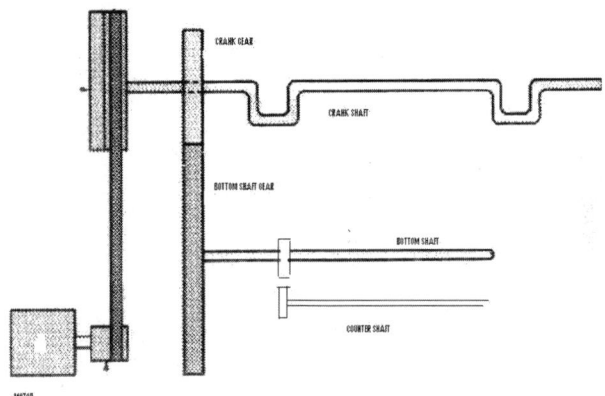

Figure 2.2   Loom gear arrangement

Gear trains are widely used in all kinds of mechanisms and machines, from can openers to aircraft carriers. Whenever a change in the speed or torque of a rotating device is needed, a gear train or one of its cousins, the belt or chain drive mechanism, will usually be used. The simplest means of transferring rotary motion from one shaft to another is a pair of rolling cylinders. They may be an external set of rolling cylinders or an internal set. Provided that sufficient friction is available at the rolling interface, this mechanism will work quite well. There will be no slip between the cylinders until the maximum available frictional force at the joint is exceeded by the demands of torque transfer.

Power of transmission

Blow room	Mixing bale opener	Monocylinder	Scutcher
Spur gear	Drive for hopper lattice, inclined lattice feed roller	Drive for other members	Drive for cages, calendar roller, shell rollers, etc., hunter & log
Worm and worm wheel	Spiked lattice movement, creeper lattice movement		Speed reduction from cone drum to feed roller
V-belt drive	Drive for stripper, u-inclined spike lattices drive for far	Drive for cylinder	Drive for lattices, stripper, lap knock-off motion, beater (2, 3 blade kirsner beater) and flat belt drive from beat-up to lap forming unit
Ratchet and pawl	-	-	Knock-off motion
Star wheel	-	-	Monitoring length of lap delivered
Cone drum	-	-	Feed regulation through control of feed roller speed
Flat belt	-	-	Driving the driven cone drum from driver
Rack and pinion	-	-	For lifting and lowering lap spindle
PIV	-	-	Feed regulation in scutcher instead of cone drums
Chain drive	-	-	Drive for cages (in some blow rooms) from feed rollers through an assembly

	Carding	Draw frame
Flat belt drive	Cylinder motion, cylinder to likerin, cross belt drive fro likerin	-
Rope drive	Drive for ancillary devices like doffer comb and flat comb etc.	-
Bevel drive	Drive from feed roller to doffer coiler region (helical bevel)	-
Spur gears	Drive for all the driving elements	Drive for drafting zone, chain drive-creel zone for self-weighed roller

Chain drive	Drive for flat comb	
Eccentric	To and fro motion for flat comb	
Spicyclic gears	Coiler motion in G2 & high production and tandom cards, crush rollers with slots – known as cross roll and India roll in modern cards	Coiler and doffing mechanism Bewel gears for top roll cleaner (old D/F)
Rubber apron	Condensing device in doffer region found on MMC cards	-
Worm and worm wheel	Drive for flats	Drive to can (old D/F)
Quadrant gear	-	Drafting system zone
V-belt drive	-	Drafting system zone from motion to drafting rollers

2.2   Classification of gear drives

All the gears are classified into

1. Spur gear.
2. Helical gear.
3. Herringbone gear.
4. Bevel gear.
5. Worm gear.
6. Rack and pinion.
7. Planetary gear.

2.3   Spur gears

Spur gears are simple, easily manufactured gears and are usually the first choice when exploring gear options. Transmitting power between parallel axes, the teeth project radially on the disc. Spur gears have straight teeth cut parallel to the rotational axis. The tooth form is based on the involute curve, and generated during gear machining processes using gear cutters with straight sides. Spur gears are the least expensive to manufacture and the most commonly used, especially for drives with parallel shafts. Spur gears are the most common type of gears. Spur gears are known as slow speed gears. If noise is not a serious design problem, spur gears can be used at any speed. Majority of the gear drive in any textile machine whether it is a spinning or weaving, it is found that spur gears are used to a greater extent (Figures 2.3–2.5).

Figure 2.3   Spur gear drive for ring frame

Figure 2.4   Showing the gear drive

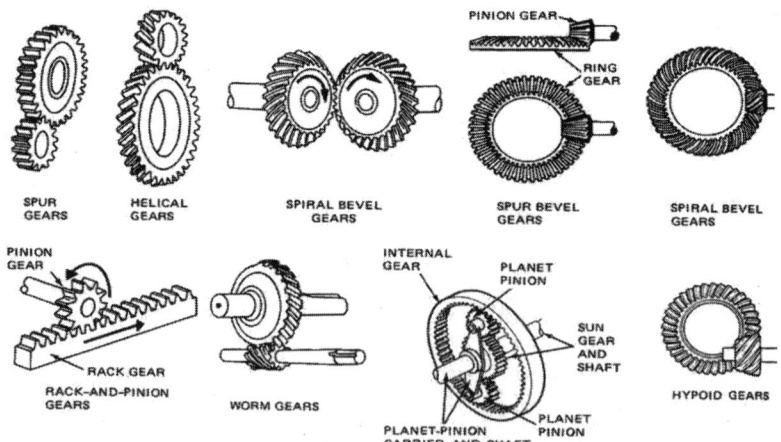

Figure 2.5   Classification of gear drives

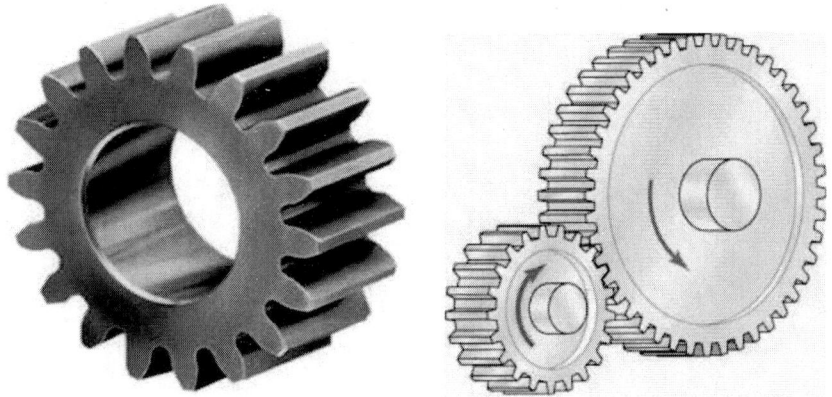

Figure 2.6 Showing the spur gear

The main classes of spur gears are internal and external. Parallel and co-planer shafts connected by spur gears. Spur gears have straight teeth and are parallel to the axis of the wheel. Spur gears are the most common type of gears. Spur gears are regularly used for speed reduction or increase, torque multiplication, resolution and accuracy enhancement for positioning systems. The teeth run parallel to the gear axis and can only transfer motion between parallel-axis gear sets. Spur gears mate only one tooth at a time, resulting in high stress on the mating teeth and noisy operation.

Internal (ring) gears – Ring gears produce an output rotation that is in the same direction as the input, Figure 2.7. As the name implies, teeth are cut on the inside surface of a cylindrical ring, inside of which are mounted a single external-tooth spur gear or set of external-tooth spur gears, typically consisting of three or four larger spur gears (planets) usually surrounding a smaller central pinion (sun). Normally, the ring gear is stationary, causing the planets to orbit the sun in the same rotational direction as that of the sun. For this reason, this class of gear is often referred to as a planetary system. The orbiting motion of the planets is transmitted to the output shaft by a planet carrier.

External-tooth gears – The most common type of spur gear (Figure 2.6), has teeth cut on the outside perimeter of mating cylindrical wheels, with the larger wheel called the gear and the smaller wheel the pinion. The simplest arrangement of spur gears is a single pair of gears called a single reduction stage, where output rotation is in a direction opposite that of the input. In other words, one is clockwise while the other is counter-clockwise.

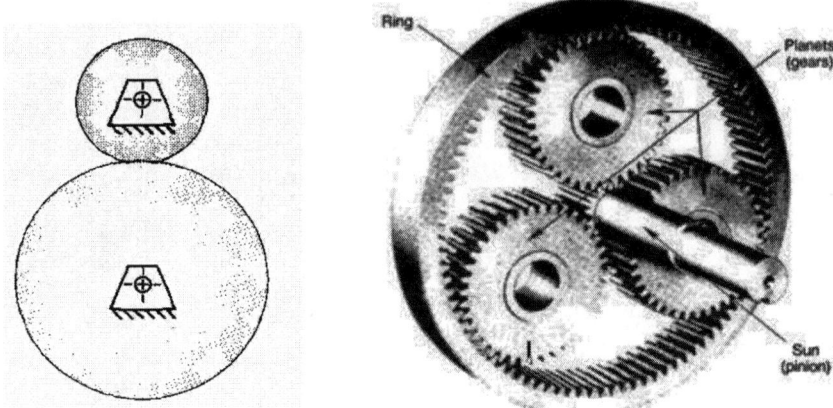

Figure 2.7   Internal and external gears

2.3.1   Spur gear materials

Gear composition is determined by application, including the gear's service, rotation speed, accuracy and more. Cast iron provides durability and ease of manufacture. Alloy steel provides superior durability and corrosion resistance. Minerals may be added to the alloy to further harden the gear. Cast steel provides easier fabrication, strong working loads and vibration resistance. Carbon steels are inexpensive and strong, but are susceptible to corrosion. Aluminium is used when low gear inertia with some resiliency is required. Brass is inexpensive, easy to mould and corrosion resistant. Copper is easily shaped, conductive and corrosion resistant. The gear's strength would increase if bronzed. Plastic is inexpensive, corrosion resistant, quiet operationally and can overcome missing teeth or misalignment. Plastic is less robust than metal and is vulnerable to temperature changes and chemical corrosion – nylon and polycarbonate plastics are common. Other material types like wood may be suitable for individual applications (Figure 2.8).

The advantages of spur gears are their simplicity in design, economy of manufacture and maintenance, and absence of end thrust. They impose only radial loads on the bearings. Spur gears are known as slow speed gears. If noise is not a serious design problem, spur gears can be used at almost any speed.

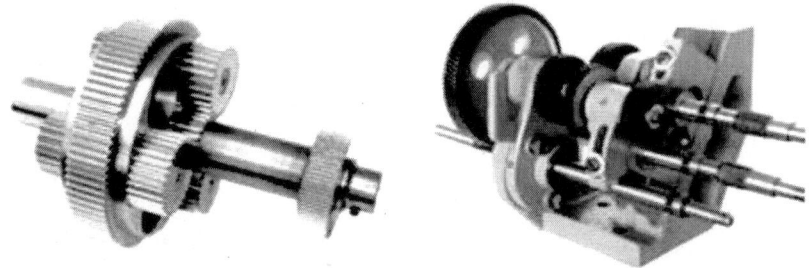

Figure 2.8 Spur gear assemble found in comber and drafting roller driving in simplex and ring frame

2.4   Helical gears

Helical gearing differs from spur in that helical teeth are cut across the gear face at an angle rather than straight. Thus, the contact line of the meshing teeth progresses across the face from the tip at one end to the root of the other, reducing the noise and vibration characteristic of spur gears (Figure 2.9).

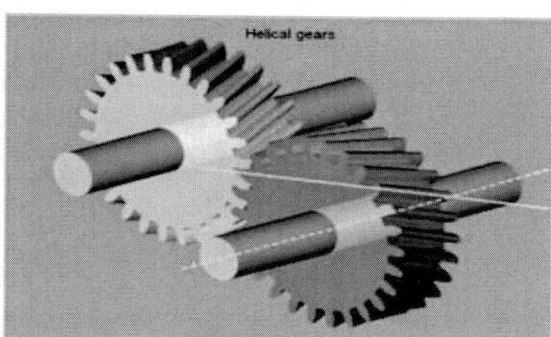

Figure 2.9 Simplest helical gear

Helical gears are produced with longer and stronger teeth. Due to greater surface contact area on teeth, these gears are able to effectively carry heavy loads. These helical gears are quiet and make fewer vibrations. These are available in right as well as left-hand configuration choices. Also, several teeth are in contact at any one time, producing a more gradual loading of the teeth that reduces wear substantially. The increased amount of sliding action between helical gear teeth, however, places greater demands on the lubricant to prevent metal-to-metal contact and resulting premature gear failure. Also, since the teeth mesh at an angle, a side thrust load is produced along each gear shaft. Thus, thrust bearings must be used to absorb this load so that the gears are held in proper alignment (Figure 2.10a–c).

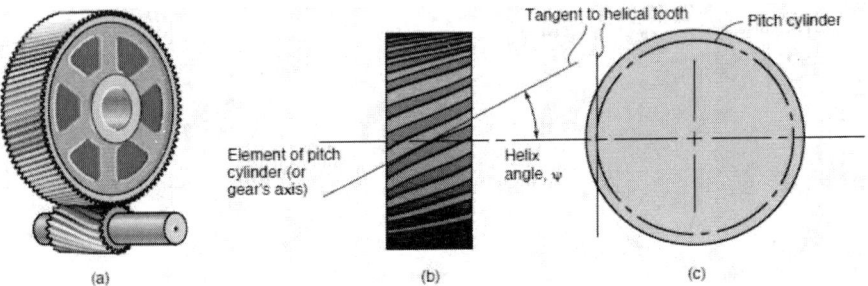

Tangent to helical tooth   Pitch cylinder

Element of pitch cylinder (or gear's axis)

Helix angle, ψ

(a)   (b)   (c)

Figure 2.10   (a) Meshing helical gears; (b) front view; (c) side view

2.4.1   Parallel helical gears

In these the meshing gears are mounted on parallel shafts. The hands of the meshing gears are opposite. For example, a left hand gear drives a right hand gear, or vice versa. The helix angle of the meshing gears must be same. Parallel helical gears find application in high speed and high power transmission or where noise control is important. as compared to spur gears. This is because their precision and power transmission efficiencies are good, but lower in comparison to spur gears. A helical gear is smaller in size compared to spur gear, for the same number of teeth, speed reduction ratio, power transmission and speed. The three other principle classes of helical gears are double-helical, herringbone, and cross-helical **(Figure 2.11)**.

Figure 2.11   Showing the set of drafting gears in draw frame / roving / ring frame

2.4.2   Double helical gears

Double helical gear is of cylindrical form. It has two sections of teeth, right hand and the other left hand. They operate by engaging themselves simultaneously with the teeth of an identically designed mating gear. In these helical gear thrust loading is eliminated by using two pairs of gears with tooth angles opposed to each other, In this way, the side thrust from one gear cancels the thrust from the other gear. These opposed gears are usually manufactured with a space between the opposing sets of teeth.

Types of double helical gears

These helical gears are generally found to be of two kinds. One type is the one that has a gap between the helices. In the other one, opposite hand helices get cut without giving a gap. This is called the herringbone gears. Both the type of helical gears uses a perfect combination of right and left hand helices. This makes the thrust load fully balanced. This balance is a conclusive advantage for selection of supporting bearings. These helical gear can offer the same type of advantage and operating smoothness as that of a single helical gear. Further, it offers the benefit of a greater strength in contact of the teeth and elimination of sideways force. Double helical gears provides an efficient and smooth transfer of torque and motion even at considerably high rotational velocities. As there is absence of axial thrust, helix angles are maintained around 30°, thereby offering the advantage of a large face overlap. Double helical gears give an efficient transfer of torque and smooth motion at very high rotational velocities.

Herringbone gears

Teeth in these gears resemble the geometry of a herring spine, with ribs extending from opposite sides in rows of parallel and slanting lines. Herringbone gears have opposed teeth to eliminate side thrust loads the same as double helical, but the opposed teeth are joined in the middle of the gear circumference. This arrangement makes herringbone gears more compact than double-helical. However, the gear centres must be precisely aligned to avoid interference between the mating helixes.

Before explaining herringbone gear observe Figure 2.12 to know the disadvantage of normal helical gear. A normal helical gear exerts some axial force on the shaft in which it is rotating, this axial force results thrust on the bearing, which is not desired.

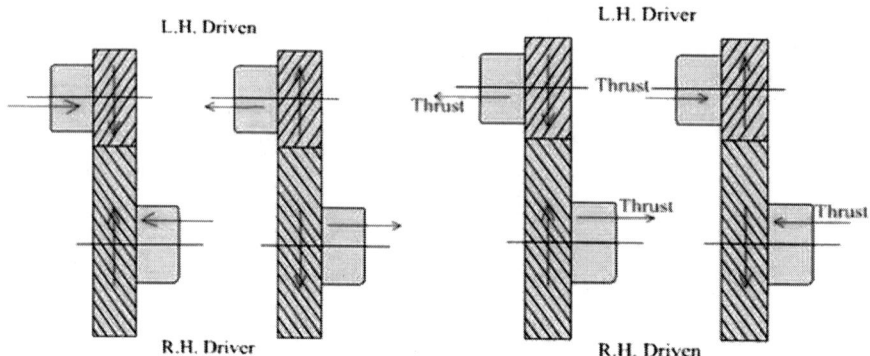

Figure 2.12 Showing the difference between normal helical gear
and herringbone gear

To eliminate this axial force, the axial force is to be countered by another axial force acting in opposite directions, this is achieved by physically
attaching a helical gear with opposite helix angle, as shown in Figure 2.12.
This double helix gear is called a herringbone gear, used mostly in precision drives. Yet the disadvantage lies in the fact that manufacturing of double
helix gear is tough, because of manufacturing problems.

A herringbone gear (Figure 2.13), a specific type of double helical
gear, is a special type of gear that is a side-to-side (not face-to-face) combination of two helical gears of opposite hands. From the top, each helical groove of this gear looks like the letter V, and many together form
a herringbone pattern (resembling the bones of a fish such as a herring).
Unlike helical gears, herringbone gears do not produce an additional axial
load. Like helical gears, they have the advantage of transferring power
smoothly because more than two teeth will be in mesh at any moment
in time. Their advantage over the helical gears is that the side-thrust of
one half is balanced by that of the other half. This means that herringbone gears can be used in torque gearboxes without requiring a substantial thrust bearing. Because of this herringbone gears were an important
step in the introduction of the steam turbine to marine propulsion. Precision herringbone gears are more difficult to manufacture than equivalent spur or helical gears and consequently are more expensive. They are
used in heavy machinery. Where the oppositely angled teeth meet in the
middle of a herringbone gear, the alignment may be such that tooth tip
meets tooth tip, or the alignment may be staggered, so that tooth tip meets
tooth trough. The latter alignment is the unique defining characteristic of
a Wuest type herringbone gear, named after its inventor.

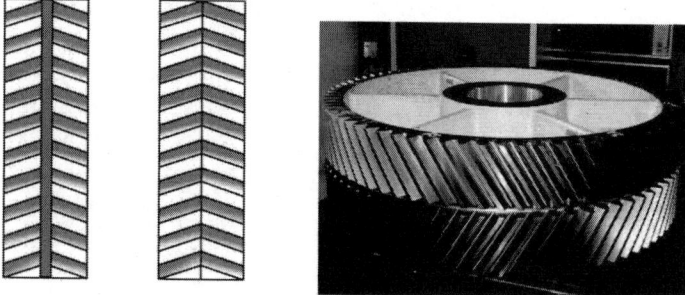

Figure 2.13   Herringbone gears. Left – with centre space;
Right – without centre space

Materials

Gear composition is determined by application, including the gear's service, rotation speed, accuracy and more. Cast iron provides durability and ease of manufacture. Alloy steel provides superior durability and corrosion resistance. Minerals may be added to the alloy to further harden the gear. Cast steel provides easier fabrication, strong working loads and vibration resistance. Carbon steels are inexpensive and strong, but are susceptible to corrosion. Aluminium is used when low gear inertia with some resiliency is required.

Brass is inexpensive, easy to mould and corrosion resistant. **Copper** is easily shaped, conductive and corrosion resistant. The gear's strength would increase if bronzed. **Plastic** is inexpensive, corrosion resistant, quiet operationally and can overcome missing teeth or misalignment. Plastic is less robust than metal and is vulnerable to temperature changes and chemical corrosion. Nylon, and polycarbonate plastics are common. **Other** material types like wood may be suitable for individual applications.

What is the difference between herringbone and double helical gears?

The first figure (Figure 2.14) is a normal helical gear, while second one is a herringbone gear or a double helical gear.

Figure 2.14   Herring bone and double helical gear

Double helical gears, or herringbone gears, overcome the problem of axial thrust presented by "single" helical gears, by having two sets of teeth that are set in a V-shape. A double helical gear can be thought of as two mirrored helical gears joined together. This arrangement cancels out the net axial thrust, since each half of the gear thrusts in the opposite direction resulting in a net axial force of zero. This arrangement can remove the need for thrust bearings. However, double helical gears are more difficult to manufacture due to their more complicated shape. For both possible rotational directions, there exist two possible arrangements for the oppositely-oriented helical gears or gear faces. One arrangement is stable, and the other is unstable. In a stable orientation, the helical gear faces are oriented so that each axial force is directed toward the centre of the gear. In an unstable orientation, both axial forces are directed away from the centre of the gear. In both arrangements, the total (or net) axial force on each gear is zero when the gears are aligned correctly. If the gears become misaligned in the axial direction, the unstable arrangement will generate a net force that may lead to disassembly of the gear train, while the stable arrangement generates a net corrective force. If the direction of rotation is reversed, the direction of the axial thrusts is also reversed, so a stable configuration becomes unstable, and vice versa. Stable double helical gears can be directly interchanged with spur gears without any need for different bearings.

2.4.3   Cross-helical gears

This type of gear is recommended only for a narrow range of applications where loads are relatively light. Because contact between teeth is a point instead of a line, the resulting high sliding loads between the teeth requires extensive lubrication. Thus, very little power can be transmitted with cross-helical gears (Figure 2.15).

Figure 2.15 Showing the use of crossed helical gears on conventional card and spindle drive in simplex

The cross-helical gears are used to (Figure 2.15) transmit power between non-parallel and, non-intersecting shafts. Alternately they are also called "spiral gears". If two helical gears are to operate as crossed helical gears, they must have the same normal pitch and normal pressure angle. During the performance of these gears it has been observed that the crossed helical gears are having point contact and after a long period of usage the point of contact get converted into contact. Compared to other types of gears, these gears exhibit poor precision and require good lubrication. For this reason crossed helical gears are used for transmission of light loads at low speeds. They also have limited speed reduction capacity.

2.5   Bevel gears

Bevel gears are used to change the direction of a shaft's rotation. Bevel gears have teeth that are available in straight, spiral, or hypoid shape. Straight teeth have similar characteristics to spur gears and also have a large impact when engaged. Like spur gears, the normal gear ratio range for straight bevel gears is 3:2 to 5:1 (Figure 2.16).

Figure 2.16   Showing the use of bevel gear

Bevel gears unlike spur and helical gears with teeth cut from a cylindrical blank, bevel gears have teeth cut on an angular or conical surface. Bevel gears are used when input and output shaft centrelines intersect. Teeth are usually cut at an angle so that the shaft axes intersect at 90°, but any other angle may be used. Often there is no room to support bevel gears at both ends because the shafts intersect. Thus, one or both gears overhang their supporting shafts. This overhung load (OHL) may deflect the shaft, misaligning gears, which causes poor tooth contact and accelerates wear. Shaft deflection may be overcome with straddle mounting in which a bearing is placed on each side of the gear where space permits.

Zero bevel gear – Zero bevel gears are similar to (Figure 2.17) straight bevel gears, but their teeth are curved lengthwise. These curved teeth of zero bevel gears are arranged in a manner that the effective spiral angle is zero.

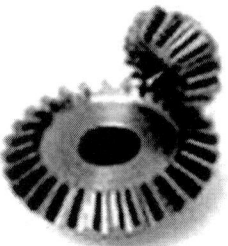

Figure 2.17   Zero bevel gear

There are two basic classes of bevels – straight-tooth and spirals. Straight-tooth bevels –-These gears, also known as plain bevels (Figure 2.18), have teeth cut straight across the face of the gear. They are subject to much of the same operating conditions as spur gears in that straight tooth bevels are efficient but somewhat noisy. They produce thrust loads in a direction that tends to separate the gears. Spiral-bevels – Curved teeth (Figures 2.19 and 2.20) provide an action somewhat like that of a helical gear. This produces smoother, quieter operation than straight-tooth bevels. Thrust loading depends on the direction of rotation and whether the spiral angle at which the teeth are cut is positive or negative.

Figure 2.18   Straight bevel gear

Figure 2.19   Spiral bevel gears used in carding and spinning machines

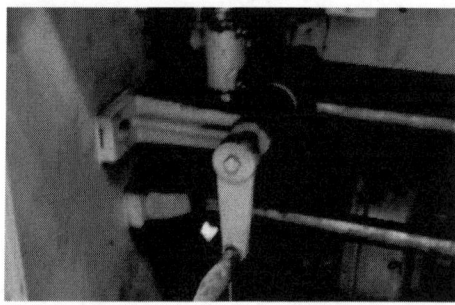

Figure 2.20   Spiral bevel gears on roving machine to release bobbin rail

2.6   Hypoid gears

Hypoid gears resemble spiral bevel gears except the shaft axes do not intersect. The pitch surfaces appear conical but, to compensate for the offset shaft, are in fact hyperboloids of revolution. Hypoid gears are almost always designed to operate with shafts at 90°. Depending on which side the shaft is offset to, relative to the angling of the teeth, contact between hypoid gear teeth may be even smoother and more gradual than with spiral bevel gear teeth, but also have a sliding action along the meshing teeth as it rotates and therefore usually require some of the most viscous types of gear oil to avoid it being extruded from the mating tooth faces, the oil is normally designated HP (for hypoid) followed by a number denoting the viscosity. Also, the pinion can be designed with fewer teeth than a spiral bevel pinion, with the result that gear ratios of 60:1 and higher are feasible using a single set of hypoid gears. This style of gear is most common in motor vehicle drive trains, in concert with a differential. Whereas a regular (non-hypoid) ring-and-pinion gear set is suitable for many applications, it is not ideal for vehicle drive trains because it generates more noise and vibration than a hypoid does (Figure 2.21).

Figure 2.21   Hypoid gears

Miter and angular bevel gears

In bevel gear drives, the shafts of the gears (Figure 2.22) mesh at 90° to each other. The number of teeth of the two bevel gear may or may not be the same. For example in conventional card coiler, we can observe the bevels on the vertical shaft and also on the coiler calendar roller with the same number of teeth. Miter gear are of the type of bevel gear when the angles between the shafts are 90°, and the two gears of a pair are having the same number of teeth.

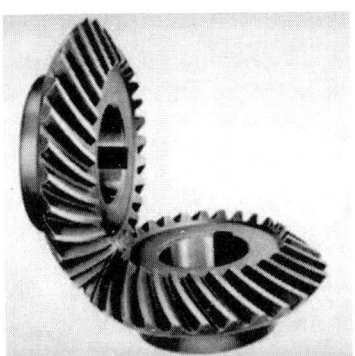 

Figure 2.22   Showing the miter and angular gear

On the other hand angular bevel gears are those in which the angles between the shafts may not be 90°, but either more or less than 90°.

Applications of bevel gear in textile production

Following are the applications

1. They are found in mixing bale opener / bale blender.
2. The drive from Doffer to feed roller in conventional card.
3. The vertical shaft has three to four sets of bevel gears with the same number of teeth in conventional card.
4. Bevel gears in driving the cam shaft in projectile weaving machine.
5. Bevel gear in positive let-off in auto loom.
6. Bevel gear in sectional warping machine for driving the traverse motion and measuring wheel.
7. In driving the spindle in semiautomatic pirn winder and also for providing the rotary traverse for spindle.

2.7   Worm and worm wheel gears

By listening the pair one can guess that when it is necessary to reduce the speed of the member under consideration it is better to choose worm and worm wheel and they are also known as reduction gear assembly. We can list the areas where it is necessary to reduce the speed like; Feed roller drive in Scutcher, Flats drive in Carding, Can drive, Weavers beam drive in loom, Linear traverse in horizontal spindle winder, Conveyer belting system in all rotoconers, etc. In this system it is to be noted that worm is always used as driver and worm wheel as driven. The meshing of teeth occurs with a sliding action resulting in very quiet operation. The sliding friction may produce overheating, which must be dissipated to the surroundings by lubrication (Figure 2.23).

Figure 2.23   A simple worm and worm wheel system

Worm gears are used to transmit power between two non-intersecting shafts, which are right angles to each other. Worm gear drives are used for large speed reduction single stage.

Figure 2.24 Showing the use of worm and worm wheel in loom for let-off motion and in cloth feeding machine

Worm and worm wheel can either be used as individuals or in combination with others. Figure 2.24 shows one such arrangement where in worm is used with bevel. Worms may be considered as cylindrical type gears with screw threads. Generally, the mesh has a right angle. The number of threads in the worm is equivalent to the number of teeth in a gear of a screw type gear mesh. Thus, a one-thread worm is equivalent to a one-tooth gear; and two-thread equivalent to two-teeth, etc., and each rotation of the worm makes the thread advance one lead.

There are four worm tooth profiles as per the manual of worm gear and are listed below.

Type I worm: This worm tooth profile is trapezoid in the radial or axial plane.

Type II worm: This tooth profile is trapezoid viewed in the normal surface.

Type III worm: This worm is formed by a cutter in which the tooth profile is trapezoid form viewed from the radial surface or axial plane set at the lead angle. Examples are milling and grinding profile cutters.

Type IV worm: This tooth profile is involute as viewed from the radial surface or at the lead angle. It is an involute helicoid, and is known by that name.

Type III worm is the most popular. Worm gear drives consists of a worm and a worm gear. Figure 2.24 show the use of worm and worm wheel in loom. Figure 2.25 show the use of worm and worm wheel with bevel in sizing machine.

The worm is similar to a screw. The threads of the worm have an involute helicoid profile. The power transmission efficiency of worm gears is lower compared to spur gears, parallel helical gears, and bevel gears; but higher than that of crossed helical gears. Worm and worm gears produce thrust load on shaft bearings. Worm gears can also be used for self-locking operation. The worm wheel in general made from phosphor–bronze alloy. The worm is usually made of hardened alloy steel. The worm is usually cut on a lathe, whereas the gear is hobbed. All the worm gears must be carefully mounted to ensure proper operation. The efficiency of power transmission varies somewhat with the conditions of assembly and lubricant, but is generally 30%–90%.

Figure 2.25   Showing the use of worm and worm wheel with bevel

Materials of construction and representation

They can be made up of steel, stainless steel, bronze or plastic as shown along with different colours like grey (Figure 2.26).

Figure 2.26   Showing the different material of construction of worm and worm wheels

K W G DL 2 – R1

Where K – Material of construction, W – Type of worm, G – Other type of gear or ground gear, DL – Duplex worm, 2 – Module, R1 – Head thread of starts (right-hand single tread).

Types of worm or classification of worm and worm wheels

Worm and worm wheels can be either right or left hand as shown in Figure 2.27.

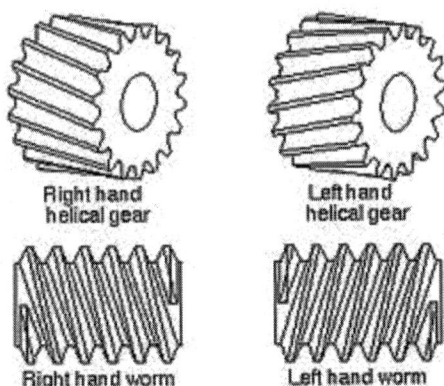

Figure 2.27   Right and left hand worm and worm wheel

Helical and worm handedness

A right hand helical gear or right hand worm is one in which the teeth twist clockwise as they recede from an observer looking along the axis. The designations, right hand and left hand, are the same as in the long established practice for screw threads, both external and internal. Two external helical gears operating on parallel axes must be of opposite hand. An internal helical gear and its pinion must be of the same hand. A left hand helical gear or left hand worm is one in which the teeth twist counter clockwise as they recede from an observer looking along the axis.

Forms of worm and worm wheel

As described above a screw (worm) is said to have one start if it advances one groove (in linear direction), in one complete revolution. It is said to have two starts if it advances two grooves (in linear direction) in one revolution.

Worm can be of different types

The first are *non-throated* worm gears. These don't have a *throat*, or groove, machined around the circumference of either the worm or worm wheel. The second are single-throated worm gears, in which the worm wheel is throated. The final type are double-throated worm gears, which have both gears throated. This type of gearing can support the highest loading. An enveloping (hourglass) worm has one or more teeth and increases in diameter from its middle portion toward both ends. Double-enveloping worm gearing comprises enveloping worms mated with fully enveloping worm gears. It is also known as globoidal worm gearing (Figure 2.28).

Advantages of worm drives

1. Worm gear drives operate silently and smoothly.
2. They are self-locking.
3. They occupy less space.
4. They have good meshing effectiveness.
5. They can be used for reducing speed and increasing torque.
6. High velocity ratio of the order of 100 can be obtained in a single step.

Disadvantages of worm drives

1. Worm gear materials are expensive.
2. Worm drives have high power losses and low transmission efficiency
3. They produce a lot of heat.

Non-enveloping worm gear sets do not have concave features and the straight plane of contact between gears places the highest level of stress on the gear teeth.

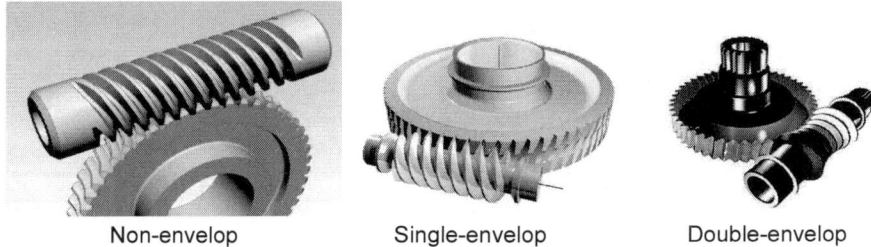

Non-envelop   Single-envelop   Double-envelop

Figure 2.28   Types of worm and worm wheel

Single-envelop worm gear sets contain a worm gear with a concave tooth width, allowing the worm drive to nestle into the gear and increasing efficiency.

Double-envelop worm gear sets contain both a word gear with a concave tooth width, and a worm drive with a concave profile. This design maximizes efficiency.

Crowning of the worm gear tooth

Crowning is critically important to worm gears (worm wheels). Not only can it eliminate abnormal tooth contact due to incorrect assembly, but it also provides for the forming of an oil film, which enhances the lubrication effect of the mesh. This can favourably impact endurance and transmission efficiency of the worm mesh.

Self-locking of worm mesh

As described above a worm can be of self-locking type and it is a unique characteristic of worm meshes that can be put to advantage. It is the feature that a worm cannot be driven by the worm gear. It is very useful in the design of some equipment, such as lifting, in that the drive can stop at any position without concern that it can slip in reverse. However, in some situations it can be detrimental if the system requires reverse sensitivity, such as a servomechanism. It is here by noted that self-locking does not occur in all worm meshes, since it requires special conditions. Factors affecting the self-locking feature include not only lead angle but also the materials of the worm and worm wheel, lead angle, precision of manufacture, types of bearings, lubricant, etc. But, in general, self-locking can occur when the lead angle in a single thread worm is less than 4°.

Multiple pawls ratchet

It has been found in some of the modern looms with take-up motion that a ratchet is provided with multiple pawls for proper gripping.

2.7.1   Applications of worm and worm wheel in textile industry

Spinning
1. In conventional blow room they are used for pre-determined lap length and driving the feed roller.
2. In card they are used to drive the flats.
3. In conventional draw frame for speed reduction and also for Hank meter.
4. As a driving tool for lap in comber.
5. In simplex the drive for Hank meter.
6. In ring spinning the drive is used for reduction in speed.

Weaving
1. In double flanged bobbin winding the drive is used to impart the traversing motion.
2. In semi-auto pirn winding they are used for speed reduction.
3. In automatic pirn winding they are used for driving the spindle and also to drive the eccentric.
4. In sectional and beam warping they are used in full beam stop motion device (Figure 2.29).

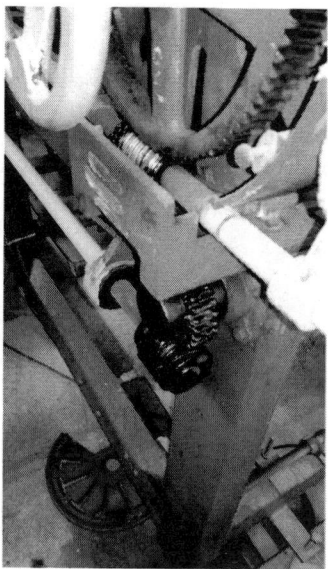

Figure 2.29   Worm and worm wheel in sectional warping

5. In loom they are used in take-up (worm and worm wheel take-up), let-off (for beam driving), pick counter, driving the heald shafts in cam driven dobby, for driving the cylinder in Damask loom, for driving the small end disc in Magzine of auto loom (Figure 2.30a, b).

(a)   (b)

Figure 2.30 a, b   Worm and worm wheel in auto loom

Figure 2.31 Worm and worm wheel in water jet loom

6. The use of worm and worm wheel is very well appreciated in water jet loom as shown in Figure 2.31.

Mangle wheel

Mangle wheel and pinion so called converts continuous rotary motion of pinion into reciprocating rotary motion of wheel. The shaft of pinion has a vibratory motion, and works in a straight slot cut in the upright stationary bar to allow the pinion to rise and fall and work inside and outside of the gearing of the wheel. The slot cut in the face of the Mangle wheel and following its outline is to receive and guide the pinion-shaft and keep the pinion in gear (Figure 2.32).

Significance of Mangle wheel in precision winding

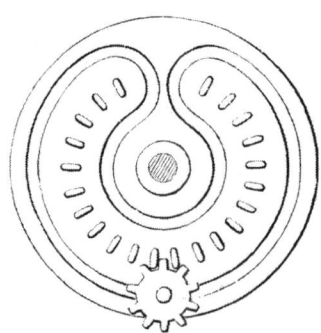
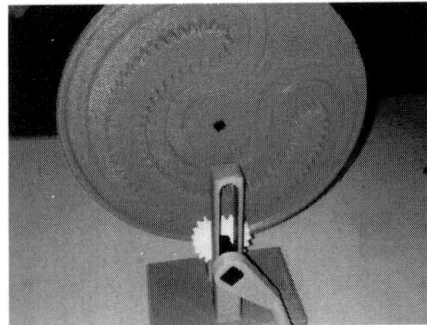

Figure 2.32 Showing the line diagram and the photograph of Mangle wheel

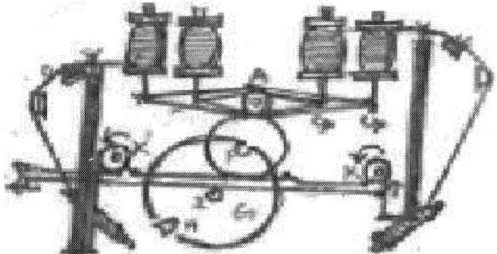

Figure 2.33   Mangle wheel in barrel shaped bobbin production

Figure 2.33 shows the use of Mangle wheel for production of barrel bobbins in precision winder of weaving preparatory.

The Mangle wheel has a pinion which move from internal to external and vice versa and this causes the rack pinion to move either from left to right or right to left. Further, the movement of traverse bar will be resulting in either upward or downward directions. But however one should note that during the pinion of the Mangle wheel movement from inside to outside to inside and vice versa, virtually the traverse bar will not move and hence the yarn is laid maximum at one place forming the barrel shaped.

2.7.2   Rack and pinion gear

Rack and pinion gears are used to convert rotation into linear motion. The flat, toothed part is the rack and the gear is the pinion. A piston coaxial to the rack provides hydraulic assistance force, and an open centred rotary valve controls the assist level. A rack and pinion gears system is composed of two gears. The normal round gear is the pinion gear and the straight or flat gear is the rack. The rack has teeth cut into it and they mesh with the teeth of the pinion gear (Figure 2.34).

Figure 2.34   Rack and pinion gear

A rack or ring and pinion gear is the differential's critical point of power transfer. A ring and pinion gear set is one of the simplest performance modifications that can be performed on a vehicle. The most common reason to change ring and pinion ratios from the original equipment is to retain power when bigger tires are put on a vehicle. The torque can be increased by a ratio change when there is enhanced pulling or higher take-off power from a dead start. A well-designed mechanism such as the rack and pinion gears save effort and time.

2.7.3   Types of rack and pinion gears

Rack and pinion gears are available in three variations.

Straight teeth have the tooth axis parallel to the axis of rotation. Straight teeth that run parallel to the axis of the gear. Load movement or transfer is manual or walk-behind. *Helical teeth* gears provide continuous engagement along the tooth length and are often quieter and more efficient than straight tooth gears. Helical tooth gears resemble spur gears in the plane of rotation, but include teeth that are twisted along a helical path in the axial direction. *Roller pinion* drives use bearing supported rollers that mesh with the teeth of that rack in order to provide minimal to no backlash (Figure 2.35).

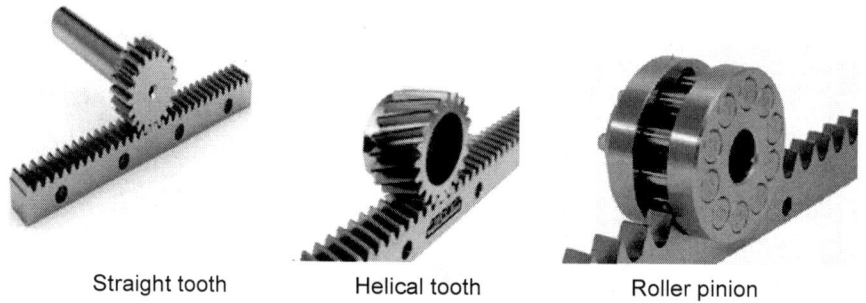

Straight tooth   Helical tooth   Roller pinion

Figure 2.35   Types of rack and pinion

2.7.4   Applications

Rack and pinion is found in many applications (Figures 2.36 and 2.37).

1. In old platt blow room the lap spindle motion is supplemented by rack and pinion.

Figure 2.36   Application in spinning machines. Use of rack and pinion in old blow room and in simplex machine for box of tricks

2. In simplex the motion to the bobbin rail is including the rack and pinion.

3. In precision winding the barrel shaped machine includes the use of Mangle wheel and the set up will also include the use of rack and pinion.

4. In modern rapier loom the flexible principle has the driving for rapier with rack and pinion.

Figure 2.37   Use of ratchet with rack and pinion for transferring the motion

2.8   Geneva

2.8.1   Introduction

The Geneva is one of the earliest of all intermittent motion mechanisms and when input is in the form of continuous rotation, it is probably still the most commonly used. The Geneva drive is also commonly called a Maltese cross mechanism. The Geneva mechanism translates a continuous rotation into an intermittent rotary motion. The driven member, or star wheel, contains evenly spaced slots into which the roller of the driving crank slides into. The number of slots determines the ratio between the dwell (stationary) and motion periods of the driven shaft. The mechanism requires a minimum of three slots to function and additional slots can be added. As the drive wheel turns it enters the slots of the driven, or star, wheel and the mechanism enters its motion period. As the pin leaves the slot the driven wheel has been indexed and it continues to stay stationary until the pin rotates back around. The drive wheel also has a raised circular blocking disc that locks the driven wheel in position between steps Genevas are available in a variety of sizes. They are cheaper than cams or star wheels and have adequate-to-good performance characteristics, depending on load factors and other design requirements (Figure 2.38).

2.8.2   Advantages of Geneva

Geneva may be the simplest and least expensive of all intermittent motion mechanisms. As mentioned before, they come in a wide variety of sizes, ranging from those used in instruments, to those used in machine tools to index spindle carriers weighing several tons. They have good motion-curve characteristics compared to ratchets, but exhibit more "jerk", or instantaneous change in acceleration, than do better cam systems. The Geneva maintains good control of its load at all times, since it is provided with locking ring surfaces to hold the output during dwell periods. In addition, if properly sized to the load, the Geneva generally exhibits very long life.

2.8.3   Disadvantages of Geneva

The Geneva is not a versatile mechanism. It can be used to produce no less than three, and usually no more than 18 dwells per revolution of the output shaft. Furthermore, once the number of dwells has been selected, the designer is well-locked into a given set of motion curves. The ratio of dwell period to motion period is also established once the number of dwells per revolution has been selected. Also, all Geneva acceleration curves start and end with finite acceleration and deceleration.

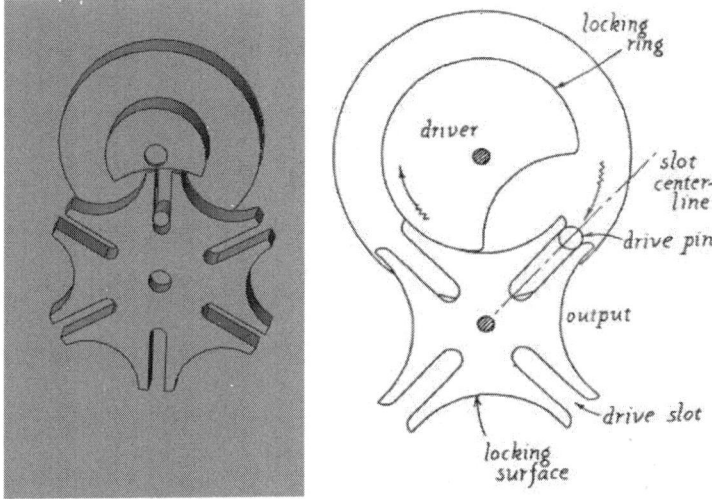

Figure 2.38   Types of Geneva

2.8.4   Types of Genevas

There are three types of Genevas: (1) external, which is the most popular; (2) internal, which is also very common and (3) spherical, which is extremely rare. Genevas are also combined with a wide variety of other mechanism, such as four-bar linkages; clutch-brake combinations; non-circular gears, etc., to modify the motion curves and dwell-motion ratios obtained from a pure Geneva (Figure 2.39).

Figure 2.39   Internal Geneva drive

In an internal Geneva drive the axis of the drive wheel of the internal drive is supported on only one side. The angle by which the drive wheel has to rotate to effect one step rotation of the driven wheel is always smaller than 180° in an external Geneva drive and is always greater than 180° in an internal one. The external form is the more common, as it can be built smaller and can withstand higher mechanical stresses.

Motion curves

A few general comments can be made about these motion curves.

1. For an external Geneva, the dwell period always exceeds the motion period.

2. For an external Geneva, the dwell period always lasts longer than the time required for 180° of motion of the input driver.

3. For an internal Geneva, the motion period always exceeds the dwell period.

4. For an internal Geneva, the dwell period is always shorter than the time required for 180° of input.

5. For a spherical Geneva, the dwell period equals the motion period, and equals the time required for exactly 180° of input.

6. The magnitude of peak acceleration and deceleration, velocity, etc., obtained with a Geneva, is a function of the number of slots or dwells. This is true of all types.

As the ratio between the diameter of the wheel and the diameter of the driver gets larger (more slots or dwell periods per revolution of the output), maximum accelerations and velocities decrease (for a given driver speed). This makes sense because the indexing angle of the output decreases as the number of slots increases.

2.9   Ratchet and Pawl

2.9.1   General features

Leonardo Da Vinci's notebooks are full of ratchet applications, and they were probably in use centuries before he came along. Most of these early applications, however, seem to have used the ratchet for mechanical advantage rather than to produce intermittent motion, the latter really being a product or need of the industrial revolution. Leonardo's ratchets, for example, are usually used to control engines of war; a man winds a catapult or crossbow, and the ratchet allows him to do this in short, easy steps, resting (dwell!) between exertions. Simplicity is one of the big advantages of the ratchet. Other related advantages include low cost and reliability. The ratchet is also noted for its ability to carry a large load in relation to its size. It is also a versatile device and is used in an amazing number of applications ranging from moderately heavy-duty machinery to high-sped instruments (Figure 2.40).

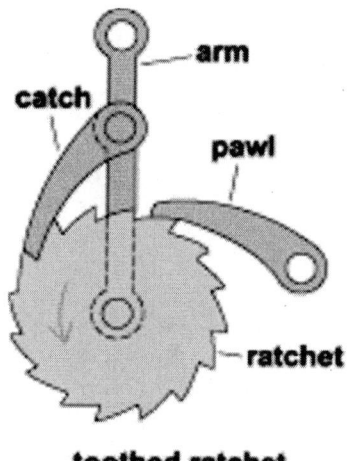

toothed ratchet

Figure 2.40   Ratchet and pawl

Disadvantages of the ratchet include the fact that it is an impacting mechanism. There are ways to reduce the impacts in certain versions, but impact will almost always be present to some extent, and can lead to wear, control, and stability problems unless the rest of the system is properly designed. The basic problem, of course, is that impacts produce forces throughout the mechanism that are well in excess of the subsequent drive forces. Impact also results in noise, which is very undesirable in most applications, and "noise pollution" should soon get a lot more attention from machine and instrument designers than it does at present. It is interesting to note about pawl and wheel-tooth geometries when designing a ratchet. It is important to shape ratchet wheel and pawl teeth properly, to reduce impact stress and wear. Proper shape also helps the pawl engage and drive the load. The forces on the pawl as it enters the wheel tooth include a normal force and a friction force. The friction force opposes the pawl motion, of course, and so must be drawn in a direction away from the wheel, and the resultant force on the pawl is in a direction to rotate it clockwise. Yet it must rotate counter-clockwise to pick up the wheel tooth. This tendency to disengage must be overcome by providing a spring that urges the pawl against the wheel; or the pawl and wheel teeth must be redesigned to place the resultant pawl force on the correct side of the pawl bearing. This is usually a better solution to the disengagement problem that a pawl spring, but pawl springs are also required in many applications to get the pawl down into engagement with the wheel more rapidly than it would under its own weight. And, of course, in many designs the ratchet is not oriented in such a way as to engage the drive pawl by weight.

Multiple pawls ratchet

It has been found in some of the modern looms with Take up motion that a ratchet is provided with multiple pawls for proper gripping.

2.9.2   Applications of ratchet and pawl in textile industry

Spinning

1. Blow room: The old type of blow room used to have the ratchet and pawl type of lap knock-off motion.

2. In simplex the builder motion also known as "Box of Tricks" has ratchet and pawl (Figure 2.41).

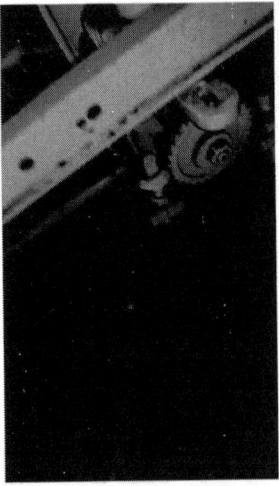

Figure 2.41   Showing the use of ratchet in Box of Tricks of simplex and as part of builder motion in ring frame

3. In ring spinning the builder motion has ratchet and pawl.

Weaving

1. In semi-auto pirn winder the rotary motion is provided by ratchet and pawl
2. In auto pirn winder the drive is given for projection wheel and rotary traverse
3. In Take up for the motion transfer from sley to cloth roller
4. In Dobby for driving the pattern cylinder
5. In beam feeler less modern let off for driving the beam (Figure 2.42a–c) for driving the cylinder in Damask loom, for driving the small end disc in Magzine of auto loom
6. For driving the fabric roller in circular weft knitting machine.

| a. | b. | c. |
| In semi-auto pirn winder | In auto pirn winder | Ratchet with slip catch |

Figure 2.42 a–c

2.9.3   Compound gear train

When there is more than one gear on a shaft, it is called a compound train of gear. These gears are useful in bridging over the space between the driver and the driven. But whenever the distance between the driver and the driven or follower has to be bridged over by intermediate gears and at the same time a great (or much less) speed ratio is required, then the advantage of intermediate gears is intensified by providing compound gears on intermediate shafts. In this case, each intermediate shaft has two gears rigidly fixed to it so that they may have the same speed. One of these two gears meshes with the driver and the other with the driven or follower attached to the next shaft as shown in Figure 2.43.

Figure 2.43 Compound Gear Train

1. A 60 teeth spur gear (Figure 2.44) which is mounted on one end of a shaft rotating at 900 rpm drives a gear having 20 teeth through a carrier gear having 120 teeth. The 20 teeth gear is compounded with another gear having 10 teeth. This 10 teeth wheel drives a spin gear which is mounted on a shaft. This is having 15 teeth. Find the rpm of the shaft.

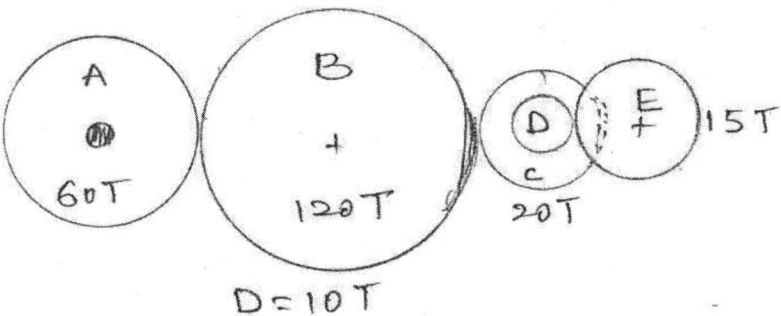

Figure 2.44 Compound Gear Train

$$\frac{A}{B} \times \frac{B}{C} \times \frac{D}{C} = e$$

$$\therefore e = 2$$

$$\frac{n}{m} = 2$$

$$\therefore \frac{n}{m} \cong 2 \qquad\qquad \therefore n = 1800$$

m = Speed of first wheel

n = Speed of second wheel

$$\frac{n}{m} = \frac{A}{B} \times \frac{B}{C} \times \frac{D}{E}$$

$$\frac{n}{900} = \frac{60}{120} \times \frac{120}{20} \times \frac{10}{15}$$

n = 1800 rpm.

2. A gear wheel having 55 teeth which is mounted on a shaft running of 1800 rpm drives another gear wheel having 30 teeth through two carriers 120 teeth and 60 teeth. This 30 teeth gear drives a double worm which is mounted on a shaft. Find the speed of the last shaft (Figure 2.45).

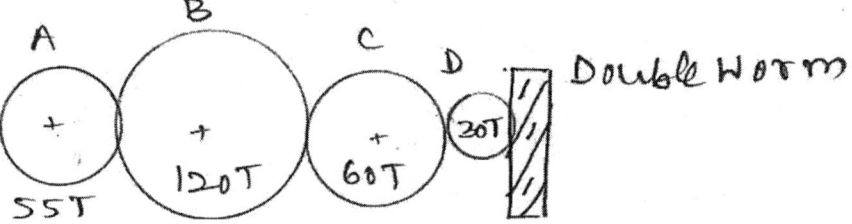

Figure 2.45   Compound Gear Train

For a single worm one teeth of the worm will rotate one teeth of D.

∴ For a double worm two teeth of the worm will rotate one teeth of D.

Let "m" rpm = speed of the first wheel i.e., wheel A

Let "n" rpm = speed of the last wheel i.e., wheel D

$$\frac{n}{m} = \frac{A}{B} \times \frac{B}{C} \times \frac{D}{E}$$

$$\frac{n}{1800} = \frac{55}{120} \times \frac{120}{60} \times \frac{60}{30}$$

$$n = \frac{55 \times 1800}{30}$$

∴ n = 3300 rpm

Speed of double worm = n × 2

$$= 3300 \times 2$$
$$= 6600 \text{ rpm.}$$

3. Shaft "A" running at 12 rpm drives shaft "B" throughout the following arrangement. Find the speed of shaft B (Figure 2.46).

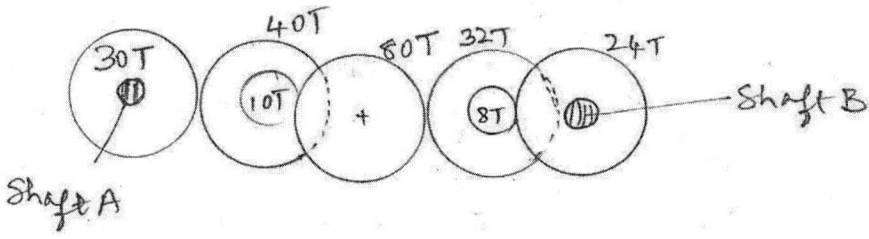

Figure 2.46   Compound Gear Train

$$\frac{n}{m} = \frac{30}{40} \times \frac{10}{60} \times \frac{60}{32} \times \frac{8}{24}$$

$$\frac{n}{m} = \frac{3}{4} \times \frac{10}{4} \times \frac{1}{24}$$

$$= \frac{30}{24 \times 16}$$

$$\frac{n}{m} = \frac{15}{24 \times 8}$$

$$\frac{n}{m} = \frac{15}{24 \times 8}$$

$$\therefore \frac{n}{12} = \frac{15}{24 \times 8}$$

$$n = \frac{15 \times 12}{24 \times 8}$$

$$\therefore n = 0.9375 \text{ rpm}$$

$$\therefore \text{ Speed of shaft B is } 0.9375 \text{ rpm.}$$

4. In a gear train, wheel "A" having 50 teeth and running at 150 rpm. drives wheel "B". What will be the speed of wheel "B" having teeth 150. If the speed of "B" is required to be 300 rpm. Find the number of teeth gear wheel.

$$e = \frac{n}{m} = \frac{50}{150}$$

$$\frac{n}{150} = \frac{50}{150}$$

$$\therefore n = \frac{50 \times 150}{150}$$

$n = 50$ rpm.

$\therefore$ Speed of the wheel "B" = 50 rpm.

If $n = 300$ rpm, $\therefore$ Number of teeth of "B" = ?

$$e = \frac{n}{m} = \frac{50}{T}$$

$$\frac{300}{150} = \frac{50}{T}$$

$$\therefore T = \frac{50}{2}$$

$$T = 25$$

5. A compounded train gear consists of 6 gears. A wheel "A" is fitted on a shaft running at 200 rpm. Find the speed of "F" (Figure 2.47).

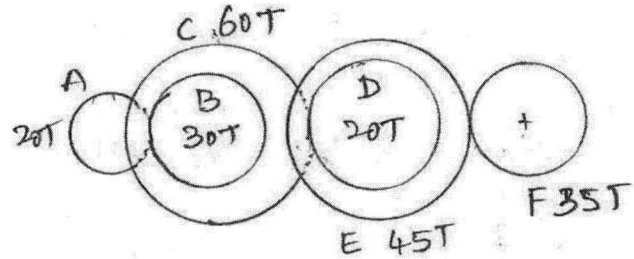

Figure 2.47   Compound Gear Train

$$e = \frac{n}{m} = \frac{n}{200}$$

$$\therefore \frac{n}{200} = \frac{20}{30} \times \frac{60}{20} \times \frac{45}{35}$$

$$\therefore n = 2.57 \times 200$$

Speed of F = 514.29 rpm.

$$\text{Circular pitch} = \frac{\text{circumference of pitch circle}}{\text{Number of teeth}}$$

$$\text{Module} = \frac{\text{Pitch diameter}}{\text{Number of teeth}}$$

$$\text{Diameter of pitch} = \frac{\text{Number of teeth}}{\text{Pitch circle diameter}}$$

6. Find the details of two spur wheels each with a pitch of 0.75 cm so as to have a velocity ratio of 4. The approximate distance between the shafts being 30 cm.

$$e = \frac{d_1}{d_2} = 4$$

$$d_1 = 4\,d_2$$

$$\frac{d_1 + d_2}{2} = 30$$

$$\frac{4d_2 + d_2}{2} = 30$$

$$5d_2 = 60$$

$\therefore \qquad d_2 = 12$ cm $\qquad$ & $\qquad d_1 = 48$ cm

For first wheel

$$\text{Circular pitch} = \frac{\text{Circumference of pitch circle}}{\text{Number of teeth}}$$

$$\therefore \text{Number of teeth} = \frac{\pi d}{pitch} = \frac{\pi \times 48}{0.75} = 201.06$$

Number of teeth = 202

For second wheel

$$\therefore \text{Number of teeth} = \frac{\pi d}{0.75} = \frac{\pi \times 12}{0.75} = 50.28\ T$$

Number of teeth $\simeq 51$

7. Find the details of two spur wheels each with a pitch of 0.6 cm. The diameter of the first wheel is 15 cm and the diameter of "B" is 3 times the diameter of "A".

$d_1 = 15$ cm

$d_2 = 45$ cm

$\overline{A}'$ Number of teeth $= \dfrac{\pi d}{pitch} = \dfrac{\pi \times 15}{0.8} = 58.93$

$= 59T.$

$\overline{B}'$ Number of teeth $= \dfrac{\pi d}{pitch} = \dfrac{\pi \times 45}{0.8} = 176.7$

$= 177T.$

8. Two parallel shafts are to be connected by a gear drive. They are approximately 1 m apart. If velocity ratio is to be exactly $\frac{9}{2}$ and the pitch of the gears is 57 mm. Find the number of teeth on two wheels and the exact distance between the shafts.

$$e = \frac{d_1}{d_2} = \frac{9}{2}$$

$$\therefore d_1 = \frac{9}{2}d_2$$

$$\frac{d_1 + d_2}{2} = 100$$

$$\therefore \frac{\frac{9}{2}d_2 + d_2}{2} = 100$$

$$\therefore \frac{11}{2}d_2 = 200$$

$$\therefore 11\, d_2 = 400$$

$$d_2 = 36.36 \text{ cm}$$

$$d_1 + d_2 = 200$$

$$\therefore d_1 = 200 - 36.36$$

$$d_2 = 163.64 \text{ cm}$$

"A" Number of teeth $= \dfrac{\pi \times 163.64}{5.7} = 90.18\, T$

$= 90T$

"B" Number of teeth $= \dfrac{\pi \times 36.26}{5.7} = 20.04\, T$

$= 20T$

$$\bar{A'}\ \text{Diameter} = \frac{\text{Number of teeth} \times \text{pitch}}{\pi} = \frac{90 \times 5.7}{\pi} = 163.3\,cm$$

$$\text{"B" Diameter} = \frac{20 \times 5.7}{\pi} = 36.29\ cms$$

$$\text{Exact centre distance} = \frac{163.3 + 36.29}{2}$$
$$= 99.8\ \text{cm}$$
$$= 0.998\ \text{m.}$$

9. The driving arrangement in ERM cleaner is shown in following gearing diagram, find the speeds of all members (Figure 2.48).

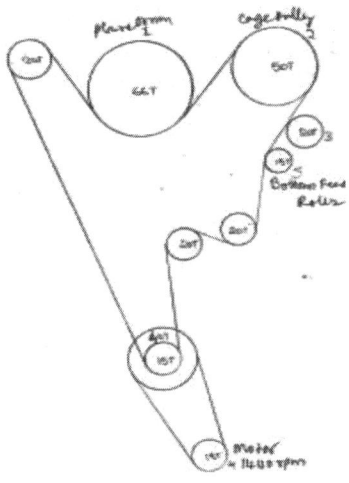

Figure 2.48 Gearing in ERM cleaner

Circumference of motor pulley $= 11''$
Circumference of disc beater $= 27''$
Speed of the motor = 1440 rpm

$$\text{Speed of the disc beater} = \frac{1440 \times 11}{27} = 586.6\ \text{rpm.}$$

Speed of feed rollers at different motor speeds
  AT motor speed is 9 rpm

$$\text{Speed of bottom feed roller} = \frac{18 \times 9 \times 15}{40 \times 15} = 4.05\ \text{rpm.}$$

$$\text{Speed of cage feed roller} = 9 \times \frac{18}{40} \times \frac{15}{20} = 3.037\ \text{rpm.}$$

Speed of the cage $= 9 \times \dfrac{18}{40} \times \dfrac{15}{20} = 1.215$ rpm.

Speed of plain drum $= 4.05 \times \dfrac{15}{50} = 0.92$ rpm.

AT motor speed is 25 rpm

Speed of bottom feed roller $= \dfrac{25 \times 18}{40} = 11.25$ rpm.

Speed of cage feed roller $= \dfrac{25 \times 18}{40} = 8.43$ rpm.

Speed of cage $= 11.25 \times \dfrac{15}{50} = 3.375$ rpm.

Speed of plain drum $= 3.375 \times \dfrac{50}{66} = 2.56$ rpm.

Speed of motor rpm	Speed of feed roller rpm	Speed of 2nd feed roller rpm	Speed of cage rpm	Speed of plain drum rpm
9	4.05	3.037	1.215	0.92
25	11.25	8.437	3.375	2.56

10. Two parallel shafts are to be connected by a gear drive. They are approximately 50 cm apart. If the velocity ratio is to be exactly "3" and the pitch of the gears is 30 mm. Find out the number of teeth and the exact distance between the shafts.

$$\dfrac{d_1 + d_2}{2} = 50 \; cm$$

$$\therefore \dfrac{d_1}{d_2} = 3$$

$$d_1 = 3d_2$$
$$3d_2 + d_2 = 100$$

$$\therefore \; d_2 = 25 \text{ cm}$$
$$d_1 = 75 \text{ cm}$$

"A"   Number of teeth $= \dfrac{\pi d}{p} = \dfrac{\pi \times 75}{3} = 78.539 T$

$$= 79T$$

"B"   Number of teeth $= \dfrac{\pi d}{p} = \dfrac{\pi \times 25}{3} = 26.179 \, T$

$$= 26\,T$$

$\overline{A'}$ $\qquad$ Diameter $= \dfrac{\text{Number of teeth} \times \text{pitch}}{\pi}$

$$= \dfrac{79 \times 3}{\pi} = 75.44\,cm$$

$\overline{B'}$ $\qquad$ Diameter $= \dfrac{\text{Number of teeth} \times \text{pitch}}{\pi}$

$$= \dfrac{26 \times 3}{\pi} = 24.83\,cm$$

2.9.4   Reverted gear train

When axis of the first wheel is co-axial with last wheel, the gear train is said to be reverted. Let N_1, N_2, N_3, N_4 be speed of gear 1,2,3,4 and T_1, T_2, T_3, T_4 be the teeth of gear 1, 2, 3, 4, let r_1, r_2, r_3, r_4 be radius of gears 1, 2, 3, 4.

We know that velocity ratio $\dfrac{N_1}{N_2} = \dfrac{T_2}{T_1}$, $\dfrac{N_3}{N_4} = \dfrac{T_4}{T_3}$, $\dfrac{N_1 \times N_3}{N_2 \times N_4} = \dfrac{T_2 \times T_4}{T_1 \times T_3}$

2nd and 3rd wheel are mounted on the same shaft.

$\therefore\ N_2 = N_3$

$$\dfrac{N_1}{N_4} = \dfrac{T_2 \times T_4}{T_1 \times T_3}$$

Pitch $= \dfrac{2\pi r}{T}$ or $\pi .m$ where $m = $ module $= \dfrac{2r}{T}$

$$r = \dfrac{m\,T}{2}$$

$$\dfrac{m\,T_1}{2} + \dfrac{m\,T_2}{2} = \dfrac{m\,T_3}{2} + \dfrac{m\,T_4}{2} = \quad (m - \text{constant})$$

$$T_1 + T_2 = T_3 + T_4$$

2.10   Sun and Planet gear or epicyclic gear or planetary gear – Special types of gear

These are the special types of gear wheels used specially to get high velocity ratio. These are extensively used in textile industry. They are found basically in card coiler, speed frame bobbin drive as differential gear, in roper let-off motion in weaving, etc. Figures 2.50 and 2.51 show the arrangement found in card coiler. In a simple epicyclic gear system two wheels known as Sun (A) and Planet (B) will be connected by an arm (C) (Figure 2.49).

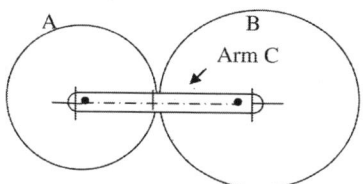

Figure 2.49   Principle of Sun and Planet Gear

Figure 2.50   Showing the top view of planetary gear of card coiler

Figure 2.51   Showing the arrangement of Sun and Planet gear (near view) in card coiler

In an epicyclic gear train the axis of the shaft over which the gears are mounted may move relative to the fixed axis. Epicyclic gear trains are useful in transmitting larger velocity ratios with moderate sizes of gears in a comparatively lesser space.

	Rotation of element		
A B Arm C Condition of motion	Arm C	Gear A	Gear B
Arm C fixed, gear A rotates ("+" 1 revolution clockwise)	0	+1	$-\dfrac{T_A}{T_B}$
Arm C fixed, gear A rotates about + "x" revolution clockwise	0	+x	$-x \times \dfrac{T_A}{T_B}$
Arm rotates "y" revolution clockwise, gear A rotates "x" revolution clockwise	y	+x	$y - x \times \dfrac{T_A}{T_B}$

Speed of gear A in relation to the movement of arm C $= N_A - N_C$
Speed of gear B in relation to the movement of arm C $= N_B - N_C$
Then the epicyclic gear ratio "e" is given by

$$e = \frac{N_B - N_C}{N_A - N_C}$$

$$e = \frac{n - a}{m - a}$$

m = Speed of first driven
n = Speed of last driven
a = Speed of arm
$n = a + e\,(m - a)$

An alternative name for planetary gear is epicyclic. In Figure 2.52 the reason for the two alternative terms is shown. The wheels A and B are free rotate on axes in the arm C. If the arm is rotated about the axis of wheel A called the sun wheel and the planet wheel B rolls round the periphery of A. The centre of B describes a circle but a point P on its periphery generates an epicycle or an epicyclic curve.

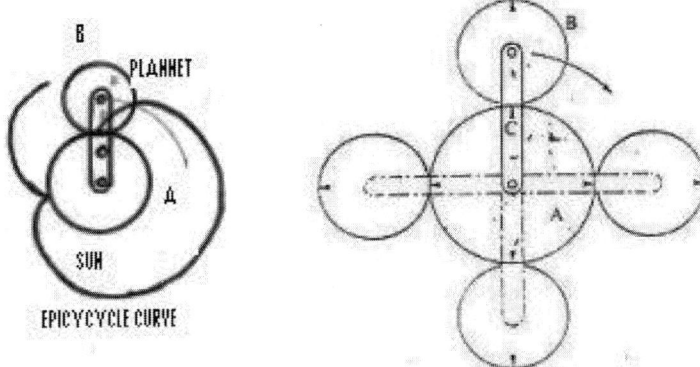

Figure 2.52   Principle of epicyclic gear

In this epicyclic system there is relative motion between two or more of the axes or shafts about which the gear wheels rotate. Consider wheels A and B in Figure. If arm C is stationary and wheel A is rotated the wheel B would rotate A/B times the revolutions of A if A and B represent the number of teeth on them. If wheel A and B are now imagined as being fixed solidly in the arm C and if C rotates one revolution, then both wheels A and B would rotate through one revolution when the system will work as simple gear train and when arm rotates the axes of the wheels have relative motion and the resulting motion of the last wheel in the system is combination of motions. If wheel A were stationary and the arm rotated then wheel B would receive two motions at the same time, one movement due to gearing with A and other movement due to C. If A and B are of the same number of teeth then B will receive two revolutions.

An epicyclic gear train is a very general term as the concept is very versatile. Basically, it involves three gears: a sun gear, a planet gear and a ring gear, the underlying concept being many gear ratios can be obtained from a small volume as compared to other types of gear trains which take up more space. Each of these three components can be the input, the output or can be held stationary. Choosing which piece plays which role determines the gear ratio for the gearset.

Sun gear	Arm	Ring gear	Speed	Power
Input	Output	Fixed	Decreased	Increased
Output	Input	Fixed	Increased	Decreased
Fixed	Output	Input	Decreased	Increased
Fixed	Input	Output	Increased	Decreased

For example let a ring gear be with 72 teeth and a sun gear with 30 teeth. Observe the different gear ratios out of this gearset.

	Input	Output	Stationary	Calculation	Gear ratio
A	Sun (S)	Planet carrier (C)	Ring (R)	1 + R/S	3.4:1
B	Planet carrier (C)	Ring (R)	Sun (S)	1 / (1 + S/R)	0.71:1
C	Sun (S)	Ring (R)	Planet carrier (C)	R/S	2.4:1

Also, locking any two of the three components together will lock up the whole device at a 1:1 gear reduction. Notice that the first gear ratio listed above is a reduction – the output speed is slower than the input speed. The second is an overdrive – the output speed is faster than the input speed. The last is a reduction again, but the output direction is reversed. There are several other ratios that can be found out of this planetary gear set. Unlike simple gear trains, an epicyclic gear train requires defining more than one input to obtain a specific output, hence making the analysis a little difficult and non-intuitive. Their advantages include space efficiency, low transmission losses and flexibility. For example, the ring gear can be made stationary by using a brake, input given to the sun gear, and output taken from the Planet gear by using a spider arrangement. This arrangement is shown below in Figure 2.53a–e.

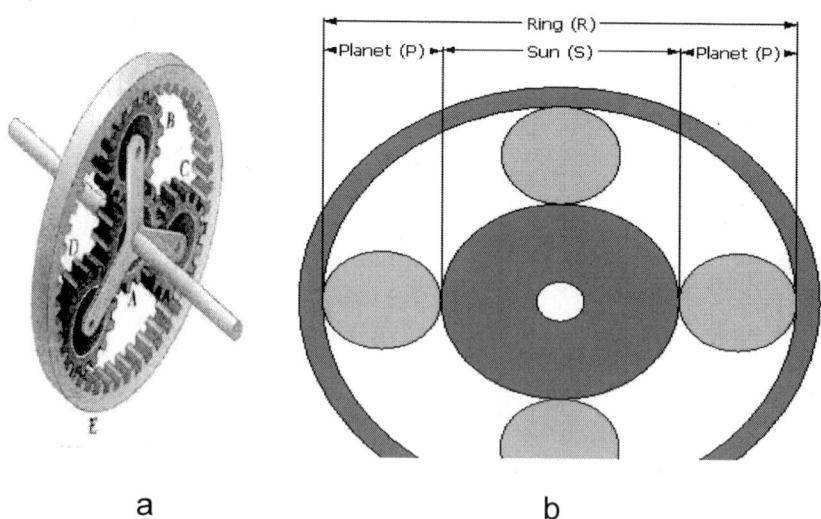

a   b

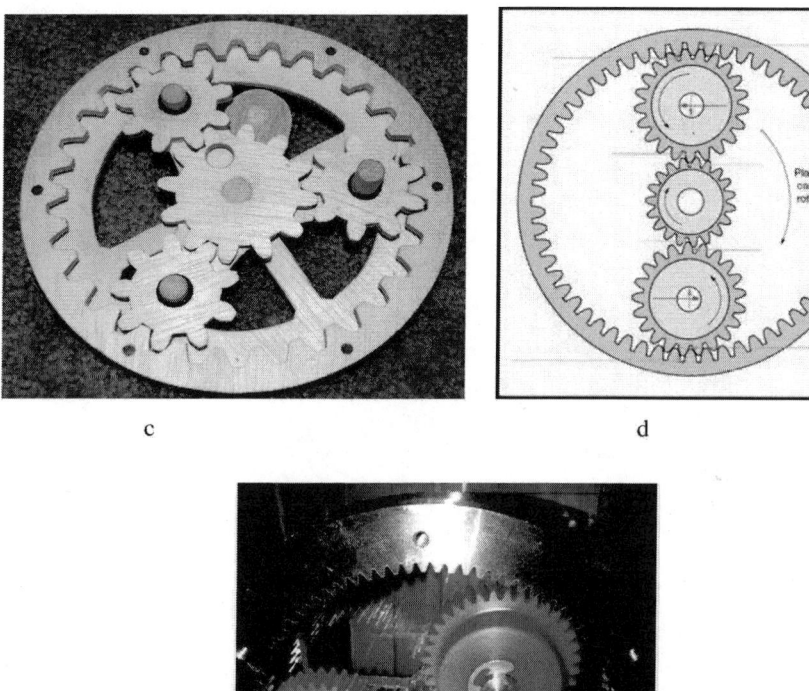

c   d

e

Figure 2.53 a- e- Examples of Sun and Planet Gear

In a planetary train at least one of the gears must revolve around another gear in the gear train. A planetary gear train is very much like our own
solar system, and that's how it gets its name. In the solar system the planets
revolve around the sun. Gravity holds them all together. In a planetary gear
train the sun gear is at the centre. A planet gear revolves around the sun gear.
The system is held together by the planet carrier. In some planetary trains,
more than one planet gear rotates around the sun gear. The system is then
held together by an arm connecting the planet gears in combination with a
ring gear (Figure 2.54).

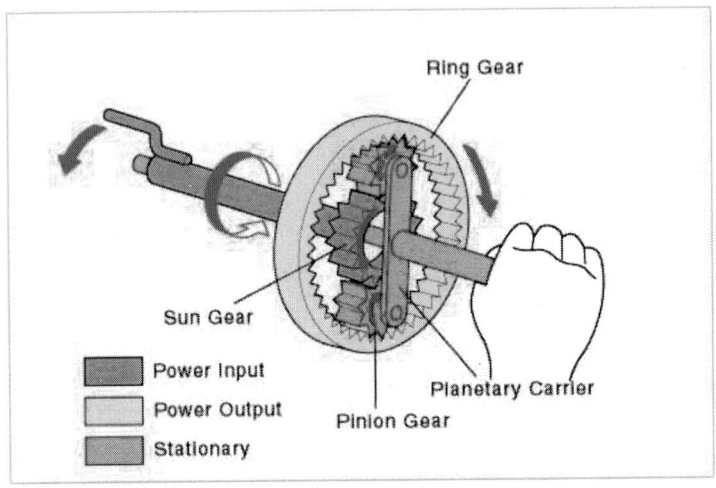

Figure 2.54   Example of Ring and Planetary Gear

Planetary gears, supply a lot of speed reduction and torque in a small package, have operating characteristics beyond those of fixed-axis gear trains planetary gearing, with its inherent in-line shafting and cylindrical casing, is often recognized as the compact alternative to standard pinion-and-gear reducers. The most basic form of planetary gearing involves three sets of gears with different degrees of freedom. Planet gears rotate around axes that revolve around a sun gear, which spins in place. A ring gear binds the planets on the outside and is completely fixed. The concentricity of the planet grouping with the sun and ring gears means that the torque carries through a straight line. In a simple planetary setup, input power turns the sun gear at high speed. The planets, spaced around the central axis of rotation, mesh with the sun as well as the fixed ring gear, so they are forced to orbit as they roll. All the planets are mounted to a single rotating member, called a cage, arm, or carrier. As the planet carrier turns, it delivers low-speed, high-torque output. Planet gears, for their size, engage a lot of teeth as they circle the sun gear – therefore they can easily accommodate numerous turns of the driver for each output shaft revolution. Simple planetary gears generally offer reductions as high as 10:1. Compound planetary systems, which are far more elaborate than the simple versions, can provide reductions many times higher. There are obvious ways to further reduce (or as the case may be, increase) speed, such as connecting planetary stages in series. The rotational output of the first stage is linked to the input of the next, and the multiple of the individual ratios represents the final reduction. Another option is to introduce standard gear reducers into a planetary train

On the other hand in differential systems every member rotates. Planetary arrangements like this accommodate a single output driven by two inputs, or a single input driving two outputs. Compound (as opposed to simple) planetary trains have at least two planet gears attached in line to the same shaft, rotating and orbiting at the same speed while meshing with different gears.

Compounded planets can have different tooth numbers, as can the gears they mesh with.

11. The speed ratio of the reverted gear train, as shown in Figure is to be 12. The module pitch of gears A and B is 3,125 mm and of gears C and D is 2.5 mm. Calculate the suitable numbers of teeth for the gears. No gear is to have less than 24 teeth (Figure 2.55).

Given: Speed ratio, $N_A / N_D = 12$; $m_A = m_B = 3.125$ mm; $m_C = m_D = 2.5$ mm.

Let
N_A = Speed of gear A,
T_A = Number of teeth on gear A,
r_A = Pitch circle radius of gear A,
N_B, N_C, N_D = Speed of respective gears,
T_B, T_C, T_D = Number of teeth on respective gears, and
r_B, r_C, r_D = Pitch circle radii of respective gears.

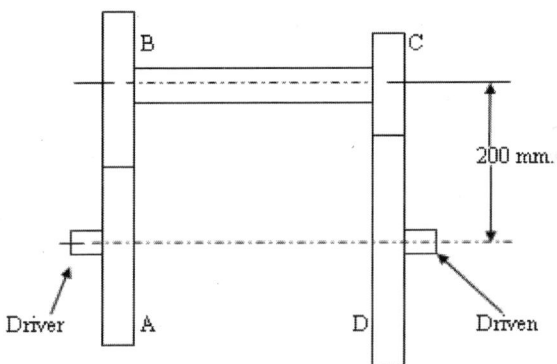

Figure 2.55   Reverted Gear train

Since the speed ratio between the gears A and B and between the gears C and D are to be same. Therefore,

$$\text{*We know that speed ratio} = \frac{\text{Speed of first driver}}{\text{Speed of last driven}} = \frac{N_A}{N_D} = 12$$

$$\text{Also} \quad \frac{N_A}{N_D} = \frac{N_A}{N_B} \times \frac{N_C}{N_D} \qquad ...(N_B = N_C, \text{ being on the same shaft})$$

For $\dfrac{N_A}{N_B}$ *and* $\dfrac{N_C}{N_B}$ to the same, each speed ratio should be $\sqrt{12}$ so that

$$\frac{N_A}{N_D} = \frac{N_A}{N_B} \ X \ \frac{N_C}{N_D} = \sqrt{12} \ X \ \sqrt{12} = 12$$

$$*\frac{N_A}{N_B} = \frac{N_C}{N_D} = \sqrt{12} = 3.464$$

Also the speed ratio of any pair of gears in mesh is the inverse of their number of teeth, therefore,

$$\frac{T_B}{T_A} = \frac{T_D}{T_C} = 3.464 \qquad\qquad ....(i)$$

We know that the distance between the shafts

$$x = r_A + r_B = r_C + r_D = 200 \text{ mm}$$

or

$$\frac{m_A . T_A}{2} + \frac{m_B . T_B}{2} = \frac{m_C . T_C}{2} + \frac{m_D . T_D}{2} + = 200 \qquad ......\left(\because r = \frac{m.T}{2} \right)$$

$$3.125 \ (T_A + T_B) = 2.5 \ (T_C + T_D) = 400 \qquad ....(\because \ m_A = m_B, \text{ and } m_C = m_D)$$

$$\therefore T_A + T_B = 400 / 3.125 = 128 \qquad\qquad\qquad (ii)$$

and $\qquad T_C + T_D = 400 / 2.5 = 160 \qquad\qquad\qquad (iii)$

From equation (i) $T_B = 3.464 \ T_A$. Substituting this value of T_B in equation (ii),

$\qquad T_A + 3.464 \ T_A = 128 \qquad$ or $\quad T_A = 128 / 4.464 = 28.67$ say 28 $\quad$ Ans.

and $\qquad\qquad T_B = 128 - 28 = 100 \qquad\qquad\qquad\qquad\qquad\qquad$ Ans.

Again from equation (i), $T_D = 3.464 \ T_C$. Substituting this value of T_D in equation (iii),

$\qquad T_C + 3.464 \ T_C = 160 \qquad$ or $\quad T_C = 160 / 4.464 = 35.84$ say 36 $\quad$ Ans.

and $\qquad\qquad T_D = 160 - 36 = 124 \qquad$ Ans.

Note: The speed ratio of the reverted gear train with the calculated values of number of teeth on each gear is

$$\frac{N_A}{N_D} = \frac{T_B \times T_D}{T_A \times T_C} = \frac{100 \times 124}{28 \times 36} = 12.3$$

12a. In an epicyclical gear train, an arm carries two gears A and B having 36 and 45 teeth, respectively. If the arm rotates at 150 rpm in the anti-clockwise direction about the centre of the gear A which is fixed, determine the speed of gear B. If the gear A instead of being fixed, makes 300 rpm in the clockwise direction, what will be the speed of gear B? (Figure 2.56).

Given: $T_A = 36$; $T_B = 45$; $N_C = 150$ rpm (anti-clockwise).

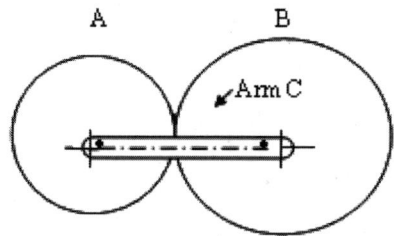

Figure 2.56   Simple Epicyclic Gear train

The gear train is shown in Figure 2.56.

Let N_A Speed of gear A, N_B Speed of gear B, and, N_C Speed of gear C
Assuming the arm C to be fixed, speed of gear A relative to arm C
$$= N_A - N_C$$
And speed of gear B relative to arm C $= N_B - N_C$
Since the gears A and B revolve in opposite directions, therefore,

$$\frac{N_B - N_C}{N_A - N_C} = -\frac{T_A}{T_B} \qquad \qquad ...(i)$$

Speed of gear B when gar A is fixed
When gear A is fixed, the arm rotates at 150 rpm in the anti-clockwise direction i.e.,

$$N_A = 0, \quad \text{and} \quad N_C = +150 \text{ rpm}$$
$$\therefore \frac{N_B - 150}{0 - 150} = -\frac{36}{45} = -0.8$$

Of $N_B = -150 \times -0.8 + 150 = 120 + 150 = 270$ rpm   Ans.

Speed of gear B when gear A makes 300 rpm clockwise
Since he gear A makes 300 rpm clockwise, therefore,
$N_A = -300$ rpm.
$$\therefore \frac{N_B - 150}{-300 - 150} = -\frac{36}{45} = -0.8$$

or $N_B = -450 \times -0.8 + 150 = 360 + 150 = 510$ rpm. Ans.

12b. Diagram shows the driving arrangement for wrap block. Gear "A" is fixed on shaft "O". Gear "B" is meshed with gear "A" and carried by an arm "D" which is freely revolving above shaft "O". The gear with gear "C", prove that for one revolution of arm "D" the wrap block which is connected with gear "C" is making two revolutions when wheel A is blocked (Figure 2.57).

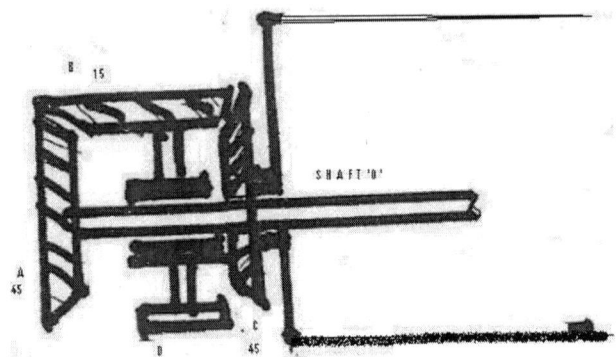

Figure 2.57   Gearing of Wrap Block

$$e = \frac{n-a}{m-a}$$

$$n = a + p(m-a)$$

$$l = \frac{T_A}{T_B} \times \frac{T_B}{T_C} = \frac{T_A}{T_C}$$

$$= \frac{45}{45}$$

$e = .1$

$m = 0$

$a = 1$

$n = a + e(m-a)$

$\quad = 1 + (-1)[0-1]$

$n = 2.$

12c. In the epicyclic or planetary arrangement shown in Figure 2.58 the wheel "A" is fixed on shaft "O". Wheel "B" is meshing with wheel A and compounded with wheel "C" on the shaft left. "C" and "D" are engaged and wheel "B" is freely revolving around shaft "O". The compound shaft F is carried by arm E, which also is freely revolving about shaft "O". If wheel,

1. If wheel A is fixed calculate the reduction ratio: $= \dfrac{Out\ put}{input}$

2. Calculate the speed of wheel "B" when wheel A is revolving around at 1000 rpm clockwise and arm E is revolving around.
    i) 250 rpm – clockwise
    ii) 250 rpm – anti-clockwise

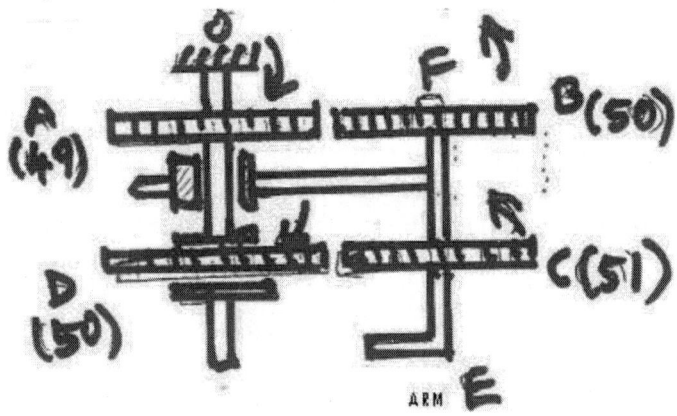

Figure 2.58   Planetary Gear

$$\text{Reduction ration} = \frac{Output}{in \; put}$$

$m = 0$

$$e = \frac{T_A}{T_B} \times \frac{T_C}{T_D}$$

$$= \frac{49}{50} \times \frac{51}{50}$$

$$e = \frac{2499}{2500}$$

$a = 1$ (assume)

$$n = 1 + \frac{2499}{2500} \; (0-1)$$

$$= 1 - \frac{2499}{2500}$$

$$n = \frac{1}{2500}$$

$N = a + e \, (m - a)$

Given: $m = 1000$ rpm.

250 rpm – clockwise
$a = +250$ rpm
$e = +0.9996$
n = 250 + (0.9996) (1000 – 250)
$n = 999.7 rpm$

250 rpm – anti-clock wise
$a = -250$ rpm
$n = -250 + (0.9996) (1000 + 250)$
$n = 999.5 rpm$

Hence we can conclude that this particular differential gearing arrangement is not influenced by the direction of movement of arm.

12d. In the figure the ratchet is moving 6 teeth per pick when the beam is filled. Find out the delivery rate of warp when the beam diameter is 60 cm. Assume loom speed as 180 ppm and wheel A is clocked (Figure 2.59).

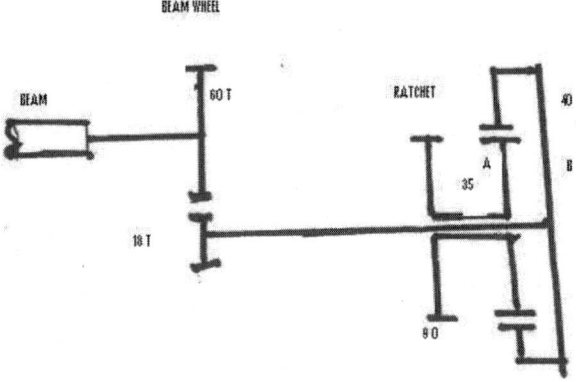

Figure 2.59   Gear in Roper Letoff

Given:
    Speed of loom = 180 rpm
    Number of teeth in ratchet = 90^T
    Number of teeth per pick = 6
    No. of revolution of ratchet =

$$\frac{\text{Crank speed}}{\text{No. of teeth in ratchet}} \text{No. of teeth movement per pick}$$

$$= \frac{180}{90} \times 6$$

$$A = 12 \text{ rpm.}$$

$$e = \frac{n-a}{m-a}$$

$$n = a + e\,(m-a)$$

$$e = \frac{T_A}{T_B} = \frac{35}{40}$$

$$e = +\,7/8$$

A is locked, so $n = 0$

$$n = a + e\,(m-a)$$
$$0 = 12 + 0.875\,(m-12)$$
$$-m = \frac{12-10.5}{0.875}$$
$$-m = 1.7$$
$$m = -1.714 \text{ rpm.}$$
$$m = -1.714 \text{ rpm}$$

$$\text{Speed of beam} = 1.714 \times \frac{18}{60}$$

$$= 0.514 \text{ rpm}$$

Delivery rate is calculated as follows

$$\text{Surface speed of beam} = \pi \times D \times N$$
$$= \pi \times 60 \times 0.514$$
$$= 96.84 \text{ cm/min.}$$

12e. In Figure 2.60 wheels A and B are internal gears, and B is free to rotate independently about centre O. Wheels C and D form a compound carrier with centre Q, carried by the arm E and meshing with A and B. Wheel A is fixed and has 22 teeth; wheel B has 23 teeth. Wheels C and D have 19 and 20 teeth, respectively. The arm E is rotated by the source power (Figure 2.60).

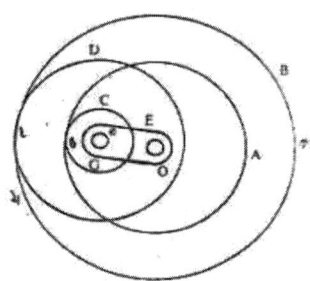

Figure 2.60 Compound Epicyclic gear

The value of the train is given by

$$e = +\frac{22}{19} \times \frac{20}{23} = +\frac{440}{437}$$

The value of "e" is positive because clockwise rotation of the first wheel, A, results in clockwise rotation of the last wheel B. Now let the source of movement rotate the arm E through one revolution. The evolutions will the last wheel, B, make is given by:

$$e = \frac{l-a}{f-a}$$

Hence,

$$\frac{440}{437} = \frac{l-1}{0-1},$$

i.e.,

$$-\frac{440}{437} = l - 1,$$

and

$$1 - \frac{440}{437} = l = -\frac{3}{437}.$$

The last wheel, B, rotates in the opposite direction to that of the arm, and the reduction ratio is $\dfrac{437}{3} \cong 146{:}1$.

12f. Figure 2.61 shows a planetary mechanism using four gears. Wheel A is fixed and has 49 teeth, wheel B has 50 teeth, and compounded with it is wheel C, with 51 teeth. Wheel C drives the last wheel, D, which has 50 teeth. The arm E rotates about the fixed shaft of wheel A.

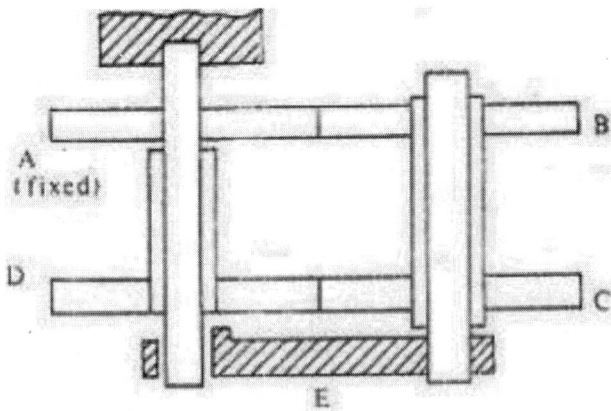

Figure 2.61   Planetary mechanism

The value of the train is

$$e = +\frac{49}{50} \times \frac{51}{50} = +\frac{2499}{2500}$$

Let the arm rotate through 1 revolution. Then:

$$e = \frac{2499}{2500} = \frac{l-1}{0-1},$$

i.e.:

$$-\frac{2499}{2500} = l - 1$$

and

$$l - \frac{2499}{2500} = l = \frac{1}{2500}.$$

13. In a reverted epicyclic gear train, the arm A carries two gears B and C and a compound gear D – E. The gear B meshes with gear E and the gear C meshes with gear D. The number of teeth on gears B, C and D are 75, 30 and 90, respectively. Find the speed and direction of gear C when gear B is fixed and the arm A makes 100 rpm clockwise (Figure 2.62).

Given: $T_B = 75$; $T_C = 30$; $T_D = 90$; $N_A = 100$ rpm (clockwise)

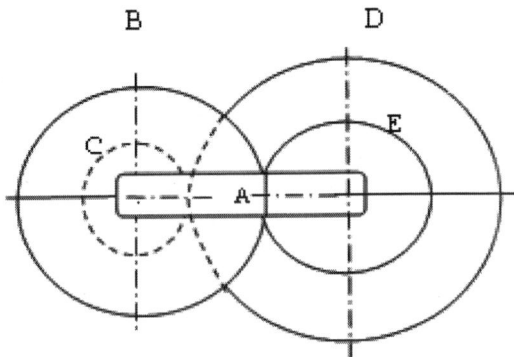

Figure 2.62   Showing the Epicyclic arrangement

First of all, let us find the number of teeth on gear E (T_E). Let d_B, d_C, d_D and d_E be the pitch circle diameters of gears B,C, D and E, respectively. From the geometry of the figure,

$$d_B + d_E = d_C + d_D$$

Since the number of teeth on each gear, for the same module, are proportional to their pitch circle diameters, therefore,

$$T_B + T_E = T_C + T_D$$

$$\therefore T_E = T_C + T_D - T_B = 30 + 90 - 75 = 45$$

The table of motions is drawn as follows:

Step No.	Conditions of motion	Revolutions of elements			
		Arm A	Compound gear D–E	Gear B	Gear C
1	Arm fixed-compound gear D–E rotated through + 1 revolution (i.e., 1 revolution anti-clockwise)	0	+ 1	$-\dfrac{T_E}{T_B}$	$-\dfrac{T_D}{T_C}$
2	Arm fixed-compound gear D–E rotated through + x revolutions	0	+ x	$- x \times \dfrac{T_E}{T_B}$	$- x \times \dfrac{T_D}{T_C}$
3	Add + y revolutions to all elements	+ y	+ y	+ y	+ y
4	Total motion	+ y	x + y	$y - x \times \dfrac{T_E}{T_B}$	$y - x \times \dfrac{T_D}{T_C}$

Since the gear B is fixed, therefore from the fourth row of the table.

$$y - x \times \frac{T_E}{T_B} = 0 \quad \text{or} \quad y - x \times \frac{45}{75} = 0$$

$$\therefore y - 0.6x = 0 \tag{i}$$

Also the arm A makes 100 rpm clockwise, therefore,

$$y = -100 \tag{ii}$$

Substituting $y = -100$ in equation (i), we get

$$-100 - 0.6x = 0 \quad \text{or} \quad x = -100 / 0.6 = -166.67$$

From the fourth row of the table, speed of gear C,

$$N_C = y - x \times \frac{T_D}{T_C} = -100 + 166.67 \times \frac{90}{30} = +400 \; rpm$$

$$= 400 \text{ rpm (anti-clockwise)} \qquad \text{Ans.}$$

14. An epicyclic gear consists of three gears A, B and C as shown in Figure 2.63. The gear A has 72 internal teeth and gear C has 32 external teeth. The gear B meshes with both A and C and is carried on an arm EF which rotates about the centre of A at 18 rpm. If the gear A is fixed, determine the speed of gears B and C.

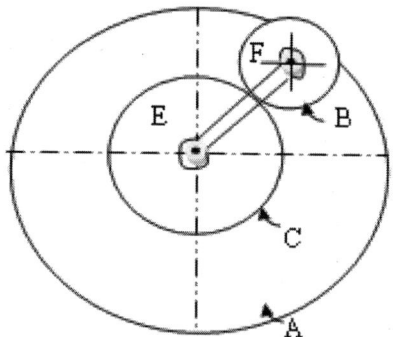

Figure 2.63   Epicyclic with internal gear

Given: $T_A = 72$; $T_C = 32$; Speed of arm EF = 18 rpm.

Considering the relative motion of rotation as shown in the following table.

Step No.	Conditions of motion	Revolutions of elements			
		Arm EF	Gear C	Gear B	Gear A
1	Arm fixed–gear C rotates through + 1 revolution) i.e., 1 revolution anti-clockwise)	0	+1	$-\dfrac{T_C}{T_B}$	$-\dfrac{T_C}{T_B} \times \dfrac{T_B}{T_A} = -\dfrac{T_C}{T_A}$
2	Arm fixed-gear C rotates through + x revolutions	0	+ x	$-x \times \dfrac{T_C}{T_B}$	$-x \times \dfrac{T_C}{T_A}$
3	Add + y revolutions to all elements	+ y	+ y	+ y	+ y
4	Total motion	+ y	$x + y$	$y - x \times \dfrac{T_C}{T_B}$	$y - x \times \dfrac{T_C}{T_A}$

Speed of gear C

We know that the speed of the arm is 18 rpm, therefore,

$$y = 18 \text{ rpm}$$

And the gear A is fixed, therefore,

$$y - x \times \frac{T_C}{T_A} = 0 \quad \text{or} \quad 18 - x \times \frac{32}{72} = 0$$

$$\therefore x = 18 \times 72 / 32 = 40.5$$

$\therefore$ *Speed of gear C*

$= x + y = 40.5 + 18$

$= + 58.5$ rpm.

$= 58.5$ rpm in the direction of arm.

Speed of gear B

Let d_A, d_B and d_C be the pitch circle diameters of gears, A, B and C, respectively. Therefore, from the geometry of Figure.

$$d_B + \frac{d_C}{2} = \frac{d_A}{2} \quad or \quad 2d_B + d_C = d_A$$

Since the number of teeth are proportional to their pitch circle diameters, therefore,

$$2\,T_B + T_C = T_A \quad or \quad 2\,T_B + 32 = 72 \quad or \quad T_B = 20$$

$\therefore$ *Speed of gear B*

$$= y - x \times \frac{T_C}{T_B} = 18 - 40.5 \times \frac{32}{20} = -46.8\ rpm$$

$= 46.8$ rpm in the opposite direction of arm.

15. Speed of A = 160 rpm "m", Speed of arm "a" = 80 rpm. Find the corresponding speed of the last wheel with train (Figure 2.64).

Also   A = 42T,   B = 60T,   C = 54T and D = 46T

$$(n - a) = (m - a) \times \frac{A}{B} \times \frac{C}{D}$$

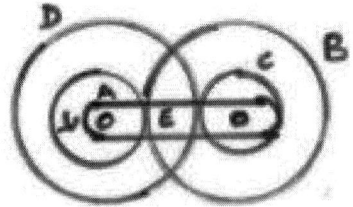

Figure 2.64   Epicyclic Gear

$$e = \frac{(n - a)}{(m - a)} = \frac{A}{B} \times \frac{C}{D} = \frac{\text{Driver wheel}}{\text{Driven wheel}}$$

$$\therefore e = \frac{A}{B} \times \frac{C}{D} = \frac{42}{60} \times \frac{54}{46}$$

$$\frac{n-80}{160-80} = \frac{42}{60} \div \frac{54}{46}$$

$$n - 80 = (0.82)\ 80$$
$$n - 80 = 65.73$$
$$n = 65.73 + 80$$
$$\therefore\ n = 145.73 \text{ rpm.}$$

This is the case if the arm "A" rotates in the same direction as that of first wheel.

Case II: If the arm rotates in the opposite direction to that of first wheel.

$$e = 0.82$$

$$\frac{n+80}{160+80} = \frac{42}{60} \times \frac{54}{46}$$

$$n + 80 = (0.82)\ 240$$
$$n = 196.8 - 80$$
$$n = 116.8 \text{ rpm.}$$

16.   A = 40 T, B = 60 T, C = 55 T and D = 45 T

Speed of first wheel = m = 108 rpm, Speed of arm "A" = 54 rpm. Find speed of last wheel ("n") in the train (Figure 2.65).

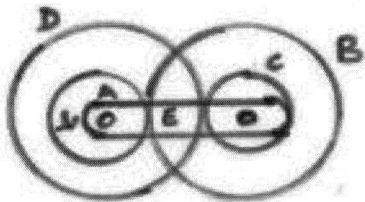

Figure 2.65   Epicyclic Gear

Case (I):   When arm rotates in the same direction as first wheel.

$$e = \frac{n-a}{m-a} = \frac{\text{Driver wheel}}{\text{Driven wheel}}$$

$$= \frac{A}{B} \times \frac{C}{D}$$

$$= \frac{40}{60} \times \frac{55}{45}$$

$$= \frac{2}{3} > \frac{11}{9}$$

$$= \frac{22}{27}$$

$$\frac{n-a}{m-a} = \frac{22}{27}$$

$$\frac{n-54}{108-54} = \frac{22}{27}$$

$$n-54 = \frac{22}{27} \times 54$$

$$n = 44 + 54$$

$$\therefore n = 98 \text{ rpm.}$$

Case (II):   When the arm rotates in the direction opposite to 1st wheel.

$$\frac{n+a}{m+a} = \frac{22}{27}$$

$$\frac{n+54}{108+54} = \frac{22}{27}$$

$$(n+54) = \frac{22}{27} \times 162$$

$$n = 132 - 54$$
$$n = 78 \text{ rpm.}$$

17. In the Figure 2.66 shown "A" and "B" are internal gears. "A" is fixed in the shaft "E" and "B" is freely revolving about shaft "E". Wheel C is engaged with wheel A and compounded with wheel D which is meshed with wheel B. The compound shaft is carried by arm F which is freely revolving about shaft. Calculate the speed of wheel B when wheel A is revolving about 250 rpm clockwise and arm "F" is given a variable speed of 500 rpm clockwise to 500 rpm anti-clockwise.

Speed of arm

$$e = \frac{n-a}{m-a}$$

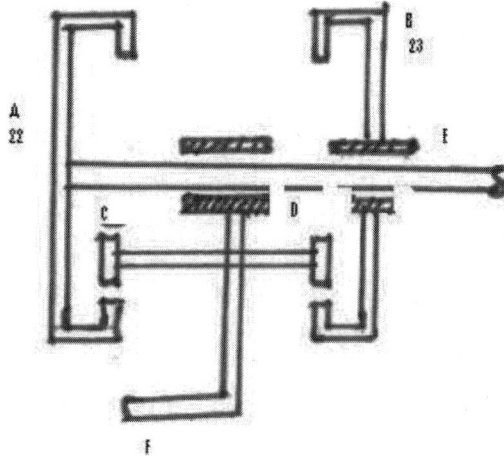

Figure 2.66   Differential Gear

$n = a + e\,(m - a)$

$e = \dfrac{T_A}{T_C} \times \dfrac{T_D}{T_B}$

$= \dfrac{22}{19} \times \dfrac{20}{25}$

$e = +1.0061$

Internal gear A and C so both clockwise A and B in same shaft

$m = +250$ rpm

$a = +500$ to -500

$a = 0$

$\qquad n = 251.5$ rpm

(i) $a = +500$ rpm

$\qquad n = 248.5$ rpm

(ii) $a = +400$ rpm

$\qquad n = 249.1$ rpm

(iii) $a = +300$ rpm

$\qquad n = a + e\,(m - a)$

$\qquad n = 249.7$ rpm

(iv) $a = +200$ rpm

$\qquad n = 250.3$ rpm

(v) $a = +100$ rpm

$\qquad n = 250.9$ rpm

(vi) $a = -100$ rpm
         $n = 252.1$ rpm

(vii) $a = -200$ rpm
         $n = 252.7$ rpm

(viii) $a = -300$ rpm
         $n = 253.3$ rpm

(ix) $a = -400$ rpm
         $n = 253.9$ rpm

(x) $a = -500$ rpm
         $n = 254.5$ rpm

2.10.1   Applications of epicyclic or planetary gears in textile industry

Use of epicyclic gear in warp block (Figure 2.67)

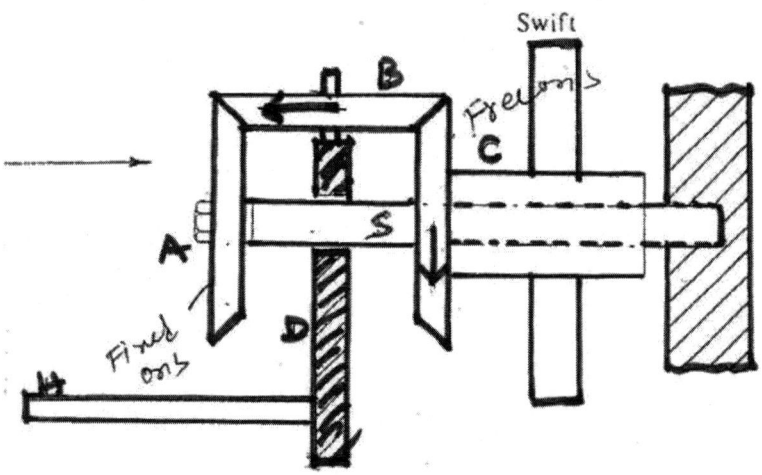

Figure 2.67   Showing the use of epicyclic gear in wrap block

18. Diagram shows the driving arrangement for wrap block. Gear "A" is fixed on shaft "O". Gear "B" is meshed with gear "A" and carried by an arm "D" which is freely revolving above shaft "O". The gear with gear "C", prove that for one revolution of arm "D" the wrap block which is connected with gear "C" is making two revolutions when wheel A is blocked.

$$e = \frac{n-a}{m-a}$$

$$n = a + p\,(m-a)$$

$$l = \frac{T_A}{T_B} \times \frac{T_B}{T_C} = \frac{T_A}{T_C}$$

$$= \frac{45}{45}$$

$e = 1$

$m = 0$

$a = 1$

$n = a + e\,(m-a)$

$= 1 + (-1)\,[0-1]$

$n = 2$

Roper let-off

19. In the figure the ratchet is moving 6 teeth per pick when the beam is filled. Find out the delivery rate of warp when the beam diameter is 60 cm. Assume loom speed as 180 ppm and wheel A is clocked (Figure 2.68).

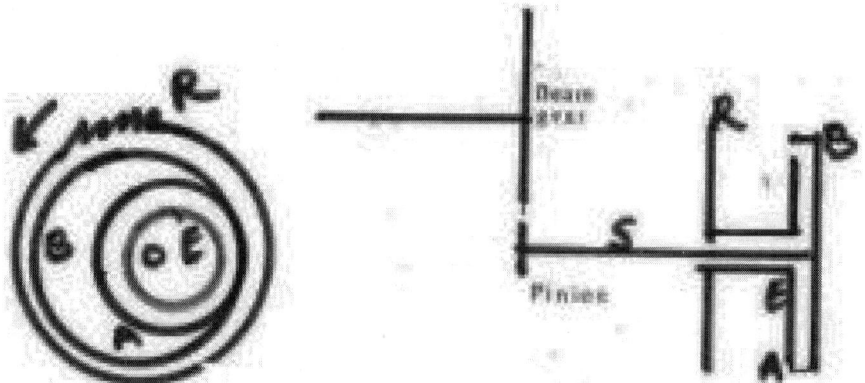

Figure 2.68   Sun and plane gear in positive letoff

Given: Speed of loom = 180 rpm, Number of teeth in ratchet = 90, Number of teeth per pick = 6, Number of revolution of ratchet =

$$\frac{\text{Crank speed}}{\text{No. of teeth in ratchet}} \text{No. of teeth movement per pick}$$

$$= \frac{180}{90} \times 6$$

$$A = 12 \text{ rpm.}$$

$$e = \frac{n-a}{m-a}$$

$$n = a + e(m-a)$$

$$e = \frac{T_A}{T_B} = \frac{35}{40}$$

$$e = +7/8$$

A is locked, so $n = 0$

$$n = a + e(m-a)$$

$$0 = 12 + 0.875(m - 12)$$

$$-m = \frac{12 - 10.5}{0.875}$$

$$m = 1.7$$

$$m = 1.714 \text{ rpm.}$$

$$m = 1.714 \text{ rpm.}$$

$$\text{Speed of beam} = 1.714 \times \frac{18}{60}$$

$$= 0.514 \text{ rpm}$$

Delivery rate is calculated as follows

$$\text{Surface speed of beam} = \pi \times D \times N$$
$$= \pi \times 60 \times 0.514$$
$$= 96.84 \text{ cm/min.}$$

Speed frame differential

20. From the given differential motion diagram calculate the spindle speed and bobbin speed when there is no input from bottom cone drum (BCD), while the main shaft is running at 705 rpm. Suppose that at some stage of bobbin build if the excess revolutions required from bobbin at the starting stage of bobbin build these 250 rpm and at final (Figure 2.69).

$$e = \frac{n-a}{m-a}$$

$$n = a + e(m-a)$$

$$e = \frac{T_A}{T_B} \times \frac{T_C}{T_D}$$

$$= \frac{32}{16} \times \frac{15}{33}$$

$$e = +0.909$$

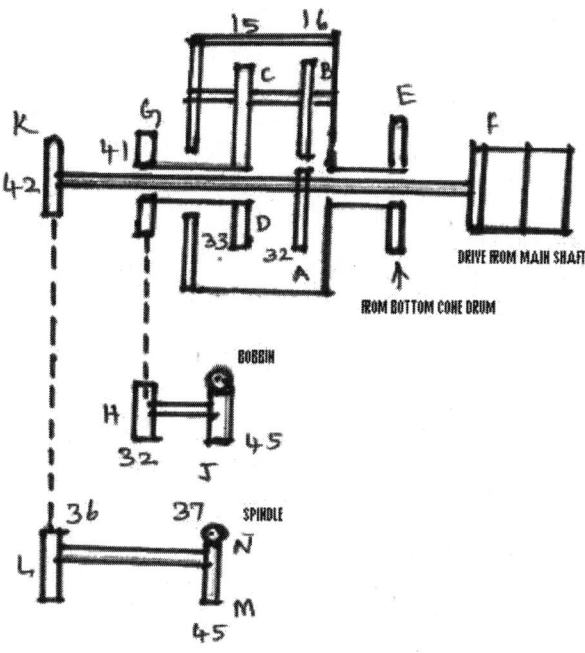

Figure 2.69   Speed frame Differential

Spindle speed = Speed of main shaft × $\dfrac{T_K}{T_L} \times \dfrac{T_M}{T_N}$

$$= 705 \times \frac{42}{36} \times \frac{45}{37}$$

$$= 1000.3 \text{ rpm.}$$

Bobbin speed $= 705 \times \dfrac{T_A}{T_B} \times \dfrac{T_C}{T_D} \times \dfrac{T_G}{T_H} \times \dfrac{T_I}{T_J}$

$$= 705 \times \frac{32}{16} \times \frac{15}{33} \times \frac{41}{32} \times \frac{45}{37}$$

$$= 998.7 \text{ rpm.}$$

Initial stage

Bobbin speed = spindle speed + Excess speed required

$$= 1000.3 + 250$$

$$= 1250.3 \text{ rpm}$$

$$= 1250.3 \times \frac{T_J}{T_1} \times \frac{T_H}{T_G}$$

$$\text{Speed of wheel G} = 1250.3 \times \frac{37}{45} \times \frac{32}{41}$$

$$= 802.4 \text{ rpm}$$

$$e = +0.909$$
$$m = 705 \text{ rpm}$$
$$n = 802.4 \text{ rpm}$$
$$n = a + e\,(m - a)$$
$$802.4 = a + 0.909\,(705 - a)$$
$$802.4 - 640.85 = a - 0.909\,a$$
$$a\,(1 - 0.909) = 802.4 - 640.85$$
$$a = \frac{802.4 - 640.85}{0.091}$$
$$a = 1775.27 \text{ rpm.}$$

Final stage

$$\text{Bobbin speed} = \text{Spindle speed} + \text{Excess speed required}$$
$$= 1000.3 + 80$$
$$= 1080.3 \text{ rpm}$$

$$\text{Speed of wheel G} = 1080.3 \times \frac{37}{45} \times \frac{32}{41}$$

$$= 693.27 \text{ rpm}$$

$$n = a + e\,(m - a)$$
$$693.27 = a + 0.909\,(705 - a)$$

$$693.27 - 640.85 = a - 0.909\,a$$
$$a = 576.10 \text{ rpm}$$

Smith differential

21. From the following gearing diagram find the necessary data (Figure 2.70).

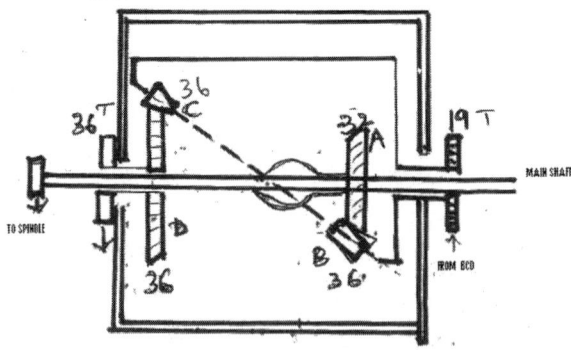

Figure 2.70 Smith Differential

$$e = \frac{T_A}{T_B} \times \frac{Tc}{T_D}$$

$$= \frac{32}{36} \times \frac{36}{36}$$

$$= \frac{+8}{9}$$

$$e = \frac{n-a}{m-a}$$

$$n = a + e\,(m - a)$$

$$= a + \frac{8}{9}\,(m - a)$$

$$= a + \frac{8}{9}\,m - \frac{8}{9}\,a$$

$$= \frac{8}{9}\,m + \left(a - \frac{8}{9}\,a\right)$$

$$a - \frac{8a}{9} = \frac{9a - 8a}{9}$$

$$N = \frac{8}{9}\,m + \frac{1}{9}\,a$$

T and S differential motion (Figure 2.71)

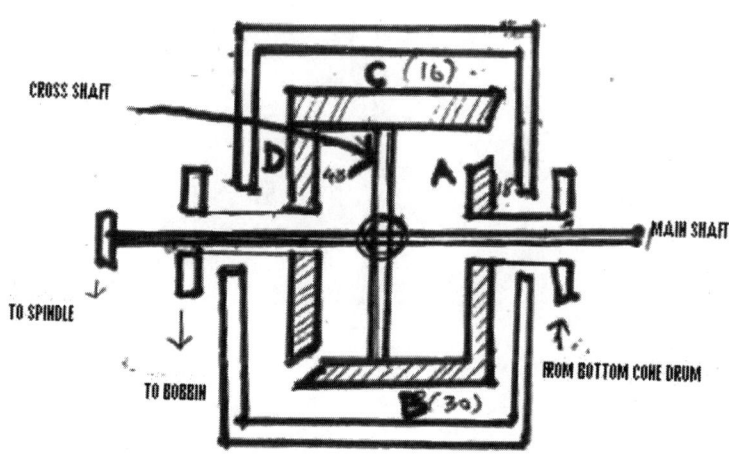

Figure 2.71   T & S Differential motion

$$e = \frac{T_A}{T_B} \times \frac{T_C}{T_D}$$

$$= \frac{18}{30} \times \frac{16}{48}$$

$$= +\frac{1}{5}$$

$$n = a + e\,(m - a)$$

$$n = a + \frac{1}{5}(m - a)$$

$$= a + \frac{1}{5}m - \frac{1}{5}a$$

$$= \frac{1}{5}m + \left(a - \frac{1}{5}a\right)$$

$$= \frac{1}{5}m + \left(\frac{5a - 1a}{5}\right)$$

$$n = \frac{1}{5}m + \frac{4}{5}a$$

Speed frame differential (Figure 2.72)

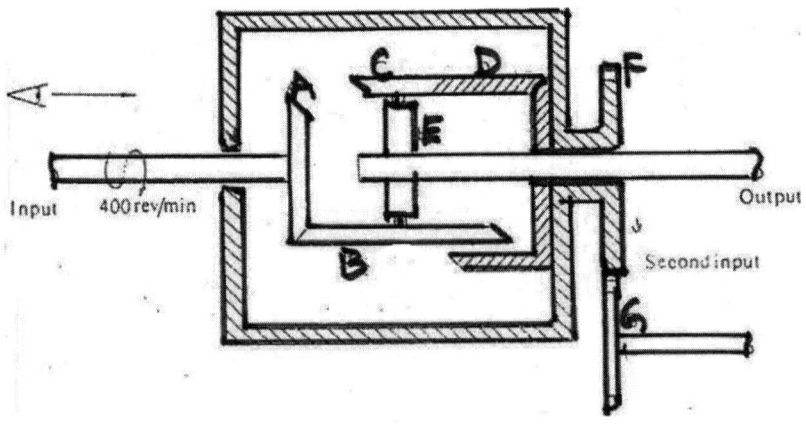

Figure 2.72   Brookes Differential Motion

Here A – 18, B – 30, C – 16, D – 48, E – arm of gear
The value of the system is:

$$e = +\frac{18}{30} \times \frac{16}{48} = +\frac{1}{5}.$$

$$e = \frac{l-a}{f-a}, \text{ when } f = 400$$

$$\frac{1}{5} = \frac{0-a}{400-a},$$

so that

$$a = -\frac{400}{4} = -100 \ rev/\text{min}.$$

Hence the output shaft rotates anti-clockwise at 100 rev/min.

$$\frac{1}{5} = \frac{20-a}{400-a}.$$

From which

$$a = -\frac{300}{4} = -75 \ rev/\text{min}.$$

If, instead of a clockwise 20 rev/min., we had fed in an anti-clockwise 20 rev/min., the output speed would be given by

$$\frac{1}{5} = \frac{-20-a}{400-a}.$$

From which

$$a = -\frac{500}{4} = -125 \ rev \ / \ min.$$

Table 2.1 shows how the resultant speed can be determined by the tabular method. The example for a positive speed of 20 rev/min. for the second input is illustrated.

Table 2.1 Analysis of differential gear

Conditions	A	D	E
Gears fixed on arm	a	a	a
Arm fixed	$400 - 4$	$\frac{1}{5}(400 - a)$	0
Resultants	400	20	a

Hence

$$20 = \frac{1}{5}(400 - a) + a$$

$$= 80 + \frac{4}{5}a.$$

i.e.

$$-60 = \frac{4}{5}a$$

and thus,

$$a = -\frac{5 \times 60}{4} = -75 \ rev/\text{min}.$$

Follows differential gear found on simplex (Figure 2.73)

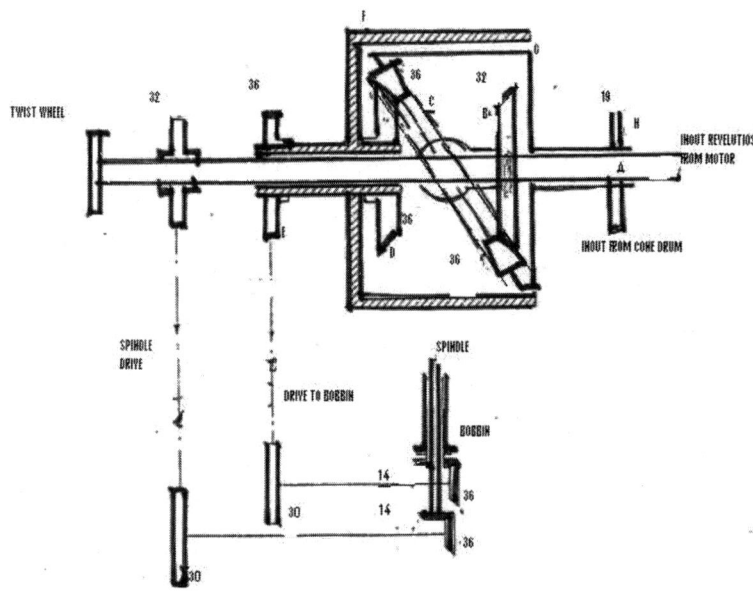

Figure 2.73   Follows Differential

In this train, the value of e is:

$$+\frac{32}{36} \times \frac{36}{36} = \frac{8}{9}$$

Consider the train operating with the arm (housing) G stationary and the shaft A rotating at 365 rev/min. Then

$$\frac{8}{9} = \frac{l - 0}{f - 0} = \frac{l - 0}{365 - 0}$$

Hence

$$8 \times 365 = 9 \times l,$$

From which the output rev/min. at the wheel E would be

$$\frac{8 \times 365}{9} \cong 325.$$

Under these conditions, the bobbin speed be

$$325 \times \frac{36}{30} \times \frac{36}{14} = 1000 \; rev/\text{min}.$$

Belt, rope and chain drives

3.1 History of belt drives

Power transmission belting has been used for more than 200 years. The first belts were flat and ran on flat pulleys. Later, cotton or hemp rope was used with V-groove pulleys to reduce belt tension. This led to the development of the vulcanised rubber V-belt in 1917. The need to eliminate speed variations led to the development of synchronous or toothed belts about 1950 and development of fabric-reinforced elastomer materials. Today, flat, V, and synchronous belting is still being used in power transmission. When compared to other forms of power transmission, belts provide a good combination of flexibility, low-cost, simple installation and maintenance, and minimal space requirements.

Belts, pulleys, and sheaves for high-volume produced products such as home appliances and passenger car engines are usually custom designed and manufactured by the thousands for specific functions and operating conditions. Standard belts, include flat, classical V, narrow V, double V, V-ribbed, joined V, and synchronous designs. Advantages of belt drives include:

1. No lubrication is required, or desired.
2. Maintenance is minimal and infrequent.
3. Belts dampen sudden shocks or changes in loading.
4. Quiet, smooth operation.
5. Sheaves (pulleys) are usually less expensive than chain drive sprockets and exhibit little wear over long periods of operation.

Drawbacks of belt drives that are more important in some applications than others are:

1. Endless belts usually cannot be repaired when they break. They must be replaced.
2. Slippage can occur, particularly if belt tension is not properly set and checked frequently. Also, wear of belts and sheaves, and bearings can reduce tension, which makes retensioning necessary.
3. Adverse service environments (extreme temperature ranges, high moisture, oily or chemically-filled atmospheres, etc.) can damage belts or cause severe slipping.
4. Length of endless belts cannot be adjusted.

3.2   Belt types

All power transmission belts are either friction drive or positive drive. Friction drive belts rely on the friction between the belt and pulley to transmit power. They require tension to maintain the right amount of friction. Flat belts are the purest form of friction drive while V-belts have a friction multiplying effect because of wedging action on the pulley. Positive drive or synchronous belts rely on the engagement of teeth on the belt with grooves on the pulley. There is no slip with this belt except for ratcheting or tooth jumping. Flat belts modern flat belts are made with reinforced, rubberised fabric that provides strength and high friction levels with the pulley (Figure 3.1). This eliminates the need for high tension, lowering shaft and bearing loads. Flat belts can transmit up to 150 hp/in at speeds exceeding 20,000 fpm.

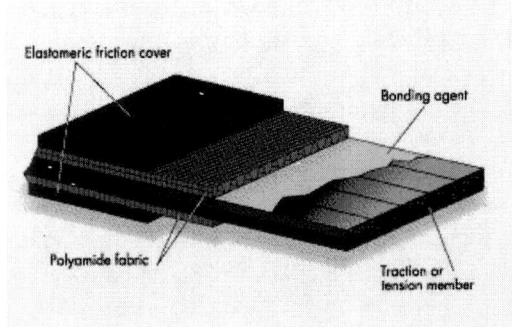

Figure 3.1 Flat belts have thin cross-sections and wrap around pulleys easily

A significant advantage of flat belts is efficiency of nearly 99%, about 2.5%–3% better than V-belts. Good efficiency is due to lower bending losses from a thin cross-section, low creep because of friction covers and high modulus of elasticity traction layers, and no wedging action into pulleys. Pulley alignment is important to flat belts. Belt tracking is improved by crowning at least one pulley, usually the larger one. Flat belts are forgiving of misalignment; however, proper alignment improves belt life. Different flat belt surface patterns serve various transmission requirements. In high-horsepower (hp) applications and outdoor installations, longitudinal grooves in the belt surface reduce the air cushion flat belts generate. The air cushion reduces friction between the pulley and belt. The grooves nearly eliminate the effects of dirt, dust, oil, and grease and help reduce the noise level. Flat belts operate most efficiently on drives with speeds above 3000 fpm. Continuous, smooth-running applications are preferred. Speed ratios usually should not exceed 6:1. At higher ratios, longer centre distances or idlers placed on the slack side of the belt create more wrap around the smaller pulley to transmit the required load.

3.2.1   V-belts

V-belts are commonly used in industrial applications because of their relative low-cost, ease of installation, and wide range of sizes (Figure 3.2). The V-shape makes it easier to keep fast-moving belts in sheave grooves than it is to keep a flat belt on a pulley. The biggest operational advantage of a V-belt is the wedging action into the sheave groove. This geometry multiplies the low tensioning force to increase friction force on the pulley sidewalls (Figure 3.3).

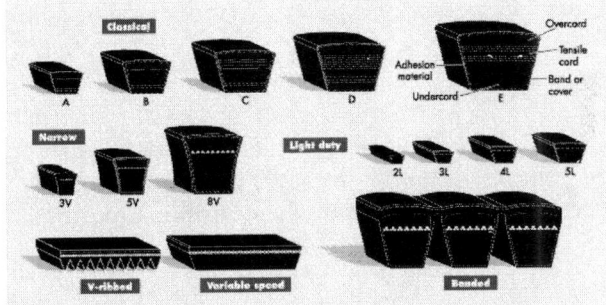

Figure 3.2 V-belts come in different types

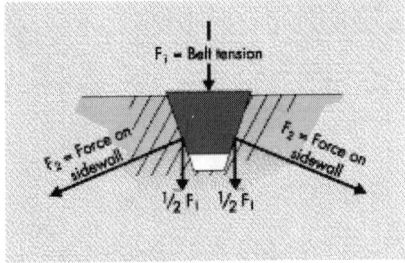

Figure 3.3 V- Belt Geometry

3.2.2   Classical V-belts

Classical V-belts are frequently used individually, particularly in A and B sizes. The larger C, D, and E sizes generally are not used in single-belt drives because of cost penalties and inefficiencies. Multiple A or B belts are economical alternatives to using single-belt C, D, or E sections. Narrow V-belts, for a given width, offer higher power ratings than conventional V-belts. They have a greater depth-to-width ratio, placing more of the sheave under the reinforcing cord. These belts are suited for severe duty applications, including shock and high starting loads.

Banded V-belts solve problems conventional multiple V-belt drives have with pulsating loads. The intermittent forces can induce a whipping action in multiple-belt systems, sometimes causing belts to turn over. The joined configuration avoids the need to order multiple belts as matched sets. Banded V-belts should not be mounted on deep-groove sheaves, which are used to avoid turnover in standard V-belts. Such sheaves have the potential for cutting the band of joined belts. Extremely worn sheaves produce the same result – ribbed belts combine some of the best features of flat belts and V-belts. The thin belt operates efficiently and can run at high speeds. Tensioning requirements are about 20% higher than V-belts. The ribs ensure the belt tracks properly, making alignment less critical than it is for flat belts.

3.2.3   Synchronous belts

Synchronous belts have a toothed profile that mates with corresponding grooves in the pulleys, providing the same positive engagement as gears or chains. They are used in applications where indexing, positioning, or a constant speed ratio is required.

The first tooth profile used on synchronous belts was the trapezoidal shape (Figure 3.4). It is still recognized as standard. Recent modifications to tooth profiles have improved on the original shape. The full-rounded profile distributes tooth loads better to the belt tension members. It also provides greater tooth shear strength for improved load capacity.

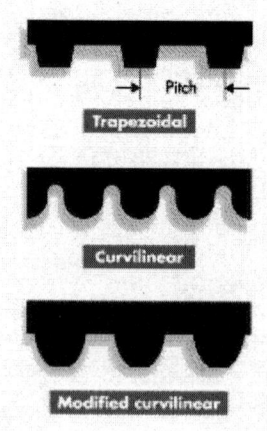

Figure 3.4   Synchronous belts have several tooth shapes

A modified curvilinear tooth design has a different pressure angle, tooth depth, and materials for improved load/life capacity and non-ratcheting resistance. Synchronous belts can wear rapidly if pulleys are not aligned properly, especially in long-centre-distance drives, where belts tend to rub against pulley flanges. To prevent the belt from riding off the pulleys, one of them is usually flanged. A recent development has produced a belt and pulley that use a V-shaped, instead of straight, tooth shape. It runs quieter than the other shapes and doesn't require pulley flanges. Under tensioning causes performance problems. The drive may be noisy because belt teeth do not mate properly with pulley grooves or the belt may prematurely wear from ratcheting. High forces generated during belt ratcheting are transmitted directly to shafts and bearings and can cause damage.

3.2.4   Link belts

Link-type V-belts consist of removable links that are joined to adjacent links by shaped ends twisted through the next link (Figure 3.5). With this design, belts can be made up of any length, reducing inventory. The belts are available in 3L, A/4L, B, C, and D widths in lengths from 5 to 100 ft.

Figure 3.5   Link-type belts are used to make instant V-belt replacements

These belts can transmit the same horsepower as classic V-belts. The links are made of plies of polyester fabric and polyurethane that resist heat, oil, water, and many chemicals. Advantages of link belts include quickly making up matched sets, fast installation because machinery doesn't have to be disassembled, and vibration dampening. Disadvantages include cost and the possible generation of static charges. The belt should be grounded when used in high-dust applications.

3.3 Alignment

Misalignment is one of the most common causes of premature belt failure (Figure 3.6). The problem gradually reduces belt performance by increasing wear and fatigue. Depending on severity, misalignment can destroy a belt in a matter of hours. Sheave misalignment on V-belt drives should not exceed 1/2° or 1/10" of centre distance. For synchronous belts it should not exceed 1/4° or 1/16" of centre distance.

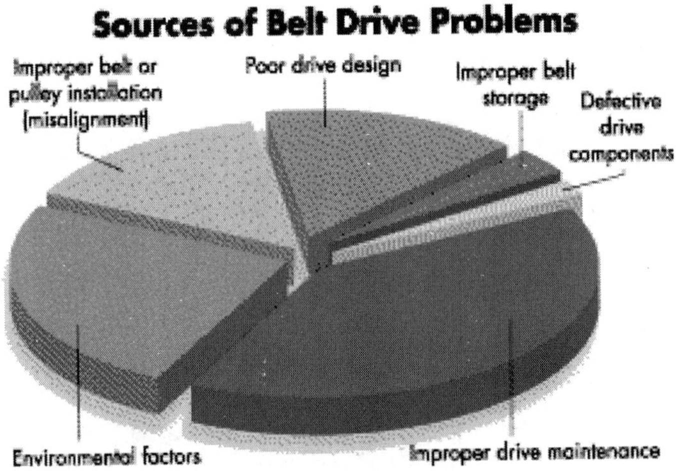

Sources of Belt Drive Problems

Improper belt or pulley installation (misalignment)

Poor drive design

Improper belt storage

Defective drive components

Environmental factors

Improper drive maintenance

Figure 3.6 Improper drive maintenance is the biggest source of belt drive problems

Angular misalignment (Figure 3.7) results in accelerated belt/sheave wear and potential stability problems with individual V-belts. A related problem, uneven belt and cord loading, results in unequal load sharing with multiple belt drives and leads to premature failure. Angular misalignment has a severe effect on synchronous belt drives. Symptoms such as high belt tracking forces, uneven tooth/land wear, edge wear, high noise levels, and potential failure due to uneven cord loading are possible. Wide belts are more sensitive to angular misalignment than narrow belts.

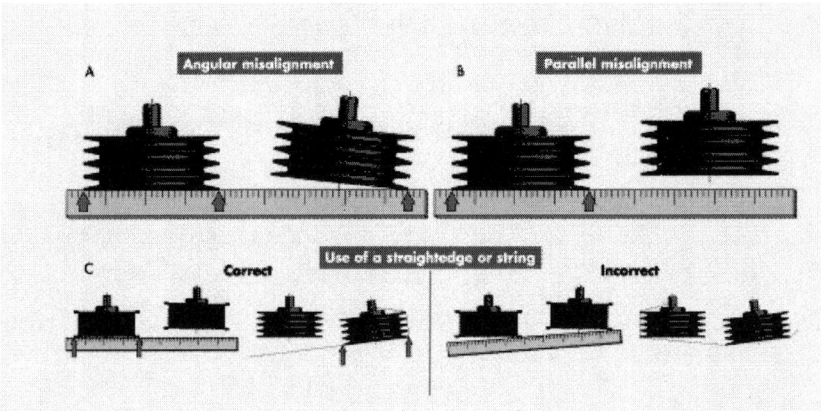

Figure 3.7 Misalignment causes belt wear, noise and excessive temperatures

Parallel misalignment also results in accelerated belt/sheave wear and potential stability problems with individual belts. Uneven belt and cord loading is not as significant a concern as with angular misalignment. Parallel misalignment is typically more of a concern with V-belts. They run in fixed grooves and cannot free float between flanges to a limited degree as synchronous belts can. Parallel misalignment is generally not a critical concern with synchronous belts as long as the belt is not trapped or pinched between opposite sprocket flanges and tracks completely on both sprockets.

3.3.1   Advantages belt drives

- Cleanliness, lubrication-free, wide selection of speed ratios.
- Can provide variable speeds.
- Quiet operation, efficiency over 95%, transmits power between widely spaced shafts.
- They don't require parallel shafts.
- Belts drives are provided with overload and jam protection.
- Noise and vibration are damped out. Machinery life is increased because load fluctuations are shock-absorbed.
- They require less maintenance cost.
- They are very economical when the distance between shafts is very large.

3.3.2   Disadvantages of belt drives

- Need to retension periodically.
- Deterioration from exposure to lubricants or chemicals.
- Cannot be repaired, must be replace.
- In belt drives, angular velocity ratio is not necessarily constant or equal to the ratio of pulley diameters, because of slipping and stretching.
- Heat build-up occurs. Speed is limited to usually 35 m/s. Power transmission is limited to 370 kilowatts.
- Operating temperatures are usually restricted to −35°C–85°C.
- Some adjustment of centre distance or use of an idler pulley is necessary for wearing and stretching of belt drive compensation.

3.3.3   Belts used with gear drive

Figure 3.8 Belts with gear drive

Most of the textile spinning machineries use this type of drive to get the positive drive and one can observe the use of tension pulley. Most of the modern bale openers or bale blenders are provided with this type of drive (Figure 3.8).

Design considerations – Belt type, belt materials, belt and sheave construction, power requirements of the drive, speeds of driving and driven sheaves, sheave diameters, and sheave centre distance are key belt drive design considerations.

Belt creep and slip – All belts (except synchronous) creep, but creep must be differentiated from slip. For example, a V-belt under proper tension creeps about 0.5% because of its elasticity and the changes in cross-section and length taking place as a section of the belt moves from the tight side to the slack side of the drive and back. That cyclical stressing, plus the bending action of the belt as it travels around the sheaves, causes only a slight increase in belt temperature. Most of that heat will be dissipated by the sheaves so that they will be only slightly warm if touched. Slip, which is a movement greater than the 0.5% creep, can create enough heat to be very uncomfortable if the sheaves are touched (again, when the drive is stopped). Another way to check for slip is to touch the belt (when it is stopped). If the belt is uncomfortable to the touch (over 140°F), it probably needs to be tightened.

3.3.4   Belt tension

Total tension required in a belt drive depends on the type of belt, the design horsepower, and the drive rpm. Since running tensions cannot be measured, it is necessary to tension a drive statically. The force/deflection method is most often used. Once a calculated force is applied to the centre of a belt span to obtain a known deflection, the recommended static tension is established. Most design catalogues provide force and deflection formulas. With too little tension in a V-belt drive, slippage can occur and lead to spin burns, cover wear, overheating of the belt, and possibly overheating of bearings. Not enough tension in a synchronous belt causes premature tooth wear or possible ratcheting that will destroy the belt and could break a shaft. When installing a new belt, installation tension should be set higher. Generally 1.4–1.5 times the normal static tension. This is necessary because drive tension drops rapidly during the seating-in process. This extra initial tension does not affect bearings because it decays rapidly. Checking operating temperatures of sheaves and belts (the touch test) is but one of several ways experienced machine operators and plant maintenance people check belt tension without the need for complicated measurements and calculations. Other equally simple and useful belt tensioning approaches involve visual and sound techniques. The bow or sag on the slack side of belt drives increases as a drive approaches full load. Undulations and flutter on the slack side of belts can be very informative to the experienced eye. Proper tensioning can reduce or eliminate these conditions. Tension adjustment based on belt sound can be a useful technique. Loads such as industrial fans require peak torque at starting. If belts squeal as the motor comes on or at some subsequent peak load, experienced belt people say the belts should be tightened until the squeal disappears.

Calculating V-belt tension – Represents a belt drive arrangement which can be used to study factors affecting V-belt tensions and stresses. At standstill, the belt strand tensions T_1 and T_2 are equal. When load is applied to the driving pulley, tension T_1 increases and T_2 decreases. T_1, tight side tension, divided by T_2 slack side tension, is the tension ratio of a belt drive. The formula for tension ratio is as follows:

Where

$$\frac{T_1}{T_2} = e^{kf\varphi}$$

$e = 2.718$ (the base of natural logarithm).
k = Wedging factor, which is considered constant for a given drive.
f = Dynamic coefficient of friction between belt and sheave.
f = Wrap angle (are of contact on the smaller sheave in radians).
k applies only to V-belt drives. A practical tension ratio of 5 for V-belts allows for such variations as lack of rigidity of mountings, low initial tension, moisture, and load fluctuations. For rubber flat belt drives, a tension ratio of 2.5 is considered most efficient. The difference between tight-strand and slack-strand tensions is effective tension, or $T_e = T_1 - T_2$. Effective tension is the actual force that turns the driven sheave.

Stress and power ratings – The usable strength of a belt is the tensile stress the belt withstands for a specified number of stress cycles, usually the equivalent of 3 years of continuous operation, or about 25,000 h. The relationships of effective tension, T_0 (lb), belt stress ratings, S_p (psi), and cross-sectional area, A (in.^{2}), of the belt are given in the formula $T_0 = AS_p$.

For belt drive transmitted horsepower:
Where

$$hp = \frac{T_e V}{33,000}$$

V = Belt velocity, fpm = 0.262DN
D = Pulley diameter, in
N = Pulley speed, rpm

Great advancements have been made in various materials used in V-belts in recent years. Horsepower (hp) ratings may vary considerably from one belt to another. It is important to provide greater tension for belts carrying higher power. Otherwise the newly designed belts will not run properly; they may slip, vibrate too much, or turn over in the grooves. All torque is transmitted through the belts, and torque does not change for a given horsepower and speed. This basic concept is best revealed in a multiple-belt drive. For example, a multiple V-belt drive formerly requiring seven belts may need only five higher-rated belts with the same cross-section. For five belts to deliver the same horsepower as seven belts, each of the five new belts must carry more load. Increased tension in the individual belts will not overload the bearings, as some engineers may fear, because total tension in the drive should be exactly the same value. Each belt has a higher tension, but there are fewer belts to transmit the tension to bearings.

3.3.5   Materials of belt

In general the belting materials are cotton and synthetic yarns, both spun filament and continuous filament. Belt carcasses can be impregnated and coated
with elastomers or synthetic resin. If an endless belt less than 0.010″ thick is
needed, specify one made of polyester film. Flat belts with tension members
made of nylons and polyesters are popular because they offer high strength-
to-weight ratios and negligible permanent stretch. More favourable friction
characteristics can be achieved by laminating nylon or polyester flat belting
with a friction surface of chrome leather, polyurethane, rubber, poly vinyl
chloride (PVC), or other material. Laminated belts are used widely in industrial drives ranging from fractional horsepower to more than 6,000 hp at belt
speeds to 20,000 fpm.

Leather belts are very largely used, but woven belts have replaced
leather to a considerable extent. Woven belts are generally made of cotton,
though other materials are used, and the belts can be made of any length,
width, and thickness without joints. Generally the belt is woven direct
of the required width and thickness, but sometimes the canvas is woven
of single thickness, then folded up, cemented and stitched. Leather belts
are generally made by joining together the 4 ft or 5 ft lengths in which
leather is obtainable, the joints being made in the form of a cemented and
laced splice which does not materially increase the belt thickness. Leather
strips are also built up into endless belts of any width and thickness, but
such belts are expensive. The ends of the belt have generally to be joined
by some temporary fastener, so that the belt may be shortened and tightened up if it stretches. If the pulley centres can be adjusted, an endless
belt, or one with a permanent fastener, may be used, which gives better
results, since the permanent joint may have a strength of about 70% of
the belt strength as compared with 25%–40% for a temporary fastener.
Temporary fasteners are made in much variety, but not many are quite satisfactory. Such a fastener should be strong and flexible and should be free
from projecting parts, such as nuts, which are dangerous. The fastener
ought also to be easy to put in place. For leather belts the old-fashioned
butt joint with a leather thong lacing is as good as most, though it takes
some considerable time to make.

Metal belts – Flat belts made of metal offer lightweight, compact drives
with little or no stretch. Endless belts are made by butt welding the ends with
laser or electron-beam methods. Belts are generally available in thicknesses
ranging from 0.002″ to 0.030″ and width from 0.030″ to 24″. Circumferential
length ranges from 6″ to about 100 ft. Usually made of stainless steel, metal
belts have a high strength-to-weight ratio, low-creep, high accuracy, and resistance to corrosion and high temperature. Some applications take advantage

of the electrical conductivity of metal belts to ground conveyed parts that are sensitive to static electricity. Attachments, such as pins and brackets, can be provided on metal belts to carry, position, or locate parts being transported on the belt.

Endless round belts (tangential belt drive) – An elastomeric O-belt is a seamless, circular belt that features a round cross section and an ability to stretch. Although O-belts look like O-rings, they are designed for power transmission applications. The elasticity of O-belts simplifies design problems and reduces costs. However, they are limited to sub-fractional horsepower applications such as recorders, projectors, and business machines. O-belt materials include natural rubber and four polymers – neoprene, urethane, ethylene-propylene-terpolymer (EPT), and ethylene-propylene-diene-monomer (Figure 3.9).

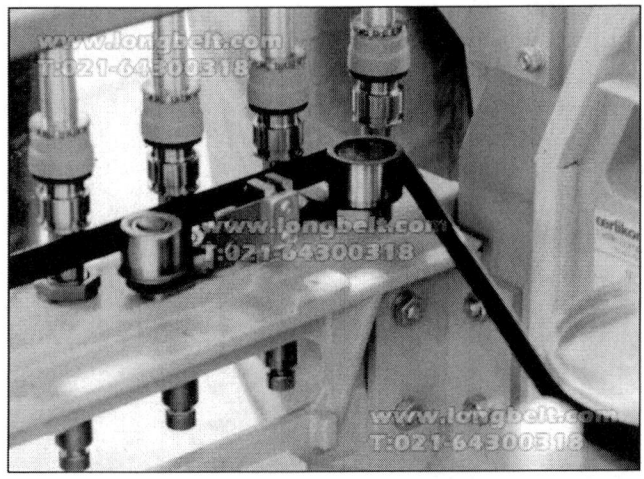

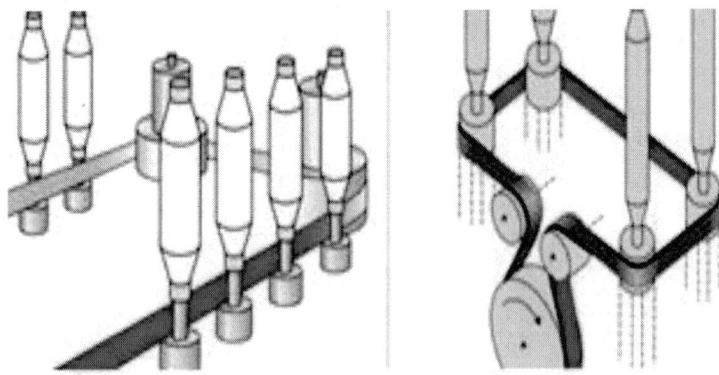

Figure 3.9   Tangential belt drive in Zinser ring frame

3.4   Applications of belt drives in textile machineries

Spinning
Blow room
Mixing bale opener (V and flat belt), Mono cylinder (V-belt), ERM cleaner (V-belt), Hopper feeder (V and flat belt), Scuthcer (V and flat belt).

Carding
Cylinder drive (flat belt), Licker in crossed-flat belt drive, Drive to flats from cylinder (Ols Card) flat belt
Draw frame – V-belt
Simplex – V-belt
Ring frame – V-belt and tangential belt.

Winding
Precision winder – Double flanged bobbin winder – Flat belt
Cone winder – V-belt for motor drive and flat belt for conveyer system.

Warping
Beam warping – V-belt for driving the drum
Sectional warping – Flat belt for drum drive and V-belt for motor drive.

Sizing
Flat belt for beam drive.

V-belts – Available from virtually all power transmission components distributors, standard V-belts are adaptable to practically any drive, although sometimes they may not be optimal in terms of life-cycle cost or compactness. Besides their wide availability, V-belts are often used in industrial and commercial applications because of their relative low-cost, ease of installation and maintenance, and wide range of sizes. The V-shape obviously makes it easier to keep fast-moving belts in sheave grooves than it is to keep a flat belt on a pulley. Probably the biggest operational advantage of a V-belt is that it is designed to wedge into the sheave groove, which multiplies the frictional force it produces in tension and, in turn, reduces the tension required to produce equivalent torque. Naturally, wedging action requires adequate clearance between the bottom of the belt and the bottom of the sheave groove. The effect of the wedging factor, k, on the belt tension ratio discusses V-belt tension calculations.

When V-belts first appeared in industrial applications to replace wide flat belts, it was not unusual to sue 10–15 belts between a single pair of shafts. Thus term "multiple" belts originated, which today is referred to as "classical multiple" or "heavy-duty conventional" belts.

Classical V-belts and mating sheaves have been standardised with letter designations from A through E, small to large cross-sections. Those standard sizes are recognised worldwide. A and B sizes are frequently used individually but not the C, D, and E sizes because of cost and efficiency penalties. Belts with cogged or notched bases permit more severe bends, which allows operation over smaller diameter sheaves. Although classical V-belts can be used in some applications individually, they tend to be over designed for a number of light duty applications. Thus, a special category of belts has evolved under the description *single V-belts*; they are also denoted as fractional horsepower and light duty. Cross-sectional size designations run from 2L (the smallest) to 5L (the largest). The 4L and 5L sections are dimensionally similar to A and B classical belts and can operate interchangeably on A and B sheaves.

Narrow V-belts are the latest step in the evolution of a single-belt configuration. For a given belt width, narrow belts offer higher power ratings than conventional V-belts. Narrow belt size designations are standardised as 3V, and 5V, and 8V. They are also available in notched designs to maximize bending capability.

Cogged, raw-edge belts have no cover, thus, the cross-sectional area normally occupied by the cover is used for more load-carrying cord. Cogs on the inner surface of the belt increase air flow to enhance cooler running. They also increase flexibility, enabling the belt to operate with smaller sheaves. Cogged belts are available in both AX, BX, and CX Classical V-shapes plus 3VX and 5VX narrow V configurations.

Double sided or hexagonal belts come in AA, BB, and CC narrow V-belt cross-sections. These belts transfer power from either side in serpentine drive configurations where a single belt operates with multiple pulleys.

Joined V-belts solve special problems for conventional multiple V-belt drives produced by pulsating loads, such as those generated by internal combustion engines driving compressors. The intermittent forces can produce a whipping action in multiple belt systems, which sometimes cause belts to turn over in the grooves. The basic belt element can be either classical or narrow. The joined configuration avoids the need to order multiple belts as matched sets. Joined 5V and 8V belts are available with aramid fibre reinforcement which offer extremely high power capacity – up to 125 hp per inch of width.

V-ribbed belts combine some of the best features of flat belts and V-belts. Tensioning requirements are not as high as flat belts but are about 20% more than V-belts. Five standard configurations are available, with designations H, J, K, L, and M. The M section is capable of transmitting up to 1,000 hp. The power density permits compact permits compact drive configurations, but the belt is usually applied only in mass-produced products.

Belts for variable-speed drives require special care in selection. Special wide belts have been developed for variable-speed applications that accommodate speed variations of up to 9:1.

Synchronous (timing) belts are used where input and output shafts must be synchronised. Trapezoidally shaped teeth of the belt mate with matching grooves in the pulleys to provide the same positive, no-slip engagement of chain or gears. Because stable belt length is essential for synchronous belts, they were originally reinforced with steel. Today, glass fibre reinforcement is common and aramid is used if maximum capacity is required. Modifications of traditional trapezoidal tooth profiles to more circular forms offer more uniform load distribution, increased capacity, and smoother, quieter action. These newer synchronous belts incorporate a rounded curvilinear tooth design to handle the higher torque capabilities normally associated with chain drives.

Link-type V-belts consist of removable links that are joined by T-shaped rivets or interlocking tabs. These belts offer application advantages such as installation without dismantling drive components, reduced belt inventory (no need for different lengths), wide temperature range, and resistance to chemicals, abrasion, and shock loads. A matrix of polyester fabric and polyurethane elastomer enables link-type belts to meet the horsepower ratings of classical V-belts. They are available in 3/8–7/8″ width for speed up to 6,000 rpm.

Materials of construction

The composition of a conveyor belt can be considered in two parts:

a)   The carcass, whether ply type (textile) or steel cord construction, which must have sufficient strength to handle the operating tensions and to support the load.

b)   The covers, which must have the required physical properties and chemical resistance to protect the carcass and give the conveyor belt an economical life span.

The general properties and the application usage of the more economical available reinforcement fabrics and rubber compounds are discussed here.

3.5   Reinforcements

3.5.1   Fabrics

Fabrics that are commonly used as reinforcement in conveyor belts are shown in Table 3.1.

The fabric designation indicates the material used in both warp and weft, e.g., PN signifies that the fabric has polyester warp fibres and nylon weft fibres.

The ultimate strength of the belt in kilo newton per metre (kN/m) width is shown along with the number of piles. PN1000/4 designates a belt with four piles of polyester warp, nylon weft fabric and an ultimate full-belt tensile strength of 1000 kN/m. Alternatively, the belt can be often described as 4 ply PN250 where the strength of the individual plies is shown.

Table 3.1 Reinforcement fabrics

Carcass type	Carcass materials		Strength range	Feature and applications
	WARP (longitudinal)	WEFT (transverse)	kilo newton per metre width	
PN Plain weave (DINcode EP)	Polyester	Nylon	315–2000 kN/m (150–400 kN/m/ply)	Low elongation. Very good impact resistance. Good fastener holding. An excellent general purpose fabric
PN Crow's foot weave	Polyester	Nylon	630–2500 kN/m (315–500 kN/m/ply)	Low elongation. Good impact resistance. Very good fastener holding. Excellent rip resistance. For high abuse installations
PN Double weave	Polyester	Nylon	900 &1350 kN/m (450 kN/m/ply)	Low elongation. Excellent impact resistance. Excellent fastener holding. For high abuse installations
PP Plain weave	Polyester	Polyester	Up to 900 kN/m (120 & 150 kN/m/ply)	Used in special applications where acid resistance is needed. Contact us for information
NN Plain weave	Nylon	Nylon	Up to 2000 kN/m (150–450 kN/m/ply)	High elongation, mostly replaced by polyester-nylon. Used in special applications where low modulus needed or in high pH environment
CC Plain weave	Cotton	Cotton	Up to 400 kN/m (65–70 kN/m/ply)	Used in special applications such as plasterboard belting and hot pellet handling

SW Solid woven	Nylon/Cotton or Polyester/ Cotton	Nylon/Cotton	600–1800 kN/m	Main use in underground coal mining. Good fastener holding and impact resistance. Used for bucket elevators
ST Steel cord	Steel cord	None (special reinforcement available)	500–7000 kN/m	Very low elongation and high strength. Used for long haul and high-tension applications
AN Aramid nylon (Kevlar)	Polyaramide	Nylon	630–2000 kN/m	Low elongation, high strength, low weight. Used on high-tension applications and on equipment conveyors

3.5.2   Belt capacities

For maximum haulage efficiency, conveyors should be operated fully loaded at maximum recommended speed.

Belt capacity is dependent upon these inter-related factors:

Belt width

Minimum belt width may be influenced by loading or transfer point requirements, or by material lump size and fines mix– refer to Table 3.4. Trough ability and load support restrictions will also influence final belt width selection.

Belt speed

Possible belt speed is influenced by many factors, importantly the loading, discharge and transfer arrangements, maintenance standards, lump sizes, etc. Typical belt speeds are shown in Table 3.2.

Material bulk density and surcharge angle

Due to undulations of the belt passing over the conveyor idlers, the natural angle of repose of the material is decreased. This decreased angle known as ANGLE OF SURCHARGE is one of the most important characteristics in determining carrying capacity as it directly governs the cross-sectional area of material on the belt and hence the "volume" being conveyed.

Table 3.4 shows bulk density and surcharge angles for some common materials. With materials which slump readily, such as fine powders or dust, or on long conveyors where the load may settle, consideration should be given to using a reduced surcharge angle for capacity determination, and may require the compensatory use of other factors (such as greater belt width or speed) to provide the required capacity.

Inclination angle

The angle of inclination of a conveyor changes the carrying capacity. The load cross-section area of an inclined load is reduced when viewed in a vertical plane as the surcharge angle is reduced perpendicular to the belt. An approximation of the reduced capacity can be determined by multiplying the horizontal capacity by the cosine of the inclination angle (see Table 3.2). Table 3.4 shows maximum inclination angle for some common materials. Effectively the capacity reduction is usually less than 3%.

Troughing angle

For standard 3 roll idlers, the most common trough angle is 35° although trough angles from 20° to 45° are not uncommon. Steeper trough angles give increased capacity but can have consequences for convex and concave curves and transition zones.

Idler configuration

The most common configuration for idler rollers is 3 rolls of equal length. This configuration and normal clearances are shown in Figure 3.10.

Other configurations from flat belt to 5 roll idlers and with unequal roll lengths are sometimes used. For wide belts, 5 roll suspended idlers are not uncommon and Table 3.2 shows capacity for different trough and surcharge angles.

Clear edge distance $= 0.055B + 23$ mm

TROUGHED BELT CROSS-SECTION

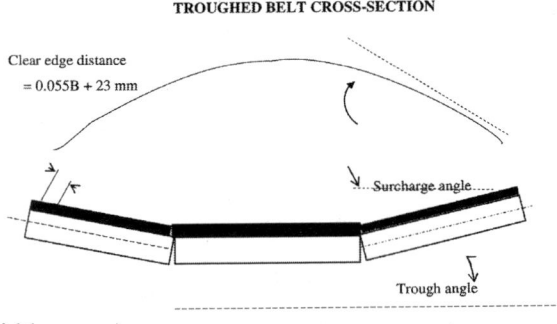

Clear edge distance
$= 0.055B + 23$ mm

Surcharge angle

Trough angle

(B = belt width – mm)

Figure 3.10   Troughed belt cross-section

3.5.3   Capacity calculations

General formula

The general formula for capacity is:

Capacity - tonnes per hour =
    $3.6 \times$ Load cross-section area * (m²) × Belt speed (m/s) × Material density (kg/m³) (* perpendicular to the belt)

For common idler configurations

Capacity can be determined using the tables provided.

Table 3.2 has been designed for quick reference and give the capacities of conveyors from 400 mm to 2200 mm wide, assuming the use of 3 roll equal length idlers at 35° troughing angle and an average material surcharge angle of 20° and bulk density of 1000 kg/m³. Capacities for conveyors using other troughing angles or materials can be obtained by multiplying the capacity shown in Table 3.3 by the appropriate CAPACITY FACTOR obtained from Table 3.2.

3 equal roll idlers
    Capacity tonnes per hour =
    Capacity × Material density (kg/m³) × Capacity factor × Belt speed (m/s) / 1000

5 equal roll idlers
    Capacity - tonnes per hour =
    Capacity × Material density (kg/m³) × Belt speed (m/s) / 1000.

Table 3.2   Capacity factor – three equal roll trough idlers

Surcharge angle	Idler troughing angle				
	20°	25°	30°	35°	45°
0°	0.43	0.53	0.61	0.69	0.81
5°	0.52	0.61	0.69	0.77	0.88
10°	0.61	0.70	0.77	0.84	0.94
15°	0.70	0.78	0.86	0.92	1.04
20°	0.79	0.87	0.94	1.00	1.08
25°	0.88	0.96	1.03	1.08	1.15

For inclined or declined conveyors, multiply the above values by the cosine of the inclination angle given in Table 3.3.

Table 3.3   Cosines

Inclination angle	0°	5°	10°	15°	17.5°	20°	22.5°	25°
Cosine	1.000	0.996	0.985	0.966	0.954	0.940	0.924	0.906

3.5.4   Properties of materials

Typical densities, angles of repose and surcharge angles for various materials are shown in Table 3.4.

For many materials these factors are subject to considerable variation, depending on the moisture content, lump size, cohesive properties, etc. Unless otherwise stated, the Tables refer to dry weight conditions, based usually on broken materials in sizes most commonly found in conveyor systems.

The physical characteristics of the material affect the operating parameters of the belt in other ways, for example, typical belt speeds, recommended maximum lump sizes, maximum slope if the belt is inclined, etc.

Where the material to be conveyed has unusual slumping characteristics, or where sufficient water is present to provide lubrication between the belt cover and the material, the slope angles to be used would be appreciably below those listed and should be determined by test or from experience in the field.

Moulded cleats can be used to raise permissible slope angles where otherwise slipping of the load on the belt would be experienced – refer to FENNER DUNLOP for advice.

Table 3.4   Properties of materials

Material	Density (kg/m³)	Angle of repose	Angle of surcharge
Acid phosphate	1540	*	*
Alumina	800–960	22°	5°
Alum – lump	800–960	27°	*
- Pulverised	720–800	35°	*
Ashes, boiler-house – dry, loose	560–690	38°–45°	25°
Asphalt	1280–1360	*	*
Bagasse – fresh, moist	120	*	25°
- Dry, loose	80	*	25°
Barytes – 50–75 mm lumps	2320–2400	30°	25°
- 15 mm screenings	2080–2320	30°	20°
- dust	1760–2080	30°	15°
Basalt – 50–75 mm lumps	1680–1760	*	25°
- 15 mm screenings	2080–2320	*	20°
- dust	1760–2080	*	15°
Bauxite – crushed	1200–1360	30°–35°	5°–15°
Borax, solid – 50–100 mm lumps	960–1040	40°	*
- 40–50 mm lumps	880–960	30°- 45°	*
Brewers grain – dry	400–480	45°	*
- wet	880–960	45°	*
Brick – hard	2000	30°- 45°	*
- soft	1600	30°- 45°	*
Carbon black – powder	80	*	*

- pellets	400	40°	*
Cement, Portland – loose	1200–1360	40°	20°
- clinker	1280–1520	33°	25°
- slurry	1440	*	5°
Chalk – 50–75 mm lumps	1280–1360	45°	*
- 40–50 mm lumps	1200–1280	40°–45°	*
Char – sugar refinery	720	*	*
Chips, paper mill – softwood	190–480	*	25°
- yellow pine	320–400	*	25°
Clay – dry, loose	1010–1440	40°–45°	15°–25°
- brick, ground fine	1760	35°	15°
Coal – 150 mm domestic sizes	830–900	*	25°
- run-of-mine	720–880	35°	25°
- slack	690–800	37°	25°
- pulverized for coking	480–590	*	10
- lignite, broken	720–880	*	25°
Cocoa	480–560	*	*
Coke – run of oven	400–480	30°	25°
- breeze	380–560	30°–45°	20°
Concrete, wet, on conveyor	1760–2400	*	5°
Copper ores, crushed	2080–2400	*	25°
Copra	350	*	*
Corn grits	670	*	*
Cryolite – 50–75 mm lumps	1600–1680	*	20°
- 15 mm screenings	1440–1600	*	15°
- dust	1200–1440	*	5°
Dolomite – lump	1440–1600	See limestone	*
Earth – as excavated, dry	1120–1280	30°–45°	20°–25°
- wet, mud	1600–1760	*	5°
Foundry refuse, old sand, cores,	960–1280	*	15°
etc.	800	*	*
Garbage – household	1680	*	*
Glass – batch	1280–1600	*	*
- broken	1360–1440	25°	*
Granite – 40–50 mm lumps	1280–1440	*	*
- 15 mm screenings	1520–1600	*	*
- broken	1440–1600	30°–45°	25°
Gravel – dry, sharp	1600–1920	32°	25°
- wet	960	*	*
Gutta percha	1200–1280	30°	20°
Gypsum – 50–75 mm lumps	1120–1280	40°	15°
- 15 mm screenings	960–1120	42°	*
- dust	560	30°–45°	*
Hops – brewery and moist	640	*	*
Ice – crushed	2000	*	*
Iron borings – machine shop	1600–3200	35°	25°
Iron ores, depends on iron	2160–2320	*	20°
percentage	1920–2160	*	15°
Iron pyrites – 50–75 mm lumps	1680–1920	*	5°
- 15 mm screenings	3200–4320	30°	15°
- dust	1440–1520	30°–40°	25°
Lead ores, depends on lead	1280–1440	*	15°
percentage	1200–1280	*	5°
Limestone – 50–75 mm lumps	760–780	*	*
-15 mm screenings	2000–2240	39°	*
- dust	570–640	*	*
Linseed cake – crushed	700	*	*
Manganese ore	640–960	*	5°
Malt meal	560–640	*	*
Meal			
Paper pulp			
Petroleum coke			

Phosphate rock	1360	*	*
Pitch	1150	*	*
Quartz, solid – 50–75 mm lumps	1440–1520	35°	*
- 40–50 mm lumps	1360–1440	35°	*
- dust	1120–1280	40°	*
Rock, soft, excavated with shovel	1600–1760	*	20°
Rubber	930	*	*
Rubber – reclaim	560	*	*
Salt – coarse	640–900	*	25°
- fines	720	25°	5°
- lump for stock	1600	*	25°
Sand – beach or river, wet	1600–2080	15°- 30°	5°- 15°
- dry	1440–1600	34°- 45°	15°
- foundry, loose	1280–1440	*	15°
- foundry, rammed lump	1600–1760	*	10°
Sandstone	1360–1440	*	*
Sawdust	160–200	35°	*
Shale – broken	1440–1600	*	*
- crushed	1360–1440	39°	*
Slag – blast furnace, crushed	1280–1440	25°	25°
- granulated, dry	960–1040	25°	10°
-granulated, wet	1440–1600	45°	10°
Slate – 40–75 mm lumps	1360–1520	*	*
- 15 mm screenings	1280–1440	28°	*
Soda ash	800–1040	32°	*
Sugar cane stalks	400	*	*
Sugar – raw	880–1040	37°	*
- refined	880	*	*
Sulphur – 50–75 mm lumps	1360–1440	35°	25°
- 15 mm screenings	1200–1360	*	15°
Talc – solid	2640	*	*
- 50–75 mm lumps	1440–1520	*	*
- dust	1220–1280	*	*
Turf – dry	480	*	*
Wheat	720–770	28°	10°
Zinc ores, crushed	2400–2560	38°	20°
Zinc oxide – light	160–480as	*	10°
- heavy	480–560	*	10°

3.6   Belt power and tensions

3.6.1   Belt power calculation formulae

Power requirements for belt conveyors may be calculated from the following formulae:

$$\text{Power} = \frac{F_c\left(L + t_f\right)\left(C + 3.6\,QS\right)}{367} \pm \frac{CH}{367}\,(kW)$$

$$\text{Or} \quad \text{Power} = \frac{F_e\left(L + t_f\right)3.6\,QS}{367} + \frac{F_l\left(L + t_f\right)}{367} \pm \frac{CH}{367}\,(kW)$$

Where

F_c, F_e, F_l = Equipment friction factors – refer item (1) below.

L = Horizontal centre to centre distance (m).

t_f = Terminal friction constant expressed in metres – refer item (2) below.

C = Capacity (t/h).

Q = Mass of moving parts expressed in kilograms per metre of centre to centre distance.

S = Belt speed (m/s).

H = Net change in elevation (m).

The values of the main factors and constants are as follows:

(1) Equipment friction factors

(a) On short centre conveyors using best quality equipment, it is often more convenient to use an average equipment friction factor (F_c) of 0.0225 for horizontal and inclined conveyors and 0.0135 on decline regenerative systems.

(b) On many systems, such as portable conveyors, "Chevron" steep angle conveyors and temporary installations using anti-friction bearings, the following value for equipment friction will apply:

$F_c = 0.030$.

(c) On longer centre conveyors and individual component tension calculations as detailed below, equipment friction factors F_e and F_l are used for empty and loaded belt conditions respectively, thus,

(i) Horizontal and elevating conveyors

$F_e = 0.020$ for empty calculations

$F_l = 0.025$ for load calculations

(ii) Regenerative decline conveyors

$F_e = 0.010$ for empty calculations

$F_l = 0.017$ for load calculations

(2) Terminal friction constant, T_f, expressed in metres of centre to centre distance

(a) Horizontal and elevating conveyors

(i)   Up to 300 m centres = 60 m

(ii) From 300 m to 1200 m = 45 m

(iii) From 1200 m to 1800 m = 30 m

(iv) Above 1800 m this influence is disregarded.

(b) Regenerative decline conveyors

$T_f = 90$ m

(c) On the systems as described, i.e., where

$F_c = 0.030$

T_f = 45 m (except for "Chevron" belt conveyors, where values as in (2) (a) are used.

(d) On rare occasions, it is possible to find long centre belts that are slightly regenerative under loaded conditions, but require more power for the empty belt. When this condition is met the terminal friction, allowance is varied as follows for empty tension calculations only,

$$T_f = 260 \text{ m.}$$

Additional friction considerations. Where the number of pulleys is large in relation to the length of the conveyor, e.g., multi-tripper conveyors, it is necessary to make allowance for the additional tension required to overcome these pulley frictions.

Also where skirt board lengths are long in relation to conveyor length, allowance should be made for the additional friction involved.

Component frictions and tensions. Should it be necessary to calculate individual component tensions for the assembly of tension diagrams on multiple slope conveyors, or for assessment of the effects of acceleration or deceleration on a particular system, the following formulae will apply:

Return side friction

$$= F_e \times Q \times L \times 0.4 \times (9.81 \times 10^{-3}) \text{ (kN)}$$

Total empty friction

$$= F_e (L + T_f) Q \times (9.81 \times 10^{-3}) \text{ (kN)}$$

Carrying side empty friction

$$= \text{Total empty friction} - \text{return side friction (kN)}$$

Load friction

$$= F_l \left(L + t_f\right) \frac{C}{3.6S} \times \left(9.81 \times 10^{-3}\right) \quad (kN).$$

Load slope tension

$$= \pm \frac{CH}{3.6S} \times \left(9.81 \times 10^{-3}\right) \quad (kN)$$

Belt slope tension

$$= \pm B \times H \times \left(9.81 \times 10^{-3}\right) \quad (kN).$$

where B is the belt weight per lineal metre, and all other symbols are as given previously.

Effective tension, T_e = Total empty friction + load friction + load slope tension.

$$= \left[F_e \left(L + t_f\right) Q + F_l \left(L + t_f\right) \frac{C}{3.6S} \pm \frac{CH}{3.6S} \right] 9.81 \times 10^{-3} \quad (kN)$$

Slack side tension $T_2 = T_e \times k$ (kN)

where k is the drive factor dependent on pulley surface, arc of contact and method of tensioning.

Power is simply calculated from Te by:

$$\text{Power} = T_e \times S \ (\text{kW})$$

and

$$T_e = \frac{kW}{S} \ (kN)$$

The above values are calculated from the basic tension relationship formula:

$$\frac{T_1}{T_2} = e^{\mu\theta}$$

Where:

T_1 = Tight side tension
T_2 = Slack side tension
e = Naperian logarithm base
μ = Coefficient of friction
θ = Arc of contact in radians

The "k" values are calculated from:

$$k = \frac{1}{e^{\mu\theta} - 1}$$

Coefficient of friction, bare pulley, dry conditions:
$\mu = 0.30$
Coefficient of friction, lagged pulley, dry conditions:
$\mu = 0.35$
Coefficient of friction, lagged and grooved pulley, wet or dry conditions:
$\mu = 0.35$

The mass of betting and idler spacing obviously affect the above values considerably and therefore this table gives conservative and average figures only. For closer determination, each individual installation should be checked using the following formula, and taking into consideration belt mass, type of idler, and idler spacing:

$$Q = 2B + \frac{W_t}{S_1} + \frac{W_r}{S_2} \ (kg \ / \ m)$$

Where:

B = Belt mass in kilograms per lineal metre.
W_t = Mass of moving parts of troughing idler (kg).
W_r = Mass of moving parts of return idler (kg). from idler manufacturers
S_1 = Spacing of troughing idlers (m).
S_2 = Spacing of return idlers (m).

"Q" values for steel cord applications should always be calculated accurately using above formula.

3.6.2   Calculation of maximum tensions

Because of the different types of belt arrangement which may be required to suit a particular application, various basic formulae are needed to determine the maximum tension, expressed in kilo newtons.

The main components of the maximum operating tension, T_{max}, are

T_e = Effective tension (kN)

T_2 = Slack side tension (kN)

T_{sag} = is the tension to prevent excess Belt sagging

1. Horizontal belts

$$T_{max} = T_e + T_2$$

2. Inclined belts

*(a) Drive at head pulley

$T_{max} = T_e + T_2$ or

$T_{max} = T_e$ + belt slope tension – return side friction + T_{sag}.

(b) Drive at tail pulley

$T_{max} = T_e + T_2$ or

$T_{max} = T_e + T_2$ + belt slope tension - return side friction.

3. Decline belts

*(a) Regenerative belt – Tai drive

$T_{max} = T_e + T_2$. Or

$T_{max} = T_e$ + belt slope tension + return side friction + T_{sag}.

(b) Regenerative belt – Head drive

$T_{max} = T_e + T_2$ + belt slope tension + return side friction.

(c) Partially regenerative – Head drive

$T_{max} = T_e + T_2$ Or

$T_{max} = T_2$ + belt slope tension + return side friction.

(d) Partially regenerative – Tail drive

$T_{max} = T_e + T_2$ or

T_{max} = Belt slope tension + return side friction + T_{sag}.

3.6.3   Graduated idler spacing

On long centre, heavily loaded, high tension conveyor systems, it is possible to use graduated idler spacing. The sag will vary inversely with the tension in the belt. Since the tension varies along the length of the belt the spacing can be graduated, being smallest at the zone of low tension and increasing as the belt tension increases. Savings can thus be effected on both the carrying and return runs. The spacing at any point can be obtained from the formula:

$$\text{Idler spacing} = \frac{8 \times T \times sag}{M_I \times \left(9.81 \times 10^{-3}\right)} \quad (m)$$

where

M_I = Mass of belt and live load expressed in kilograms per metre.

T = Tension at the point being investigated (kN).

sag = A percentage of the idler spacing expressed as a decimal and usually 0.02 (2%).

$$\text{Live load} = \frac{C}{3.6S} \quad (kg / m)$$

where

S = Belt speed (m/s)

C = Capacity (t/h)

Supplementary notes

(1) Return idler spacing = approximately 3.0 m.

(2) Impact idler spacing = approximately ¼–½ carrying idler spacing.

(3) Convex curve idler spacing – at most ½ carrying and return idler spacing.

(4) Self aligning idlers – one or two sets for the return side of belt approaching tail pulley at 6–9 m intervals from the pulley. Also useful at times on the carrying side approaching head pulleys and along the whole carrying and return runs at approximately 120 m and 60 m, respectively.

3.6.4   Feeder belt calculations

The following applies to fully skirted feeder belts.

Belt speeds. For feeder belts supported by idlers, belt speeds should not exceed 0.25 m/s with abrasive materials and 0.5 m/s with non-abrasive materials. For slider bed support it is usual not to exceed 0.13 m/s.

Feeder belt capacity. For width of skirted load = 80% of belt width, depth of skirted load = 40% of skirted load width (flat belts).

$$\text{Capacity} = \frac{W^2 \times M \times S}{1.085} \quad (t / h)$$

where

W = Belt width (m)

M = Material density (kg/m³)

S = Belt speed (m/s)

Feeder belt tensions and power. The effective height of load in the hopper supported by the belt can be assumed as twice the loaded belt width for most lumped bulk materials, thus the mass of the load supported by and to be moved by the belt is, approximately

$$\text{Mass} = 2W_l^2 \times L_l \times M \,(\text{kg})$$

where

W_l = Hopper opening width (m)
L_l = Hopper opening length (m)
M = Material density (kg/m³)

Effective belt tension, $Te = \mu_o \times W_l^2 \times L_l \times M \times (9.81 \times 10^{-3})(\text{kN})$

where

μ_o = Overall friction coefficient
(0.4 for idler operation)
(0.6 for slider bed operation)
(up to 1.0 for difficult flow materials)

Maximum belt tension, $T_{max} = \{1 + k\}\, T_e \,(\text{kN})$.

Where

k = Drive factor
(0.97 for 180° wrap, screw take-up, bare steel pulley);
(0.90 for 180° wrap, screw take-up, lagged pulley);
Belt power = $T_e \times S \,(\text{kW})$.

Where

S = Belt speed (m/s)

Note: Height of hopper opening above belt should not be less than three times maximum lump size.

Feeder belt specification. For carcass design, the following considerations apply:

(a) Carcass: low-elongation, high strength carcass preferred – such as KN (Special weave construction) or PN (Plain weave construction) constructions,

(b) Top covers: usually from 5 mm thick for lumps up to 50 mm, to 10 mm thick for lumps up to 150 mm,

(c) Bottom covers: 2–4 mm thick depending on degree of abrasion and impact.

3.7   Acceleration and deceleration

3.7.1   Accelerating belt conveyors

The ideal starting arrangement for a belt conveyor is one which provides a gradual steeples increase in torque, which rises to a value just sufficient to put the belt in motion. Once this is achieved there should be a slight pause to allow shock tensions within the system to dampen out. Following this the drive should continue the steeples increase in torque at a faster rate until full speed is reached.

The question of cost must always be considered and generally there is some compromise in choice of motor and controller. The following table lists most, but not necessarily all, methods that can be regarded as acceptable, and classifies them in respect of ideal starting arrangements for the belt.

With direct on line (D.O.L.) starting of normal squirrel cage motors, it is often possible to start a conveyor and not harm the belt in any way, particularly on short centre installations, because the inertia of the driving components is in excess of the load and thus the torque transmitted to the belt is quite often below the customary limit of 150%–160% of the running torque.

As necessary, D.O.L. starting characteristics are modified with the substitution of primary resistance or auto-transformer starters.

3.7.2   Decelerating belt conveyors

Regenerative decline conveyors. If an induction motor is driven by its load in the same direction as the rotation of its flux, its speed rises above synchronous speed. The motor acts as an induction generator, taking magnetising current from the line and absorbing mechanical power through its shaft. Electric power is then fed back into the power system. The motor will restrain the load with little rise in speed above synchronous as long as the load torque does not exceed the maximum torque of the motor. If maximum torque is exceeded the motor becomes unstable and runs away.

Since the motor draws magnetising current from the line it follows that no dynamic braking is possible when the connection between the motor starter and the power line is interrupted. A decline conveyor is an excellent illustration of this effect when the speed of the belt and load are restrained by the motor. The motor serves as a generator feeding power back into the system without special control.

Regenerative braking at speeds below synchronous is not possible. Retardation must be affected by other braking methods and the brake selected for a regenerative decline conveyor must have the following characteristics:

(i) It must always fail safe.

(ii) It must stop the conveyor in a reasonable time so that there will be no damage to the belt, the driving components or the brake itself.

(iii) When the brake stops the conveyor, or it fails safe, it must have sufficient torque rating to hold the conveyor motionless under fully loaded conditions of operation.

Decline conveyors may therefore be stopped by any one of the following means:

i) Dynamic braking.

ii) Eddy current braking.

iii) Hydraulically.

iv) Friction braking.

It is, of course, possible to use a combination of the above methods but the only brake that will fail safe and hold the load is the friction type. As a result, the most common type of brake used as a control on decline conveyors is the gravity or spring actuated thrust or brake.

Many factors have to be considered in the selection of brake size of sufficient torque to decelerate the load and hold it motionless. Problems such as gradual deceleration and the heating effect on the drum shoes have also to be considered. The inertia of the driving components will also have a marked effect and therefore these brakes must always be designed on a most liberal basis.

Horizontal conveyors. When power is interrupted on long centre horizontal conveyors, "drift" or "coasting" will take place because of the inertia of the system. Normally this condition is not troublesome unless the long centre belt is one of a multiple system or is feeding onto a shorter inclined belt with vastly different natural deceleration characteristics.

Unless such conveyors are braked, material may pile up at transfer points with the possibility of damage to both belt and machinery. If the natural deceleration time is excessive for the system, retardation may be affected by the use of dynamic or eddy current braking.

3.7.3   Hold back or anti-run devices

When an inclined conveyor is stopped under load, the force of gravity will tend to drive the belt in the reverse direction. A hold-back device is necessary to prevent the belt backing down the incline. A reversal of motion would cause spillage and, in many cases, belt and mechanical damage. The following list details the most commonly used types:

(i) Clutch type hold-backs. These are among the safest devices and consist of an over-running clutch with its outer race held stationary between two ever arms rigidly mounted to the outer housing. These clutch-type holdbacks are considerably more expensive than other devices but they require the lowest maintenance and operate with zero back-lash on the drive shaft. They come in a wide variety of sizes, have high capacities and cannot be contaminated by dust.

(ii) Roller type hold-back. These consist of hardened steel rollers which lie in wedge-shaped slots as long as the hold-back rotor is running in the direction of normal travel. Once the direction of rotation is reversed, the rollers jam in the narrow end of the slots holding the rotor against the stationary housing.

(iii) Ratchet and pawl hold-backs. These are the simplest and least expensive type of hold-back. Their simplicity of design promotes easy maintenance but unless serviced regularly, dust and lack of lubrication could make the device ineffective. Such units should always be covered if mounted in dusty locations.

(iv) Differential brake hold-backs. This device is also comparatively cheap but as it relies on friction, is not as positive as the other devices listed above. This hold-back consists of a brake wheel, brake band and base-mounted cam. Movement in normal direction of travel actuates the cam through friction between the brake wheel and the band in such a manner that the band perimeter is increased, allowing free rotation. Movement in the opposite direction decreases the perimeter of the band, binding it against the brake wheel and preventing movement.

Differential band brakes are subject to failure from over-greasing, wear and icing. If used in dusty locations, they should always be covered. Such brakes require careful adjustment in the field.

(v) Magnetic brake. This friction brake is normally actuated by a thruster or solenoid, and is always located on the high speed side of the reducer. It is generally used to supplement the action of more conventional types on high lift conveyors.

3.7.4   Counterweight reaction – Accelerating and braking

Today's long centre, terrain following conveyors quite frequently, during acceleration or deceleration, generate forces which sometimes affect the amount of counterweight tension required in the system. If the counterweight is not sufficiently heavy to resist such forces, then it will move inwards and inevitably cause an accumulation of slack belt at some point of lower tension in the system. This in turn can cause severe spillage, damage to belt and/or idlers and in some cases, when the acceleration or deceleration phase is complete and the belt gives up its energy, the generation of longitudinal waves in the system, producing a phenomenon analogous to water hammer, which can severely damage terminal equipment such as pulleys, bearings, structures and even the belt itself.

The accelerating and decelerating force calculations are based on the simple assumption that the entire belt starts or stops at a uniform rate of acceleration or deceleration. This assumption is not completely accurate because of the elastic qualities of the belt itself. Textile belts tend to elongate more with the result that some sections of the belt reach either full speed or standstill much quicker than other sections. On the other hand, steel cord belts, with relatively high moduli of elasticity, and therefore inherently lower stretch characteristics, tend to behave in a manner much closer to the single mass assumption mentioned earlier. Depending on the magnitude of acceleration/deceleration forces, and the load condition of the system, the following general conditions will apply.

3.7.5　　Accelerating and braking forces

The belt tension during acceleration or deceleration can be calculated for any critical point in the system (e.g., vertical curve). This tension is equal to the normal operating tension at that specific point in the system plus the additional tension caused by the forces of acceleration or deceleration. Each significantly different condition of loading should be investigated.

The three basic formulae used in such calculations are as follows:

$F = ma$ where,　　　m = Mass to be accelerated or decelerated (kg)
a = Acceleration or deceleration (m/s²)
F = Force or tension (N)

Where time required for acceleration or deceleration has to be calculated:

$$t = \frac{S}{a}$$

where　　S = Belt speed (m/s)
a = Average acceleration or deceleration (m/s²)
t = Time (seconds)

Where coasting is involved and distance is required:

$$d = \frac{at^2}{2}$$

a = Deceleration (m/s²)
t = Decelerating time (s)
d = Coasting or decelerating distance (m)

also

$$f = \frac{F}{m} \text{ where}$$

a, F and m are described earlier.

When combinations of incline/decline become involved, e.g., on a terrain following conveyor where there could be multiple slopes involved, then careful consideration to acceleration and deceleration forces should be given.

3.7.6   Application of forces

Accelerating and decelerating forces are distributed around conveyor systems in direct proportion to the mass involved. These masses are as follows:

(1) Conveyor carrying side

Belt mass + material mass + mass of carrying idler rotating parts.

Belt mass $= B$ (kg/m)

Material mass $= \dfrac{C}{3.6S}$ $(kg\,/\,m)$

Mass of trough idler parts $\dfrac{w_t}{S_1}$ (kg/m)

(2) Conveyor return side

Belt mass + mass of return idler rotating parts

Belt mass $= B$ (kg/m)

Mass of return idler parts $= \dfrac{w_r}{S_1}$ (kg/m)

3.7.7   Coasting

Unless a conveyor is regenerative, it will coast to a gradual halt due to the inertia in the system when the drive power is shut-off.

It may be necessary to calculate this natural coasting time (or distance) to determine if extra retardation in the form of a brake is required.

The three basic formulae described earlier under "Accelerating and braking forces" are again used in this respect, viz.,

$$a = \frac{Fi}{m}$$

$$t = \frac{S}{a}$$

$$d = \frac{at^2}{2}$$

where

$a =$ Coasting deceleration (m/s²)

$Fi =$ Decelerating gravitational force (N)

$m =$ Mass to be decelerated (kg)

$t =$ Time to coast to halt (s)

$S =$ Normal operating belt speed (m/s)

$d =$ Distance to coast to half (m)

In the application of these formulae the terms (Fi) and (m) need further elaboration.

The decelerating gravitational force (Fi) is composed of two elements

(a) The effective tension (T_e) for the load condition being considered (i.e., conveyor empty or fully loaded for the whole length or perhaps for only one particular section of the conveyor), and

(b) The friction losses of the drive. As the speed reducer portion of the drive will have significantly greater friction losses than the motor portion, the latter may be neglected for simplicity. If the efficiency of the reduction unit is not known it may be assumed to be 97% or 98% (i.e., the friction losses may be assumed to be 2.5%). To convert these losses to units of force (newton) calculate the following expression:

$$\frac{\text{Redn. Unit Power Rating (kW)} \times 1000 \times \text{No. of Redn Units} \times (\% \text{ Loss})}{\text{Belt Speed} \times 100}$$

Thus, the total decelerating gravitational force (Fi) is then the sum of (T_e) and the value obtained from the above formula is expressed in newton.

The mass to be decelerated (m) is also composed of two elements:

(a) The total mass of the material load, the belt and the rotating idler parts for the condition being considered:

$$\left(Q + \frac{C}{3.6S} \right) \times L \text{ (kg)}$$

(b) The equivalent mass of the drive system. To calculate this, the inertia (WK^2), expressed in kilogram metre squared, is substituted in the following expression:

$$\frac{(WK^2) \times (\text{Reduction ratio})^2 \times \text{No. of drives}}{(\text{Radius of drive pulley in metres})^2} \text{ (kg)}$$

Values of WK2 for the reduction units, motors and couplings should be obtained from the equipment manufacturers.

From the above, values for (Fi) and (m) may now be substituted into the original basic formulae to determine (t) time taken and (d) distance required for the conveyor to coast to halt.

3.8   Belt carcass selection

3.8.1   Belt construction requirements

To select the optimum plied belt carcass, five properties must be considered:

- The belt width.
- The service conditions under which the belt will operate.

- The maximum operating tension (T_{max}) – both steady state condition and peak.
- The minimum number of plies required to support the load.
- The maximum number of plies beyond which transverse flexibility is reduced and the troughing efficiency is affected. This varies with the belt width, trough angle and the idler roll arrangement.

3.8.2   Considerations

Operating conditions

The allowable working tensions are applicable for reasonably well maintained conveyors operating with moderate impact, infrequent starts and good loading. Peak tension – on starting or braking, should not exceed 140% of the allowable working tension.

For more severe operating conditions, moderate maintenance, short time cycles, frequent DOL or loaded starts, poor loading or severe impact, hot materials handling etc., reduce the tabled figures by 15%. Tension on starting or braking should not exceed 150% of the resulting rated tension.

For severe service conditions, poor maintenance, very hot materials, chemically aggressive environment, severe impact and short time cycles etc., reduce the tabled figures by 30%. Tension on starting or braking should not exceed 160% of the resulting rated tension.

Safety factors

The working tensions shown on these tables are based on the application of a safety factor of 6.7:1 on the strength of the belt at the splice or fastened join. The safety factor is increased for more difficult operating conditions with further restrictions applying for starting and braking.

Starting and braking tensions

A check should always be made comparing the acceleration or braking tension with the allowable peak tension for the belt, i.e., 140% of rated working tension. If the peak tension exceeds the latter figure, a stronger belt can be selected or the choice of control must be changed to reduce peak tension.

Mechanical fasteners

Belt manufacturers always recommend vulcanised splices for plain weave plied belting. Other constructions including the Crows Foot Weave, Double Weave and Solid Woven PVC can be successfully operated at close to vulcanised joint tensions for long periods of time whereas plain weave constructions generally operate at reduced tensions when fitted with mechanical fasteners.

Recommended precautions including frequent inspection and monitoring, any local authority restrictions and greater than normal care should always be observed when using mechanical fasteners. Belt cleaners should only be fitted if specifically suited to operation with mechanical fasteners. If a conveyor belt is to be operated for any length of time with mechanical fasteners then the selected combination of belt and fastener should be statically tensile tested and a working tension of not more than 15% of that result. Table 3.1 lists general recommendations for its common range of belts.

Troughability and load support
This table provides a guide to the maximum width of belt that will support the load when carrying material with the bulk density shown. This table provides a guide to the minimum width of belt that will trough satisfactorily at the trough angle shown. The widths shown above are a guide only and experience may dictate the selection of a ply more or a ply less. Some factors that may influence the choice are:

- Partially filled belt.
- Idler trough angle.
- Convex or concave curve radius and idler pitch.
- Lump size of material.
- Installed pulley diameters.

3.9   Cover gauges and qualities

3.9.1   Considerations

There are a number of factors that must be taken into account when selecting the belt grade or cover material, such as follows:

- Fire resistance or anti-static properties.
- Resistance to oils or chemicals.
- Temperature of the operating environment or conveyed material.
- Resistance to ageing, weathering and ozone.
- The type of material being conveyed.
- The lump size and shape of the material being conveyed.
- The mix of lumps and fines in the material.
- The abrasiveness of the material.
- The method of loading the belt.
- The fall height of material to the belt.
- The cycle time of the conveyor for a single revolution of the belt.

- Performance or experience in a similar application.
- For replacement belts – the performance of previous belts on the same installation.
- Availability and cost.

3.9.2   Selection

Previous experience will always be the best guide to the optimum selection of both the type and thickness of belt cover, however if this information is not available as will be the case for new installations, the following steps should be followed.

- In some cases, statutory requirements or the operating conditions will limit selections to one or two possibilities.
- Calculate the time cycle of the conveyor $= \dfrac{(2 \times L)}{S}$.

Where

L = conveyor centres (m)
S = belt speed (m/s)

- Use Table 3.1 as a guide to select the appropriate thickness of top cover. Consideration should be given to the applicable properties of the cover in making this selection.

For difficult applications such as belt feeders, or impact belts, heavier covers may be required.

3.10   Pulley side cover

As a guide, pulley side cover should generally be not less than 1/4 of carry side cover for covers up to 9 mm and about 1/3 of carry cover thickness for covers heavier than 9 mm. Operating conditions can dictate that heavier pulley side covers are required.

For long centre, long time cycle conveyors, pulley side cover can be up to 1/2 of carry side cover.

3.10.1   Pulley diameters

The minimum pulley diameter recommended for a particular belt depends upon three factors:

- Carcass thickness – The wire rope diameter in the case of steel cord belts.

— The overall thickness of all plies plus the rubber skims between plies in the case of ply type belts.

— The overall thickness of the thick woven fabric separating the top and bottom covers in the case of solid-woven belts.

- Operating tension   – The relationship of the operating tension of the belt at the particular pulley to the belt's allowable working tension.

- Carcass modulus   – The relationship between elongation of the carcass and the resulting stress.

Whatever the carcass type, steel cord, ply type or solid woven, when the belt is bent around a small radius, tension stresses are developed in the outer fibres while compression stresses are built up in the inner fibres. At a given tension, if the radius is too small the elastic limit of the outer fibres may be exceeded and fracture, and at the same time, the compression of the inner fibres may cause severe crinkling and eventual ply separation.

Since the elastic properties of the rubber or PVC cover material is so much greater than the carcass material, the cover thickness of the belting is not a factor in determining minimum pulley size, and may be ignored.

The tables of recommended pulley diameters in the FENNER DUNLOP handbook for ply type belting are based on the three classes of pulleys defined in ISO 3684. viz.;

- Type "A" – High tension / tight side pulleys ($T1$) e.g., head, drive, tripper and shuttle pulleys.

- Type "B" – Low tension or slack side pulleys ($T2$) such as tail and take up pulleys.

- Type "C" – Low tension snub or bend pulleys with wrap angle of less than 30°.

Two sets of pulley diameter tables follow:

- For belts operating at over 60% of allowable working tension.
  Table 1A – for standard belt constructions.
  Table 1B – for coal master series belts.

- For belts operating at 30%–60% of allowable working tension.
  Table 2A – for standard belt constructions.
  Table 2B – for coal master series belts.

For belts operating at less than 30% of the allowable working tension, the diameter of Type "A" pulleys can be reduced to the same as Type "B".

3.10.2   Parallel face pulleys

With just a few special exceptions, all pulleys used with modern high strength, high modulus fabric belts should be parallel face types. It is absolutely mandatory that all pulleys used in conveyors fitted with Steel Cord belting be parallel face type. One notable exception to this rule is in the case of Bucket Elevators which, lacking any other means of tracking the belt centrally, may benefit from Crown Faced Pulleys.

3.10.3   Crown face pulleys

A crown faced pulley can have the effect of centering the tracking of the belt, but only in the case where there is a long unsupported length of belt leading into the pulley, as the belt must be able to bend longitudinally along its centreline to benefit from the crown. High modulus ply type belts have very little ability to bend longitudinally and steel cord belts have virtually no ability at all. Solid woven belts are not quite so rigid but still need an unsupported distance of something like 4–6 times the belt width to be able to react.

Apart from not serving much purpose in troughed conveyor systems, crowned pulleys can seriously damage the belt by severely overstressing the carcass in the centre of the belt, particularly in the case of steel cord belts.

The few special cases where crowned pulleys are useful include, Bucket Elevators, the Take- up pulley in long gravity Take-up arrangements and for some short centre – wide belt, reversing conveyors. In cases like this where there are no supporting idlers to train the belt, some benefit may be obtained from the installation of crowned pulleys.

It is fairly common practice to crown a pulley by machining a taper of 1 in 100 from each pulley edge towards the centre over a distance of 1/4 pulley face. It is more correct to relate the amount of pulley crown to the pulley diameter, not to its face width viz., $d = D - 0.008 \times D$.

3.10.4   Pulley face width

As all belts tend to wander a bit in operation, the overall face width of the pulleys should exceed the belt width by the following minimum amounts, if serious edge damage is to be avoided;
- Belts up to 650 mm wide   100 mm.
- Belts 750 to 1400 mm wide   150 mm.
- Belts over 1400 mm wide   200 mm.

For conveyors built on unstable ground, as in underground coal mines and very long overland conveyors, the above allowances should be increased by 50 mm.

3.11   Design considerations

3.11.1   Multiple slope and vertical curve conveyors

Conveyors with grade variations, and particularly decline regenerative systems with concave vertical curves, require special consideration. In their design a thorough analysis is necessary, particularly at points of low tension. The effect of varying load conditions as well as acceleration and deceleration must be considered carefully.

Determination of vertical curves

(1) Concave vertical curves
(a) Selection of radius to prevent belt lifting off idlers
    (i) Squirrel-cage motor, D.O.L start:

$$R = \frac{^* \ (2.0 \ to \ 2.5) \times T_i}{B \times 9.81 \times 10^{-3}} \ (m)$$

    (ii) Squirrel-cage motor, D.O.L start with traction type coupling or centrifugal clutch:

$$R = \frac{^* \ (1.7 \ to \ 2.0) \times T_i}{B \times 9.81 \times 10^{-3}} \ (m)$$

    (iii) Squirrel-cage motor, D.O.L start with scoop type fluid coupling or eddy-current clutch; slip-ring motor, electrically controlled start:

$$R = \frac{^* \ (1.3 \ to \ 1.6) \times T_i}{B \times 9.81 \times 10^{-3}} \ (m)$$

    where
    R = Radius of curvature (m).
    T = Maximum operating tension (total belt width) at the point of tangent intersection (kN).
    B = Belt mass (kg/m).
    * = The difference and range in these values apply to the motor and starter characteristics, and should be varied according to the type of control chosen for the installation under consideration.
    (iv) On decline regenerative conveyors even with the best of control, the very minimum radius at the foot of a steep decline should always be calculated from the following:

$$R = \frac{1.8 \ T_i}{B \times 9.81 \times 10^{-3}} \ (m)$$

Consideration should then be given to closer idler spacing at locations immediately following such steep declines.

(b) Selection of radius to prevent belt edge tension from dropping below zero and causing possible buckling:

The formula to give a radius of curvature which will result in zero edge tension in the belt is:

$$R = \frac{\sin\theta \times W \times E \times N}{4.5\left(T_{c1} - 0\right)} \text{ (m) where}$$

W = Belt width (m).
E = Elastic modulus (kN/m/ply).
N = Number of plies.
Tc = Maximum operating tension at point of tangent intersection (kN/m).
θ = Carrying idler troughing angle.
* For steel cord belts, $N = 1$ and E has units of kilo newton per metre.

As positive tension is preferable in the belt edges, the value of that tension should be substituted in the denominator of the formula. As a general rule, 4.4 kN is used as the desired minimum tension. With that minimum tension value, the formula will read:

$$R = \frac{\sin\theta \times W \times E \times N}{4.5\left(T_{c1} - 4.4\right)} \text{ (m).}$$

Other minimum tension values could be substituted to suit particular installations.

Note: (i) In calculations to determine the radius of concave curves to prevent the belt lifting off the idlers, design tonnage figures should be used.

(ii) In calculations to determine the radius of concave curves below which negative edge tensions (and hence edge bucking) will occur, actual tonnage rates should be used, particularly when design tonnage rates are significantly above the rates which will occur in practice.

(2) Convex vertical curves

(a) Selection of radius to keep belt edge tension within acceptable limits:

$$R = \frac{\sin\theta \times W \times E \times N}{4.5\left(T_a - T_{c2}\right)} \text{ (m)}$$

$$.. 4.5\left(T_a - T_c\right).$$

(b) Selection of radius to keep tension at centre of belt above zero and thereby prevent possible buckling:

The formula to give zero tension in the centre of the belt is

$$R = \frac{\sin\theta \times W \times E \times N}{9\left(T_{c2} - 0\right)} \text{ (m)}$$

where

θ = Troughing angle of carrying idlers.

W = Bet width (m).

*E = Elastic modulus (kN/m/ply).

*N = Number of plies.

T_a = Recommended allowable working tension for the belt reinforcement used, (kN/m).

T_{c2} = Maximum operating tension at point of tangent intersection (kN/m).

*For steel cord belts, $N = 1$ and E has units of kilo newton per metre.

As a general rule, positive tension should be provided, and 4.4 kN is usually used as the desired minimum tension, in which case the formula becomes:

$$R = \frac{\sin \theta \times W \times E \times N}{9\left(T_{c2} - 4.4\right)} \text{ (m).}$$

Note: (i) In calculations to determine radius of convex curves to keep edge tension within acceptable limits, design tonnage rates should be used.

(ii) In calculations to determine the radius of convex curves below which negative tension at the centre of the belt will occur, actual tonnage rates should be used particularly when design tonnage rates are significantly above rates which will occur in practice.

The equipment supplier should always check radial force component caused by the belt on the idlers in the transition area. If the curve is calculated by means of either (a) or (b) (formula) above, the general result is that method (a) determines the radius in most cases. For idler load limitations, check ratings, etc., with the idler supplier.

(3) Horizontal curves

In recent years horizontal curves have been used in long centre terrain – following conveyors as a means of eliminating transfer points and so reducing costs and increasing efficiency. No attempt is made here to detail all the variables involved in designing such curves, except to say that the largest possible radius should always be used and the minimum allowable radius is 900 × belt width in metres.

Idlers in the transition area should be somewhat wider than usual and not less than 35′ inclination on the carrying side and 15′ inclination on the return. In some cases, normal trough idlers are also used on the return run.

In addition to forward tilting of the carry idlers in the direction of belt travel, it is also helpful to tilt the idler frames to provide a reverse camber around the curve. That is, starting at the beginning of the curve, packing should be provided under the end of the idler frames at the inside of the curve. The thickness of this packing should be progressive from say 1 mm at the beginning of the curve to about 75 mm at the centre of the curve, then progressively decreasing to 1 mm again at the end of the curve. This reverse camber effect assists in preventing the belt climbing the idlers towards the centre of the curve (Figure 3.11).

 * Pack under inside end of idler frames progressing to maximum at centre of curve

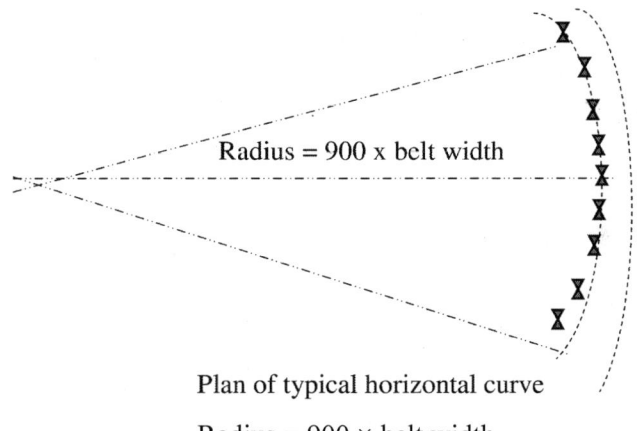

Plan of typical horizontal curve

Radius = 900 × belt width

Figure 3.11   Reverse Camber effect

3.12   Terminal troughing idler arrangements

It is recommended that the first or last standard troughing idler over which the belt passes under high tension as it leaves or approaches a terminal pulley, be mounted on a level which will average the belt edge stresses.

If the terminal pulley is set so that a tangent line from its rim is above the top of the centre roll by an amount equal to half the height of the troughing idler, the belt edge stresses are minimised and the optimum level of the last standard troughing idler is obtained.

Where the belt is not operating at high tension and very large lumps are carried, the best position for the terminal pulley with respect to the last standard troughing idler is to have the pulley rim set so that the tangent line from it will be tangent to the top of the centre roll. This position increases belt edge stress but lessens the chances of belt injury due to the impinging effect of large lumps against the pulley.

At tail end loading points with low to moderate belt tensions, this position is also used, and ensures that the belt cannot lift at the loading point and interfere with the skirting.

Always locate first the standard troughing idler prior to the loading chute, and incorporate intermediate troughing angle idler sets between the terminal pulleys and the steep-angle-troughing idler run.

3.13   Take-up arrangements

Gravity take-ups - Vulcanised splices. The take up travel requirements shown in Table 3.5 are applicable to belts operating at between 75% and 100% of the allowable working tension, and with starting tension limited to 150% of allowable working tension.

For belts operating at between 50% and 75% of the allowable working tension, the travel distances shown in Table 3.5 can be reduced by 25%, whilst the travel distance for belts operating at less than 50% of the allowable tension can be reduced by 50%.

Belts operating at high temperatures or under very wet conditions may require take up travel distances up to 50% longer than shown in Table 3.5.

For long to very long centre belts using low elongation Kurlon/Nylon and Polyester/Nylon carcass constructions, take-up travel can be progressively reduced as necessary to suit available space, down to as little as 0.25%–0.5% of centre to centre length of conveyor (contact Apex Belting for recommendations). Provided accelerating and braking forces are kept to reasonable limits, belt stretch after some initial elongation becomes minimal.

Screw take-ups - Vulcanised splices and all fastened joints. Travels can generally be reduced to approximately half those shown in Table 3.5.

However, for take-ups with vulcanised splices, always provide sufficient travel to permit re-splice of the belt if required without having to insert a new piece.

(A) Fabric belting

Table 3.5 Gravity take-up travel

Centre to centre length of conveyor (m)	Travel in percent of conveyor centre distance (minimum)		
	CC	NIN	KN, PN, PP
Up to 30	2.0	4.0	2.0
31–60	1.9	3.5	1.7
61–180	1.8	3.4	1.6
181–300	1.8	3.3	1.5
Over 300	1.8	3.3	1.3

Note: A practical minimum take-up travel should not be less than 1 m.

Initial location of take-ups. To eliminate all belt sag on the installation of a new belt, pull the belt ends together against the installed counterweight. Take-ups can be initially located as follows:

(1) For nylon/nylon carcass belts, hard against the inner stop,

(2) For all other synthetic fabric carcass constructions, from hard against the inner stop to 20% of total travel from inner stop,

(3) With cotton/cotton carcasses, from 20% up to 50% travel from inner stop for operating conditions ranging from dry to wet.

(B) Steel cord belting

The high modulus characteristic of the steel wire ropes used in the construction of steel cord conveyor belting results in extremely low elongation. This high strength, low stretch characteristic gives steel cord belting a tremendous advantage over fabric belts on long single flight systems.

With proper tensioning of the belt during the closing splice operation, take-up as low as 0.25% of conveyor centre distance is possible. Where the belt is designed to operate at tensions up to 100% rated tension a take-up travel of 0.5% is normally recommended.

3.14   General data

3.14.1   Belt mass and thickness

On the following there are three tables:
Table 1
> Belt carcass mass and thickness for standard belt constructions (other than coal master belting for underground coal mining).

Table 2
> Belt carcass mass and thickness for coal master belting (for underground coal mining).

Table 3
> Carcass mass factor and cover mass factor.

To calculate belt mass

- Look up belt carcass mass (kg/m²) from either Table 1 – Standard belt constructions or Table 2 – Coal master belting as applicable.

- To obtain the actual belt carcass mass, multiply this by the belt carcass mass factor from Table 3.

- To obtain the mass of the belt covers, add together the top and bottom cover thickness and multiply this by the belt cover mass factor, also from Table 3.

- Add together the mass of the belt carcass and covers for the belt mass per square metre (kg/m²) and multiply by the belt width (in metres) for the belt mass per metre run (kg/m).

To obtain belt thickness

- Look up belt carcass thickness (mm) from either Table 1 – Standard belt constructions or Table 2 – Coal master belting as applicable.
- Add to this the thickness of both the top and bottom covers.

3.14.2   Shipping dimensions and roll sizes

Roll diameter

Roll diameter for belts can be determined from the below formula given. The diameters shown are for a belt wrapped on a 600 mm-diameter centre. For belts supplied on enclosed drums, an additional 0.15 m should be added for clearance and slats where fitted.

Alternatively, the diameter can be calculated from the following formula:

$$D = \sqrt{d^2 + \left(0.001273 \times L \times G\right)}$$

Where D = Overall diameter (m)
d = core diameter (m)
L = Belt length (m)
G = Belt thickness (mm)

3.15   Solid woven belting

3.15.1   Belt construction

Fenaplast conveyor belting consists of three main components

- Textile solid woven carcass.
- PVC impregnation.
- Cover material.

3.15.2   The textile solid woven carcass

The solid-woven carcass is typically woven with nylon or polyester load bearing warp fibres and nylon or nylon/cotton weft. Synthetic binder yarns follow a complex pattern to give the carcass its solid-woven properties. Various combinations of synthetic and natural fibres are chosen, together with the fabric design to meet the requirements of impact resistance, belt elongation, flexibility for troughing and small diameter pulleys, load support and fastener retention. The patented Fenaplast PVC impregnation method also renders the carcass impervious to attach from moisture, dirt, chemicals, bacteria, and oils. Cotton pile warp yarns may be included for improved impact resistance and special edge reinforcement can be included where these are particular problems. The Fenaplast carcass design facility enables users to choose the properties of a custom-built belt.

All Fenaplast belts have a solid woven carcass where all layers of yarn are mechanically interlocked during the weaving process and bound together by a self-binding warp yarn interweave, thus making subsequent delamination impossible. High tenacity continuous filament synthetic yarns are used for the warp fie length direction), such yarns also provide most of the necessary strength in the weft (transverse/width) direction.

3.15.3   PVC impregnation

After weaving the roll of carcass is vacuum impregnated with PVC plastisol containing a careful blend of polymer, plasticisers, stabilisers, fire retardants, and special additives, with special attention being given to viscosity control in order to ensure full impregnation of the woven structure.

Whilst the textile elements fix many of the belts properties such as tensile strength and elongation in service, the properties of the plastisol are equally important, and its formulation will influence not only the fire performance properties but also operational factors such as troughability and the ability to hold fasteners.

3.15.4   Cover material

PVC covers to meet numerous fire resistance specifications or for other properties such as resistance to oils, chemicals, fertiliser etc., are generally available up to 3 mm thick per side. They can also be compounded to give improved abrasion resistance or coefficients of friction.

Rubber covers to a specified safety standard may be applied on one or both sides of a PVC impregnation parent belt up to a maximum of 6 mm + 2 mm, dependent on belt width, tensile and construction. SR wear-resistant nitrile rubber covers are also available, single or double sided, up to 6 mm + 2 mm maximum, dependent on belt construction.

Belt and cover thickness

When considering cover thickness the user should be aware of the thick, high textile content of Fenaplast and the special solid-woven carcass properties. Consequently, thinner covers may generally be chosen than normal with rubber, plied belting; the Fenaplast carcass being more substantial and providing the necessary load support and impact resistance.

Operating temperature range
Above 90°C PVC softens and the belt properties change, therefore, Fenaplast is not recommended for conveying materials above this temperature. Standard Fenaplast can be used in cold climates at minus 15°C and special cover compounds are available for operation down to minus 40°C. Cold weather details should be supplied to ensure a belt with suitable coefficient and flexibility characteristics is supplied.

3.16   Belt joints

3.16.1   Vulcanised spliced joint

Fenaplast belting can be vulcanised using the Fenaplast finger splice method on a variety of polymeric materials with conventional rectangular presses. Vulcanised joints achieve strengths close to original strength of the parent belt.

3.16.2   Mechanical fasteners

The unique PVC impregnated solid woven Fenaplast carcass provides superior belt clip holding capabilities, improving joint efficiency and increased service life

Note! Mechanical fasteners are not recommended for long-term trunk or short centered high trip conveyor installations.

Safety
Fenaplast is used extensively in underground coal mines and as such it exhibits excellent fire resistant, anti-static properties.
    Australia – manufactured to AS1332 and tested to AS1334, which meets and exceeds all of the requirements of AS 4606-2000.

Steel cord belting
Steel cord belting is used on conveyors where high belt tension, available take-up travel or the requirement for a high elastic modulus dictates that a fabric belt is unsuitable. Steel cord belts also provide excellent splice efficiency for difficult pulley configurations and are able to accommodate smaller pulleys than plied fabric belts of similar working tension.

Construction
Steel cord belts are comprised of zinc coated steel cables embedded in a bonding rubber matrix with covers top and bottom as shown below:

Cord spacing and diameter are varied to provide the required strength. There are various international and industry standards for belt construction and some of these are shown below. Available belt covers and belt grades are shown above.

3.17   Operating conditions

3.17.1   Allowable working tension

The accepted safety factor for steel cord belting is 6.7:1, however, considerably lower safety factors can be accommodated on installations with well controlled starts and long-time cycles, and where transient or peak tensions are low.

Conversely, adverse operating conditions may dictate that a higher safety factor is applied.

3.17.2   Pulley diameters

Pulley diameters for belts operating at over 60% of allowable working tension and for belts operating at 30%–60% of allowable working tension. Where belts are operating at less than 30% of allowable working tension, the diameter of type A high tension pulleys can be reduced to the same diameter as type B pulleys. Crowned pulleys should NOT be used with steel cord belting.

3.17.3   Transition lengths

Terminal pulleys at head and discharge ends of a conveyor or at a tripper are most commonly located at half the trough depth above the centre idler roller. The tail end pulley is most often located in line with the centre idler roller.

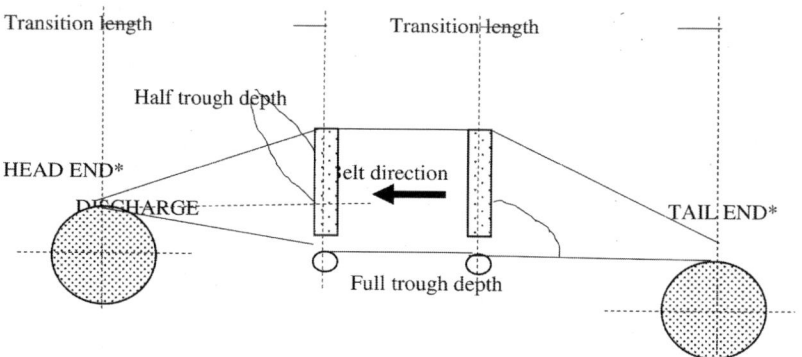

Figure 3.12 Transition arrangements

Most common arrangement. Circumstances may dictate that either DIS-CHARGE / HEAD or TAIL pulleys are located at different trough depths (Figure 3.12).

Table 3.2 lists the transition lengths required for the most common trough angles for 3 roll equal length idlers. The transition length can be determined by multiplying the belt width by the factor below. Longer transitions are required at increased belt tension, however at very low tensions; longer transitions are also required to maintain positive tension in the centre of the belt.

3.17.4   Load support and troughability

Load support is normally not an issue however for steeply troughed, wide belts or light belts with heavy materials or very large lumps, please contact your Fenner Dunlop representative. Fitment of a transverse breaker can provide necessary increased stiffness.

Carrying idler spacing, should limit sag between idlers to 1.5% for lumps up to 200 mm and light materials. For very heavy materials and lumps over 200 mm, sag should be limited to 1% of idler spacing. The gap between idler rollers should not exceed 15 mm to avoid pinching and belt wear at the idler junction. Minimum belt width for troughing is influenced by belt construction and thickness of covers and the following table provides a guide. Where a belt incorporates a transverse breaker, add 150 mm to the belt width for a single breaker and 300 mm where there are two breakers. If in doubt, contact your Fenner Dunlop representative.

3.17.5   Take-up travel

Recommended take-up travel allowances for most circumstances are shown in Table 3.6 below.

Table 3.6 Recommended take-up travel allowance

Conveyor centres (metres)	Take-up travel (% of centres)
150	0.80
300	0.70
500	0.60
750	0.50
1000	0.45
1500	0.40
2000	0.30%
3000	0.25%

With proper tensioning of the belt during the closing splice operation and careful control of starting and stopping, take-up travel can be as small as 0.25% of the conveyor centre distance, which will accommodate:

0.05% thermal expansion and contraction.

0.05% permanent stretch.

0.15% elastic elongation.

3.18 Belt covers

3.18.1 Cover types

In addition to this range of cover compounds, Fenner Dunlop offer a complete range of specialised "*Low Rolling Resistance*" pulley cover compounds under the brand name "PowerPlus". On long centre horizontal conveyors these compounds help reduce total system power consumption and operating costs.

3.18.2 Cover thickness

As a guide, top cover thickness may be determined by adding the plied belt cover thickness to the minimum cover thickness.

Pulley cover thickness may be increased from the minimum thickness by 1 – 3 mm depending on the application and the condition of the installation. In any event, the pulley cover should be no less than one third of the thickness of the carrying cover.

3.19 Belt protection

Steel cord belting is more vulnerable to longitudinal ripping than fabric belting as there is no weft tension member or reinforcement.

3.19.1 Transverse reinforcement

Various types of transverse reinforcement can be incorporated into either the top or bottom belt covers, or both covers. These may include woven breakers, designed to prevent penetration, or transverse cords to prevent ripping and eject any tramp material that has penetrated the belt, or both.

3.19.2 Rip detection

Electronic and electromagnetic systems can be incorporated in the belt to enable a rip to be detected and the conveyor stopped to prevent further damage.

Fenner Dunlop's "**EagleEye**" system is available to also provide continuous condition monitoring which will detect and monitor such defects as damaged or corroded cords while at the same time providing rip detection.

For details of available options, please contact your Fenner Dunlop representative.

Splicing
Steel cord belts are made endless by a hot vulcanised splice with cords over-lapped from each end in a symmetrical pattern which is determined by the cord spacing and cord diameter.

Stepped pulleys (Figure 3.13)

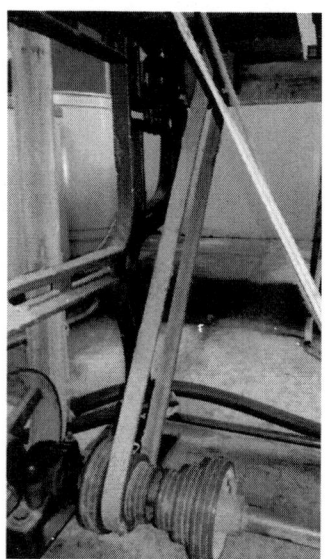

Figure 3.13 Stepped pulleys

1. A motor having an rpm of 1200 and motor diameter is 8″ and machine pulley diameter is 6″. What is the speed of machine shaft? If motor pulley diameter is increased by 2″, what is the machine shaft speed?

The machine shaft speed is same what is the change you are going to make in the machine tool?

Solution :

i)
$$n_2 = \frac{n_1 d_1}{d_2} = \frac{1200 \times 8 \times 2.54}{6 \times 2.54}$$

$$n_2 = 1600 \text{ rpm}$$

(ii)   If $d_1 = 10''$, $n_2 = ?$

$$n_2 = \frac{n_1 d_1}{d_2} = \frac{1200 \times 10 \times 2.54}{6 \times 2.54}$$

$n_2 = 2000$ rpm

iii) $n_2 = 1600$ rpm, $d_2 = ?$

$$d_2 = \frac{n_1 d_1}{d_2}$$

$$= \frac{1200 \times 10}{1600}$$

$d_2 = 7.5''$

2. A motor running at 1200 rpm having its pulley diameter 10″. The motor pulley is driving the machine pulley of diameter 12″. What is the speed of the machine shaft? If you want to increase the machine shaft speed by about 200 rpm what is the motor pulley diameter you suggest?

Solution :

We know that $d_2 = \dfrac{n_1 d_1}{d_2}$

$n_2 = 1200 \times 10 \times 2054 / 12 \times 2054$ and $n_2 = 1000$ rpm.

If $n_2 = 1200$ rpm, $d_2 = 1200 \times 12 / 1200 = 12''$, therefore, the motor pulley dia is 12″.

3. "A" pulley on a lined belt drive shaft 30″ diameter drives a pulley "B" 12″ diameter on a counter shaft. Pulley "C" from the counter shaft which is of 22″ diameter drives the machine pulley "D" of 10″ diameter. What is the speed of the machine if the pulley on the line shaft makes 200 revolutions per minutes? (Figure 3.14)

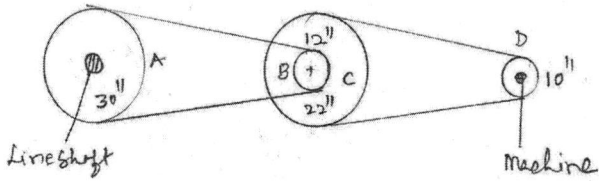

Figure 3.14   Belt Drive

Solution :

$$\text{Speed of counter shaft} = \frac{200 \times 30}{12} = 500 \text{ rpm.}$$

$$\text{Speed of machine shaft} = \frac{500 \times 22}{10} = 1100 \text{ rpm.}$$

4. A pulley on a line shaft having 20″ diameter drives a pulley on a counter shaft which is having a diameter of 24″. One more pulley mounted on the counter shaft having diameter 8″ drive a beater pulley having diameter 18″. Find the speed of the beater if the line shaft makes 30 rpm. Find the speed of the beater with the following drive arrangement.

Solution :

$$\text{Speed of crank shaft} = \frac{30 \times 20}{24} = 25 \; rpm.$$

$$\text{Speed of beater} = \frac{25 \times 8}{18} = 11.11 \; rpm.$$

5. Find the speed of the Kirchner beater with the following driving diagram (Figure 3.15).

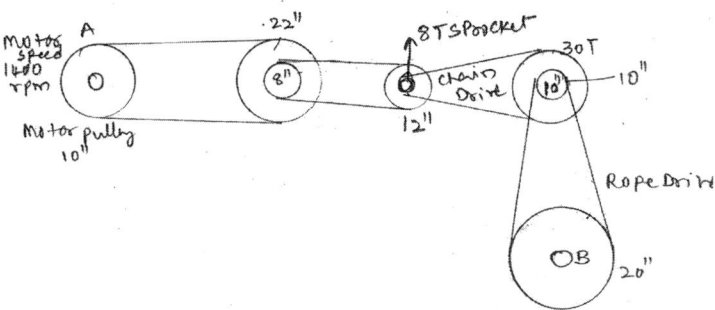

Figure 3.15   Belt Drive

Solution :

Belt drive

Motor pulley = 10″ diameter, Motor speed = 1400 rpm.

$$\text{Speed of counter shaft (I)} = \frac{1400 \times 10}{22}$$

= 636.36 rpm.

$$\text{Speed of counter shaft (II)} = \frac{636.36 \times 8}{12}$$

= 424.24 rpm.

$$\text{Speed of counter shaft (III)} = \frac{424.24 \times 8}{30}$$

= 113.13 rpm.

$$\text{Speed of "B"} = \frac{113.13 \times 10}{20}$$

= 56.57 rpm.

$$\text{Speed of shaft B} = 1400 \times \frac{10 \times 8 \times 8 \times 10}{22 \times 12 \times 30 \times 20}$$

= 56.57 rpm.

6. A machine is driven by leather belt, ¼" thick from an overhead line shaft rotating at 140 rpm. The line shaft and the machine pullies are 15.5" and 12" diameter, respectively. Calculate the machine speed (allows 2% slip).

Solution :

Thickness = 0.25"

∴ Effective diameters d_1 = 15.75 and d_2 = 12.25

$$e = \frac{d_1}{d_2} = \frac{15.5 + 0.25}{12 + 0.25} = \frac{n}{m}$$

$$\therefore n = \frac{15.75}{12.25} \times 140 = 180 \; rpm.$$

2% slip therefore speed of machine   = 180 × 0.98
= 176.4 rpm.

7. A beater is driven by a flat belt 0.3" thick from a line shaft rotating at 1000 rpm. The line shaft and the beater pullies are 18" and 16" diameter, respectively. Calculate the beater speed ignoring the slip and also taken into account 3% slip.

Solution :

$$d_1 = 18''$$
$$\text{and} \quad d_2 = 16''$$

∴   Effective diameter = 18.3" and 16.3"

(ignore slip) $\text{Speed of beater} = 1000 \times \dfrac{18.3}{16.3} = 1122.699 \; rpm.$

(with slip)   Speed of beater = 1122.699 × 0.97 = 1089.02 rpm.

8. A shaft is driven by a rope drive from an individual motor, the speed of which is 1000 rpm. The pulley diameters are 4″ and 12″ for motor and shaft, respectively. Find the speed of shaft.

Solution :

Motor		Shaft
$n_1\, d_1$	$=$	$n_2\, d_2$
4×1000	$=$	$n_2 \times 12$
n_2	$=$	$\dfrac{4 \times 1000}{12}$

$$= 333.33 \text{ rpm speed of shaft.}$$

9. A loom is driven by a V-rope drive from an individual motor, the motor speed is 960 rpm and the effective pulley diameters, when the ropes are new are 3″ and 16″, respectively (motor and loom). (i) Find out the loom speed. (ii) If wear causes the ropes to sink 1/20″ into grooves the pulley diameters will be reduced by how much. What will the loom speed be?

Solution :

(i)

	Motor		Loom
	$n_1\, d_1$	$=$	$n_2\, d_2$
	3×960	$=$	$n_2 \times 16$
	n_2	$=$	$\dfrac{3x\,960}{16}$
$\therefore n_2$		$=$	180 rpm

Speed of loom is 180 rpm.

(ii) $\qquad 960 \times \dfrac{2.9}{15.9} = 175.09 \; rpm.$

10. A roller is driven by a motor by the following rope drive arrangement. Find out the speed of the roller (Figure 3.16).

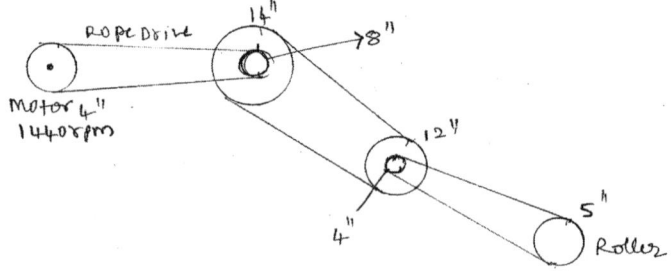

Figure 3.16   Belt Drive

Solution :

$$\frac{\text{Speed of roller}}{\text{speed of motor}} = \frac{n_2}{n_1} = \frac{4}{8} \times \frac{14}{12} \times \frac{4}{5}$$

$$n_2 = 1440 \times 0.466$$
$$= 672 \text{ rpm.}$$

11. A carding engine requires 1 hp to keep it running steadily at 180 rpm. If the pulley is having 18″, what is effective tension on the belt? A 3″ wide belt is used and is loaded to 225 lbs (maximum on the tight side). What maximum hp is the belt capable of transmitting of slipping ratio is 3?

Solution :

$$D = 18'' \qquad N = 180 \text{ rpm.}$$

$$\frac{T_t}{T_s} = 3$$

$$\therefore 3T_s = T_t.$$

(i) Effective tension $= T_t - T_s$

$$(T_t - T_s) = \frac{hp \times 33,000}{\pi DN} = \frac{1 \times 33000}{\pi \times \frac{18}{12} \times 180}$$

$$3\,T_s - T_s = 38.9$$
$$2\,T_s = 38.9$$
$$T_s = 19.45 \text{ lbs} \qquad T_t = 58.35 \text{ lbs.}$$

(ii) $T_t = 225$

$$\frac{T_t}{T_s} = 3$$

$$\therefore \frac{225}{3} = T_s$$

$$\therefore T_s = 75 \text{ lbs}$$

$$\text{Maximum hp} = \frac{(T_t - T_s)\pi DN}{33000}$$

$$= \frac{(225 - 75)\pi \times 1.5 \times 180}{33000}$$

$$\therefore \text{ Maximum hp} = 3.85.$$

12. The width of a belt is 150 mm and the maximum tension/mm width is not to exceed 1.6 kg/mm. The ratio of the tension on the two sides is 2¼ (2.25), the diameter of the driver is 1 m and it makes 220 rpm. Find the hp that can be transmitted.

Solution :

Here,

$T_1 = 1.6 \times 150$ 1 mm = 1.6 kg
$T_1 = 240$ kg   150 mm = 1.6×150
$T_1 = 240$ kg

Ratio of tension is $\dfrac{T_1}{T_2} = 2.25$

Width of belt $= \dfrac{\text{Tight side } T_1}{\text{Maximum tenion/mm}}$

$= \dfrac{240}{T_2} = 2.25$

$T_1 = W \times \text{Max} = 1.6 \times 150$

$T_2 = \dfrac{240}{2.25} = 106.7/\text{kg}$

$d = 1$ md

$n = 220$

$\text{hp} = \dfrac{Pv}{4500}$

$= \dfrac{P \times \pi dn}{4500}$

$\text{hp} = \dfrac{(T_1 - T_2) \times \pi dn}{4500}$

$= \dfrac{240 - 106.7 \times \pi \times 1 \times 220}{4500}$

$\text{hp} = 20.5.$

13. A pump consuming 5 hp running at 200 rpm. The diameter of the pulley is 0.375 m. What is the tension in the tight side of the belt? if it is 3 times the slack side tension.

Solution :

We know that

$$\text{hp} = \frac{(T_1 - T_2) \times \pi \, d \, n}{4500} \qquad \text{(i)}$$

Given $T_1 = 3 \, T_2$

hp = 5

Substitute T_1 in equation (ii)

$d = 0.375$ m $n = 200$

$$5 = \frac{(3T_2 - T_2) \, \pi \times 0.375 \times 200}{4500}$$

$$2T_2 = \frac{4500 \times 5}{\pi \times 0.375 \times 200}$$

$T_2 = 47.7$ kg

$\therefore T_1 = 3 \, T_2 = 3 \times 47.7$

$T_1 = 143.1$ kg.

Effective pull

$P = T_1 - T_2$

$P = 143.1 - 47.7 = 95.4$ kg.

14. The tension in the two sides of a belt is 50 kg and 25 kg, respectively. If the speed of the driver is 200 rpm and its diameter is 1 m, find the hp transmitted by the belt.

Solution :

Given

$T_1 = 50$ kg $T_2 = 25$ kg

    = 200 rpm $d = 1$ m

$$\text{hp} = \frac{(T_1 - T_2) \times \pi \, d \, n}{4500}$$

$$= \frac{50 - 25 \times \pi \times 1 \times 200}{4500}$$

hp = 3.49 » 3.5 hp.

15. The speed of a belt is 500 m/min and it transmits a power of 80 hp. Find the difference of tension in two sides of the belt and also if the tension in the tight side is 2½ times the tension in the slack side, find the two tensions and the width of the belt required if maximum tension is not to exceed 2.5 kg/mm width of the belt.

Solution :

$v = 500$ m/min   hp = 80

$T_1 = 2.5\ T_2$ $T_1 - T_2 = ?$

$T_1 = ?$ $T_2 = ?$

Width = ?   Maximum tension = 2.5 kg/mm.

$$hp = \frac{(T_1 - T_2)v}{4500}$$

$$80 = \frac{(T_1 - T_2)500}{4500}$$

$$T_1 - T_2 = \frac{4500 \times 80}{500}$$

$T_1 - T_2 = 720$ kg   (1)

We know that $T_1 = 2.5\ T_2$

Substitute T_1 in equation (1)

2.5 $T_2 - T_2 = 720$

1.5 $T_2 = 720$

$$T_2 = \frac{720}{1.5} = 480 \text{ kg}$$

$T_1 = 2.5\ T2$

$T_1 = 2.5 \times 480$

$T_1 = 1200$ kg

Width of the belt $= \dfrac{\text{Tight side } T_1}{\text{Maximum tenion in kg/mm}}$

$$= \frac{1200 kg}{2.5 kg / mm}$$

Width = 480 mm.

16. In a modern weaving shed the Air Jet looms have a compressor engine running at 150 rpm, drives a line shaft by means of a belt. The engine pulley is 750 mm diameter and the pulley on the line shaft being 450 mm. A 900 mm diameter pulley on the line shaft drives a 150 mm diameter pulley keyed to a crank shaft. Find the speed of the crank shaft, (a). Without slip and (b) with a slip of 2% at each drive (Figure 3.17).

Solution :

Given: N_1 = 150 rpm; d_1 = 750 mm; d_2 = 450 mm; d_3 = 900 mm; d_4 = 150 mm

The arrangement of belt drive is shown in Figure.

Let N_4 = Speed of the crank shaft.

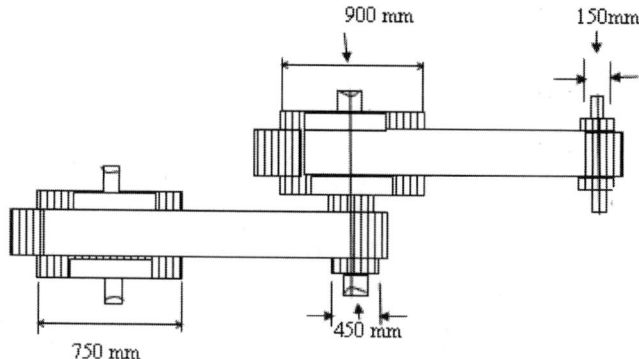

900 mm   150mm

450 mm

750 mm

Figure 3.17   Belt Drive in Air jet loom

(a) Without slip

We know that $\dfrac{N_4}{N_1} = \dfrac{d_1 \times d_3}{d_2 \times d_4}$ or $\dfrac{N_4}{150} = \dfrac{750 \times 900}{450 \times 150} = 10$

$\therefore N_4 = 150 \times 10 = 1500$ rpm.

(b) With a slip of 2% at each drive

We know that $\dfrac{N_4}{N_1} = \dfrac{d_1 \times d_3}{d \times x \, d_4}\left[1 - \dfrac{s_1}{100}\right]\left[1 - \dfrac{s_2}{100}\right]$

$\dfrac{N_4}{150} = \dfrac{750 \times 900}{450 \times 150}\left[1 - \dfrac{2}{100}\right]\left[1 - \dfrac{2}{100}\right] = 9.6$

$\therefore N4 = 150 \times 9.6 = 1440$ rpm.

17. The blow room of cotton processing has some machineries with conventional belt drive. The power is transmitted from a pulley 1 m diameter running at 200 rpm to a pulley 2.25 m diameter by means of a belt. Find the speed lost by the driven pulley as a result of creep, if the stress on the tight and slack side of the belt is 1.4 MPa and 0.5 MPa, respectively. The Young's modulus for the material of the belt is 100 MPa.

Solution :

Given: $d_1 = 1$ m; $N_1 = 200$ rpm; $d_2 = 2.25$ m; $\sigma_1 = 1.4$ MPa $= 1.4 \times 10^6$ N/m2; $\sigma_2 = 0.5$ MPa $= 0.5 \times 10^6$ N/m²; $E = 100$ MPa $= 100 \times 10^6$ N/m²

Let N_2 = Speed of the driven pulley

Neglecting creep, we know that

$\dfrac{N_2}{N_1} = \dfrac{d_1}{d_2}$ or $N_2 = N_1 \times \dfrac{d_1}{d_2} = 200 \times \dfrac{1}{2.25} = 88.9\,rpm.$

Considering creep, we know that

$$\frac{N_2}{N_1} = \frac{d_1}{d_2} \times \frac{E + \sqrt{\sigma_2}}{E + \sqrt{\sigma_1}}$$

$$\therefore \ N_2 = 200 \times \frac{1}{2.25} \times \frac{100 \times 10^6 + \sqrt{0.5 \times 10^6}}{100 \times 10^6 + \sqrt{1.4 \times 10^6}} = 88.7 \, rpm.$$

$\therefore$ Speed lost by driven pulley due to creep
$$= 88.9 - 88.7 = 0.2 \text{ rpm.}$$

18. The precision winder of cotton has stepped pulleys of two. The sum of the diameters of two pulleys A and B connected by a belt is 600 mm. If they run at 1400 and 2800 rpm, respectively, determine the diameters of each pulley.

Solution :

Given $\qquad d_A + d_B = 600$ $\hfill$ (1)
$$d_A \, n_A = d_B \, n_B$$
$$d_A \times 1400 = d_B \times 2800$$
$$\therefore d_A = 2 \, d_B$$
Substituting the value of d_A in equation (1)
$$d_A + d_B = 600$$
$$2 \, d_B + d_B = 600$$
$$3 \, d_B = 600$$
$$\therefore d_B = \frac{600}{3} = 200 \text{ mm.}$$
We know that $d_A + d_B = 600$
$$\therefore d_A = 600 - d_B = 600 - 200$$
$$d_A = 400 \text{ mm.}$$

19. A bottom shaft running at 200 rpm is required to drive a power generator at 300 rpm by means of a belt. The pulley on the driving belt is 500 mm diameter. Determine the diameter of generator pulley if thickness of the belt is 8 mm. Assume a slip of 4%.

Solution :

$$\frac{n_2}{n_1} = \frac{d_1 + t}{d_2 + t} \times \frac{100 - s}{100}$$
$$t = \text{thickness of the belt}$$
$$s = \text{Total percentage of slip}$$
$$n_1 = 200$$
$$n_2 = 300$$

$$t = 8 \text{ mm}$$
$$s = 4\%$$
$$d_1 = 500 \text{ mm}$$
$$d_2 = ?$$

$$\frac{n_2}{n_1} = \frac{d_1 + t}{d_2 + t} \times \frac{100 - s}{100}$$

$$\frac{300}{200} = \frac{500 + 8}{d_2 + 8} \times \frac{100 - 4}{100}$$

$$d_2 + 8 = \frac{200 \times 508 \times 96}{300 \times 100}$$

$$d_2 + 8 = 325$$
$$d_2 = 325 - 8 = 317 \text{ mm.}$$

20. In H & B carding machine the belt drive is arranged for driving the 51″ cylinder. Find the power transmitted by a belt running over a pulley of 600 mm diameter at 200 rpm. The coefficient of friction between the belt and the pulley is 0.25, angle of lap 160o and maximum tension in the belt is 2500 N.

Solution :

Given: $d = 600 \text{ mm} = 0.6 \text{ m}$; $N = 200 \text{ rpm}$; $\mu = 0.25$;
$\theta = 160° = 160 \times \pi/180 = 2.793 \text{ rad}$; $T_1 = 2500 \text{ N}$.

We know that velocity of the belt,

$$v = \frac{\pi d.N}{60} = \frac{\pi \times 0.6 \times 200}{60} = 6.284 \, m/s$$

Let T_2 = Tension in the slack side of the belt.

We know that $2.3 \log \left(\dfrac{T_1}{T_2} \right) = \mu.\theta = 0.25 \times 2.793 = 0.6982$

$$\log \left(\frac{T_1}{T_2} \right) = \frac{0.6982}{2.3} = 0.3036$$

$$\left(\frac{T_1}{T_2} \right) = 2.01$$

and $$T_2 = \frac{T_1}{2.01} = \frac{2500}{2.01} = 1244 \, N$$

We know that power transmitted by the belt,

$$P = (T_1 - T_2) v = (2500 - 1244) \, 6.284 = 7890 \, W$$
$$= 7.89 \text{ kW.}$$

21. In the condenser carding machine the rope drive is meant for driving the workers and strippers. A casting weighing 9 kN hangs freely from a rope which makes 2.5 turns round a drum of 300 mm diameter revolving at 20 rpm. The other end of the rope is pulled by a typical system. The coefficient of friction is 0.25. Determine (i) The force required by the man, and (ii) The power to raise the casting.

Solution :

Given: $W = T_1 = 9$ kN $= 9000$ N; $d = 300$ mm $= 0.3$ m; $N = 20$ rpm; $\mu = 0.25$

(i) Force required by the man

Let $T_2 =$ Force required by the man

Since the rope makes 2.5 turns round the drum, therefore angle of contact,
$$\theta = 2.5 \times 2\,\pi = 5\,\pi \text{ rad}$$

We know that $2.3 \log \left(- \right) = \mu.\theta = 0.25 \times 5\,\pi = 3.9275$

$$\log \left(\frac{T_1}{T_2} \right) = \frac{3.9275}{2.3} = 1.71 \quad or \quad \frac{T_1}{T_2} = 51$$

$$\therefore\ T_2 = \frac{T_1}{51} = \frac{9000}{51} = 176.47\,N$$

(ii) Power to raise the casting

We know that velocity of the rope,
$$v = \frac{\pi d.N}{60} = \frac{\pi \times 0.3 \times 20}{60} = 0.3142\,m\,/\,s$$

$\therefore$ Power to raise the casting,
$$P = (T_1 + T_2)\,v = (9000 - 176.47)\,0.3142 = 2772 \text{ W.}$$
$$= 2.772 \text{ kW.}$$

22. In the conventional Cimmco Muller cone winding machine the driving arrangement has two pulleys of 450 mm and 200 mm diameter and are on parallel shafts 1.95 m apart which use a crossed-belt. Find the length of the belt required and the angle of contact between the belt and each pulley. If the larger pulley rotates at 200 rev /min, if the maximum permissible tension in the belt is 1 kN, and the coefficient of friction between the belt and pulley is 0.25, find the power expended in driving the arrangement.

Solution :

Given: $d_1 = 450$ mm $= 0.45$ m or $r_1 = 0.225$ m; $d_2 = 200$ mm $= 0.2$ m or
$r_2 = 0.1$ m; $x = 1.95$ m; $N_1 = 200$ rpm; $T_1 = 1$ kN $= 1000$ N; $\mu = 0.25$.

We know that speed of the belt,

$$v = \frac{\pi d_1 N_1}{60} = \frac{\pi \times 0.45 \times 200}{60} = 4.714 \, m/s$$

Length of the belt, we know that length of the crossed-belt,

$$L = \pi \, (r_1 + r_2) + 2x + \frac{(r_1 + r_2)^2}{x}$$

$$= \pi \, (0.225 + 0.1) + 2 \times 1.95 + \frac{(0.225 + 0.1)^2}{1.95} = 4.975 \, m.$$

Angle of contact between the belt and each pulley

Let θ = Angle of contact between the belt and each pulley

We know that for a crossed-belt drive,

$$Sin \, \alpha \frac{r_1 + r_2}{x} = \frac{0.225 + 0.1}{1.95} = 0.1667 \quad or \quad \alpha = 9.6^o$$

$$\therefore \, \theta = 180o + 2\alpha = 180o + 2 \times 9.6o = 199.2o.$$

$$= 199.2 \times \frac{\pi}{180} = 3.477 \, \text{rad.}$$

Power transmitted

Let T_2 = Tension in the slack side of the belt.

We know that

$$2.3 \log \left(\frac{T_1}{T_2} \right) = \mu . \, \theta = 0.25 \times 3.477 = 0.8692$$

$$\text{Log} \left(\frac{T_1}{T_2} \right) = \frac{0.8692}{2.3} = 0.378 \quad or \quad \frac{T_1}{T_2} = 2.387$$

$$\therefore \quad T_2 = \frac{T_1}{2.387} = \frac{1000}{2.387} = 419 N$$

We know that power transmitted,

$$P = (T1 - T2) \, v = (1000 - 419) \, 4.714 = 2740 \, W = 2.74 \, kW.$$

23. In a cotton ginning mill, a shaft rotating at 200 rpm drives another shaft at 300 rpm and transmits 6 kW through a belt. The belt is 100 mm wide and 10 mm thick. The distance between the shafts is 4 m. The smaller pulley is 0.5m in diameter. Calculate the stress in the belt, if it is (i). An open belt drive, and (ii). A cross-belt drive. Given the coefficient of friction $\mu = 0.3$.

Solution :

Given: N_1 = 200 rpm; N_2 = 300 rpm; P = 6 kW = 6×10^3 W; b = 100 mm;

$t = 10$ mm; $x = 4$ m; $d_2 = 0.5$ m; $\mu = 0.3$.
Let $\quad \sigma =$ Stress in the belt.

(i) Stress in the belt for an open belt drive

First of all, let us find out the diameter of larger pulley (d_1), we know that

$$\frac{N_2}{N_1} = \frac{d_1}{d_2} \quad or \quad d_1 = \frac{N_2.d_2}{N_1} = \frac{300 \times 0.5}{200} = 0.75m$$

and velocity of the belt, $v = \dfrac{\pi \, d_2.N_2}{60} = \dfrac{\pi \times 0.5 \times 300}{60} = 7.855 \; m \,/\, s$

Now let us find the angle of contact on the smaller pulley. We know that, for an open belt drive,

$$\sin \alpha = \frac{r_1 - r_2}{x} = \frac{d_1 - d_2}{2x} = \frac{0.75 - 0.5}{2 \; x \; 4} = 0.03125 \quad or \quad \alpha = 1.8^{o}$$

$\therefore$ Angle of contact, $\theta = 180o - 2 \; \alpha = 180 - 2 \times 1.8 = 176.4°$
$= 176.4 \times \pi/180 = 3.08$ rad.

Let $\qquad T_1 =$ Tension in the tight side of the belt, and
$\qquad\qquad T_2 =$ Tension in the slack side of the belt

We know that

$$2.3 \log \left(\frac{T_1}{T_2} \right) = \mu. \; \theta = 0.3 \times 3.08 = 0.924.$$

$$\therefore \quad \log \left(\frac{T_1}{T_2} \right) = \frac{0.924}{2.3} = 0.4017 \; or \; \frac{T_1}{T_2} = 2.52 \qquad\qquad \text{(i)}$$

Power transmitted (P),

$$6 \times 103 = (T_1 - T_2) \, v = (T_1 - T_2) \, 7.855$$
$$\therefore \qquad (T_1 - T_2) = 6 \times 103 \, /7.855 = 764 \; N \qquad\qquad \text{(ii)}$$

From equations (i) and (ii)

$$T_1 = 1267 \; N, \text{ and } T_2 = 503 \; N$$

Maximum tension in the belt (T_1),

$$1267 = \sigma.b.t = \sigma \times 100 \times 10 = 1000 \, \sigma$$
$$\sigma = 1267 \, /1000 = 1.267 \; N/mm^2 = 1.267 \; MPa$$

(ii) Stress in the belt for a cross-belt drive

We know that for a cross-belt drive

$$\sin \alpha = \frac{r_1 + r_2}{x} = \frac{d_1 + d_2}{2x} = \frac{0.75 + 0.5}{2 \; x \; 4} = 0.1562 \, or \; \alpha = 9^{o}$$

$\therefore \quad$ Angle of contact, $\theta = 180° + 2 \; \alpha = 180 + 2 \times 9 = 198°$
$$= 198 \times \pi \,/180 = 3.456 \text{ rad.}$$

We know that

$$2.3 \log \left(\frac{T_1}{T_2}\right) = \mu.\theta = 03 \times 3.456 = 1.0368$$

$$Log\left(\frac{T_1}{T_2}\right) = \frac{1.0368}{2.3} = 0.4508 \; or \; \left(\frac{T_1}{T_2}\right) = 2.82 \qquad \text{(iii)}$$

From equations (ii) and (iii),

$T_1 = 1184$ N and $T_2 = 420$ N

We know that maximum tension in the belt ($T1$),

$$1184 = \sigma.b.t = \sigma \times 100 \times 10 = 1000 \, \sigma$$
$$\sigma = 1184 / 1000 = 1.184 \text{ N/mm2} = 1.184 \text{ MPa.}$$

24. In a waste spinning machine (the machine named DEVIL) has a leather belt and is required to transmit 7.5 kW from a pulley 1.2 m in diameter, running at 250 rpm. The angle of contact is 165o and the coefficient of friction between the belt and the pulley is 0.3. If the safe working stress for the leather belt is 1.5 MPa, density of leather 1 mg /m3 and thickness of belt 10 mm, determine the width of the belt considering the centrifugal tension into account.

Solution :

Given: $P = 7.5$ kW $= 7500$ W; $d = 1.2$ m; $N = 250$ rpm;

$\theta = 165° = 165 \times \pi/180 = 2.88$ rad; $\mu = 0.3$; $\sigma = 1.5$ MPa $= 1.5 \times 10^6$ *N/m²;

$\rho = 1$ mg/m³ $= 1 \times 10^6$ g/m³ $= 1000$ kg/m³; $t = 10$ mm $= 0.01$ m.

Let $b = $ Width of belt in metres,

$T_1 = $ Tension in the tight side of the belt in N, and

$T_2 = $ Tension in the slack side of the belt in N.

We know that velocity of the belt.

$$v = \pi \, d.N /60 = \pi \times 1.2 \times 250 /60 = 15.71 \text{ m/s}$$

And power transmitted (P),

$$7500 = (T_1 - T_2) \, v = (T_1 - T_2) \, 15.71$$
$$\therefore \quad T_1 - T_2 = 7500 /15.71 = 477.4 \text{ N} \qquad \text{(i)}$$

We know that

$$2.3 \log \left(\frac{T_1}{T_2}\right) = \mu.\theta = 0.3 \times 2.88 = 0.864$$

$$Log\left(\frac{T_1}{T_2}\right) = \frac{0.864}{2.3} = 0.3756 \; or \; \frac{T_1}{T_2} = 2.375$$

From equations (i) and (ii),

$T_1 = 824.6$ N, and $T_2 = 347.2$ N

We know that mass of the belt per metre length,

$$m = \text{Area} \times \text{length} \times \text{density} = b.t.l.\rho$$
$$= b \times 0.01 \times 1 \times 1000 = 10\ b\ \text{kg}.$$
$\therefore$ Centrifugal tension,
$$TC = mv2 = 10\ b\ (15.71)^2 = 2468\ b\ \text{N}$$
And maximum tension in the belt,
$$T = \sigma.b.t = 1.5 \times 10^6 \times b \times 0.01 = 15000\ b\ \text{N}$$
We know that
$$T = T_1 + TC \text{ or } 15,000\ b = 824.6 + 2468\ b$$
$$15,000\ b - 2468\ b = 824.6 \text{ or } 12,532\ b = 824.6$$
$$\therefore \qquad b = 824.6\ /12,532 = 0.0658\ \text{m} = 65.8\ \text{mm}.$$

25. A conventional cotton ginning machine has a belt of 9.75 mm thick leather belt required to transmit 15 kW from a motor running at 900 rpm. The diameter of the driving pulley of the motor is 300 mm. The driven pulley runs at 300 rpm and the distance between the centres of two pulleys is 3 m. The density of the leather is 1000 kg/m³. The maximum allowable stress in the leather is 2.5 MPa. The coefficient of friction between the leather and pulley is 0.3. Assume open belt drive and neglect the sag and slip of the belt. Work out the width of the belt used.

Solution :

Given: $t = 9.75$ mm $= 9.75 \times 10^{-3}$ m; $P = 15$ kW $= 15 \times 10^3$ W; $N_1 = 900$ rpm;
$d_1 = 300$ mm $= 03$ m; $N_2 = 300$ rpm; $x = 3$ m; $\rho = 1000$ kg/m³;
$\sigma = 2.5$ MPa $= 2.5 \times 106$ N/m²; $\mu = 0.3$.

First of all, let us find out the diameter of the driven pulley ($d2$). We know that

$$\frac{N_2}{N_1} = \frac{d_1}{d_2} \text{ or } d_2 = \frac{N_1 \times d_1}{N_2} = \frac{900 \times 0.3}{300} = 0.9m$$

and velocity of the belt, $v = \dfrac{\pi d_1 N_1}{60} = \dfrac{\pi \times 0.3 \times 900}{60} = 14.14\ m/s$

For an open belt drive,

$$\sin \alpha = \frac{r_2 - r_1}{x} = \frac{d_2 - d_1}{2x} = \frac{0.9 - 0.3}{2 \times 3} = 0.1 \qquad \qquad ....(\because d_2 > d_1)$$

or $\alpha = 5.74$o
$\therefore$ Angle of lap $\theta = 180° - 2\alpha = 180 - 2 \times 5.74 = 168.52°$
$$= 168.52 \times \pi/180 = 2.94 \text{ rad}.$$
Let T_1 = Tension in the tight side of the belt,
T_2 = Tension in the slack side of the belt.
We know that

$$\log \left(\frac{T_1}{T_2}\right) = \mu.\theta = 0.3 \times 2.94 = 0.882$$

$$\log \left(\frac{T_1}{T_2}\right) = \frac{0.882}{2.3} = 0.3835 \quad or \quad \frac{T_1}{T_2} = 2.42 \qquad (i)$$

We also know that power transmitted (P),

$$15 \times 10^3 = \left(T_1 - T_2\right) v = \left(T_1 - T_2\right) 14.14$$

∴ $\qquad T_1 - T_2 = 15 \times 10^3 / 14.14 = 1060$ N $\qquad (ii)$

From equations (i) and (ii),

$T_1 = 1806$ N

Let $\qquad b$ = Width of the belt in metres.

We know that mass of the belt per metre length.

m = Area × length × density = $b.t.l.\rho$

$\qquad = b \times 9.75 \times 10^{-3} \times 1 \times 1000 = 9.75\ b$ kg.

∴ Centrifugal tension

$\qquad T_C = mv^2 = 9.75\ b\ (14.14)^2 = 1950\ b$ N

Maximum tension in the belt,

$\qquad T = \sigma.b.t = 2.5 \times 10^6 \times b \times 9.75 \times 10^{-3} = 24,400\ b$ N

We know that $\qquad T = T_1 + T_C$ or $T - T_C = T_1$

$\qquad 24,400\ b - 1950\ b = 1806$ or $22,450\ b = 1806$

∴ $\qquad b = 1806 / 22,450 = 0.080$ m = 80 mm.

26. In the driving arrangements for Crigthon opener of a conventional blow room has a pulley driven by a flat belt, subtending the angle of lap 120°. The belt is 100 mm wide by 6 mm thick and density 1000 kg/m³. If the coefficient of friction is 0.3 and the maximum stress in the belt is not to exceed 2 MPa, find the greatest power transmitted by belt and its speed.

Solution :

Given: θ =120° = 120 × π /180 = 2.1 rad; b = 100 mm = 0.1 m; t = 6 mm = 0.006 m;

$\qquad \rho$ = 1000 kg/m³; μ = 0.3; σ = 2 MPa = 2 × 10⁶ N/m²

Speed of the belt for greatest power, we know that maximum tension in the belt.

$\qquad T = \sigma.b.t = 2 \times 10^6 \times 0.1 \times 0.006 = 1200$ N

and mass of the belt per metre length, m = Area × length × density = $b.t.l.\rho$.

$\qquad = 0.1 \times 0.006 \times 1 \times 1000 = 0.6$ kg/m

∴ $\qquad$ Speed of the belt for greatest power,

$$v = \frac{\sqrt{T}}{3m} = \frac{\sqrt{1200}}{3 \times 0.6} = 25.82 \text{ m/s.}$$

Greatest power which the belt can transmit

We know that for maximum power to be transmitted, centrifugal tension,

$T_C = T/3 = 1200/3 = 400$ N,

and tension in the tight side of the belt,

$T_1 = T - T_C = 1200 - 400 = 800$ N

Let T_2 = tension in the slack side of the belt.

We know that

$$2.3 \log \left(\frac{T_1}{T_2}\right) = \mu.\theta = 0.3 \times 21 = 0.63$$

$$\text{Log} \left(\frac{T_1}{T_2}\right) = \frac{0.63}{2.3} = 0.2739 \ or \ \frac{T_1}{T_2} = 1.88$$

and $T_2 = \dfrac{T_1}{1.88} = \dfrac{800}{1.88} = 425.5 \, N$

∴ Greatest power which the belt can transmit,

$P = (T_1 - T_2) v = (800 - 425.5) \, 25.82 = 9670$ W $= 9.67$ kW.

27. The Techmech make 2 m drum diameter sectional warping machine is driven by an open belt drive connecting two pulleys, 1.2 m and 0.5 m in diameter, on parallel shafts 4 m apart. The mass of the belt is 0.9 kg per metre length and the maximum tension in not to exceed 2000 N. The coefficient of friction is 0.3. The 1.2 m pulley, which is the driver, runs at 200 rpm. Due to belt slip on one of the pulleys, the velocity of the driven shaft is only 450 rpm. Calculate the torque on each of the two shafts, the power transmitted, and power lost in friction. Also work out the efficiency of the drive?

Solution :

Given: $d_1 = 1.2$ m or $r_1 = 0.6$ m; $d_2 = 0.5$ m or $r_2 = 0.25$ m; $x = 4$ m; $m = 0.9$ kg/m; $T = 2000$ N; $\mu = 0.3$; $N_1 = 200$ rpm; $N_2 = 450$ rpm.

We know that velocity of the belt,

$$v = \frac{\pi d_1 N_1}{60} = \frac{\pi \times 1.2 \times 200}{60} = 12.57 m/s$$

and centrifugal tension, $T_C = mv^2 = 0.9 \, (12.57)^2 = 142$ N

∴ Tension in the tight side of the belt,

$T_1 = T - T_C = 2000 - 142 = 1858$ N

We know that for an open belt drive,

$$\sin \alpha = \frac{r_1 - r_2}{x} = \frac{0.6 - 0.25}{4} = 0.0875 \quad or \quad \alpha = 5.02°$$

∴ Angle of lap on the smaller pulley,

$\theta = 180\text{o} - 2\alpha = 180° - 2 \times 5.02° = 169.96°$

$= 169.96 \times \pi / 180 = 2.967$ rad.

Let T_2 = Tension in the slack side of the belt.

We know that

$$2.3 \log\left(\frac{T_1}{T_2}\right) = \mu.\theta = 0.3 \times 2.967 = 0.8901$$

$$\text{Log}\left(\frac{T_1}{T_2}\right) = \frac{0.8901}{2.3} = 0.387 \ or \ \frac{T_1}{T_2} = 2.438$$

$$\therefore T_2 = \frac{T_1}{2.438} = \frac{1858}{2.438} = 762 \, N$$

Torque on the shaft of larger pulley

We know that torque on the shaft of larger pulley,

$$T_L = (T_1 - T_2)\, r_1 = (1858 - 762)\, 0.6 = 657.6 \text{ N-m.}$$

Torque on the shaft of smaller pulley

We know that torque on the shaft of smaller pulley

$$T_S = (T_1 - T_2)\, r_2 = (1858 - 762)\, 0.25 = 274 \text{ N-m.}$$

Power transmitted

We know that the power transmitted,

$$P = (T_1 - T_2)\, v = (1858 - 762)\, 12.57 = 13780 \text{ W}$$
$$= 13.78 \text{ kW.}$$

Power lost in friction

We know that input power,

$$P_1 = \frac{T_L \times 2\pi N_1}{60} = \frac{657.6 \times 2\pi \times 200}{60} = 13,780 \, W = 13.78 \, kW.$$

and output power, $P_2 =$

$$\frac{T_S \times 2\pi N_2}{60} = \frac{274 \times 2\pi \times 450}{60} = 12,910 \, W = 12.91 \, kW$$

∴ Power lost in friction $= P1 - P2 = 13.78 - 12.91 = 0.87$ kW.

Efficiency of the drive

We know that efficiency of the drive

$$\eta = \frac{\text{output power}}{\text{input power}} = \frac{12.91}{13.78} = 0.937 \ or \ 93.7\%$$

28. The Porcupine beater of a blow room has a flat belt drive with the initial tension is 2000 N. The coefficient of friction between, the belt and the pulley is 0.3 and the angle of lap on the smaller pulley is 150°. The smaller pulley has a radius of 200 mm and rotates as 500 rpm, workout the power in kW transmitted by the belt.

Solution :

Given: $T_0 = 2000$ N; $\mu = 0.3$; $\theta = 150° = 150° \times \pi/180 = 2.618$ rad.;

$r_2 = 200$ mm or $d_2 = 400$ mm $= 0.4$ m; $N_2 = 500$ rpm.

We know that velocity of the belt,

$$v = \frac{\pi d_2 N_2}{60} = \frac{\pi \times 0.4 \times 500}{60} = 10.47 m/s$$

Let $\quad T_1 =$ Tension in the tight side of the belt, and

$\quad T_2 =$ Tension in the slack side of the belt.

We know that initial tension ($T0$)

$$2000 = \frac{T_1 + T_2}{2} \quad or \quad T_1 + T_2 = 4000 \ N$$

We also know that

$$2.3 \log\left(\frac{T_1}{T_2}\right) = \mu.\theta = 0.3 \times 2.618 = 0.7854$$

$$\text{Log}\left(\frac{T_1}{T_2}\right) = \frac{0.7854}{2.3} = 0.3415 \ or \ \frac{T_1}{T_2} = 2.2 \qquad \text{(ii)}$$

From equations (i) and (ii),

$\quad T_1 = 2750$ N; and $T_2 = 1250$ N

$\therefore$ Power transmitted, $P = (T_1 - T_2) \ v = (2750 - 1250) \ 10.47 = 15700$ W
$$= 15.7 \ kW.$$

29. A drive is available for driving the fabric folding machine through two parallel shafts whose centre lines are 4.8 m apart, are connected by open belt drive. The diameter of the larger pulley is 1.5 m and that of smaller pulley 1 m. The initial tension in the belt when stationary is 3 kN. The mass of the belt is 1.5 kg/m length. The coefficient of friction between the belt and the pulley is 0.3. Taking centrifugal tension into account, calculate the power transmitted, if the smaller pulley rotates at 400 rpm.

Solution :

Given: $x = 4.8$ m.; $d_1 = 1.5$ m; $d_2 = 1$ m; $T_0 = 3$ kN $= 3000$ N;

$m = 1.5$ kg/m; $\mu = 0.3$; $N_2 = 400$ rpm.

We know that velocity of the belt,

$$v = \frac{\pi d_2 N_2}{60} = \frac{\pi \times 1 \times 400}{60} = 21 \ m/s$$

And centrifugal tension,

$$T_C = mv^2 = 1.5\,(21)^2 = 661.5 \text{ N}$$

Let $\quad T_1$ = Tension in the tight side, and

$\quad\quad T_2$ = Tension in the slack side.

We know that initial tension (T_0)

$$3000 = \frac{T_1 + T_2 + 2T_C}{2} = \frac{T_1 + T_2 + 2 \times 661.5}{2}$$

$\therefore \quad T_1 + T_2 = 3000 \times 2 - 2 \times 661.5 = 4677 \text{ N}$

For an open belt drive,

$$\sin \alpha = \frac{r_1 - r_2}{x} = \frac{d_1 - d_2}{2x} = \frac{1.5 - 1}{2 \times 4.8} = 0.0521 \ or \ \alpha = 3^o$$

$\therefore$ Angle of lap on the smaller pulley,

$$\theta = 180° - 2\alpha = 180° - 2 \times 3° = 174°$$
$$= 174° \times \pi/180 = 3.04 \text{ rad.}$$

We know that

$$2.3 \log \left(\frac{T_1}{T_2} \right) = \mu.\theta = 0.3 \times 3.04 = 0.912$$

$$\log \left(\frac{T_1}{T_2} \right) = \frac{0.912}{2.3} = 0.3965 \quad or \quad \frac{T_1}{T_2} = 2.5 \qquad \text{(i)}$$

From equations (i) and (ii)

$$T_1 = 3341 \text{ N; and } T_2 = 1336 \text{ N}$$

$\therefore$ Power transmitted,

$$P = (T_1 - T_2) v = (3341 - 1336) 21 = 42,100 \text{ W}$$
$$= 42.1 \text{ kW.}$$

30. In the Slasher sizing shed an open flat belt drive connects two parallel shafts 1.2 m apart. The driving and the driven shafts rotate at 350 and 140 rpm, respectively and the driven pulley is 400 mm in diameter. The belt is 5 mm thick and 80 mm wide. The coefficient of friction between the belt and pulley in 0.3 and the maximum permissible tension in the belting is 1.4 MN/m². Find the diameter of the driving pulley.

Solution :

Given: $x = 1.2$ mm; $N_1 = 350$ rpm; $N_2 = 140$ rpm; $d_2 = 400$ mm = 0.4 m;
$\quad\quad\quad t = 5$ mm = 0.005 m; $b = 80$ mm = 0.08 m; $\mu = 0.3$;
$\quad\quad\quad \sigma = 1.4$ MN/m² = 1.4×10^6 N/m².

Diameter of the driving pulley

Let d_1 = Diameter of the driving pulley.

We know that $\quad\quad \dfrac{N_2}{N_1} = \dfrac{d_1}{d_2} \quad or \quad d_1 = \dfrac{N_2 d_2}{N_1}$

$$= \frac{140 \times 0.4}{350} = 0.16m.$$

31. Power is transmitted using a V-belt drive. The included angle of V-groove is 30°. The belt is 20 mm deep and maximum width is 20 mm. If the mass of the belt is 0.35 kg per metre length and maximum allowable stress in 1.4 MPa, determine the maximum power transmitted when the angle of lop is 140°. $\mu = 0.15$.

Solution

Given: $2\beta = 30°$ or $\beta = 15°$; $t = 20$ mm $= 0.02$ m; $b = 20$ mm $= 0.02$ m; $m = 0.35$ kg/m;

$\sigma = 1.4$ MPa $= 1.4 \times 106$ N/m2; $\theta = 140o = 140o \times \pi /180 = 2.444$ rad; $\mu = 0.15$.

We know that maximum tension in the belt,

$$T = \sigma.b.t = 1.4 \times 106 \times 0.02 \times 0.02 = 560 \text{ N},$$

and for maximum power to be transmitted, velocity of the belt,

$$v = \sqrt{\frac{T}{3m}} = \sqrt{\frac{560}{3 \times 0.35}} = 23.1 m/s$$

Let T_1 = Tension in the tight side of the belt, and
                T_2 = Tension in the slack side of the belt.

We know that

$$2.3 \log \left(\frac{T_1}{T_2} \right) = \mu.\theta \operatorname{cosec} \beta = 0.15 \times 2.444 \times \operatorname{cosec} 15° = 1.416$$

$$\log \left(\frac{T_1}{T_2} \right) = \frac{1.416}{2.3} = 0.616 \quad or \quad \frac{T_1}{T_2} = 4.13 \qquad \text{(i)}$$

Centrifugal tension $T_C = \dfrac{T}{3} = \dfrac{560}{3} = 187 \text{ N}$

And $T_1 = T - T_C = 560 - 187 = 373 \text{ N}$

$$T_2 = \frac{T_1}{4.13} = \frac{373}{4.13} = 90.3 \text{ N}$$

We know that maximum power transmitted,

$$P = (T_1 - T_2) v = (373 - 90.3) \, 23.1 = 6530 \text{ W} = 6.53 \text{ kW}.$$

32. In a twisting machine two pulleys 240 mm and 600 mm diameter connects two parallel shafts 3 m apart and are connected by flat open belt drive and transmits 4 kW from the smaller pulley that rotates at 300 rpm. Coefficient of friction between the belt and the pulley is 0.3 and the safe working tension is 10 N per mm width, determine minimum width of the belt, initial belt tension, and, length of the belt required.

Solution :

Given: d_2 = 240 mm = 0.24 m; d_1 = 600 mm = 0.6 m; x = 3 m; P = 4 kW = 4000 W;

N_2 = 300 rpm; μ = 0.3; T_1 = 10 N/mm width.

Minimum width of belt: We know that velocity of the belt,

$$v = \frac{\pi d_2.N_2}{60} = \frac{\pi \times 0.24 \times 300}{60} = 3.77 m/s$$

Let T_1 = Tension in the tight side of the belt, and
           T_2 = Tension in the slack side of the belt.

$\therefore$ Power transmitted (P),

$$4000 = (T_1 - T_2) v = (T_1 - T_2) 3.77$$

or $T_1 - T_2 = 4000 / 3.77 = 1061$ N

We know that for an open belt drive,

$$\sin \alpha = \frac{r_1 - r_2}{x} = \frac{d_1 - d_2}{2x} = \frac{0.6 - 0.24}{2 \times 3} = 0.06 \ or \quad \alpha = 3.44°$$

and angle of lap on the smaller pulley,

$$\theta = 180° - 2\alpha = 180° - 2 \times 3.44° = 173.12°$$
$$= 173.12 \times \pi / 180 = 3.022 \text{ rad.}$$

We know that

$$2.3 \log \left(\frac{T_1}{T_2} \right) = \mu.\theta = 0.3 \times 3.022 = 0.9066$$

$$\text{Log} \left(\frac{T_1}{T_2} \right) = \frac{0.9066}{2.3} = 0.3942 \quad or \quad \frac{T_1}{T_2} = 2.478$$

From above equations

$$T_1 = 1779 \text{ N and } T_2 = 718 \text{ N}$$

Since the safe working tension is 10 N per mm width, therefore minimum width of the belt,

$$b = \frac{T_1}{10} = \frac{1779}{10} = 177.9 mm. \qquad Ans.$$

Initial belt tension

We know that initial belt tension

$$T_0 = \frac{T_1 + T_2}{2} = \frac{1779 + 718}{2} = 1248.5 N \ Ans$$

Length of the belt required
We know that length of the belt required

$$L = \frac{\pi}{2}(d_1 + d_2) + 2x + \frac{(d_1 - d_2)^2}{4x}$$

$$= \frac{\pi}{2}(0.6 + 0.24) + 2 \times 3 + \frac{(0.6 - 0.24)^2}{4 \times 3}$$

$$= 1.32 + 6 + 0.01 = 7.33.$$

33. The ring frame drive consists of V-belt drive of two V-belts in parallel, on grooved pulleys of the same size. The angle of the groove is 30°. The cross-sectional area of each belt is 750 mm² and $\mu = 0.12$. The density of the belt materials is 1.2 mg/m³ and the maximum safe stress in the material is 7 MPa. Calculate the power that can be transmitted between pulleys 300 mm. diameter rotating at 1500 rpm. Find also the shaft speed is rpm at which the power transmitted would be a maximum.

Solution :

Given: $2\beta = 30°$ or $\beta = 15°$; $a = 750$ mm² $= 750 \times 10^{-6}$ m²; $\mu = 0.12$;
$\rho = 1.2$ mg/m³ $= 1200$ kg/m³; $\sigma = 7$ MPa $= 7 \times 10^6$ N/m²;
$d = 300$ mm $= 0.3$ m;
$N = 1500$ rpm.

Power transmitted
We know that velocity of the belt,

$$v = \frac{\pi d.N}{60} = \frac{\pi \times 0.3 \times 1500}{60} = 23.56 m/s$$

and mass of the belt per meter length,
$m =$ Area × length × density $= 750 \times 10^{-6} \times 1 \times 1200 = 0.9$ kg/m.

∴ Centrifugal tension,
$$T_C = mv^2 = 0.9 \, (23.56)^2 = 500 \text{ N}$$

We know that maximum tension in the belt,
$T =$ Maximum stress × cross-sectional area of belt $= \sigma \times a$
$$= 7 \times 10^6 \times 750 \times 10^{-6} = 5250 \text{ N}$$

∴ Tension in the tight side of the belt,
$$T_1 = T - T_C = 5250 - 500 = 4750 \text{ N}$$

Let $T_2 =$ Tension in the slack side of the belt.

Since the pulleys are of the same size, therefore, angle of contact, $\theta = 180° = \pi$ rad.

We know that

$$2.3 \log\left(\frac{T_1}{T_2}\right) = \mu.\theta \cos ec\,\beta = 0.12 \times \pi \times \cosec 15° = 1.457$$

$$\log \left(\frac{T_1}{T_2}\right) = \frac{1.457}{2.3} = 0.6334 \qquad or \qquad \frac{T_1}{T_2} = 4.3$$

and $T_2 = \dfrac{T_1}{4.3} = \dfrac{4750}{4.3} = 1105\,N$

We know that power transmitted,

$P = (T_1 - T_2)\, v \times 2$ ...($\because$ No. of belts = 2)

$= (4750 - 1105)\,23.56 \times 2 = 17{,}1752\ W = 171.752\ kW.$

Shaft speed

Let N_1 = Shaft sped in rpm and v_1 = Belt speed in m/s.

We know that for maximum power, centrifugal tension

$$T_C = T/3 \text{ or } m\,(v_1)^2 = T/3 \text{ or } 0.9\,(v_1)^2 = 5250\,/3 = 1750$$

$$\therefore \quad (v_1)^2 = 1750\,/0.9 = 1944.4 \text{ or } v_1 = 44.1 \text{ m/s.}$$

We also know that belt speed (v_1),

$$44.1 = \frac{\pi d N_1}{60} = \frac{\pi \times 0.3 \times N_1}{60} = 0.0157\,N_1$$

$\therefore\ N1 = 44.1\,/0.0157 = 2809$ rpm.

34. Find the hp transmitted by a belt running over a pulley of 60 cm diameter at 200 rpm. The coefficient of friction between the belt and pulley is 0.25, angle of lap 160° and maximum tension in the belt is 250 kg.

Solution :

Given: $d = 60$ mm $= 0.6$ m, $N = 200$ rpm, $\mu = 0.25$, $\theta = 160°$

$$= 160 \times \frac{\pi}{180} = 2.793\ radians$$

$T_1 = 250$ kg

Let $T_2 = ?$

hp = ?

$$hp = \frac{(T_1 - T_2)\,v}{4500} \qquad \text{If } n \text{ is in m/min}$$

$$= \frac{(T_1 - T_2)\,v}{75} \qquad \text{If } n \text{ is in m/s}$$

First step:

We know that $v = \dfrac{\pi d N}{60}$

$$= \frac{\pi \times 0.6 \times 200}{60}$$

$$v = 6.283 \text{ m/s.}$$

We know that

$$\log \frac{T_1}{T_2} = \mu\theta$$

$$\log\left(\frac{T_1}{T_2}\right) = 0.25 \times 2.793$$

$$\therefore \log \frac{T_1}{T_2} = 0$$

$$\frac{T_1}{T_2} = 2.01$$

$$\therefore T_2 = \frac{2.50}{2.01} = 124.4 \text{ kg}$$

We know that

$$\text{HP} = \frac{(T_1 - T_2)\,v}{75}$$

$$= \frac{(250 - 124.4)\,6.283}{75}$$

$$= 10.52 \text{ hp.}$$

35. The tension on the tight side of a belt is 300 kg and angle of lap is 160°. If the coefficient of friction is 0.3, find the tension on the slack side of the belt.

Solution :

$$T_1 = 300 \text{ kg}$$

$$\theta = 160° = 160 \times \frac{\pi}{180} = 2.792$$

$$\mu = 0.3$$

$$T_2 = ?$$

$$\log \frac{T_1}{T_2} = 0.4343\,\mu q$$

$$= 0.4343 \times 0.3 \times 2.792$$

$$= \log \frac{T_1}{T_2} = 0.3637$$

$$\frac{T_1}{T_2} = \text{antilog}\,(0.3637)$$

$$\frac{T_1}{T_2} = 2.310$$

$$\frac{300}{T_2} = 2.310$$

$$\therefore T_2 = \frac{300}{2.310}$$

$$T_2 = 129.8 \text{ kg}.$$

36. A belt transmits power from a pulley of 75 mm diameter running at 200 rpm. If the power transmitted is 10 hp and the angle of lap on the pulley is 160°, find the necessary initial tension in the belt and the belt width, if the pull is not to exceed 2 kg/mm width. Take coefficient of friction as 0.28.

Solution :

$$D = 75 \text{ mm} \quad = \frac{75}{100} mt$$

$N = 200$ rpm.

hp $= 10$

$$q = 160° = 160 \times \frac{\pi}{180} = 2.792 \; radians$$

$\mu = 0.28$

Maximum tension $= 2$ kg/mm

$$T_0 = ?$$

$$\log \frac{T_1}{T_2} = 0.4343 \; \mu q$$

$$\log \frac{T_1}{T_2} = 0.4343 \times 0.28 \times 2.792$$

$$\log \frac{T_1}{T_2} = 0.9700$$

$$\log \frac{T_1}{T_2} = \text{antilog} \, (0.3395)$$

$$\frac{T_1}{T_2} = 32.19$$

So $T_1 = 2.19 \, T_2$

$$\text{hp} = \frac{(T_1 - T_2)\pi\,d\,n}{4500}$$

$$10 = \frac{(T_1 - T_2)\pi \times 75 \times 200}{4500 \times 1000}$$

$$T_1 - T_2 = \frac{4500 \times 10 \times 1000}{\pi \times 75 \times 00}$$

$$T_1 - T_2 = 954.93 \text{ kg.}$$

Substitute the value of T_1

$$2.19\, T_2 - T_2 = 954.93$$

$$1.19\, T_2 = 954.93$$

$$T_2 = \frac{954.93}{1.19}$$

$$T_2 = 802.46 \text{ kg}$$

$$\therefore T_1 = 2.19\, T_2$$

$$= 2.19 \times 802.46$$

$$T_1 = 1757.39 \text{ kg.}$$

We know that $T_0 = \dfrac{(T_1 - T_2)}{2}$

$$= \frac{1757.39 + 802.46}{2}$$

$$T_0 = 1279.92 \text{ kg.}$$

$$\text{Width of belt} = \frac{T_1}{Maximum\ Tension}$$

$$= \frac{1757.39}{2}$$

$$= 878.69 \text{ mm.}$$

37. Find the power transmitted by a belt running over a pulley of 600 mm. diameter at 200 rpm. The coefficient of friction between the belt and the pulley is 0.25, angle of lap 160° and maximum tension in the belt is 2500 N.

Solution :

Given: $d = 600$ mm $= 0.6$ m; $N = 200$ rpm; $\mu = 0.25$;
       $\theta = 160 = 160 \times \pi/180 = 2.793$ rad; $T_1 = 2500$ N.

We know that velocity of the belt,

$$v = \frac{\pi d.N}{60} = \frac{\pi \times 0.6 \times 200}{60} = 6.284\, m/s$$

Let T_2 = Tension in the slack side of the belt.

We know that $2.3 \log \left(\dfrac{T_1}{T_2} \right) = \mu.\theta = 0.25 \times 2.793 = 0.6982$

$$\log \left(\frac{T_1}{T_2} \right) = \frac{0.6982}{2.3} = 0.3036$$

$$\left(\frac{T_1}{T_2} \right) = 2.01$$

and $T_2 = \dfrac{T_1}{2.01} = \dfrac{2500}{2.01} = 1244\,N$

We know that power transmitted by the belt,

$$P = \left(T_1 - T_2 \right) v = (2500 - 1244)\,6.284 = 7890\ W$$
$$= 7.89 \text{ kW.}$$

3.20   Effect of centrifugal tension (T_c)

Since the belt continuously runs over the pulleys, therefore some centrifugal force is caused, whose effect is to increase the tension on both tight as well as the slack sides. The tension caused by the centrifugal force is called *centrifugal tension* (T_c).

When a belt is running at high speed and moving in a circular path while passing through a pulley, a centrifugal force acts on it, this force tries to remove the belt away from the pulley. In order to keep the belt in position, an additional tension T_C acts on both sides of the belt drive.

Consider a small portion of the belt xy subtending an angle $d\theta$ at the centre of the pulley as show in Figure.

Let W = Width of the belt / unit length (speed)

n = Linear velocity of the belt

r = Radius of the pulley

T_c = Centrifugal tension acting tangentially at x and y

$\therefore$ Centrifugal tension T_c is given by

$$T_c = \frac{w v^2}{g}\ \text{kg.}$$

Where w = Width of the belt / unit kg

n = Speed of the belt in m/s

g = Acceleration due to gravity (9.81 m/s² or 981 cm/s²)

It may be noted that T_C exits in a belt in addition to T_1 and T_2. Thus, when the centrifugal tension is considered the total tension in the tight side "T_t" and that on the slack side "T_s" are given by

$T_t = T_1 + T_C$

$T_s = T_2 + T_C$

Maximum tension in the belt T_M is also equal to the total tension on the tight side of the belt (T_t) i.e., $T_M = T_1$

Let f = maximum stress in the belt is kg/cm²

b = width of the belt in cm and

t = thickness of the belt in cm.

Maximum tension in the belt is given by

T_M = Maximum stress × Cross-sectional area of the belt

$T_M = f \times b \times t$

When centrifugal tension Tc is considered then

Tension on the tight side $T_M = T_t = T_1 + T_c$ in kg.

Or $T_1 = T_M - T_1$

When centrifugal tension is neglected then

$T_M = T_t = T_1$

Maximum HP transmitted by a belt if 1/3 of the maximum tension is utilised as centrifugal tension.

We know that

$T_M = T_1 + T_C$

Let $T_1 = 2\ T_C$ (Assumption)

∴ $TM = 3\ T_C$

Problems

1. A pulley is driven by a flat belt at the angle of lap being 120°. The belt is 20 cm width and 0.5 cm thick and weighs 0.0012 kg/cm². If the coefficient of friction is 0.3 and the maximum stress in the belt is not 10 exceed 20 kg/cm², find the maximum hp transmitted by the belt and the corresponding speed of the belt.

Solution :

$\theta = 120° = 120 \times \dfrac{\pi}{180} = 2.09\ radians$

Width = 20 cm = b

Thickness = 0.5 cm = t

Weight = 0.0012 kg/cm²

$\mu = 0.3$

$f = 20$ kg/cm²

Max hp = ?   v = ?

We know that

$$T_c = \frac{w v^2}{g}$$

Maximum tension in the belt is given by

$$T_M = f \times b \times t$$
$$= 20 \times 20 \times 0.5$$
$$T_M = 200 \text{ kg}$$

Total weight of the belt /mt length

$$W = 20 \times 0.5 \times 0.0012 \times 100 \qquad \because 100 \text{ cm} = 1 \text{ m}$$
$$W = 1.2 \text{ kg.}$$

But we know that for maximum hp transmitted

$$T_M = 3 \, TC$$
$$T_C = \frac{T_M}{3}$$
$$T_C = \frac{200}{3} = 66.67 \, kg.$$

We know that

$$T_C = \frac{w v^2}{g}$$

Or

$$66.67 = \frac{1.2 \, v^2}{9.81}$$
$$\therefore v^2 = \frac{9.81 \, x \, 66.67}{1.2}$$
$$v^2 = 545$$
$$v = \sqrt{545}$$
$$v = 23.34 \text{ m/s.}$$

To find maximum hp we have to assume that $T_1 = 2 \, T_C$

$$= 2 \times 66.67 = 133.34 \text{ kg.}$$

To find T_2, we know that

$$\log\left(\frac{T_1}{T_2}\right) = 0.4343 \, \mu\theta$$
$$= 0.4343 \times 0.3 \times 2.09$$
$$\log\left(\frac{T_1}{T_2}\right) = 0.2723$$
$$\therefore \left(\frac{T_1}{T_2}\right) = \text{antilog} \, (0.2723)$$

$$\left(\frac{T_1}{T_2}\right) = 1.872$$

$$\therefore T_2 = \frac{T_1}{1.872} = \frac{133.34}{1.872}$$

$T_2 = 71.23$ kg.

$$\therefore hp = \frac{\left(T_1 - T_2\right) v}{75} \quad \text{where v is in m/s.}$$

$$= \frac{\left(133.34 - 71.23\right) 23.34}{75}$$

hp = 19.32

Ratio of driving tensions for V-belts is given by

$$\text{Log } \frac{T_1}{T_2} = 0.4343 \; \mu q \text{ cosec a}$$

2. A leather belt 9 mm × 250 mm is used to drive a CI pulley 90 cm in diameter at 336 rpm. If the active arc (angle of lap) on the smaller pulley is 120° and the stress in the tight side is 20 kg/cm², find the hp coefficient of the belt which weighs 0.00098 kg/cm², coefficient of friction of leather on CI is 0.35 and also find centrifugal tension.

Solution :

Given $t = 9$ mm $= 0.9$ cm
            $b = 250$ mm $= 25$ cm
            $d = 90$ cm $= 0.9$ m
            $N = 336$ rpm.

$$\theta = 120° = 120 \times \frac{\pi}{180} = 2.1 \; radians$$

$F_1 = 20$ kg/cm²
hp = ? and T_c = ?
Density = 0.00098 kg/cm²
hp is given by

$$hp = \frac{\left(T_1 - T_2\right) v}{75} \quad \text{if n is in m/s.}$$

We know that

$$v = \frac{\pi d n}{60}$$

$$= \frac{\pi \times 0.9 \times 336}{60}$$

$$v = 15.83 \text{ m/s}.$$

Cross-section area of the belt a = $b \times t$

$$= 0.9 \times 25 = 22.5 \text{ cm}^2$$

$\therefore$ Tension on tight side of the belt T_1 is given by

$$T_1 = f \times b \times t$$
$$T_1 = 20 \times 22.5$$
$$T_1 = 450 \text{ kg}.$$

To find T_2

We know that log $\dfrac{T_1}{T_2}$ = 0.4343 μq

$$= 0.4343 \times 0.35 \times 2.1$$

$$\log \frac{T_1}{T_2} = 0.3192$$

$$\frac{T_1}{T_2} = \text{antilog } (0.3192)$$

$$T_2 = 2.085$$

$$\therefore T_2 = \frac{T_1}{2.085}$$

$$= \frac{450}{2.085}$$

$$T_2 = 215.8 \text{ kg}.$$

$\therefore$ Centrifugal tension is given by $T_C = \dfrac{w v^2}{g}$

To find weight of the belt/m length

$$w = \text{Area} \times \text{length} \times \text{density}$$
$$= 22.5 \times 100 \times 0.00098$$
$$w = 2.2 \text{ kg}$$

$$\therefore T_C = \frac{w v^2}{g}$$

$$= \frac{2.2 \times (15.83)^2}{9.81}$$

$$T_C = 56.2 \text{ kg}.$$

$$\therefore hp = \frac{\left(T_1 - T_2\right) v}{75}$$

$$= \frac{\left(450 - 215.8\right) 15.83}{75}$$

$$= 49.4 \text{ hp.}$$

3.21   Rope drives

Rope drives are the next to belt drives and finds less application. The use of rope for the transmission of power is more common in Europe than in the United States. When power is to be transmitted over long distances then belts cannot be used due to the heavy losses in power. In such cases ropes can be used. Ropes are used in elevators, mine hoists, cranes, oil well drilling, aerial conveyors, tramways, haulage devices, lifts and suspension bridges, etc.

3.21.1   Material of construction

Fibre ropes

These ropes can use at a distance up to 60 m. Fibre ropes are commonly made of hemp, manila, or cotton. The ropes made from hemp and manila are not flexible and have inferior mechanical properties compared to the cotton. In order to occlude moving of fibres during bending and thus protect it from wear, manila and hemp belt are lubricated with tar or graphite. It also makes rope waterproof.

Cotton ropes are smooth and flexible, so the lubrication is not necessary. It may lubricate to reduce the wear.

Advantages – Smooth and quiet, no precise alignment required, high efficiency.

They are further classified into

- Vegetable fibre (hemp, manila, cotton, jute)
- Synthetic fibre (nylon) – They are stronger than vegetable fibre and can use in moister conditions.

Manila rope

Manila rope – often referred to as hemp – is a natural fibre product of a type which has been in use for thousands of years. It is made from the "abaca" or "musa" textiles plant, which is grown in the Philippines and is related to the banana plant.

Manila ropes without the wire core are used, largely due to their good "hand" and ability to absorb perspiration, in gymnasium and obstacle course climbing systems, and in a number of exercise and hand strength building applications. Frequent inspection and padded landing surfaces are required for climbing or other load bearing systems. Long a standard for stage rigging, many still prefer "hemp systems" over systems using stronger, higher cost synthetics. Used in diameters 5/8″, 3/4″, 7/8″ and 1″ (mostly 3/4″), its low stretch and good "hand" are important here. The wire centre construction is sometimes used for added strength. Despite preferences, habits, and its-what-we've-always-used traditions, there are stronger, more durable synthetics for use in towing, safety, or climbing lines, or in applications where rope failure could cause damage to property or personal injury or worse. On the other hand, Manila is a popular choice for tug of war because it won't snap back – potentially causing serious injuries – as nylon rope has been known to do. Manila's good grip and ability to absorb sweat make it popular for obstacle courses and numerous strength building exercises.

Wire ropes

A wire rope is made up of stands and a strand is made up of one or more layers of wires as shown in Figure. The number of strands in a rope denotes the number of groups of wires that are laid over the central core. For example a 6 × 19 construction means that the rope has 6 strands and each strand is composed of 19 (12/6/1) wires. The central part of the wire rope is called the core and may be of fibre, wire, plastic, paper or asbestos. The fibre core is very flexible and very suitable for all conditions (Figure 3.18).

The points to be considered while selecting a wire rope are

- Strength
- Abrasion resistance
- Flexibility
- Resistance of crushing
- Fatigue strength
- Corrosion resistance.

Figure 3.18   Folds of Rope

Ropes having wire core are stronger than those having fibre core. Flexibility in rope is more desirable when the number of bends in the rope is too many.

They used for power transmission over a large distance (they can transmit power over 150 m). These are usually employed in an elevator and cranes. The wire ropes are made from twisting steel wires (other alloys used for special application).

Advantages of wire rope drives

1. These are lighter in weight
2. These offer silent operation
3. These can withstand shock loads
4. These are more reliable
5. They do not fail suddenly
6. These are more durable
7. The efficiency is high, and
8. The cost is low.

Horse power (hp) transmitted is given by

$$\text{hp} = \frac{p\,v\,n}{4500} \text{ or } \frac{p\,v\,n}{75} \text{ if } v \text{ is in m/s.}$$

Where p is the driving force in kg.

v = Velocity in m/min

n = Number of ropes in the pulley

Ratio of driving tensions is given by

$$\log \frac{T_1}{T_2} = \frac{0.4343\,\mu\,\theta}{\sin \alpha}$$

where μ = coefficient of friction between the belt and the sides of the groove.

θ = angle of contact or lap in radians

$2\,\alpha$ = angle of groove in degrees

T_1 and T_2 = tension on tight and slack sides.

1. A rope pulley with 5 ropes surface speed of 1000 m/min transmits 100 hp. Find the tensions on the tight side and slack side. If the angle of lap is 130° and the angle between the sides of the pulley is 45°. Assume coefficient of friction as 0.3.

Solution :

μ = 5 ropes

hp = 100 hp

v = 1000 m/min

$\theta = 130° = 130 \times \dfrac{\pi}{180} = 2.268$ radians

$2\alpha = 45°$

$\therefore \alpha = 22.5°$

$\mu = 0.3$

$T_1 = ?$ hp transmitted/rope $= \dfrac{100}{5} = 20 HP$

$T_2 = ?$

$$hp = \dfrac{(T_1 - T_2)v}{4500}$$

$$20 = \dfrac{(T_1 - T_2)1000}{4500}$$

$$T_1 - T_2 = \dfrac{4500 \times 20}{1000}$$

$$T_1 - T_2 = 90 \qquad (1)$$

We know that

$$\log \dfrac{T_1}{T_2} = \dfrac{0.4343\, \mu\theta}{\sin \alpha}$$

$$= \dfrac{0.4343 \times 0.3 \times 2.268}{\sin 22.5}$$

$$\log \dfrac{T_1}{T_2} = 0.7721$$

$$\dfrac{T_1}{T_2} = \text{antilog } (0.7721)$$

$$\dfrac{T_1}{T_2} = 5.9179$$

$\therefore T_1 = 5.9179\, T_2$

Substitute the value of T_1 in equation (1)

$5.9179\, T_2 - T_2 = 90$

$4.9179\, T2 = 90$

$$T_2 = \dfrac{90}{4.9179} = 18.30 \text{ kg.}$$

$\therefore T_1 = 5.9179\, T_2$

$= 5.9179 \circ 18.30 = 108.29$ kg.

2. Find the HP transmitted by a rope drive from the following data.
Angle of contact – 180° (v)

Pulley groove angle (2 α) = 60°
Coefficient of friction (μ) = 0.2
Width of rope/m length= 0.4 kg (w)
Permissible tension = 150 kg (T_M)
Velocity of rope (v) = 15 m/s

Solution :

$$\theta = 1800 = 180 \times \frac{\pi}{180} = \pi \ radians$$

2 α = 60° or a = 30°
μ = 0.2,
w = 0.4 kg/m
T_M = 150 kg
v = 15 m/s
hp = ?

$$hp = \frac{\left(T_1 - T_2\right) v}{75}$$

To find the value of T_1 and T_2.
To find T_1, we know that

$$T_C = \frac{w v^2}{g}$$

$$= \frac{0.4 \times 15^2}{9.81}$$

T_C = 9.17 kg.
We know that $T_M = T_1 + T_C$
$$\therefore T_1 = T_M - T_C$$
T_1 = 150 – 9.17 = 140.83 kg.
To find T_2 we know that

$$\log \frac{T_1}{T_2} = \frac{0.4343 \ \mu \theta}{\sin \alpha}$$

$$= \frac{0.4343 \times 0.2 \times \pi}{\sin 30^o}$$

$$\log \frac{T_1}{T_2} = 0.5457$$

$$\frac{T_1}{T_2} = antilog \ (0.5457)$$

$$\frac{T_1}{T_2} = 3.5136$$

$$\therefore T_2 = \frac{T_1}{3.5136}$$

$$= \frac{140.83}{3.5136}$$

$T_2 = 40.08$ kg.

$$\text{hp} = \frac{(T_1 - T_2)\, v}{75}$$

$$= \frac{(140.83 - 40.08)\, 15}{75}$$

$= 20.15$ hp.

3. A rope drive is required to transmit a power of 230 KW from a pulley of 1 m diameter running at 450 rpm. The safe pull in each rope is 800 N and the mass of the rope is 0.46 kg/m length. The angle of lap and the groove angle is 160° and 45°, respectively. If the coefficient of friction between the rope and the pulley is 0.3, find the number of ropes required.

Solution :

Given:

$P = 230$ KW

$d = 1$ m

$N = 450$ rpm

$T_M = 800$ N

$m = 0.46$ kg/m

$\mu = 0.3$

$\theta = 160° = 160 \times \dfrac{\pi}{180} = 2.79\ radians$

$2\,\alpha = 45°$

$\alpha = 22.5°$

Number of ropes = ?

$v = ?$

We know that power transmitted/rope

$= (T_1 - T_2)\, n$ in watts (in SI system)

$$v = \frac{\pi\, d\, n}{60}$$

$$= \frac{\pi \times 1 \times 450}{60}$$

v = 23.56 m/s

To find T_1 and T_2 we know that

Centrifugal tension $T_C = mn^2$

$$= 0.46 \times 23.56^2$$

$T_C = 255.33$ N

∴ Tension on the tight side T_1 is given by

$$T_M = T_1 + T_C$$
$$\therefore T_1 = T_M - T_C$$
$$= 800 - 255.33$$
$$T_1 = 544.67 \text{ N}$$

To find T_2 we know that

$$\log \frac{T_1}{T_2} = \frac{0.4343 \, \mu \theta}{\sin \alpha}$$

$$= \frac{0.4343 \times 0.3 \times 2.79}{\sin 22.5^0}$$

$$\log \frac{T_1}{T_2} = 0.9498$$

$$\therefore \frac{T_1}{T_2} = 8.91$$

$$\therefore T_2 = \frac{\pi}{8.91}$$

$$= \frac{544.67}{8.91}$$

$T_2 = 61.13$ N

We know that power transmitted / rope is given by

$$P = \left(T_1 - T_2\right) v$$
$$= (544.67 - 61.13) \, 23.56$$
$$= 11392 \text{ W}$$
$$P = 11.392 \text{ KW} \qquad \because 1000 \text{ W} = 1 \text{ KW}$$

$$\text{Number of ropes} = \frac{\text{Total power transmitted}}{\text{Power transmitted/rope}}$$

$$= \frac{230}{11.39^2}$$

$$= 20.18 \text{ or } 21 \text{ ropes.}$$

3.21.2   Advantages of rope drive

- A larger amount of power is transmitted.
- A rope can be run in any direction or to any distance.
- Electrical disturbances are absent.
- There is an absence of slip.
- Significant power transmission.
- It can be used for long distance.
- Ropes are strong and flexible.
- Provides smooth and quiet operation.
- It can run any direction.
- Low-cost and economic.
- Precise alignment of the shaft not required.

3.21.3   Demerits of rope drives

Internal failure of the rope has no sign on external, so it if often get unnoticed. Corrosion of wire rope.

3.21.4   Material of construction

Ropes for power transmission are generally made of cotton, as the cotton fibre stands the bending and unbending much better than hemp. The ropes vary in diameter from 1″ to 2″ for main and secondary drives, down to about ¼″ for machine-part drives. The breaking strength of the ropes is about 8000–12,000 lb per square inch of net section, which is about 0.9 times the area of the circumscribing circle. But the working stress is generally very low – from about 150 to 300 lb per square inch of net section. The low working stress ensures a long life of the ropes. The rope speed is generally about 4800 ft per minute for main drives, and for others the speed is kept as high as possible since the best speed of 4800 ft per minute is seldom practicable.

4. A number of trucks are to be hauled by a rope passed round a hydraulic capstan; assuming a coefficient of friction of 0.25 between the rope and the capstan and a constant manual pull of 30 lb. What will be the horse power exerted by the capstan under these conditions when there are 2.5 turns of rope and it is being wound-off at 100 ft/min?

Solution :

We know that

$$\frac{T_1}{T_2} = e^{\mu\theta}$$

From the question: $T_2 = 30, \mu = 0.25$ and $\theta = 2\pi\, n\, radians$
From which:

$$T_1 = 30\, e^{0.25 \times 2\pi n} = 30(4.82)^n$$

When $n = 2.5$, $T_1 = 1520\, lb.$

$$\frac{(T_1 - T_2) \times v}{550} = \frac{(1520 - 30) \times 100}{550 \times 60} = 4.52\, h.p.$$

To which must be added the power supplied manually which is:

$$\frac{30 \times 100}{550 \times 60} = 0.091\, h.p.$$

So the horse power taken by the trucks = 4.52 + 0.091 = 4.611 $h.p.$

5. A rope drive transmits 600 kW from a pulley of effective diameter 4 m, which runs at a speed of 90 rpm. The angle of lap is 160°; the angle of groove 45°; the coefficient of friction 0.28; the mass of rope 1.5 kg/m and the allowable tension in each rope 2400 N. Find the number of ropes required.

Solution :

Given: $P = 600$ kW; $d = 4$ m; $N = 90$ rpm; $\theta = 160° = 160 \times \pi/180 = 2.8$ rad;
$2\beta = 45°$ or $\beta = 22.5°$; $\mu = 0.28$; $m = 1.5$ kg/m; $T = 2400$ N.

We know that velocity of the rope

$$v = \frac{\pi d N}{60} = \frac{\pi \times 4 \times 90}{60} = 18.85 m/s$$

$\therefore$ Centrifugal tension, $T_C = mv^2 = 1.5\,(18.85)2 = 533$ N
and tension in the tight side of the rope,

$$T_1 = T - T_c = 2400 - 533 = 1867\ N$$

Let T_2 = Tension in the slack side of the rope.
We know that

$$2.3 \log\left(\frac{T_1}{T_2}\right) = \mu.\theta \cos ec\beta = 0.28 \times 2.8 \times \cos ec\, 22.5° = 2.05$$

$$\log\left(\frac{T_1}{T_2}\right) = \frac{2.05}{2.3} = 0.8913 \qquad or \qquad \frac{T_1}{T_2} = 7.786$$

......(Taking antilog of 0.8913)
and

$$T_2 = \frac{T_1}{7.786} = \frac{1867}{7.786} = 240 N$$

We know that power transmitted per rope

$= (T_1 - T_2) v = (1867 - 240) 18.85 = 30670 \text{ W} = 30.67 \text{ kW}$

$\therefore$ Number of ropes $= \dfrac{\text{Total power transmitted}}{\text{Power transmitted per rope}}$

$$= \frac{600}{30.67} = 19.56 \text{ or } 20 \quad Ans.$$

6. A pulley used to transmit power by means of ropes has a diameter of 3.6 m and has 15 grooves of 45° angle. The angle of contact is 170° and the coefficient friction between the ropes and the groove sides is 0.28. The maximum possible tension in the ropes is 960 N and the mass of the rope is 1.5 kg per metre length. What is the speed of pulley in rpm and the power transmitted if the condition of maximum power prevail.

Solution :

Given: $d = 3.6$ m; No. of grooves $= 15$; $2\beta = 45°$ or $\beta = 22.5°$;
$\theta = 170° = 170 \times \pi/180 = 2.967$ rad; $\mu = 0.28$; $T = 960$ N; $m = 1.5$ kg/m.

Speed of the pulley

Let $N =$ Speed of the pulley in rpm

We know that for maximum power, velocity of the rope or pulley,

$$v = \sqrt{\frac{T}{3m}} = \sqrt{\frac{960}{3 \times 1.5}} = 14.6 m/s$$

$$\therefore N = \frac{v \times 60}{\pi d} = \frac{14.6 \times 60}{\pi 3.6} = 77.5 r.p.m \quad Ans. \qquad ...\left(\because v = \frac{\pi d N}{60} \right)$$

Power transmitted

We know that for maximum power, centrifugal tension,

$T_C = T/3 = 960/3 = 320$ N

$\therefore$ Tension in the tight side of the rope,

$T_1 = T - T_C = 960 - 320 = 640$ N

$T_2 =$ Tension in the slack side of the rope.

We know that $2.3 \log \left(\dfrac{T_1}{T_2} \right) = \mu.\theta \cos ec\beta = 0.28 \times$

$2.967 \times \cos ec\, 22.5° = 2.17$

$$\log \left(\frac{T_1}{T_2} \right) = \frac{2.17}{2.3} = 0.9438 \qquad or \qquad \frac{T_1}{T_2} = 8.78$$

(Taking antilog of 0.9438)

And $T_2 = \dfrac{T_1}{8.78} = \dfrac{640}{8.78} = 73\,N$

∴ Power transmitted per rope

$= (T_1 - T_2)\,v = (640 - 73)\,14.6 = 8278\,W = 8.278\,kW$

Since the number of grooves are 15, therefore total power transmitted.

$= 8.278 \times 15 = 124.17\,kW$

7. Following data is given for a rope pulley transmitting 24 kW: Diameter of pulley = 400 mm; Speed = 110 rpm; angle of groove = 45°; Angle of lap on smaller pulley = 160°; Coefficient of friction = 0.28; Number of ropes = 10; Mass in kg/m length of ropes = 53 C^2; and working tension is limited to 122 C_2 kN, where C is girth of rope in meters (m). Find initial tension and diameter of each rope.

Solution :

Given: $P_T = 24$ kW; $d = 400$ mm = 0.4 m; $N = 110$ rpm; $2\beta = 45°$ or $\beta = 22.5°$;
$\theta = 160° = 160 \times \pi/180 = 2.8$ rad; $\mu = 0.28$; $n = 10$; $m = 53\ C^2$ kg/m;
$T = 122\ C^2$ kN $= 122 \times 10^3\ C^2$ N.

Initial tension

We that power transmitted per rope

$$P = \frac{\text{Total power transmitted}}{\text{No. of ropes}} = \frac{P_T}{n} = \frac{24}{10} = 2.4kW = 2400\,W$$

And velocity of the rope $v = \dfrac{\pi d.N}{60} = \dfrac{\pi \times 0.4 \times 110}{60} = 2.3m/s$

Let T_1 = Tension in the tight side of the rope, and
           T_2 = Tension in the slack side of the rope

We know that power transmitted per rope (P),

$2400 = (T_1 - T_2)\,v = (T_1 - T_2)\,2.3$

∴ $T_1 - T_2 = 2400/2.3 = 1043.5$ N   (i)

We know that

$$2.3 \log\left(\frac{T_1}{T_2}\right) = \mu.\theta \cos ec\,\beta = 0.28 \times 2.8 \times \cos ec\,22.5° = 2.05$$

$$\log\left(\frac{T_1}{T_2}\right) = \frac{2.05}{2.3} = 0.8913 \quad or \quad \frac{T_1}{T_2} = 7.786.$$

From equations (i) and (ii)

$T_1 = 1197.3$ N   and $T_2 = 153.8$ N

We know that initial tension in each rope.

$$T_0 = \frac{T_1 + T_2}{2} = \frac{1197.3 + 153.8}{2} = 675.55 N \qquad Ans.$$

Diameter of each rope

Let $\quad d_1$ = Diameter of each rope

We know that centrifugal tension

$$T_C = mv^2 = 53\ C^2\ (2.3)^2 = 280.4\ C^2\ N$$

And working tension (T),

$$122 \times 10^3\ C^2 = T1 + T_C = 1197.3 + 280.4\ C^2$$
$$122 \times 10^3\ C^2 - 280.4\ C^2 = 1197.3$$

$\therefore \qquad C^2 = 9.836 \times 10{-}3$ or $C = 0.0992$ m $= 99.2$ mm

We know that girth (i.e., circumference) of rope (C),

$$99.2 = \pi\ d_1 \text{ or } d_1 = 99.2\ /\ \pi = 31.57 \text{ mm.}$$

8. A casting weighing 9 kN hangs freely from a rope which makes 2.5 turns round a drum of 300 mm diameter revolving at 20 rpm. The other end of the rope is pulled by a man. The coefficient of friction is 0.25. Determine the force required by the man, and the power to raise the casting.

Solution :

Given: $W = T_1 = 9$ kN $= 9000$ N; $d = 300$ mm $= 0.3$ m; $N = 20$ rpm; $\mu = 0.25$.

(i) Force required by the man

Let $\quad T_2$ = Force required by the man.

Since the rope makes 2.5 turns round the drum, therefore angle of contact,

$$\theta = 2.5 \times 2\ \pi = 5\ \pi \text{ rad}$$

We know that $2.3 \log \left(\dfrac{T_1}{T_2} \right) = \mu.\theta = 0.25 \times 5\ \pi = 3.9275$

$$\log \left(\frac{T_1}{T_2} \right) = \frac{3.9275}{2.3} = 1.71 \quad or \quad \frac{T_1}{T_2} = 51$$

$$\therefore\ T2 = \frac{T_1}{51} = \frac{9000}{51} = 176.47 N$$

(ii) Power to raise the casting

We know that velocity of the rope,

$$v = \frac{\pi d.N}{60} = \frac{\pi \times 0.3 \times 20}{60} = 0.3142\ m\ /\ s$$

$\therefore$ Power to raise the casting,

$$P = (T_1 + T_2)\ v = (9000 - 176.47)\ 0.3142 = 2772 \text{ W.}$$
$$= 2.772 \text{ kW.}$$

3.22   Chain drive

Chain drive is a way of transmitting mechanical power from one place to another. It is often used to convey power to the wheels of a vehicle, particularly bicycles and motorcycles. It is also used in a wide variety of machines besides vehicles (Figure 3.20).

Figure 3.19   Simple Chain Drive

Figure 3.20   Chain drives of the stenter machine

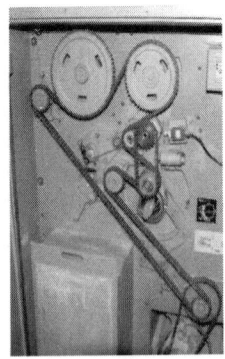

a   b

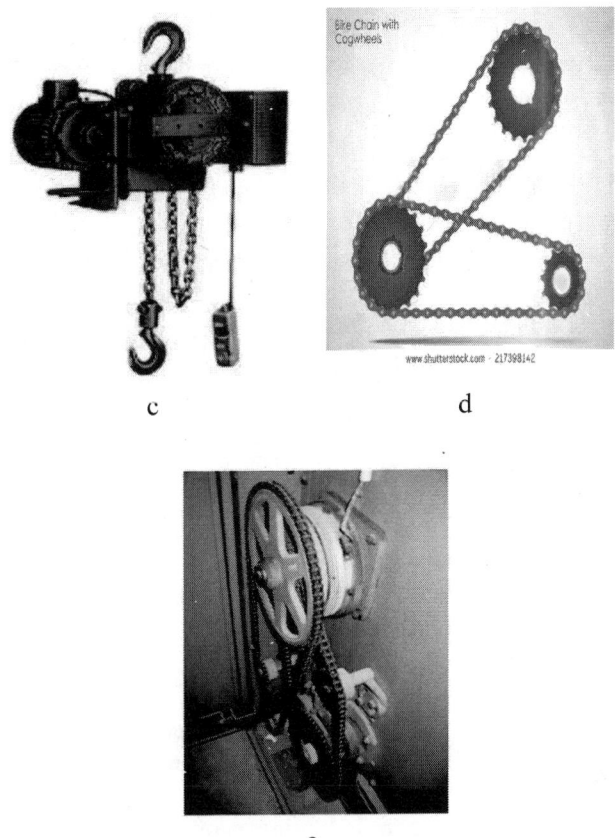

Figure 3.21   a- to e- Chain Drive in Textile Machines

3.22.1   Textile applications of chain drive (Figure 3.21a–e)

Chain drive is used to a greater extent in MBO, ERM cleaner, Cage driving in Scuther, Carding , hank meter drive in Drawframe, Creel roller drive in simplex, etc., in spinning and in yarn preparatory, weaving looms, etc., following are the few applications listed.

- Drives from beater to plain and perforated drums and feed rollers on fine cleaner are through duplex roller chains.
- Drive from inclined lattice to feed apron and creel apron of bale opener through clutch.
- Motor to feed-roller, lap winding-roller, and tuft-feeder in high production carding machine.

- Duplex roller chain transmits motion from main shaft to lap rollers via bottom calendar roller on sliver lap machine.
- Drive to shafts driving the flyers and bobbins on conventional roving machines.
- Drive to ring rail on ring spinning machines.
- Drive to creel rollers on drawing machines, and hank meters
- Drive to brush roller shaft on comber.
- Drive to drafting rollers that feed sheath fibres on friction spinning machine.

3.22.2   Merits of chain drive

(1) They can be used for short to medium distances. Gears need additional idler gears.
(2) They can be used for transmission of higher loads compared with belt drives.
(3) The power transmission efficiency of chain drives may be as high as 99%; higher compared to flat and V-belts. A chain drive does not slip and to that extent, it is positive drive.
(4) Due to the polygonal effect and wears in the chain joints, they are not suitable for precise speed control.
(5) They have longer life, no creep, and the ability of driving several shafts from a single source of power compared to belt drives.
(6) They can only be used for transmitting motion between parallel shafts. Crossed-flat belts, bevel gears, worm gears, and some crossed helical gears can be used to transmit motion between non-parallel shafts.
(7) The location (centre distance) and alignment tolerances need not be as precise as with gear drives, but require precise alignment of shafts, compared to belt drives. The best services can be expected when both the input and output sprockets lie in the same vertical plane.
(8) They require proper maintenance, particularly lubrication, when compared to gears.
(9) A number of parallel shafts with sprockets in the same plane can be driven in the same or opposite direction by a chain from a single driving sprocket. This is not possible with gears without idler. This type of drive is used to drive various rotors (feed rollers, plain and perforated rollers) on the fine cleaner in blow room.

3.22.3   Chain drives compared

Following is the table gives the comparison for power transmission.

	Shaft	Belt	Chain	Gear	Hydraulic	Electric
Required alignment accuracy	High	Medium	Medium	High	None	None
Positive drive	Yes	No (except toothed)	Yes	Yes	Yes	No
Efficiency	High	Medium	High	Variable	Low	Medium
Stiffness	High	Low	High	High	High	N/A
Strength	Medium	Low-medium	High	High	High	N/A
Ability to span large distances	Low	Medium	Medium	Low	Very high	Very high
Maintenance	Low	Medium	High	Medium	Medium	Low
Cost	Low	Low	Low	Medium	High	Medium

Three major types of chain are used for power transmission; roller, engineering steel, and silent. Roller chains are probably the most common and are used in a wide variety of low-speed to high-speed drives. Engineering steel chains are used in many low-speed, high-load drives. Silent chains are mostly used in high-speed drives. Chains can span long centre distances like belts, and positively transmit speed and torque like gears. For a given ratio and power capacity, chain drives are more compact than belt drives, but less compact than gear drives. Mounting and alignment of chain drives does not need to be as precise as for gear drives. Chain drives can operate 98%–99% efficiency under ideal conditions. Chain drives are usually less expensive than gear drives and quite competitive with belt drives.

3.22.4   Roller chains

Standard roller chains

The American National Standards Institute (ANSI) has standardised limiting dimensions, tolerances, and minimum ultimate tensile strength for chains and sprockets of 0.25–3.0 in pitch.

Multiple-strand roller chains

Multiple-strand roller chains consist of two or more parallel strands of chain assembled on common pins. They are also standardised.

Roller chain technology has evolved over the centuries. During this time new design features and production processes have been introduced. This extensive range of transmission roller chains is supplemented with ranges of attachment, double pitch, conveyor, agricultural and leaf chains.

Fenner roller chain

Increased fatigue resistance
- Shot peening of rollers and side plates imparts beneficial fatigue resisting properties, counteracting the onset of fatigue failure.
- Ball swaged side plate holes improve the whole quality and surface finish thus combating fatigue failure.
- Deed waisted side plates increase the effective cross-section of the side plate, thereby providing additional resistance to fatigue failure.
- Increased wear life
- Seam oriented bushes ensure that seams are positioned away from the critical bearing areas, extending wear life thus reducing down-time.
- Case hardened pins increase resistance to wear, further extending wear life.
- Solid rollers evenly distribute working loads, prolonging wear life.

Fit and forget reliability

Fenner chain is pre-loaded to bed in all component parts. This also acts as a final 100% quality check, enabling you to simply "fit and forget" fenner chain.

Lubrication and protection

Fenner roller chain is pre-lubricated for protection against corrosion and contamination. The light mineral oil used provides as initial lubrication, however, Fenner Power Transmission Distributor recommend all chains should be approximately lubricated prior to operation. Fenner roller chain is wrapped in a protective covering and is boxed in 5 m lengths.

Selection

(a) Service factor
Determine the service factor which is applicable to the drive.

(b) Design power
Multiply the normal running power required by the service factor. This gives the design power which is used as the basis for selecting the drive.

(c) Chain pitch is determined.

(d) Speed ratio
Divide the speed of the faster shaft by the speed of the slower shaft to obtain the speed ratio.

(e) Sprocket sizes

Select driving and driven sprockets to match the speed ratio.

(f) Power rating is done. If the power rating figure does not equal or preferably exceed the design power, single strand chain offers the most economical solution, and should be used where possible. However, for limitations in space, high speed or smooth running requirements a smaller pitch, duplex or triplex drive may be considered.

(g) Chain length

To find the chain length in pitches, use the formula below:

$$L = \frac{2C}{P} + \frac{T+t}{2} + \frac{KP}{C}$$

L = Length of chain in pitches,

C = Centre distance in mm.

P = Pitch of chain in mm.

T = Number of teeth on large sprocket.

t = Number of teeth on small sprocket.

K = Factor

The calculated number of pitches should be rounded up to an even, whole number of pitches. If the centre distance cannot be adjusted, to allow for the use of an even number of pitches, it may be necessary to use an offset or cranked link, in which case the chain power rating will need to be reduced, consult Fenner Power Transmission Distributor. Recalculate the exact centre distance required for the adjusted number of pitches, for recommended centre distance, if a jockey or tensioning sprocket is used add an extra 2 pitches. To obtain the chain length, multiply the number of pitches by the pitch of the chain.

$$\text{Length of chain in feet} = \frac{LP}{305}$$

Double-pitch roller chains: Non-standard roller chains, sprockets

Engineering steel chains

Standard engineering steel chains: The engineering steel chains designated for power transmission are heavy-duty offset sidebar chains.

Non-standard chains: Some manufacturers offer engineering steel chains in straight-sidebar and multiple-strand versions, and in pitches.

Sprockets: Machine-cut engineering-steel-chain sprockets look much like roller-chains sprockets, but they have pitch line clearance and undercut bottom diameters to accommodate the dirt and debris in which engineering-class chain drives often operate.

Silent chain – Standard silent chains

Silent (inverted-tooth) chains are standardised for pitches of 0.375 to 2.0″. Silent chain is an assembly of toothed link plates interlaced on common pins.

The sprocket engagement side of silent chain looks much like a gear rack. Silent chains are designed to transmit high power at high speeds smoothly and relatively quietly. Silent chains are a good alternative to gear trains where the centre distance is too long for one set of gears. The capacity of a given pitch of silent chain varies with its width. Standard widths of silent chain range from 0.5 to 6.0″ for 0.375-in pitch, and from 4.0 to 30.0″ for 2.0-in pitch.

Non-standard silent chains

Some manufacturers offer silent chains with special rocket-type joints. These chains generally transmit higher horsepower more smoothly and quietly than the standard joint designs. However, they generally require sprockets with special tooth forms.

Sprockets : Silent-chain sprockets have straight-sided teeth. They are designed to engage the toothed link plates of the chain with mostly rolling and little sliding action.

3.23   Roller chains: nomenclature and dimensions

3.23.1   Standard roller – chain nomenclature

Roller chain: Roller chain is an assembly of alternating roller links and pin links in which the pins pivot inside the bushings, and the rollers, or bushings, engage the sprocket teeth to positively transmit power.

Roller links: Roller links are assemblies of two bushings press-fitted into two roller link plates with two rollers free to rotate on the outside of each of the bushings.

Pin links: Pin links are assemblies of two pins press-fitted into two pin link plates.

Connecting links: Connecting links are pin links in which one of the pin link plates is detachable and is secured either by a spring clip that fits in grooves on the ends of the pins or by cotters that fit in cross-drilled holes through the ends of the pins.

Offset links: Offset links are links in which the link plates are bent to accept a bushing in one end and a pin in the other end. The pin may be a press fit in the link plates, or it may be a slip fit in the link plates and be secured by cotters.

3.23.2   Selection of roller-chain drives

General design recommendations

The following are only the more important considerations in roller-chain drive design. For more detailed information, consult Ref.[15.5] or manufacturers' catalogues.

Chain pitch

The most economical drive normally employs the smallest-pitch single-strand chain that will transmit the required power. Small-pitch chains generally are best for lighter loads and higher speeds, whereas large-pitch chains are better for higher loads and lower speeds. The smaller the pitch, the higher the allowable operating speed.

Number of sprocket teeth

Small sprocket: The small sprocket usually is the driver. The minimum number of teeth on the small sprocket is limited by the effects of chordal action.

Large sprocket: The number of teeth on the large sprocket normally should be limited to 120. Larger numbers of teeth are very difficult (expensive) to manufacture. The number of teeth on the large sprocket also limits maximum allowable chain wear elongation. The maximum allowable chain wear elongation, in percent, is 200/N2.

Hardened teeth

The fewer the number of teeth on the sprocket, the higher the tooth loading. Sprocket teeth should be hardened when the number of teeth is less than 25 and any of the following conditions exist:

1. The drive is heavily loaded.
2. The drive runs at high speeds.
3. The drive runs in abrasive conditions.
4. The drive requires extremely long life.

Angle of wrap

The minimum recommended angle of wrap on the small sprocket is 120°.

Speed ratio

The maximum recommended speed ratio for a single-reduction roller-chain drive is 7:1. Speed ratios up to 10:1 are possible with proper design, but a double reduction is preferred.

Centre distance

The preferred centre distance for a roller-chain drive is 30–50 times the chain pitch. At an absolute minimum, the centre distance must be at least one-half the sum of two sprocket outside diameters. A recommended minimum centre distance is the pitch diameter of the large sprocket plus one-half the pitch diameter of the small sprocket. The recommended maximum centre distance is 80 times the chain pitch.

Chain length

Required chain length may be estimated from the following approximate equation:

$$L = 2C + \frac{N_1 + N_2}{2} + \frac{N_2 - N_1}{4\pi^2 C}$$

Wear and chain sag

As a chain wears, it elongates. Roller-chain sprocket teeth are designed to allow the chain to ride higher on the teeth as it wears, to compensate for the elongation. Maximum allowable wear elongation normally is 3%. Where timing or smoothness is critical, maximum allowable elongation may be only 1.5%. The size of the large sprocket may also limit allowable elongation, as noted earlier.

Idlers

When the centre distance is long, the drive centres are near vertical, the centre distance is fixed, or machine members obstruct the normal chain path, idler sprockets may be required. Idler sprockets should engage the chain in the slack span and should not be smaller than the small sprocket. At least 3 teeth on the idler should engage the chain, and there should be at least 3 free pitches of chain between sprocket engagement points.

Multiple-strand chain

Multiple-stand chain may be required when the load and speed are too great for a single-strand chain, or when space restrictions prevent the use of large enough single-strand sprockets.

3.23.3   Drive arrangements

Power ratings of roller-chain drives

Conditions for ratings

The roller-chain horsepower ratings presented are based on the following conditions:

1. Standard or heavy series chain
2. Service factor of 1
3. Chain length of 100 pitches
4. Use of the recommended lubrication method
5. A two-sprocket drive, driver and driven
6. Sprockets properly aligned on parallel, horizontal shafts and chains
7. A clean, nonabrasive environment
8. Approximately 15,000 hours service life.

Horsepower rating equations

When operating under the above conditions, the maximum horsepower capacity of standard roller chains is defined by the equations shown. Depending on speed and the number of teeth on the smaller sprocket, the power capacity may be limited by link plate fatigue, roller and bushing impact fatigue, or galling between the pin and the bushing. The power capacity of the chain is the lowest value obtained from the following three equations at the given conditions.

1. Power limited by link plate fatigue:

$$HP_f = K_f N_1^{1.08} R^{0.9} P^{(3.0-0.07P)}$$

where $Kf = 0.0022$ for no. 41 chain, and 0.004 for all other numbers.

2. Power limited by roller and bushing impact fatigue:

$$HP_r = \left(K_r N_1^{1.5} P^{0.8} \right) / R^{1.5}$$

where $K_r = 29,000$ for numbers 25 and 35 chain, 3400 for no. 41 chain, and 17,000 for nos. 40 through 240 chain.

3. Power limited by galling:

$$HP_g = \left(RPN_1 / 110.84 \right) \left(4.413 - 2.073P - 0.0274N_2 \right)$$
$$- \left[ln \left(R/1000 \right) \right] \left(1.59 log\, P + 1.873 \right)$$

3.24   Lubrication and wear

In all roller and engineering steel chains, and in many silent chains, each pin and bushing joint essentially is a traveling journal bearing. So, it is vital that they receive adequate lubrication to attain full potential wear life. Even silent chains with rocking-type joints are subject to some sliding and fretting, and so they also need good lubrication to obtain optimum wear life.

3.24.1   Purpose of chain lubrication

Effective lubrication aids chain performance and life in several ways:

1. By resisting wear between the pin and bushing surfaces
2. By flushing away wear debris and foreign materials
3. By lubricating the chain-sprocket contact surfaces
4. By dissipating heat
5. By cushioning impact loads
6. By retarding rust and corrosion.

3.24.2   Lubricant properties

General lubricant characteristics

Chain lubrication is usually best achieved by a good grade of non-detergent petroleum-based oil with the following properties:

- Low enough viscosity to penetrate to critical surfaces
- High enough viscosity to maintain an effective lubricating film at prevailing bearing pressures

- Free of contaminants and corrosive substances
- Able to maintain lubricating properties in the full range of operating conditions.
- Additives that improve film strength, resist foaming, and resist oxidation usually are beneficial, but detergents or additives to improve viscosity index normally are not needed.

Recommended viscosities

The oil must be able to flow into small internal clearances in the chain, and so greases and very high-viscosity oils should not be used. The recommended viscosity for various ambient temperature ranges.

3.24.3   Types of chain lubrication

All three types of chain drives – roller, engineering steel, and silent – will work with three types of lubrication system. The type of lubrication system used is dependent on the speed and the amount of power transmitted. The three types of chain drive lubrication systems are

Type 1. Manual or drip
Type 2. Oil bath or slinger disk
Type 3. Oil stream

1. A chain drive is used for reduction of speed form 240 rpm to 120 rpm. The number of teeth on the driving sprocket is 20. Find the number of teeth on the driven sprocket. If the pitch circle diameter of the driven sprocket is 600 mm and centre to centre distance between the two sprockets is 800 mm, determine the pitch and length of the chain.

Solution :

Given: $N_1 = 240$ rpm; $N_2 = 120$ rpm; $T_1 = 20$;
$d_2 = 600$ mm or $r_2 = 300$ mm $= 0.3$ m; $x = 800$ mm $= 0.8$ m.

Number of teeth on the driven sprocket

Let T_2 = Number of teeth on the driven sprocket.

We know that

$$N_1.T_1 = N_2.T_2 \text{ or } T_2 = \frac{N_1.T_1}{N_2} = \frac{240 \times 20}{120} = 40.$$

Pitch of the chain

Let p = Pitch of the chain

We know that pitch circle radius of the driven sprocket ($r2$),

$$0.3 = \frac{p}{2} \cos ec\left(\frac{180°}{T_2}\right) = \frac{p}{2} \cos ec\left(\frac{180°}{40}\right) = 6.37\,p$$

$\therefore \qquad p = 0.3 / 6.37 = 0.0471 \text{ m} = 47.1 \text{ mm}.$

Length of the chain

We know that pitch circle radius of the driving sprocket,

$$r_1 = \frac{p}{2} \cos ec \left(\frac{180^o}{T_1} \right) = \frac{47.1}{2} \cos ec \left(\frac{180^o}{20} \right) = 150.5 mm$$

And $\qquad x = mp$ or $m = x/p = 800 / 47.1 = 16.985$

We know that multiplying factor,

$$K = \frac{(T_1 + T_2)}{2} + 2m + \frac{\left[\cos e \left(\dfrac{180^o}{T_1} \right) - \cos e \left(\dfrac{180^o}{T_2} \right) \right]^2}{4m}$$

$$= \frac{(20 + 40)}{2} + 2 \times 16.985 + \frac{\left[\cos ec \times \left(\dfrac{180^o}{20} \right) - \cos ec \left(\dfrac{180^o}{40} \right) \right]^2}{4 \times 16.985}$$

$$= 30 + 33.97 + \frac{(6.392 - 12.745)^2}{67.94} = 64.56 \ say \ 65$$

$\therefore$ Length of the chain, $L = pK = 47.1 \times 65 = 3061.5 \text{ mm} = 3.0615 \text{ m}.$

2. Select a chain drive to transmit 1.5 kW from a gearbox running at 80 rev/min and driven by a direct-on-line electric motor to a uniformly loaded conveyor drive shaft which is required to run at approximately 40 rev/min for 12 hours per day. Gearbox output shaft is 35 mm and the head shaft is 65 mm diameter.

Solution :

(a) Service factor

The service factor is 1.2.

(b) Design power

= 1.5 × 1.2 = 1.8 kW.

(c) Chain pitch

The intersection of design power and the rev/min of the faster shaft indicates a 16B 1² pitch chain.

(d) Speed ratio

$$\frac{80}{40} = 2 : 1$$

(e) Sprocket size

Sprockets of 19 and 38 teeth give a ratio of 2:1.

(f) Power rating

The required power rating from step (b) is 1.8 kW. For a 19T driver sprocket, running at 80 rev/min, the power rating for 16B-1 simplex chain is 3.79 kW. As this exceeds the required design power the selection is satisfactory.

If space limitations demand smaller sprocket dimensions, an alternative selection would be to use 12B-2 duplex chain which has a power rating of 2.11 kW at 80 rev/min.

(g) Chain length

Recommended centre distance for 16B-1 chain is 1000 mm

Therefore, the chain length as per selection step (g) (chain length) is 108 pitches including a connecting link.

3.25 Stepped pulleys and their construction

3.25.1 Cone pulley drive for lathe machine (Figure 3.22)

The cone pulley drive is one of the oldest form of the lathe machines, in which the different speeds are obtained with the help of the belt arranged on the four different steps of the cone pulley. A step cone pulley is attached with the head stock spindle. The arrangement is also known as: Step Pulleys, Multi Groove Pulleys, Multi V-Pulleys, Multi V-Belt Pulleys, Cone Pulleys, and Stepped Cone Pulleys. Stepped pulleys provide a low cost and efficient means of enabling a selection of speeds. Speed changes are made when the drive is stationary, the belt tension must first be released and the belt manually relocated in an alternative set of grooves. In many arrangements the pulley centre distance remains virtually unchanged. It is important that the belt tension is released adequately when changing speeds, and that belts are not forced or sprung over the groove flanges. Stepped pulleys are made up of different engineering materials like mild steel, cast iron, aluminium, wood or plastic, etc.

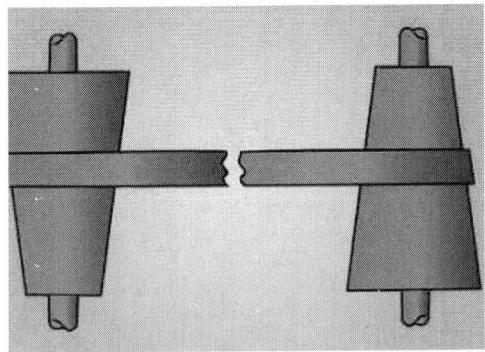

Figure 3.22 Cone pulleys

Cone pulleys (Figure 3.22) are used in pairs to obtain speed variations in a driven machine. By using a narrow flat belt and moving it across the pulley faces by means of a shifter fork, variable speed ratios are developed. This service puts extra strain first on one edge of the belt, and then on the other, as the belt passes over pulleys tapered in opposite directions. The shifting mechanism can also damage the belt edges.

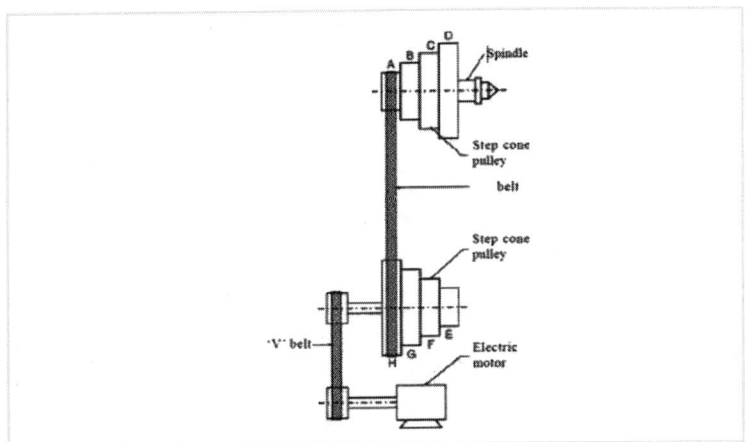

Figure 3.23   A typical Stepped cone pulley drive

As mentioned above step pulley is a system of pulleys made up of many different sizes. The typical step pulley consists of a two- to four-pulley configuration. The step pulley is always operated in pairs, and when the belt is changed on one pulley, it is also changed on the corresponding pulley on the other side of the belt. Most step pulley assemblies are cast from liquid materials; some of the very high-end pulleys, however, are forged from solid billet material such as aluminium or steel.

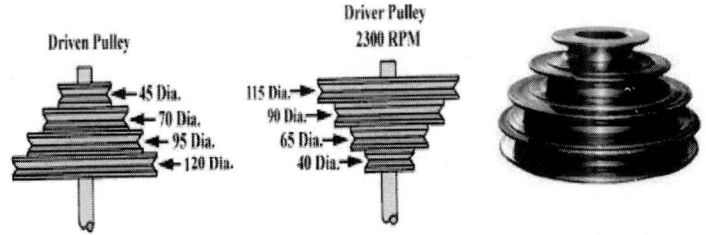

Figure 3.24   Different steps of cone

The conventional lathe machine or cone pulley lathe has four steps as in Figure 3.23 (A,B,C,D) and on the other side there is another pulley which also have four steps (E,F,G,H) which is placed parallel to the spindle cone pulley. Both the pulleys are connected with the belt and the V-belt is connected to the electric motor. The belt can be arranged between the steps A&H, B&G, C&F and D&F manually, thus the required speed is attained. The spindle speed will be maximum when the belt is arranged between A&H while the minimum speed will be attained when the belt is arranged between D&E. In this kind of lathe, operator have to slip the belt from one step to another manually. While most of the step assemblies are configured for V-belt use, there are also some step pulleys which utilise flat belts. This type of step system can be switched or changed on the fly and is found in lathe transmissions or some milling machines. The flat belt pulleys are typically always forged from billet stock and operate much more smoothly than the V-belt design. The V-belt type of pulley is favoured in higher torque applications, though, due to the ability of a V-belt to grip the pulley with more strength (Figure 3.24).

Number of pulley in a set

Virtually there is no limit or the application may call for a specific purpose and thus it may range from 2 to 10 or more so. One such piece is shown in the Figure 3.25.

Figure 3.25   Multiple steps of cone

3.25.2   Applications of stepped pulley in textile industry

Following are the applications found in textile machines.

1. Precision winding

In this machine the supply package is Hank mounted on swift and yarn is wound onto double flanged bobbin with precision winding principle. The driving arrangements are shown here for the benefit of the reader and the calculations show the use of stepped pulleys (Figure 3.26).

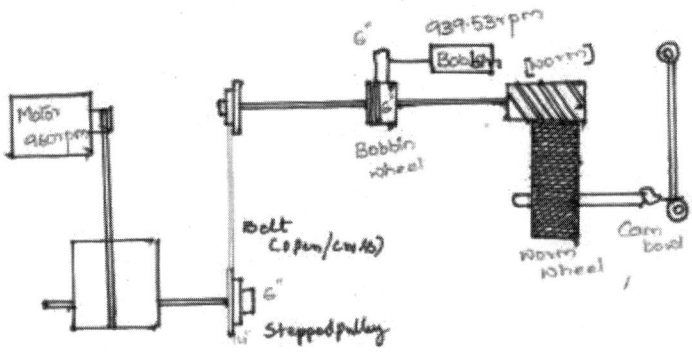

Figure 3.26   Gearing in Precision winder

Speed calculations in precision winder
Given motor speed = 960 rpm, Circumference of motor pulley = 16.5 cm, Circumference of intermediate pulley = 77.5 cm.
∴ Speed of intermediate pulley =
$$\frac{\text{Speed of motor} \times \text{Circumference of motor}}{\text{Circumference of intermediate pulley}}$$

$$= \frac{960 \times 16.5}{77} = 240.3 \; rpm.$$

<u>Drive from final step to initial step pulley:</u>
Circumference of final pulley = 22 cm, Circumference of initial stepped pulley = 35 cm.
Speed of initial final stepped pulley = 204.3 rpm.
Speed of initial pulley = $\dfrac{204.3 \times 22}{35} = 128.4 rpm.$

Speed of bobbin wheel = $\dfrac{128.4 \times 66}{15.24} = 556$ rpm.
<u>Drive from worm to worm wheel.</u>
Number of teeth of worm = 1
Speed of worm = Speed of bobbin shaft = 128.4 rpm, Number of teeth of worm wheel = 30
Speed of worn wheel = $\dfrac{128.4 \times 1}{30} \times = 4.28$ rpm.

Length of spindle $l = 18$ cm. Diameter of spindle $D0 = 7.6$ cm.
∴ Rate of winding = $\sqrt{\left(\pi DN\right)^2 + \left(2ln\right)^2}$

Where D = Diameter of spindle, N = Bobbin speed, n = Cam speed

$$\text{Rate of winding} = \sqrt{(3.14 \times 7.6 \times 556)^2 + (2 \times 18 \times 4.25)^2}$$

$$R = 132.6 \text{ rpm.}$$

Middle stepped pulley to middle stepped pulley:

Circumference of middle stepped pulley = 28 cms.

Speed of middle stepped pulley = 204.3 rpm.

Speed of bobbin shaft = 204.3 rpm.

$$\text{Speed of bobbin} = \frac{204.3 \times 66}{15.24} = 884.7 rpm.$$

Drive from worm to worm wheel:

Number of teeth of worm = 1, Speed of worm = Speed of bobbin shaft = 204.3 rpm.

Number of teeth of worm wheel = 30

$$\text{Speed of worm wheel} = \frac{325.22 \times 1}{30} = 10.83 rpm.$$

$$\text{Rate of winding} = \sqrt{(3.14 \times 7.6 \times 884.7)^2 + (2 \times 18 \times 6.8)^2}$$

$$= 210.01 \text{ rpm.}$$

Drive from initial stepped pulley to final stepped pulley:

Final stepped pulley circumference = 22 cm, Initial stepped pulley circumference = 35 cm.

Speed of initial stepped pulley = 204.3 rpm.

$$\text{Speed of final stepped pulley} = \frac{204.3 \times 35}{22} = 325.02 rpm.$$

$$\text{Speed of bobbin stepped pulley} = \frac{325.02 \times 66}{15.24} = 1407.5 \ rpm.$$

Drive from worm to worm wheel:

Number of teeth to the worm = 1, Speed of worm = Speed of bobbin shaft = 325.22 rpm.

Number of teeth to worm wheel = 30

$$\text{Speed of worn wheel} = \frac{325.22 \times 1}{30} = 10.83 rpm.$$

∴ Speed of cam = Speed of worn wheel,

∴ Speed of cam = 10.83 rpm.

∴ Traverse speed = Cam speed = 10.83 rpm.

$$\text{Rate of winding} = \sqrt{(3.14 \times 7.6 \times 1407.5)^2 + (2 \times 18 \times 10.8)^2}$$

$$= 335.9 \text{ rpm.}$$

Details	Rate of winding (rpm)
Initial step	132.6
Middle step	210.01
Final step	335.9

Now we can summarise the change of speed of stepped pulley on net rate of winding in precision winding machine.

Problems

1. Design a pair of stepped pulley system found on a Franze muller cone winding machine with the driving shaft running at 150 rpm with the smallest pulley as 30 cm in diameter. It is necessary to drive the countershaft at 450 rpm, 150 rpm and 50 rpm, respectively. The shafts are at a distance of 4 m. Mentioning the assumptions for closed belt drive select the suitable scale and design the system.

Solution :

In a stepped pulley system the sum of the radii of the pulleys are constant and the ratio of pulleys is equal to the ratio of corresponding speeds. Let r_1, r_2, r_3, ...r_6 be the radii of the pulleys. And N_1, N_2............N_6 be the radii of speeds. Then $r_1 + r_4 = r_2 + r_5 = r_3 + r_6$ and $\dfrac{r_4}{r_1} = \dfrac{N_1}{N_4}$, Given n_1, n_2, $n_3 = 150$ rpm, $r_3 = 15$ cm or $d_3 = 30$ cm.

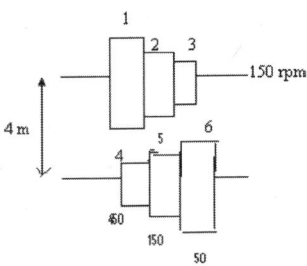

Consider last step (i.e., pulleys 3 and 6 in contact see diagram)

$$\frac{r_6}{r_3} = \frac{N_3}{N_6} = \frac{150}{50}$$

$r_6 = 3r_3 = 45$ cm
$d_3 = 30$ cm, $d_6 = 90$

Consider intermediate step (i.e., pulleys 2 and 5 in contact)

$$\frac{r_5}{r_2} = \frac{N_2}{N_5} = \frac{150}{150}$$

$$r_5 = r_2 \tag{1}$$

We also know that $r_2 + r_5 = r_3 + r_6$

By (1) $2r_2 = 60$

$r_2 = 30,$ $d_2 = 60$ cm $d_5 = 60$ cm

Consider first step (i.e., pulleys 1 and 4 in contact)

$$\frac{r_4}{r_1} = \frac{N_1}{N_4} = \frac{150}{450}$$

$r_1 = 3r_4$

$r_1 + r_4 = r_2 + r_5$

$4r_4 = 60$

$r_4 = 15 \quad d_4 = 30$ cm

$r_1 = 40 \quad d_1 = 90$ cm

Now contact a suitable scale to manipulate the diameter

$d_1 = 90$ cm or 9 m, $d_2 = 60$ cm or 6 m, $d_3 = 30$ cm or 3 m, $d_4 = 30$ cm or 3 m, $d_5 = 60$ cm or 6 m,

$d_6 = 90$ cm or 9 m, Scale 1:10, (Figure 3.27).

The distance between the centres is given as 4 m. Take scale and represent 4 m by 8 cm on drawing.

Scale = 1 m;

Actual size = 2 cm drawing size

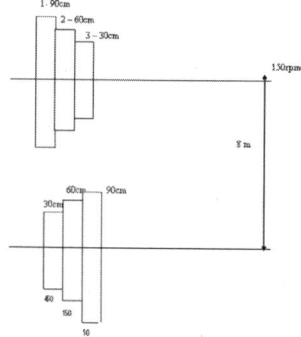

Figure 3.27   Designing of stepped pulley system

2. Design a set of stepped pulleys for driving a machine by a cross belt from a counter shaft running at 300 rpm. The machine is to run at 100, 150 and 200 rpms and the smallest pulley in the counter shaft is 300 mm. The centre distance between 2 sets of pulleys is 4 m.

Solution :

$$n_1 d_1 = n_4 d_4$$

$\therefore \ 300 \times 300 = 150 \times d_4$

$\therefore \ d_4 = 900$ mm.

$$n_2\, d_2 = n_5\, d_5$$
$$300\, d_2 = 150\, d_5$$
$$\frac{d_5}{d_2} = 2$$
$$\therefore\quad d_5 = 2d_2$$
$$n_3\, d_3 = n_6\, d_6$$
$$300\, d_3 = 200\, d_6$$
$$d_3 = \frac{2}{3}\, d_6$$

$$d_1 + d_4 = d_2 + d_5 = d_3 + d_6$$
$$d_1 = 300 \text{ mm.}$$
$$\frac{d_1}{d_4} = \frac{n_4}{n_1} = \frac{100}{300}$$

$$\frac{d_1}{d_4} = \frac{1}{3}$$

$$d_4 = 3\, d_1$$
$$\therefore d_4 = 3 \times 300$$
$$d_4 = 900 \text{ mm}$$

$$\frac{d_2}{d_5} = \frac{n_5}{n_2} = \frac{150}{300} = \frac{1}{2}$$

$$\frac{d_2}{d_5} = \frac{1}{2}$$

$$2\, d_2 = d_5 \tag{1}$$
$$\frac{d_3}{d_6} = \frac{n_6}{n_3} = \frac{200}{300}$$

$$\frac{d_3}{d_6} = \frac{2}{3}$$

$$2\, d_6 = 3\, d_3 \tag{2}$$

We know that
$$d_1 + d_4 = d_2 + d_5 = d_3 + d_6 = 1200$$
$$d_2 + d_5 = 1200$$

$$d_2 + 2\,d_2 = 1200$$
$$\therefore \quad d_2 = 400 \text{ mm } \& \, d_5 = 800 \text{ mm}$$
$$d_3 + d_6 = 1200$$
$$\frac{2}{3}\,d_6 + d_6 = 1200$$
$$= \frac{5}{3} d\,6 = 1200$$
$$\therefore \; d_6 = 720 \text{ mm}$$
$$\quad d_3 = 480 \text{ mm}$$

Ans:

$d_1 = 300$ mm	$d_2 = 400$ mm	$d_3 = 480$ mm
$d_4 = 900$ mm	$d_5 = 800$ mm	$d_6 = 720$ mm

3. Design a set of stepped pulleys for driving a lathe spindle at 100, 120 and 150 rpm from a counter shaft running at 80 rpm. The radius of the smallest pulley is to be 15 cm on counter shaft and the distance between the lathe spindle and counter shaft is 3 m. The same belt is to be used for all the three drives.

Solution :

$$n_1\,d_1 = n_4\,d_4$$
$$80 \times 300 = d_4 \times 100$$
$$\therefore \; d_4 = 240 \text{ mm.}$$
$$n_2\,d_2 = n_5\,d_5$$
$$\therefore 80 \times d_2 = 120\,d_5$$
$$2\,d_2 = 3\,d_5$$
$$\therefore \qquad d_5 = \frac{2}{3}\,d_2 \qquad\qquad (1)$$
$$n_3\,d_3 = n_6\,d_6$$
$$80\,d3 = 150\,d_6$$
$$d_6 = \frac{8}{15}\,d_3$$
$$d_6 = 0.533\,d_3$$

We know that

$$d_1 + d_4 = d_2 + d_5 = d_3 + d_6 = 540$$
$$d_2 + d_5 = 540$$
$$d_2 + \frac{2}{3}\,d_2 = 540$$
$$\frac{5}{3}\,d_2 = 540$$

$$\therefore \quad d_2 = 324 \text{ mm. } d_2$$
$$d_5 = 216 \text{ mm.}$$
$$d_3 + d_6 = 540$$
$$0.533 \, d_3 + d_3 = 540$$
$$1.533 = 540$$
$$\therefore \quad d_3 = 352.17 \text{ mm.}$$
$$d_6 = 187.8 \text{ mm.}$$

4. A shaft which rotates at a constant speed of 160 rpm is connected by belting to a parallel shaft 720 mm apart, which has to run at 60, 80 and 100 rpm. The smallest pulley on the driving shaft is 40 mm in radius. Determine the remaining radii of the two stepped pulleys for (1) a crossed-belt, and (2) an open belt. Neglect belt thickness and slip.

Solution :

Given: $N_1 = N_3 = N_5 = 160$ rpm; $x = 720$ mm; $N_2 = 60$ rpm;
$N_4 = 80$ rpm; $N_6 = 100$ rpm; $r_1 = 40$ mm (Figure 3.28).

Let r_2, r_3, r_4, r_5, and r_6, be the radii of the pulleys 2, 3, 4, 5 and 6, respectively, as shown in Figure 3.28.

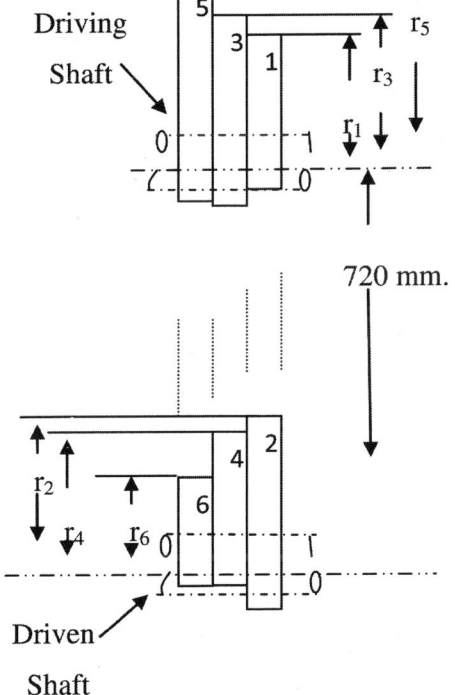

Figure 3.28   Designing of stepped pulley system

(1) For a crossed-belt

We know that for pulleys 1 and 2,

$$\frac{N_2}{N_1} = \frac{r_1}{r_2}$$

or $\quad r2 = r1 \times \dfrac{N_1}{N_2} = 40 \times \dfrac{160}{60} = 106.7 mm \qquad$ Ans .

and for pulleys 3 and 4,

$$\frac{N_4}{N_3} = \frac{r_3}{r_4} \quad \text{or } r_4 = r_3 \times \frac{N_3}{N_4} = r_3 \times \frac{160}{80} = 2\, r_3$$

We know that for a crossed-belt drive.

$$r_1 + r_2 = r_3 + r_4 = r_5 + r_6 = 40 + 106.7 = 146.7 \text{ mm.} \qquad \text{(i)}$$

∴ $\qquad r_3 + 2\, r_3 = 146.7 \text{ or } r_3 = 146.7/3 = 48.9 \text{ mm.}$

and $\qquad r_4 = 2\, r_3 = 2 \times 48.9 = 97.8 \text{ mm.}$

Now for pulleys 5 and 6,

$$\frac{N_6}{N_5} = \frac{r_5}{r_6} \quad \text{or } r_6 = r_6 \times \frac{N_5}{N_6} = r_5 \times \frac{160}{100} = 1.6\, r_5$$

From equation (i),

$$r_5 + 1.6\, r_5 = 146.7 \text{ or } r_5 = 146.7/2.6 = 56.4 \text{ mm.}$$

And $\qquad r_6 = 1.6\, r_5 = 1.6 \times 56.4 = 90.2 \text{ mm.}$

(2) For an open belt

We know that for pulleys 1 and 2,

$$\frac{N_2}{N_1} = \frac{r_1}{r_2} \quad \text{or } r_2 = r_1 \times \frac{N_1}{N_2} = 40 \times \frac{160}{60} = 106.7 \; mm. \quad Ans$$

And for pulleys 3 and 4,

$$\frac{N_4}{N_3} = \frac{r_3}{r_4} \quad \text{or } r_4 = r_3 \times \frac{N_3}{N_4} = r_3 \times \frac{160}{80} = 2 r_3$$

We know that length of belt for an open belt drive.

$$L = \pi\,(r_1 + r_2) + \frac{(r_2 - r_1)^2}{x} + 2x$$

$$= \pi\,(40 + 106.7) + \frac{(106.7 - 40)^2}{720} + 2 \times 720 = 1907 mm.$$

Since the length of the belt in an open belt drive is constant, therefore for pulleys 3 and 4, length of the belt (L),

$$1907 = \pi \ (r_3 + r_4) + \frac{(r_4 - r_3)^2}{x} + 2x$$

$$= \pi \ (r_3 + 2r_3) + \frac{(2r_3 - r_3)^2}{720} + 2 \times 720$$

$$= 9.426 \ r_3 + 0.0014 \ (r_3)^2 + 1440$$

or $\quad 0.0014 \ (r_3)^2 + 9.426 \ r_3 - 467 = 0$

$$\therefore r3 = \frac{-9.426 \pm \sqrt{(9.426)^2 + 4 \times 0.0014 \times 467}}{2 \times 0.0014}$$

$$= \frac{-9.426 \pm 9.564}{0.0028} = 49.3 mm \quad .....(\text{taking} + \text{ve sign})$$

And $\qquad r_4 \qquad = 2 \ r_3 = 2 \times 49.3 = 98.6$ mm.

Now for pulleys 5 and 6,

$$\frac{N_6}{N_5} = \frac{r_5}{r_6} \text{ or } r6 = \frac{N_5}{N_6} \times r_5 = \frac{160}{100} \times r_5 = 1.6 \ r_5$$

and length of the belt (L), $\qquad 1907 = \pi \ (r5 + r6) + \frac{(r_6 - r_5)^2}{x} + 2x$

$$= \pi \ (r_5 + 1.6 \ r_5) + \frac{(1.6r_5 - r_5)^2}{720} + 2 \times 720$$

$$= 8.17 \ r_5 + 0.0005 \ (r_5)^2 + 1440$$

Or $\qquad 0.0005 \ (r_5)^2 + 8.17 \ r_5 - 467 = 0$

$$\therefore r_5 = \frac{-8.17 \pm \sqrt{(8.17)^2 + 4 \times 0.0005 \times 467}}{2 \times 0.0005}$$

$$= \frac{-8.17 \pm 8.23}{0.001} = 60 mm \ Ans .... \text{(Taking +ve sign)}$$

And $\qquad r_6 = 1.6 \ r_5 = 1.6 \times 60 = 96$ mm.

5. Design a set of stepped pulleys to drive a machine from a counter shaft that runs at 220 rpm. The distance between centres of the two sets of the pulleys is 2 m. The diameter of the smallest step on the countershaft is 160 mm. The machine is to run at 80, 100 and 130 rpm. Consider both the cases of belt.

Solution:

Reader is advised to solve the problem.

Differences between flat and V-belt drive

S. No.	Flat belt	V-belt
1	Rectangular in cross-section	V-shaped in cross-section
2	Pulley grooved rectangular	V-shaped groove pulley
3	Limited hp can be transmitted	High hp can be transmitted
4	Slippage is more	Slippage is less
5	Less wear and tear of belt	Wear for V-belts is more
6	Available in open length	Available in fixed length
7	Cross-belt drive is possible	Not possible
8	Flat belt can be used with large centre distance	V-belts can't be used
9	There are very noisy	There are quiet
10	Flat belts are more durable	Not so durable
11	These are cheaper	V-belts are very costly

Feed regulation motion in scutcher and simplex

4.1   Significance of feed regulation in scutcher

The cotton from the Hopper feeder enter the Scutcher which may be provided with either two or three bladed beater or three bladed beater with Kirschner beater along with feed roller and cone drum assembly. It's very essential that the rate of feed to calendar roller to be maintained constantly as it directly affects the hank of lap.

The piano feed regulation motion is based on the thickness of the material passing below the feed roller and the corresponding change required so as to maintain the constant rate of feed and thus the hank of lap.

The weight per yard (wt/yd) of the lap is based on total weight of the lap, on the assumption that the length of the lap is kept constant. If there is any small variation in the yardage of lap, it is reflected in the total weight and weighing to determine the uniformity of the lap is rendered valueless. If the operator assumes that his measuring motion is accurate and discover that the weight of lap is excessive, he will change the evener motion (feed control system) to secure a lighter lap and meet his standard. However, if the additional weight was due to inaccuracy of the measuring motion and hence to excess yardage, the operative has used the wrong remedy he is now producing a lap which is less than required weight per hard (Figure 4.1).

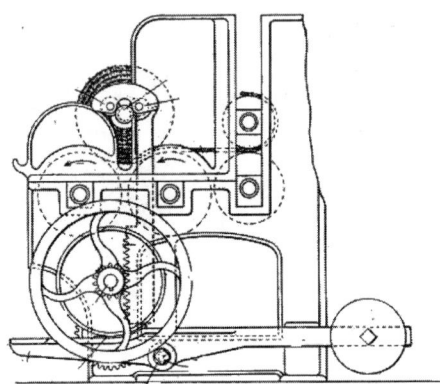

Figure 4.1   Conventional Lap knock off

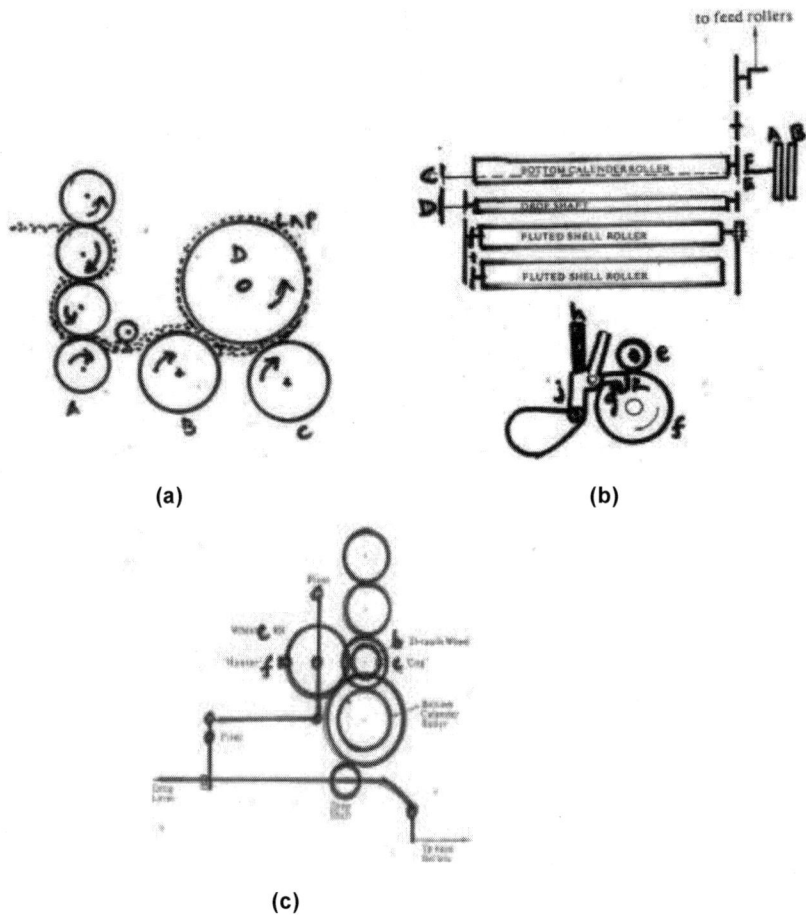

Figure 4.2 a–c   Knock-off motion arrangements in scutcher

4.2   Significance of knock-off motions in scutcher

The end product of the blow room is lap which is characterised by a parameter known as Hank. This depends on the length and weight of the lap delivered and thus it is necessary to adjust the rate of feed and to govern the length delivered to be with a specific weight. Each lap produced in scutcher is weighed and checked for the required weight and accordingly the cone drums speed are adjusted by piano feed regulation motion. Thus if the weight produced is excess or lower than the estimated, the lap is rejected. In any situations such rejected laps should not be more than 5% of the total produced. In order to avoid all these issues a knock-off motion device is fitted on to scutcher. The apparatus which ensures the required length to get the required hank is known as knock-off motions in scutcher (Figure 4.2a–c).

Object: (i) To stop the rotation of bottom calendar roller and feed roller, after a particular length of lap is completed, (ii) To measure the lap length and change the lap length as per out requirements by changing the lap length change wheel.

Figure 4.2a shows the arrangement of calendar roller set up in which the lap is wound on two shell rollers. The material coming out of the bottom calendar roller (BCR) will give the idea of the length of the lap delivered. By knowing the diameter of the BCR and the number of revolutions it makes, it is possible to calculate the hank of the lap produced.

In Figure 4.2b, the arrangement consists of pulley A and B of which A is fast and C is at the end of the shaft of A. The wheel D gets motion from C and D is a drop shaft. From here the drive goes to two areas – one side it goes to BCR through E and F and the other goes to shell roller. In the event of the drop roller falling E gets disengaged from F and thus BCR stops but shell roller continues to run, resulting in the fine cut of the lap. The drive from wheel F also reaches feed roller, in the event of E moving away from F the feed to machine and feed roller do stop. Referring to the Figure 4.2b it is said that drop shaft E is held into mesh with wheel F because it bearing is mounted on the drop lever h. The drop lever is supported by j. A worm drive from shell roller rotates f with e. A projection g on f engages with k as f continues to rotate. This causes g to drag k and thus also j (Figure 4.2c). This results in falling of h and thus E is cut-off from F.

4.3   Types of Knock-off motion in scutcher

There are basically three types of knock-off motions namely.

i) Worm and worm wheel,

ii) Ratchet and Paul,

iii) Hunter and cog.

4.3.1   Worm and Worm wheel knock-off motion

The method of knocking-off for full lap, taken from Platt bros machine, is given in the Figure 4.3.

A worm "D" on the bottom calendar roller drives a worm wheel "C" on a side shaft. A wheel "B" on this shaft gears with "A"; on the wheel "A" a pin on a stud, which during the revolution pushes a catch layer aside and allows the drop layer to fall. This takes the drop wheel immediately out of gear. Hence motion to the calendar roller is stopped whereas the lap roller still remains revolving on the lap roller thus shearing the sheet of cotton and making effective separation of the wound lap and the delivered material. After each knock-off the measuring motion is to be set back again for the next cycle of action.

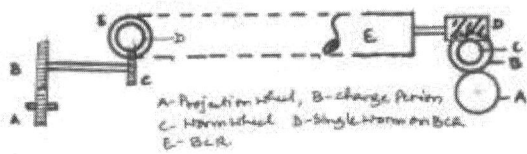

A- Projection wheel, B- change Pinion
C- Worm wheel D- Single Worm on BCR
E- BCR

Figure 4.3 Platt Knock off

A = Projection wheel   B = Change pinion

C = Worm wheel   D = Single worm on bottom calendar roller of 7″ diameter

E = Bottom calendar roller

$$\text{Actual length of the lap delivered} = \frac{A}{D} \times \frac{C}{D} \times \frac{\pi x d}{12 \times 3}$$

$$\text{Lap constant} = \frac{A \times C}{D} \times \frac{\pi \times d}{36}$$

$$\text{Length of the lap delivered} = \frac{\text{Lap constant}}{\text{No. of teeth on the change pinion}}$$

Thus, for a definite length of lap, the required wheel can be changed.

1. In a Platt's worm and worm motion BCR = 7″ diameter change wheel 22T worm wheel 24T projection wheel 48T. On the same shaft of the worm and worm wheel drives a single one. Calculate length of knock-off and weight of lap in kg if hank of lap = 0.0012.

Solution :

$$\text{Length of lap} = \frac{48}{22} \times \frac{24}{1} \times \pi \times \frac{7''}{36} = 31.98\, yd.$$

$$\sim 32 \text{ yd} = 29 \text{ m}.$$

$$\text{Weight of the lap} = \frac{\text{Length in yards}}{\text{weight in grams}} \times 8.33$$

$$= \frac{32}{7000 \times 2.2 \times 0.0012} \times 8.33 = 14.42\, kg.$$

2. Calculate the weight of full lap for a knock-off motion if hank of lap delivered is 0.00136 and tension draft 1.09. Projection wheel 50T change wheel 20T worm wheel 25T, BCR =7.

Solution :

$$\text{Length of lap} = \frac{50}{20} \times \frac{25}{1} \times \pi \times \frac{7}{36}$$

$$= 38.179 \text{ yd}$$
$$= 41.9 \text{ with tension draft.}$$

$$= \frac{42}{7000 \times 2.2 \times 0.00136} \times 8.33 = 16 \text{ kg}$$

(or) $$= \frac{42}{0.00136 \times 840 \times 2.2} = 16 \text{ kg.}$$

3. On a scutcher bottom C.R. which is 178 mm in diameter makes 14 rpm. Also the knocking-off wheel is 52T, C.P. = 181 and worm wheel = 25T. Average lap weight is 480 g. Calculate the scutcher production, per shift of 8 h when efficiency is equal to 84. What will be the weight of one full lap?

Solution :

Production / shift of 8 h =

$$= \frac{178 \times \pi \times 14 \times 480}{1000 \times 1000} \times \frac{84}{100} \times \frac{60 \times 8}{1} = 1514.88 \, kg.$$

Length of lap $$= \frac{52}{18} \times \frac{25}{1} \times \frac{\pi \times 178}{1000} m = 40.38 \text{ m.}$$

Weight of full lap $$= \frac{\text{Length in meters} \times \text{wt/m}}{1000} kg$$

$$= 40.38 \times \frac{400}{1000} = 19.38 kg.$$

4. A single worm on the end of fluted roller shaft drives a worm wheel of 20 teeth. The same shaft carries a wheel of which drives knock-off wheel of 50 teeth. If fluted roller has 25 cm diameter, calculate the length of the lap.

Solution :

For 1 revolution of knock-off wheel, the number of revolutions made by fluted roller can be found out as

$$1 \times \frac{50}{25} \times \frac{20}{1} = 40 \text{ revolutions}$$

The length of lap delivered during these revolutions is $\dfrac{40 \times 25 \times \pi}{100} = 31.42$

zm (Ans.)

5. In a worm and worm wheel knock-off motion BCR =7″ lap product hank of 0.00136 speed of calendar roller is 12.5 rpm projection wheel 48T change wheel 21T worm wheel 22T. Calculate weight of lap/knock-off number of laps produced / shift, $\eta = 91\%$.

Solution :

Length of lap $= \dfrac{48}{21} \times \dfrac{22}{1} \times \pi \times \dfrac{7}{36} = 30.7$ yd.

Weight$= \dfrac{30.7}{7000 \times 2.2 \times 0.00136} \times 8.33 = 12.2$ kg.

480 min $\longrightarrow$ 100%
436.8 min $\longrightarrow$ 91%
12.5 revolutions $\longrightarrow$ 1 min

Number of revolutions of BCR required for 1 knock-off =

$$1 \times \dfrac{48}{21} \times \dfrac{22}{1} = 50.3 \; rev.$$

To make 12.5 revolutions $\longrightarrow$ 1 min
50.3 $\longrightarrow$ 4.02 min which is at 100% η
4.024 $\longrightarrow$ 100%
4.42 min $\longrightarrow$ 91%

$= \dfrac{4.024}{91} \times 100.$

Time required at 91% η is 4.42 min.

Number of laps produced $= \dfrac{480}{4.42 \, min} = 108 \; laps.$

Problems for practice

1. In a measuring motion, a 9 fluted roller carries a single worm during 22 teeth worm wheel and on the worm wheel is 22 teeth pinion meshing with 45 teeth knock-off wheel. Calculate the length of lap made during 1 revolution of knock-off wheel. (Ans. 38.89 yards.)
2. Calculate the weight of the full lap for a knock-off, if the diameter of calendar roller is 7″ on the shaft on which is mounted a single worm geared to 25 teeth worm wheel. Change wheel on the worm wheel

shaft is 20 teeth and the projection wheel geared to it is 50 teeth. Draft between calendar roller and lap roller is 109 and hank of lap delivered is 0.0136. (Ans. 36.44 lbs.)

3. Calculate the weight of the lap if the diameter of the calendar roller is 7", the shaft of which is mounted on a single worm geared to 24 teeth worm wheel. The change wheel on the worm wheel shaft has 18.7 and the projection wheel driven by the change wheel has 50 teeth. The draft between the calendar roller and the lap roller is 1.05, hank of lap is 0.00158. (Ans. 30.55 lbs.)

4. From the Exercise No. 3, calculate knock-off constant and also change wheel needed for delivering 36-2/3 yd of lap. (Ans. 770; 21 teeth.)

5. In a Platt's worm and worm wheel knock-off motion the following particulars are given. Bottom calendar roller = 175 mm (diameter); calendar roller worm = single, worm wheel = 22 teeth, knock-off wheel 45T, change pinion=20 teeth. Calculate the length of the lap for one knock-off. (Ans. 27.27 m.)

6. Calculate the hank and gms/yard of a lap with the particulars like: Weight of the full lap = 32.5 lb, Length of the lap = 40 yd, Lap roller speed = 18 rpm. (Ans. Lap wt. – 368.5 g/yd.) (Hank – 0.00146)

4.3.2   Hunting cog

The hunting cog knock-off motion is as shown in the Figure 4.5. The set up consists of a cog e contacting the projection f on wheel of 80 teeth. The cog wheel is with 21 teeth. As the third calendar roller rotates the wheel b will move as 80 teeth wheel completes one full turn then e will come in contact with f and thus f is pushed outwards. The set-up is on the lever leading to the falling of drop lever and thus the motion to BCR stops. In Figures 4.4 and 4.5, the stages of the hunting cog are shown. First stage the beginning of the new lap. In the second stage the wheel b has moved one revolution and four teeth. In the third stage, the wheel b has completed 80 revolutions and thus e and f are facing each other causing the knock-off.

Figure 4.4   Showing the stages of the hunting cog motion

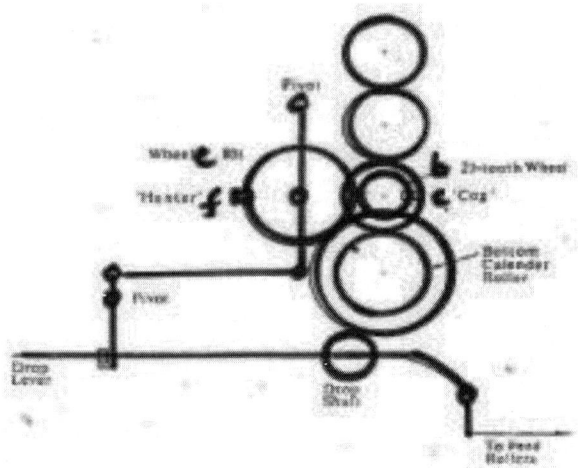

Figure 4.5 Hunting cog motion

1. In a hunting cog knock-off motion calculate the length of the lap if the hunting cog wheel is having 78 tooth and the wheel geared to it is 41 teeth. The diameter of the calendar roller is 5.5″. The draft between calendar roller and the lap roller is 1.07. Calculate the actual length of lap and also the weight of the lap if the hank of the lap is 0.00145. Knock-off takes place for every 41 revolution of cog wheel or 78 revolution of the calendar roller.

Solution :

Length of the lap for knock-off at calendar roller =

$$\frac{78}{41} \times 41 \times \frac{11}{2} \times \frac{22}{7x36} \times 1.87$$

= 39.64 yd (Ans.)

Weight of the lap = 39.64 × (13.2/16) = 32.7 lb (Ans.)

2. A certain hunting cog knock-off motion has a lap constant of 0.89, in use, is 73 teeth. Calculate the length of the lap.

Solution :

Length of the lap in yd = lap constant teeth × change wheel

= 0.89 × 73

= 64.97 yd. (Ans.)

3. What change wheel is to be used for a lap of 36.2 yd, if the calendar roller wheel has 41 teeth and the diameter of calendar roller is 5″.

Solution :

Length of lap in yards = Number of times 41T calendar roller rotates $x \times D$

$$36.2 = N \times \frac{22}{7} \times \frac{5}{12x3}$$

$$N = \frac{7 \times 36 \times 36.2}{22 \times 5} = 83 \ teeth$$

Change wheel required = 83 teeth (ans.)

4. The third C.R. (calendar roller) has diameter of 15 cm, and carries 301 pinion. This pinion drives knocking-off wheel of 91T. The draft between third and bottom calendar roller is 1.1 and that between C.R. and fluted lap roller is 1.05. Calculate the length of lap delivered for each knock-off.

Solution :

The total draft = 1.1 × 1.85 = 1.155

The number of revolutions made by third C.R. when knock-off wheel makes 1 revolution will be as: The hunter and cog will have IT difference after 3 revolutions of 30T wheel (B). Hence they will come against each other after 91 revolutions. Thus wheel B will make 91 × 3 = 273 revolutions. So C.R. will rotate through 273 revolutions.

$$\text{The length delivered will be} = \frac{273 \times \pi \times 15}{100} = 128.58 \ m$$

With 1.155 draft, this length will be = 128.58 × 1.155 = 148.5 m.

5. In a hunting cog knock-off motion cog wheel 78T and wheel geared to it is 41T diameter of BCR is 5½″ draft between calendar roller and back roller, 1.07 calculate knock-off length and weight of the lap if hank is 0.00145.

Solution :

$$\text{Length of lap} = \frac{\pi \times 5.5 \times 7.8}{36} = 34.43 \ m$$

$$\frac{37.43}{0.00145 \times 2.2 \times 840} = 13.96 \ kg$$

Problems for practice

1. What length of lap will be made if the hunting cog wheel has 81 teeth and the wheel on the 5.5″ diameter calendar roller is 41 teeth if the

hank of the lap is 0.00146. Find out the weight of the full lap for a knock-off.   (Ans. 31.3781 lb.)

2. Calculate the length of the lap and also the weight of the lap for a knock-off if the hunting cog wheel is having 106 teeth and the wheel on the calendar roller is having 74 teeth. The tension draft is 1.2 the hank of the lap delivered is and the diameter of calendar roller is 5.5″   (Ans. 3053 yd; 22.9 lb.)

3. A certain knock-off motion has the lap constant of 0.7; the hunting cog wheel is having 79 teeth. Calculate the weight of the lap for a knock-off if the hank of the lap is 0.00136.   (Ans. 46.64 lb.)

4.3.3   Ratchet and Pawl type knock-off motion

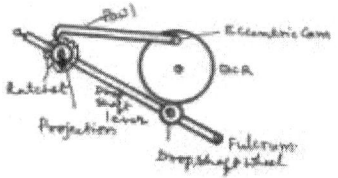

Figure 4.6   Ratchet & Pawl knock off

A = Drop shaft wheel, B = Bottom calendar roller, C=Eccentric cam, D = Pawl, E = Ratchet wheel, F = Projection, G = Drop shaft lever,

No. of revolutions required for   = No. of teeth on the Ratchet wheel ×
a knock-off from calendar roller   No. of teeth moved per revolution
                                                    of the calendar roller (Figure 4.6)

1. Workout the length and total weight of lap / knock-off if Ratchet wheel 73T BCR 7″. For every revolution of BCR 1 tooth is moved in the ratchet. Tension draft 1.09 hank 0.00136.

Solution :

$$\text{Length of the lap } = \frac{\pi \times 7 \times 73}{36} = 44.59$$

$$= 44.59 \times 1.09 = 48.6.$$

$$\text{Weight} = \frac{48.6}{0.00136 \times 2.2 \times 840} = 19.33$$

2. Find out the length and also the total weight of the lap if 72 teeth ratchet wheel is used for a knock-off motion. For every revolution of 7″ diameter calendar roller the ratchet wheel is moved by 1 teeth. The hank of the lap is 0.00119. The tension draft from the calendar roller to the lap roller is 1.03.

Solution :

Actual length of the lap delivered $= \dfrac{22}{7} \times 7 \times \dfrac{72}{12 \times 3} \times \dfrac{103}{100} = 45.37 \ yd$

Weight of the lap $= \dfrac{45.32}{16} \times 16$ (Note: 16 oz. $- 0.00119) = 45.32$ lb.

(Ans.)

3. In a single scutcher 4 laps each of 13 oz/yd are fed. The total mechanical draft is 0.8. Waste removed is 2%. Knock-off motion fitted to the machine is Ratchet and Pawl arrangement. The Pawl is fitted to the third calendar roller of 5½″ diameter; Pawl actuates 106 teeth Ratchet wheel at 2 teeth/revolution of the calendar roller. Calculate the hank of lap delivered and also the weight of the lap for a knock-off.

Solution :

Actual draft $= \dfrac{100 \times 3.8}{98}$

Weight per yard delivered $= \dfrac{4 \times 13}{3.9} = 13.33 \ oz.$

Length of the lap delivered $= \dfrac{106}{2} \times \dfrac{11}{2} \times \dfrac{22}{7} \times \dfrac{1}{36} = 25.44 \ yd.$

Weight of the lap $= \dfrac{25.44 \times 13.3}{16} = 21.158 \ lb.(Ans.)$

Hank of the lap $= \dfrac{1 \times 16}{840 \times 13.5} = 0.00141(Ans.)$

4. In a scutcher 4 laps each of 12 oz / lap are fed. Mechanical draft = 3.8, wastes 2.5%. Find the hank of lap delivered and also weight of each lap for a knock-off R = 72 BCR 7″. Every revolution of calendar roller moves the Ratchet 1 tooth workout number of laps delivered per day 16 has speed of calendar roller = 11 rpm.

(Note: Reader is advised to solve this problem)

5. In the above problem length of the lap when delivered is less than cam is shifted over to 7½″ diameter calendar roller and the length required is 42 yd. What should be the ratchet wheel teeth?

Solution :

Length of the lap delivered = p DN of calendar roller.

$$42 \text{ yd} = \frac{22}{7} \times \frac{15}{2} \times \frac{x}{36}$$

$$x = \frac{42 \times 7 \times 2 \times 36}{22 \times 15}$$

Number of teeth required on the ratchet wheel = 65 teeth (Ans.)

Problems for practice

1. What length of lap will be made if the hunting cog wheel has 81 teeth and the wheel on the 5.5" diameter calendar roller is 41 teeth if the hank of the lap is 0.00146. Find out the weight of the full lap for a knock-off.   (Ans. 31.3781 lbs.)

2. Calculate the length of the lap and also the weight of the lap for a knock-off if the hunting cog wheel is having 106 teeth and the wheel on the calendar roller is having 74 teeth. The tension draft is 1.2 the hank of the lap delivered is and the diameter of calendar roller is 5.5".   (Ans. 3053 yd; 22.9 lb)

3. A certain knock-off motion has the lap constant of 0.7; the hunting cog wheel is having 79 teeth. Calculate the weight of the lap for a knock-off if the hank of the lap is 0.00136.   (Ans. 46.64 lb.)

4.4   Design of cone drums for scutcher

1. Design of pair of cone drums for feed regulating mechanism is scutcher when belt is at centre position with cone drum dia. of 250. The cone drums are 1 m apart. The variation in the thickness of material is 50 – 150% times the normal (Refer Figure 4.7).

Solution :

Assumption:

Rate of diameter of top cone drum to bottom cone drum is constant

$$\frac{d_T}{d_B} = R \text{ constant} \quad R \times \frac{1}{T} \tag{1}$$

$$\Rightarrow \quad \frac{d_T}{d_B} \times \frac{1}{T} \quad \rightarrow (2) \text{ is thickness of material}$$

$$\Rightarrow \quad d_T + d_B = \text{Constant} \quad \rightarrow (3)$$

$$d_T + d_B = 250 + 250 \quad d_B R + d_B = \text{constant}$$
$$= 500 \text{ mm} \quad d_B (R + 1) = \text{constant}$$

$$= 19.6'' \qquad d_B = \frac{\text{Constant}}{R+1}$$

$$\Rightarrow \quad \frac{d_T}{d_B} = \frac{\text{Constant}}{T} \qquad d_B = \frac{19.67}{1+t}$$

$$d_T = 19.6 - d_B$$

$$\Rightarrow \quad d_B = \frac{19.67}{1+7}\,(cm) \qquad d_t = 19.6 - d_p\,(cm)$$

0.5	6.5	13.1
0.6	7.3	12.8
0.7	8.0	11.53
0.8	8.71	10.89
0.9	9.2	10.3
1.0	9.8	9.8
1.1	10.2	9.3
1.2	10.59	8.90
1.3	11.07	8.52
1.4	11.4	8.16
1.5	11.76	7.9

2. Design and draw the profile of cone drum used in scutchure for regulating lap thickness.

1. Lap thickness varies from 50% – 120% of the normal.
2. When the thickness is normal. The diameter of driven and driving cone drums are 100 mm and 120 mm, respectively.
3. The belt shift is 300 mm.
4. Assume suitable centre to centre distance.

Solution :

Given: $t = 1.0$

$D_1 = 100$ mm

$D_2 = 120$ mm

$$\frac{t+k}{t} = \frac{C}{D_2}$$

$$D_1 + D_2 = C$$

$$100 + 120 = 220$$

$$\frac{1.0+k}{1.0} = \frac{220}{120}$$

$$1.0 + k = 1.83 \times 1.0$$
$$k = 1.83 - 1.0$$
$$k = 0.83$$

When $t = 0.5$

$$\frac{0.5 + 0.83}{0.5} = \frac{220}{D_2}$$

$$D_2 = 82.70 \text{ mm}$$
$$D_1 = C - D_2$$
$$D_1 = 220 - 82.70$$
$$D_1 = 137.30 \text{ mm.}$$

When $t = 0.6$

$$\frac{0.6 + 0.83}{0.6} = \frac{220}{D_2}$$

$$D_2 = 92.30 \text{ mm}$$
$$D_1 = 220 - 92.30$$
$$= 127.69 \text{ mm.}$$

When $t = 0.7$

$$\frac{0.7 + 0.833}{0.7} = \frac{220}{D_2}$$

$$D_2 = 100.65 \text{ mm}$$
$$D_1 = 220 - 100.65$$
$$= 119.35 \text{ mm.}$$

When $t = 0.8$

$$\frac{0.8 + 0.83}{0.8} = \frac{220}{D_2}$$

$$D_2 = 107.97 \text{ mm.}$$
$$D_1 = 220 - 107.97$$
$$= 112.02 \text{ mm.}$$

When $t = 0.9$

$$\frac{0.9 + 0.83}{0.9} = \frac{220}{D_2}$$

$$D_2 = 114.45 \text{ mm}$$
$$D_1 = 220 - 114.45$$
$$= 105.55 \text{ mm.}$$

When $t = 1$

$$\frac{1 + 0.83}{1} = \frac{220}{D_2}$$

$D_2 = 120$ mm.
$D_1 = 220 - 120$
$\quad = 100$ mm.

When $t = 1.1$

$$\frac{1.1 + 0.83}{1.1} = \frac{220}{D_2}$$

$D_2 = 125.38$ mm
$D_1 = 220 - 125.83$
$\quad = 94.62$ mm.

When $t = 1.2$

$$\frac{1.2 + 0.83}{1.2} = \frac{220}{D_2}$$

$D_2 = 130.05$ cm
$D_1 = 220 - 130.05$
$\quad = 89.95$ mm.

S.No	T	D_2	$D_1 = C + D_2$
1	0.5	82.70	137.30
2	0.6	92.30	127.70
3	0.7	100.65	119.35
4	0.8	107.90	112.10
5	0.9	114.45	105.55
6	1	120	100
7	1.1	125.83	94.62
8	1.2	130.05	89.95

3. Design the profile of the cone drums used in scutcher to regulate lap thickness with following data,

(i) Lap thickness varies from 0.7 to 2 times to normal.

(ii) When the thickness is 1.2 times is normal, the diameter of driving and driven cone drums are equal.

(iii) Assume any other relevant data.

Solution :

$$t = 0.7\text{--}2$$

When $D_1 = D_2 = D$

$$\frac{t + k}{t} = \frac{C}{D_2}$$

$$\frac{1.2 + k}{1.2} = \frac{200}{100}$$

$$1.2 + k = 2.4$$

$$\boxed{k = 1.2}$$

When $t = 0.7$

$$\frac{0.7 + 1.2}{0.7} = \frac{200}{D_2}$$

$$D_2 = 73.68 \text{ mm.}$$
$$D_1 = 200 - 73.68 = 126.32 \text{ mm.}$$

When $t = 0.8$

$$\frac{0.8 + 1.2}{0.8} = \frac{200}{D_2}$$

$$D_2 = 80 \text{ mm.}$$
$$D_1 = 200 - 80$$
$$D_1 = 120 \text{ mm.}$$

When $t = 0.9$

$$\frac{0.9 + 1.2}{0.9} = \frac{200}{D_2}$$

$$D_2 = 85.71 \text{ mm.}$$
$$D_1 = 200 - 85.71$$
$$= 114.29 \text{ mm.}$$

When $t = 1$

$$\frac{1 + 1.2}{1} = \frac{200}{D_2}$$

$$D_2 = 90.90 \text{ mm.}$$
$$D_1 = 200 - 90.90$$
$$= 109.09 \text{ mm.}$$

When $t = 1.1$

$$\frac{1.1 + 1.2}{1.1} = \frac{200}{D_2}$$

$$D_2 = 95.65 \text{ mm}$$
$$D_1 = 200 - 95.65 = 104.35 \text{ mm.}$$

When $t = 1.2$

$$\frac{1.2 + 1.2}{1.2} = \frac{200}{D_2}$$

$$D_2 = 100 \text{ mm.}$$
$$D_1 = 100 \text{ mm.}$$

When $t = 1.3$

$$\frac{1.3 + 1.2}{1.2} = \frac{200}{D_2}$$

$$D_2 = 96 \text{ mm.}$$
$$D_1 = 200 - 96 = 104 \text{ mm.}$$

When $t = 1.4$

$$\frac{1.4 + 1.2}{1.4} = \frac{200}{D_2}$$

$$D_2 = 107.69 \text{ mm.}$$
$$D_1 = 200 - 107.69$$
$$= 92.31 \text{ mm.}$$

When $t = 1.5$

$$\frac{1.5 + 1.2}{1.5} = \frac{200}{D_2}$$

$$D_2 = 111.11 \text{ mm.}$$
$$D_1 = 200 - 111.11$$
$$= 88.88 \text{ mm.}$$

When $t = 1.6$

$$\frac{1.6 + 1.2}{1.6} = \frac{200}{D_2}$$

$$D_2 = 114.28 \text{ mm.}$$
$$D_1 = 200 - 114.28 = 85.72 \text{ mm.}$$

When $t = 1.7$

$$\frac{1.7 + 1.2}{1.7} = \frac{200}{D_2}$$

$$D_2 = 117.24 \text{ mm.}$$
$$D_1 = 82.76 \text{ mm.}$$

When $t = 1.8$

$$\frac{1.8 + 1.2}{1.8} = \frac{200}{D_2}$$

$D_2 = 120$ mm.
$D_1 = 200 - 120$
$\quad = 80$ mm.

When $t = 1.9$

$$\frac{1.9 + 1.2}{1.9} = \frac{200}{D_2}$$

$D_2 = 122.58$ mm.
$D_1 = 200 - 122.58$
$\quad = 77.41$ mm.

When $t = 2$

$$\frac{2 + 1.2}{2} = \frac{200}{D_2}$$

$D_2 = 125$ mm.
$D_1 = 200 - 125$
$\quad = 75$ mm.

S. No.	t	D_2	$D_1 = C - D_2$
1	0.7	73.68	126.31
2	0.8	80	120
3	0.9	85.71	114.29
4	1.0	90.90	109.90
5	1.1	95.65	104.35
6	1.2	100	100
7	1.3	96	104
8	1.4	107.69	92.31
9	1.5	111.11	88.88
10	1.6	114.28	85.72
11	1.7	111.24	82.76
12	1.8	120	80
13	1.9	122.58	77.41
14	2	125	75

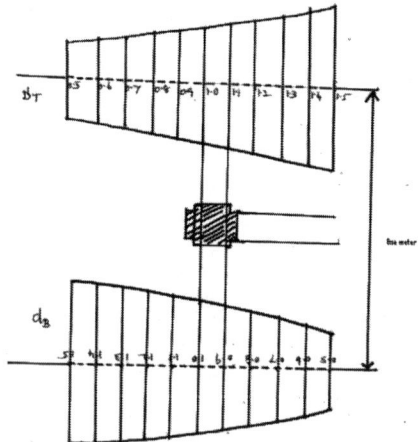

Figure 4.7   Profile of Cone drums

4.4.1   Design of profile of cone drum used for simplex

Let "L" be the delivery rate of front rollers let "B" be the bobbin diameter at any instant. Let DT and DB be the diameters of top and bottom cone drums, respectively. At any instant of time, the excess revolution per minute of bobbin over its spindle is given by

$$\frac{L}{\pi \times B} \tag{1}$$

Let "R" be the speed of top cone drum speed of bottom cone drum is given by

$$R \times \frac{D_T}{D_B} \tag{2}$$

Now equation (1) = (2) if the bottom cone drum speed is with a gearing constant $k1$ (lower case k)

$$R \times \frac{D_T}{D_B} \times k = \frac{L}{\pi \times B} \tag{3}$$

Since R, L, π, α, k are all constants for a given machine. It can be expressed as

$$\frac{D_T}{D_B} = \frac{K}{B} \tag{4}$$

Or $\dfrac{D_T}{D_B} \alpha \dfrac{1}{B}$

In a belt driven system we know that

$$D_T + D_B = k_2 \tag{5}$$

Let $D_T = 17$ cm; $D_B = 9$ cm   then $\dfrac{D_T}{D_B} = \dfrac{17}{9} = 1.889$

Referring to equation 4 if $B = 3$ and let us assume that the change in the bobbin dia from empty to full be 10 cm (i.e., from 3 cm to 13 cm).

$$1.889 = \dfrac{K}{3}$$

$$K = 5.67$$

In general we can write for the given situation as $\dfrac{D_T}{D_B} = \dfrac{5.67}{B}$

$$D_B = \dfrac{D_T \times B}{5.67}$$

Since $D_T + D_B = 17 + 9 = 26$ $\tag{6}$

$$D_T = 26 - D_B = 26 - \left(\dfrac{D_T \times B}{5.67}\right)$$

Or equation 6 can be written as

$$D_T + \left(\dfrac{D_T \times B}{5.67}\right) = 26$$

$$D_T \left(1 + \dfrac{B}{5.67}\right) = 26$$

$$D_T = \dfrac{26}{1 + \dfrac{B}{5.67}}$$

$$= \dfrac{26 \times 5.67}{5.67 + B}$$

$$= \dfrac{147.42}{5.67 + B} \tag{7}$$

And $D_B = 26 - D_T$ $\tag{8}$

Now using the above values we can get the diameters of top and bottom cone drum for different bobbin diameters from 3 to 13 cm.

Example 1: Find the top cone drum dia i.e., bobbin dia is 3.35 m

$$D_T = \dfrac{147.42}{5.67 + 3.35} = \dfrac{147.42}{9.02} = 16.34 \; cm$$

(Using equation No. 7) $D_B = 26 - 16.34 = 9.66$ cm.

Example 2: If the bobbin diameter is 10 cm what are the diameters of D_T and D_B. What is your inference.

$$D_T = \frac{147.42}{5.67 + 10} = \frac{147.42}{15.67} = 9.40 \ cm$$

$D_B = 26 - 9.40 = 16.6$ cm.

But we have assumed D_T as 11 and D_B as 9 cm the value of D_B calculated in previous case is 9.66 and thus it can be said that the bobbin diameter can be lesser than 10 cm.

Bobbin dia	$D_T = \dfrac{147.42}{5.67 + B}$	$D_B = 26 - D_T$
3	17	9
4	15.24	10.76
5	13.81	12.19
6	*	*
7	11.63	14.37
8	*	*
9	10.04	15.96
10	*	*
11	*	*
12	*	*
13	7.89	18.11

Note: Reader is directed to complete the calculation construction of profile. Procedure explained.

Consider the belt width as 10 cm and draw a horizontal line and divide into 10 equal parts and name them as 3, 4, 5 ..... 13 and call this line as axis of D_T. Now at point 3, draw a perpendicular whose total length should be equal to 17 cm above and 8.5 cm below. Like this erect perpendiculars at all points of D_T axis i.e., from 3 to 13 cm. Similar steps to be followed for D_B.

4.4.2   Speed frame – Cone drum design illustrations

1. Design the profile of a pair of cone drums for simplex if the data are: Empty and full diameter of bobbin is 4 cm and 15 cm, respectively. The cone drum diameters when bobbins empty are 25 cm and 12 cm. The centre to centre to distance is 25 cm. The belt shift is 30 cm. Find the necessary statistics when the package is two thirds full.

Solution :

Design equation:

$$\frac{d+k}{d} = \frac{C}{D_2}$$

When the bobbin is empty $d = 4$ cm.

$$\frac{4+k}{4} = \frac{25+12}{12}$$

$$4 + k = 3.08 \times 4$$
$$k = 12.33 - 4$$

$$\boxed{k = 8.33}$$

When $d = 4$ cm

$$\frac{4+8.33}{4} = \frac{37}{D_2}$$

$$D_2 = 12 \text{ cm}$$
$$D_1 = C - D_2 \text{ (because } C = 37)$$
$$= 25 \text{ cm.}$$

When $d = 5$ cm

$$\frac{5+8.33}{5} = \frac{37}{D_2}$$

$$D_2 = 13.88 \text{ cm}$$
$$D_1 = C - D_2$$
$$= 23.12 \text{ cm.}$$

When $d = 6$ cm

$$\frac{6+8.33}{6} = \frac{37}{D_2}$$

$$D_2 = 15.49 \text{ cm}$$
$$D_1 = C - D_2$$
$$= 21.51 \text{ cm.}$$

When $d = 7$ cm

$$= \frac{7+8.33}{7} = \frac{37}{D_2}$$

$$D_2 = 16.89 \text{ cm}$$

$$D_1 = C - D_2$$
$$= 20.11 \text{ cm.}$$

When $d = 8$ cm

$$\frac{8 + 8.33}{8} = \frac{37}{D_2}$$

$$D_2 = 18.13 \text{ cm.}$$
$$D_1 = C - D_2$$
$$= 18.87 \text{ cm.}$$

When $d = 9$ cm

$$\frac{9 + 8.33}{9} = \frac{37}{D_2}$$

$$D_2 = 19.22 \text{ cm}$$
$$D_1 = C - D_2$$
$$= 17.78 \text{ cm.}$$

When $d = 10$ cm

$$\frac{10 + 8.33}{10} = \frac{37}{D_2}$$

$$D_2 = 20.19 \text{ cm}$$
$$D_1 = C - D_2$$
$$= 16.81 \text{ cm.}$$

When $d = 11$ cm

$$\frac{11 + 8.33}{11} = \frac{37}{D_2}$$

$$D_2 = 21.06 \text{ cm}$$
$$D_1 = C - D_2$$
$$= 15.94 \text{ cm.}$$

When $d = 12$ cm

$$\frac{12 + 8.33}{12} = \frac{37}{D_2}$$

$$D_2 = 21.84 \text{ cm}$$
$$D_1 = C - D_2$$
$$= 15.16 \text{ cm.}$$

When $d = 13$ cm

$$\frac{13 + 8.33}{13} = \frac{37}{D_2}$$

$$D_2 = 22.55 \text{ cm}$$
$$D_1 = C - D2$$
$$= 14.45 \text{ cm.}$$

When $d = 14$ cm

$$\frac{14 + 8.33}{14} = \frac{37}{D_2}$$

$$D_2 = 23.20 \text{ cm}$$
$$D_1 = C - D_2$$
$$= 13.8 \text{ cm.}$$

When $d = 15$ cm.

$$\frac{15 + 8.33}{15} = \frac{37}{D_2}$$

$$D_2 = 23.79 \text{ cm}$$
$$D_1 = C - D_2$$
$$= 13.21 \text{ cm.}$$

S. No.	D	D_2	$D_1 = C - D_2$
1	4	12	25
2	5	13.88	23.12
3	6	15.49	21.51
4	7	16.89	20.11
5	8	18.13	18.87
6	9	19.22	17.78
7	10	20.19	16.81
8	11	21.06	15.94
9	12	21.84	15.16
10	13	22.55	14.45
11	14	23.20	13.8
12	15	23.79	13.21

When the package is 2/3rd full

$$d = \left((F - E) \times \frac{2}{3} \right) + E$$

$$= \left[(15-4) \times \frac{2}{3} \right] + 4$$

$$d = 11.33$$

When $d = 11.33$

$$\frac{11.33 + 8.33}{11.33} = \frac{37}{D_2}$$

$$\boxed{D_2 = 21.32 \text{ cm}}$$

$$D_1 = C - D_2$$

$$\boxed{D_1 = 15.68 \text{ cm}}$$

2. Design and draw the profile of cone drums used in speed frames for following curricular.

Empty bobbin diameter = 40 mm, inner diameter of flyer = 170 mm, clearance between flyer arm and full package = 5 mm. When the package is 50% full the diameters of driving and driven cone drums are equal. Assume any other relevant data (Figure 4.8).

Solution :

Given: Empty bobbin diameter = 40 mm = 4 cm. Full bobbin diameter
        =170 – (5 + 5)
        =160 mm = 16 cm

When the bobbin is 50% full,

$$D = \left[(16-4) \times 50 \right] + 4$$

$$\boxed{\frac{d+k}{d} = \frac{C}{D}}$$

Since $D_2 = D$

$$\frac{d+k}{d} = \frac{D_1 + D_2}{D}$$

Since $D_1 = D_2 = D$

$$\frac{d+k}{d} = \frac{2D}{D}$$

$$\frac{10+k}{10} = 2$$

$$\boxed{k = 10 \text{ cm}}$$

Assume $C = 10$

When $d = 4$ cm

$$\frac{4+10}{4} = \frac{10}{D_2}$$

$D_2 = 2.86$ cm.
$D_1 = C - D_2$
$\quad = 10 - 2.86$
$\quad = 7.14$ cm.

When $d = 5$ cm

$$\frac{5+10}{5} = \frac{10}{D_2}$$

$D_2 = 3.33$ cm
$D_1 = C - D2$
$\quad = 6.67$ cm.

When $d = 6$ cm

$$\frac{6+10}{6} = \frac{10}{D_2}$$

$D_2 = 3.75$ cm
$D_1 = C - D_2$
$D_1 = 6.25$ cm.

When $d = 7$ cm

$$\frac{7+10}{7} = \frac{10}{D_2}$$

$D_2 = 4.12$ cm
$D_1 = C - D_2$
$\quad = 5.88$ cm.

When $d = 8$ cm

$$\frac{8+10}{8} = \frac{10}{D_2}$$

$D2 = 4.44$ cm
$D_1 = C - D_2$
$\quad = 5.56$ cm.

When $d = 9$ cm

$$\frac{9+10}{9} = \frac{10}{D_2}$$

$D_2 = 4.74$ cm
$D_1 = C - D_2$
$\quad = 5.26$ cm.

When $d = 10$ cm

$$\frac{10 + 10}{10} = \frac{10}{D_2}$$

$D_2 = 5$ cm
$D_1 = C - D_2$
$\quad = 5$ cm.

When $d = 11$ cm

$$\frac{11 + 10}{11} = \frac{10}{D_2}$$

$D_2 = 5.24$ cm
$D_1 = C - D_2$
$\quad = 4.76$ cm.

When $d = 12$ cm

$$\frac{12 + 10}{12} = \frac{10}{D_2}$$

$D_2 = 5.45$ cm
$D_1 = C - D_2$
$\quad = 4.55$ cm.

When $d = 13$ cm

$$\frac{13 + 10}{13} = \frac{10}{D_2}$$

$D_2 = 5.65$ cm
$D_1 = C - D_2$
$\quad = 4.35$ cm.

When $d = 14$ cm

$$\frac{14 + 10}{14} = \frac{10}{D_2}$$

$D_2 = 5.83$ cm,
$D_1 = C - D_2$,
$\quad = 4.17$ cm.

When $d = 15$ cm

$$\frac{15 + 10}{15} = \frac{10}{D_2}$$

$D_2 = 6$ cm,
$D_1 = C - D_2$, $\qquad\qquad$ $D_1 = 4$ cm.

When $d = 16$ cm

$$\frac{16 + 10}{16} = \frac{10}{D_2}$$

$D_2 = 6.15$ cm,
$D_1 = C - D_2$,
$D_1 = 3.85$ cm.

S. No.	d	D_2	$D_1 = C - D_2$
1	4	2.86	7.14
2	5	3.33	6.67
3	6	3.75	6.25
4	7	4.12	5.88
5	8	4.44	5.56
6	9	4.74	5.26
7	10	5	5
8	11	5.24	4.76
9	12	5.45	4.55
10	13	5.65	4.35
11	14	5.83	4.17
12	15	6	4
13	16	6.15	3.85

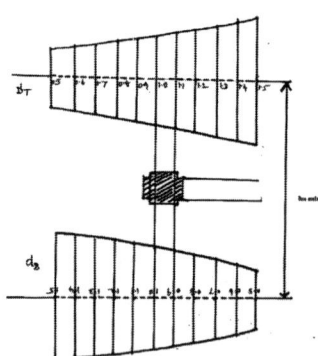

Figure 4.8   Profile of Cone drums

3. Design and draw the profile of driving and driven cone drums used in speed frame with following particulars. Empty bobbin diameter 3.5 cm, full bobbin diameter 10 cm. When the bobbin is empty the diameters of driving and driven cone drums are 17 cm, 9 cm, respectively. The centre to centre distance between driving and driven cone drum is 36 cm. Belt shift is 70 cm.

Solution :

Given
Empty bobbin diameter = 3.5 cm.
Full bobbin diameter = 10 cm.
When the bobbin is empty = D_1 = 17 cm, D_2 = 9 cm.
Centre to centre distance = 36 cm.
Belt shift   = 70 cm.

Design equation

$$\frac{d+k}{d} = \frac{C}{D_2}$$

When the bobbin is empty d = 3.5 cm.

$$\frac{3.5+k}{3.5} = \frac{17+9}{9}$$

$$k = (2.89 \times 3.5) - 3.5$$
$$\boxed{k = 6.61}$$

When d = 3.5 cm.

$$\frac{3.5+6.61}{3.5} = \frac{26}{D_2}$$

$$D_2 = 9 \text{ cm.}$$
$$D_2 = C - D_2$$
$$= 26.\ 9 \text{ cm.}$$
$$\boxed{d = 17 \text{ cm.}}$$

1	3.5	9	17
2	4	9.8	16.2
3	4.5	10.53	15.47
4	5	11.19	14.80
5	5.5	11.81	14.19
6	6	12.37	13.63
7	6.5	12.87	13.13
8	7	13.37	12.62
9	7.5	13.82	12.18
10	8	14.24	11.76
11	8.5	14.63	11.37
12	9	14.99	11.01
13	9.5	15.33	10.66
14	10	15.65	10.35

When $d = 4.0$ cm

$$\frac{4 + 6.61}{4} = \frac{26}{D_2}$$

$$D_2 = \frac{26}{2.6}$$

$D_2 = 9.8$ cm.

$D_1 = C - D_2$

$\quad = 26 - 9.8$

$\quad = 16.2$ cm.

When $d = 4.5$ cm

$$\frac{4.5 + 6.61}{4.5} = \frac{26}{D_2}$$

$$D_2 = \frac{26}{2.47}$$

$D_2 = 10.53$ cm.

$D_1 = C - D_2$

$\quad = 26 - 10.53$

$\quad = 15.47$ cm.

When $d = 5$ cm

$$\frac{5 + 6.61}{5} = \frac{26}{D_2}$$

$$D_2 = \frac{26}{2.32}$$

$D_2 = 11.19$ cm.

$D_1 = C - D_2$

$\quad = 26 - 11.19$

$\quad = 14.80$ cm.

When $d = 5.5$ cm

$$\frac{5.5 + 6.61}{5.5} = \frac{26}{D_2}$$

$$D_2 = \frac{26}{2.20}$$

$D_2 = 11.81$ cm.

$D_1 = C - D_2$

$\quad = 26 - 11.81$

$\quad = 14.19$ cm.

$$D_2 = \frac{26}{2.1} = 12.37\, cm.$$

$$D_1 = C - D_2$$
$$= 13.63 \text{ cm.}$$

When $d = 6.5$ cm

$$\frac{6.5 + 6.61}{6.5} = \frac{26}{D_2}$$

$$D_2 = \frac{26}{2.02} = 12.87 \text{ } cm.$$

$$D_1 = C - D_2$$
$$D_1 = 26 - 11.87 \text{ cm.}$$
$$D_1 = 13.13 \text{ cm.}$$

When $d = 7$ cm

$$\frac{7 + 6.61}{7} = \frac{26}{D_2}$$

$$D_2 = \frac{26}{1.94}$$

$$D_2 = 13.37 \text{ cm.}$$
$$D_1 = C - D_2$$
$$= 26 - 13.37$$
$$= 12.62 \text{ cm.}$$

When $d = 7.5$

$$\frac{7.5 + 6.61}{7.5} = \frac{26}{D_2}$$

$$D_2 = \frac{26}{1.88}$$

$$D_2 = 13.82 \text{ cm.}$$
$$D_1 = C - D_2$$
$$= 26 - 13.82$$
$$= 12.18 \text{ cm.}$$

When $d = 8$ cm

$$\frac{8 + 6.61}{8} = \frac{26}{D_2}$$

$$D_2 = \frac{26}{1.83}$$

$$D_2 = 14.24 \text{ cm.}$$
$$D_1 = C - D_2$$
$$= 26 - 14.24$$
$$= 11.76 \text{ cm.}$$

When $d = 8.5$ cm

$$\frac{8.5 + 6.61}{8.5} = \frac{26}{D_2}$$

$$D_2 = \frac{26}{1.78}$$

$D_2 = 14.63$ cm.
$D_1 = C - D_2$
     $= 26 - 14.63$
     $= 11.37$ cm.

When $d = 9$ cm

$$\frac{9 + 6.61}{9} = \frac{26}{D_2}$$

$$D_2 = \frac{26}{1.73}$$

$D_2 = 14.99$ cm.
$D_1 = C - D_2$
     $= 26 - 14.99$
     $= 11.01$ cm.

$$D_2 = \frac{26}{1.7}$$

$D_2 = 15.33$ cm.
$D_1 = C - D_2 = 26 - 14.99$
     $= 10.66$ cm.

When $d = 10$ cm

$$\frac{10 + 6.61}{10} = \frac{26}{D_2}$$

$$D_2 = \frac{26}{1.67}$$

$D_2 = 15.65$ cm.
$D_1 = C - D_2$
     $= 10.35$ cm.

(2) Design and draw the profiles of driving and driven cone drum used in a speed frame for the following particulars:

Empty bobbin diameter $= 4$ cm.

Full bobbin diameter $= 20$ cm.

When the bobbin is empty the diameters of driving and driven cone drums are 23 cm and 10 cm respectively, centre to centre distance between driving and driven cone drums is 40 cm, belt shift 80 cm.

Solution :

Given

Empty bobbin diameter – 4 cm.

Full bobbin diameter = 20 cm.

When the bobbin is empty $D_1 = 23$ cm; $D_2 = 10$ cm.

Centre to centre distance = 40 cm.

Belt shift = 80 cm.

Design equation

$$\frac{d+k}{d} = \frac{c}{D_2}$$

When the bobbin is empty $d = 4$ cm.

$$\frac{4+k}{4} = \frac{23+10}{10}$$

$$\frac{4+9.2}{4} = \frac{23}{D_2}$$

$$D_2 = \frac{23}{3.3}$$

$D_2 = 10$ cm.

$D_1 = C - D_2$

$D_1 = 23$ cm.

When $d = 4.5$ cm

$$\frac{4.5+9.2}{4.5} = \frac{33}{D_2}$$

$$D_2 = \frac{33}{3.04}$$

$D_2 = 10.84$ cm.

$D_1 = C - D_2$

$D_1 = 22.16$ cm.

When $d = 5$ cm

$$\frac{5+9.2}{5} = \frac{33}{D_2}$$

$$D_2 = \frac{33}{2.84}$$

$D_2 = 11.62$ cm.

$D_1 = C - D_2$

$D_1 = 21.38$ cm.

When $d = 6$ cm

$$\frac{6 + 9.2}{6} = \frac{33}{D_2}$$

$D_2 = 13.03$ cm.
$D_1 = C - D_2$
$D_1 = 19.97$ cm.

When $d = 7$ cm

$$\frac{7 + 9.2}{7} = \frac{33}{D_2}$$

$D_2 = 14.26$ cm.
$D_1 = C - D_2$
$D_1 = 18.74$ cm.

When $d = 8$ cm

$$\frac{8 + 9.2}{8} = \frac{33}{D_2}$$

$D_2 = 15.35$ cm.
$D_1 = C - D_2$
$D_1 = 17.65$ cm.

When $d = 9$ cm

$$\frac{9 + 9.2}{9} = \frac{33}{D_2}$$

$D_2 = 16.32$ cm.
$D_1 = C - D_2$
$D_1 = 16.68$ cm.

When $d = 10$ cm

$$\frac{10 + 9.2}{10} = \frac{33}{D_2}$$

$D_2 = 17.19$ cm.
$D_1 = C - D_2$
$D_1 = 15.81$ cm.

When $d = 11$ cm

$$\frac{11 + 9.2}{11} = \frac{33}{D_2}$$

$D_2 = 17.97$ cm.
$D_1 = C - D_2$
$D_1 = 15.03$ cm.

When $d = 12$ cm

$$\frac{12 + 9.2}{12} = \frac{33}{D_2}$$

$D_2 = 18.68$ cm.
$D_1 = C - D_2$
$D_1 = 14.32$ cm.

When $d = 13$ cm

$$\frac{13 + 9.2}{13} = \frac{33}{D_2}$$

$D_2 = 19.32$ cm.
$D_1 = C - D_2$
$D_1 = 13.68$ cm.

When $d = 14$ cm

$$\frac{14 + 9.2}{14} = \frac{33}{D_2}$$

$D_2 = 19.91$ cm.
$D_1 = C - D_2$
$D_1 = 13.09$ cm.

When $d = 15$ cm

$$\frac{15 + 9.2}{15} = \frac{33}{D_2}$$

$D_2 = 20.45$ cm.
$D_1 = C - D_2$
$D_1 = 12.55$ cm.

When $d = 16$ cm

$$\frac{16 + 9.2}{16} = \frac{33}{D_2}$$

$D_2 = 20.95$ cm.
$D_1 = C - D_2$
$D_1 = 12.05$ cm.

When $d = 17$ cm

$$\frac{17 + 9.2}{17} = \frac{33}{D_2}$$

$D_2 = 21.41$ cm.
$D_1 = C - D_2$
$D_1 = 11.59$ cm.

When $d = 18$ cm

$$\frac{18 + 9.2}{18} = \frac{33}{D_2}$$

$D_2 = 21.83$ cm.
$D_1 = C - D_2$
$D_1 = 11.17$ cm.

When $d = 19$ cm

$$\frac{19 + 9.2}{19} = \frac{33}{D_2}$$

$D_2 = 22.23$ cm.
$D_1 = C - D_2$
$D_1 = 10.77$ cm.

When $d = 20$ cm

$$\frac{20 + 9.2}{20} = \frac{33}{D_2}$$

$D_2 = 22.60$ cm.
$D_1 = C - D_2$
$D_1 = 10.4$ cm.

S. No.	D	D_2	$D_1 = C - D_2$
1	4	10	23
2	5	11.62	21.38
3	6	13.03	19.97
4	7	14.26	18.74
5	8	15.35	17.65
6	9	16.32	16.68
7	10	17.19	15.81
8	11	17.97	15.03
9	12	18.68	14.32
10	13	19.32	13.68
11	14	19.91	13.09
12	15	20.45	12.55
13	16	20.95	12.05
14	17	21.41	11.59
15	18	21.83	11.17
16	19	22.23	10.77
17	20	22.60	10.4

Mechanics of warp winding, warping and warp sizing

5.1 Kinetics of winding

Winding is a process of conversion of smaller packages (Figure 5.1) into bigger packages by clearing and cleaning the yarn. There are two main principles of winding found viz., spindle principle and drum principle. In other classification, the winding mechanisms are classified into precision and drum wound concepts.

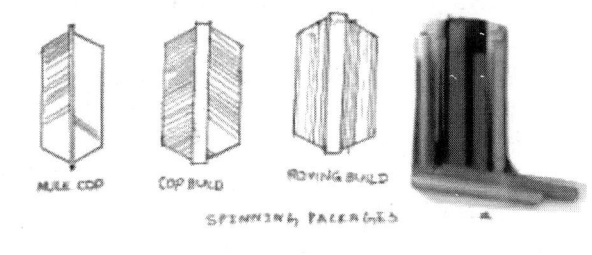

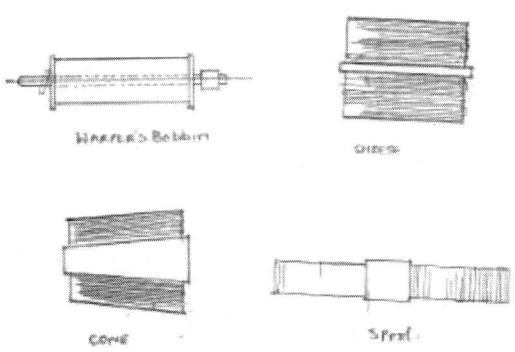

Figure 5.1 Different types of packages

All the winding machines are provided with driving and traversing mechanisms. These will vary with the principles of winding. For example, the method of driving the package may be any one of the methods as shown in Figure 5.2. But in drum wound machines, the drum is double acting in nature as it drives the package and also traverses the yarn on to the package. Therefore the rate of winding has significant effect in the process. In other words as the package diameter builds the number of coils laid per double traverses and angle of wind varies depending on the type of winding principle.

5.2   General methods of driving the package

The packages may be rotated by one of the three methods:

a)   Surface contacts between the outer surface of the yarn on the package and a drum or a roller. This gives a constant surface speed to the package and the yarn is taken up at an approximately constant speed (Figure 5.2a).

b)   Directly driving the package at a constant angular speed, this causes the yarn take-up speed to vary as the size of the package changes (Figure 5.2b).

c)   Directly driving the package at varying speed. To give constant yarn speed, it is necessary to cause the rotational speed to vary inversely with a package radius (Figure 5.2c). W = variable, R increases and V is constant.

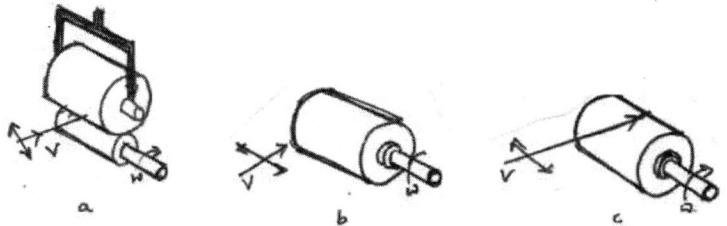

Figure 5.2 a–c   Different methods of package driving

The package build in precision winder is shown below (Figure 5.3).

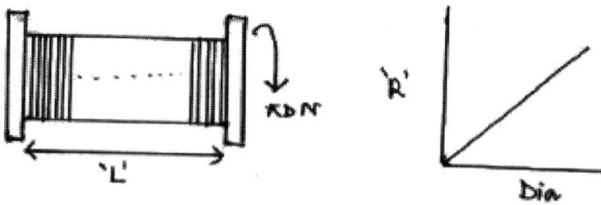

Figure 5.3   Package build in precision winder

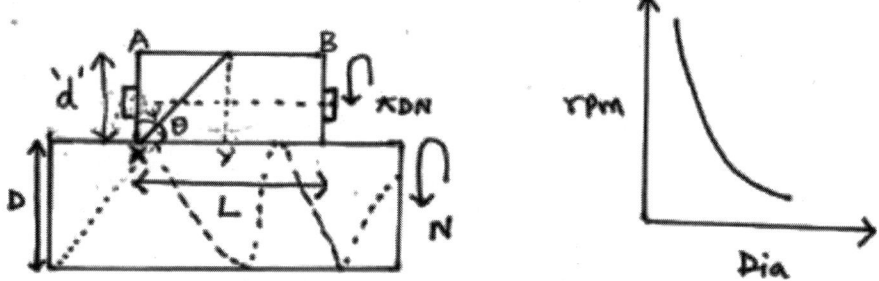

Figure 5.4 Package build in drum winder

The package build in drum winding machine is shown in Figure 5.4.

Figure 5.5(a–f) shows the methods of yarn traversing in both the principles for different types of packages. It is to be noted that yarn traversing and package driving are two important parameters of the winding process (Figure 5.6).

5.2.1   General methods of yarn traversing

The method of yarn traversing depends on the type of the package.

(i)   Parallel wound package
This comprises many threads laid parallel to one another, as in a warp. It is necessary to have a flanged package or beam, otherwise the package would not be stable and would collapse (Figure 5.5a).

Figure 5.5b show the parallel wound package with a single coil shown in thick line for reference purpose and the Figure 5.5c show the package in which the shoulders of a square ended parallel wound package are liable to be damaged if not supported by flanges.

(ii)   The near parallel wound package
The near parallel wound package comprises one or more threads which are laid very nearly parallel to the layers already existing on the package. Figure 5.5d show the package in which the successive coils are shown with angle of wind "θ" at 90°.

(iii)   Cross wound package
As the name itself indicates the yarn is wound onto the package at an angle with respect to the package axis.

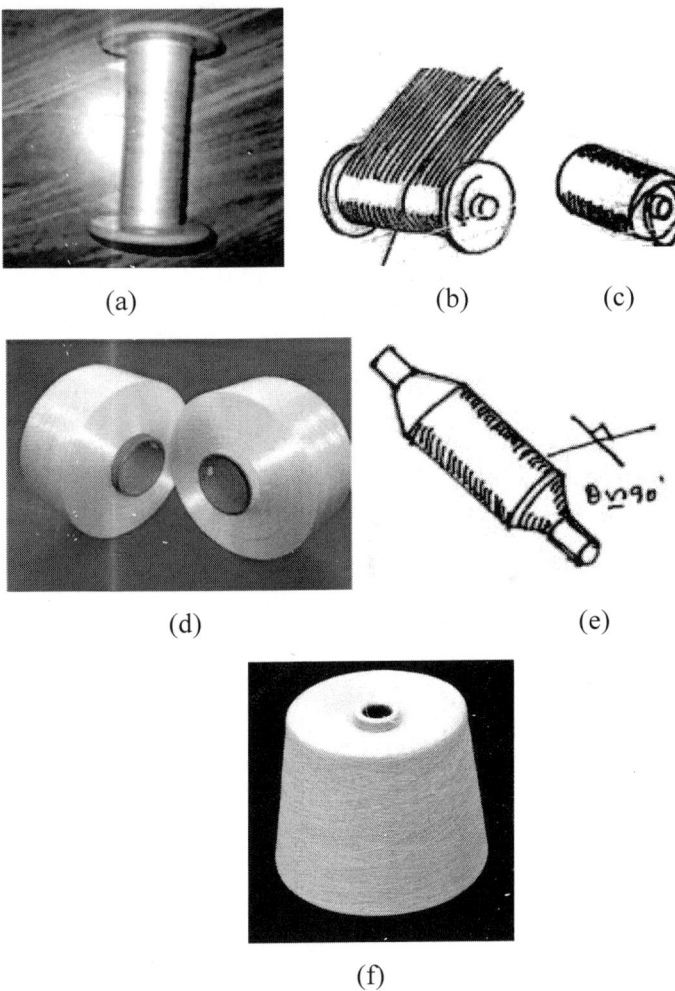

(a)   (b)   (c)

(d)   (e)

(f)

Figure 5.5   Methods of traversing in winding

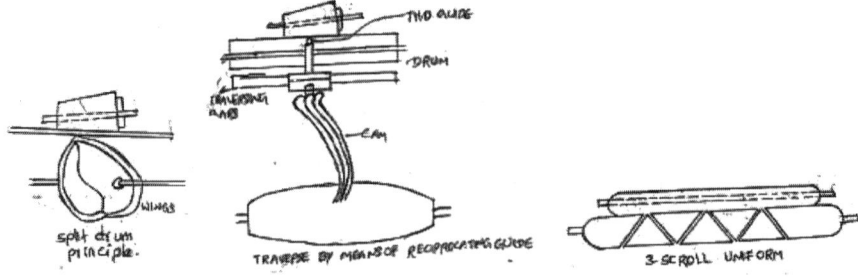

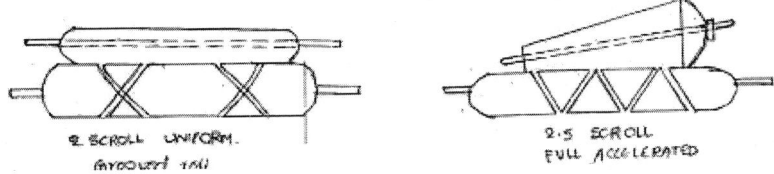

Figure 5.6 Methods of yarn traversing

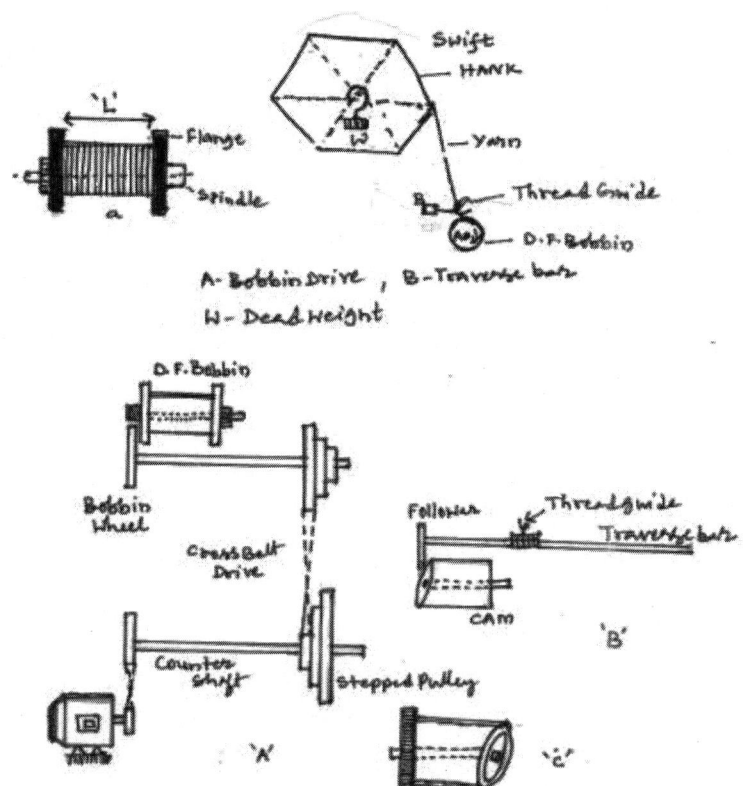

Figure 5.7 Horizontal spindle winder

5.2.2   Mechanics of winding in general (inclusive of winding concept in spinning and weaving)

When fibrous material has to be wound on a bobbin, cop, pirn, or beam, the speed of the part on which the material is wound has generally to be regulated very carefully. The best winding is usually obtained when the speed is so regulated that the material is wound at a constant surface speed and when the coils are put on in spirals of constant pitch (Figure 5.7). The surface speed of winding on a bobbin is equal to the speed of revolution × bobbin circumference. Thus, to keep surface speed constant, the speed of revolution must vary inversely as the circumference or diameter the bobbin. The speed curve, on a base of bobbin diameter, is a rectangular hyperbola, and is shown for a parallel bobbin in Figure 5.8a. For such a bobbin, if the pitch of the coils is to be kept constant, as in flyer frame winding, the speed with which the yarn is traversed across the face of the bobbin must also be varied inversely as the bobbin diameter. When the coils are not laid so closely as to touch each other, the speed of the traverse is often kept constant, which results in the pitch of the coils varying.

It is much more difficult to wind yarn at uniform surface speed in the form of a cop or pirn than it is to wind on a parallel bobbin. The diameter on which winding takes place, in the cop or pirn, is continually varying, requiring a constantly changing speed of rotation to keep constant winding speed.

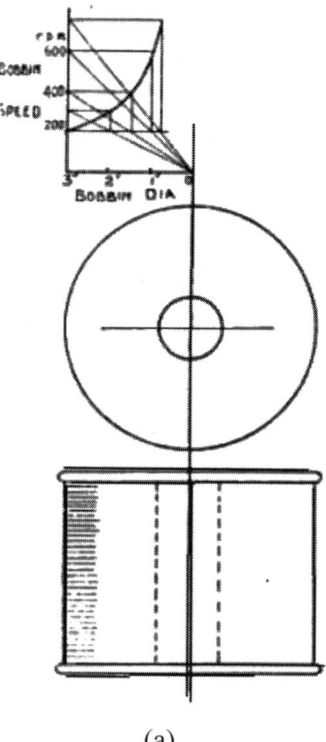

(a)

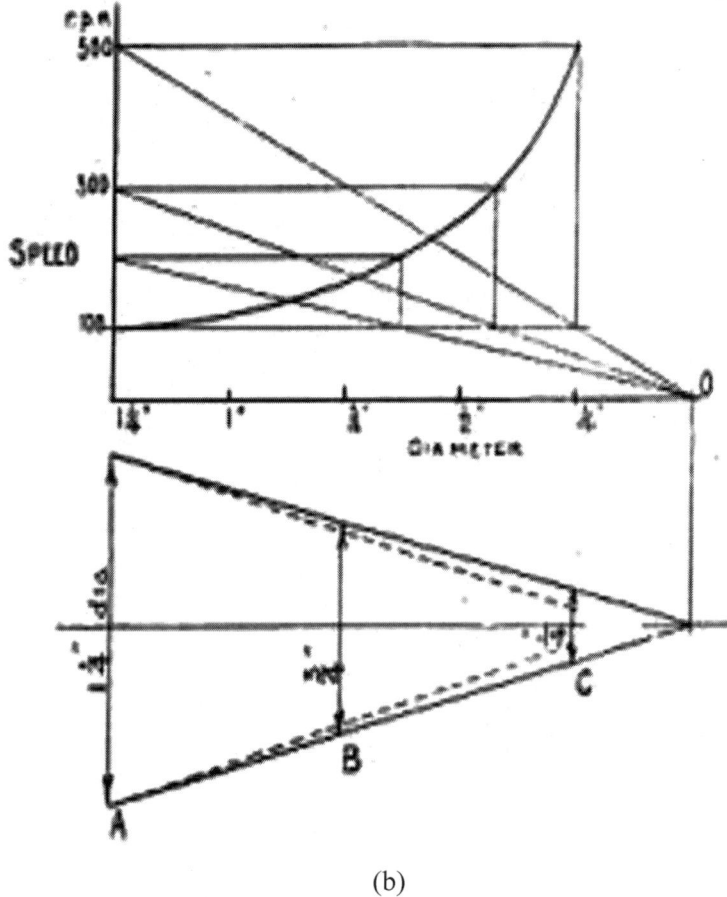

(b)

Figure 5.8 (a,b)   Precision winding
(Source: Hanton)

During one traverse, the diameter on which winding takes place changes gradually and uniformly from a maximum at A (Figure 5.8b) to a minimum at C. The diameter at any intermediate point, as B, can be obtained by calculation or by measurement from a scale drawing. Thus if B is midway between A and C, the diameter at B will be equal to half the sum of the diameters at A and B. Then if the speed of rotation when winding at A is known, the speeds at any other position can be found, remembering that speed must be inversely proportional to diameter. A speed curve can be plotted from the calculated results and is a rectangular hyperbola, which may also be got by the graphical construction shown, taking care to draw from the zero point O of speed and diameter.

Not only must the speed be changed in this way throughout one traverse, but for perfectly uniform winding speed different speeds of rotation are generally required for successive traverses. Changes are generally needed during the formation of the bottom of the cop or pirn, since the largest diameter keeps on increasing during that period. After the completion of the bottom, changes are needed on account of the tapering of the spindle, which reduces the diameters throughout the traverse, as indicated by dotted lines in Figure 5.8b. The changes in spindle speed necessary to keep a constant rate of winding on a mule spindle throughout the building of the cop are shown in Fig. 5.9. Speed curves are shown for four layers, AB, at the beginning of the cop bottom; CD, when the bottom is partly competed; EF, during the building of the body; and GH, the last layer of the cop.

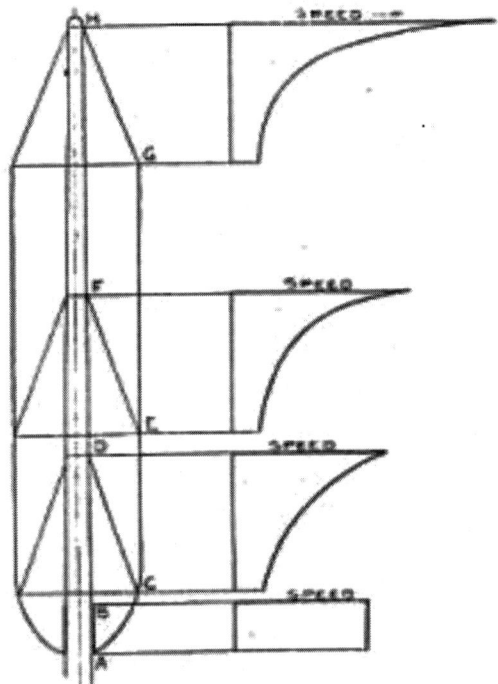

Figure 5.9   Speed curves

The first layer AB is laid on practically a constant diameter, so the speed is approximately constant. At the next layer CD, the diameter at D is nearly the same as at A and B, but the diameter at C is much greater, and consequently the speed at C must be reduced as shown. Throughout the building of the body of the cop, as at FE, the largest diameter, at E, remains constant, but the smallest diameter, at F, gets less and less for successive layers, due to the tapering of the spindle. Thus the speed at F is greater than that at D, whilst that at H is still higher, being the maximum speed reached. If the pitch of the coils is to be kept constant, then the rate at which the yarn is traversed must be varied to suit the varying speed of rotation, the rate of traversing being changed in direct proportion to the speed of rotation. Thus if the speed of rotation is doubled, two coils are wound in the same time as one at the lower speed, and evidently to keep the coils equally spaced the yarn will have to be traversed twice as far during the winding of the two coils as was needed for the coil.

The speed curves shown in Figure 5.9 are drawn on the assumption that the carriage carrying the spindles moves at constant speed during winding and that the rate at which yarn must be wound is constant; neither of these assumptions is correct, and whilst the curves show how the spindle speed would have to vary in the circumstances described, they do not show how the spindle speed in the mule does vary during winding. These ideal conditions of speed variation are not reached in many machines, as the mechanism required for the speed changes would be too complicated and expensive. Much depends on the conditions under which winding is done and the state of the material would. Thus in the mule the speed changes have to follow very closely to the theoretical, on account of the delicate material wound and the fact that the end of the yarn is practically fixed during winding. In pirn winding machines, on the other hand, the yarn is drawn freely from a hank or cop, and changes in winding speed are not so important. In the following chapters it will be seen how closely or otherwise the winding in the different machines follows the rules suggested by theory.

5.3   Package build in precision winding machines

Let the package traverse length be L cm mounted on the spindle which constantly rotates at N rpm. The cam shaft is driven from the spindle shaft by means of gear or belt and is rotating at n rpm. Usually one rotation of drum means one double traverse.

Let at any instant "r" be the diameter of the package.

Winding speed: Here also winding speed is made up of two components.

Surface speed: The surface speed is proportional to "d" so as the "d" increases the surface speed also increases: Surface speed (SS) $= \pi\, dN$

Traverse speed: Constant except at the vicinity of reversal points.

Traverse speed TS $= 2\, LN$

$\therefore$ winding speed $= \sqrt{\pi^2 d^2 N^2 + 4L^2 n^2}$

On a constant speed, precision winding machine winding speed increases as the diameter increases. But this is undesirable for the following reasons.

1. Filament yarns are easily stretched beyond their elastic limits.

2. There is of uneven compactness of packing.

Because of above facts, on precision winding machines, arrangements are provided to reduce the tension as the package builds up. In case of cone winding, the mean line as speed of winding V_m is determined by the formula:

$$V_m = \sqrt{\left(\pi N d_m\right)^2 + 4L^2 N^2}$$

Where $d_m = \dfrac{d_1 + d_2 + d_3 + d_4}{4}$ and d_1 – core diameter (ϕ) of the nose.

d_2 – core ϕ of the base, d_3 – package diameter at the nose, d_4 = package ϕ at the base.

Wind per double traverse
In precision winders the ratio of spindle speed and cam speed always remain constant and thus the number of coils per double traverse remains constant. The ratio is chosen such that no patterning occurs.

In winding or unwinding, it is necessary to have yarn guides in guiding the yarn to the desired path. The usage guide can also be linked to the type of unwinding. If the side withdrawal being used, it is possible for the yarn to pass along a smooth unvarying yarn path, but if there is some vibrating force present, yarn may vibrate between the grids. These vibrations can be controlled by the strategic placing of guides along the yarn path. If over-end unwinding is used, the yarn does not move along a fixed path, because a rotary motion is imparted as the yarn unwinds. Any given section yarn moves not only along the length of the yarn, but also in the circular fashion, this is called as ballooning. For a given yarn speed and package size, the position of the yarn guide will determine the balloon shape, this in turn determines the tension of yarn. The guide position is thus important. It is normally desired to have only one balloon and if the distance between the yarn package tip and the guide distance is more, it is found that the number of balloons formed may increase. ATIRA (Ahmadabad Textile Research Association) has developed a formulae known as Optimum Guide Distance (OGD) given by $n = G + L / L + 1$, where G is the guide distance selected, L is the lift of the bobbin and n is the integer. Further it is also to be noted that the tension during unwinding developed varies as square of speed. Guides normally are made of hard, smooth steel or ceramics. Many man-made yarns are abrasive and frequently it is essential to use ceramic guides. Guides of various shapes are used, the choice depending on the yarn motion to be controlled.

It is found in modern super speed and high speed automatic winding machines very close to the supply package. It helps in balloon breaking as the unwound coils of yarn from the supply package pass tangentially over this accelerator. It also imparts a little amount of tension to the yarn without damaging it as the thread guides are usually made of porcelain or chromium coated. There are different types of thread guides. As the name itself indicates, they guide the yarn path on the winding machine. This is the first part into which the yarn is unwound from the ring cop comes in contact with. As the yarn unwinds from the ring cop, it forms a balloon and it is undesirable to have excess tension on yarn due to ballooning. They are also called as *unwinding accelerators* because they facilitate the unwinding of yarn from the ring cop at a given speed of yarn take-up. Any thread guide shall do two important functions namely it should guide the yarn and secondly the balloon formed due to unwinding of the yarn from the ring cop is need to be controlled. The position of thread guide is however decided by the optimum guide distance. Later in the chapter, a separate discussion is given in which the type of the package and the unwinding is mentioned. It is also found that by using the guides the vibrations due to unwinding of yarn if any can be controlled. Guides are normally made of hard smooth mild steel or ceramic or chromium coated material for a stainless steel base. Thread guides used are of different shapes and may fall into two groups. The first group includes those in which the yarn end is required for threading and the second group in which the yarn end is not required for threading or self-threading in nature.

5.4   Uniform package build in drum winding

5.4.1   Cone/cheese winding

Quality package preparation is the main objective in winding as the package form the supply or feed material in warping and loom shed (if the packages are directly used on loom). Therefore it is essential to have uniform build and density. The necessary condition for uniform build can be understood if an assumption like the following is made.

"The length of yarn wound per unit surface area of the package should be constant from D_0 to D_{max}".

5.5   Calculations of spindle speed, traverse speed, number of coils/double traverse in precision winder

5.5.1   Speed calculations in precision winder (Figure 5.10)

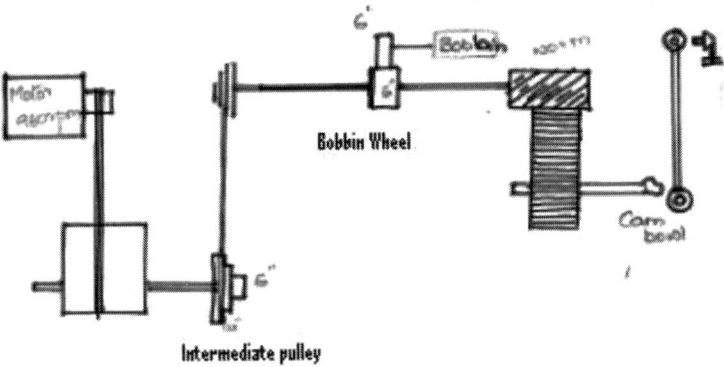

Figure 5.10   Driving arrangements for a precision winder

Given motor speed = 960 rpm
Circumference of motor pulley = 16.5 cm
Circumference of intermediate pulley = 77.5 cm
∴ Speed of intermediate pulley

$$= \frac{\text{Speed of motor} \times \text{Circumference of motor}}{\text{Circumference of intermediate pulley}}$$

$$= \frac{960 \times 16.5}{77} = 240.3 \; rpm.$$

Drive from final step to initial step pulley
Circumference of final pulley = 22 cm
Circumference of initial stepped pulley = 35 cm
Speed of initial final stepped pulley = 204.3 rpm

Speed of initial pulley $= \dfrac{204.3 \times 22}{35} = 128.4 rpm.$

Speed of initial pulley = 128.4 rpm

Speed of bobbin wheel $= \dfrac{128.4 \times 66}{15.24}$

Speed of bobbin wheel = 556 rpm.
Drive from worm to worm wheel

Number of teeth of worm = 1
Speed of worm = Speed of bobbin shaft = 128.4 rpm
Number of teeth of worm wheel = 30

Speed of worn wheel = $\dfrac{128.4 \times 1}{30}$ = 4.28 rpm

∴ Speed of cam = 4.28 rpm.
Length of spindle l = 18 cm
Diameter of spindle D_0 = 7.6 cm

∴ Rate of winding $R = \sqrt{(\pi DN)^2 + (2ln)^2}$

Where D = Diameter of spindle
         N = Bobbin speed
         n = Cam speed

Rate of winding $(R) = \sqrt{(3.14 \times 7.6 \times 556)^2 + (2 \times 18 \times 4.25)^2}$

R = 132.6 rpm.

Middle stepped pulley to middle stepped pulley
        Circumference of middle stepped pulley = 28 cm
        Speed of middle stepped pulley = 204.3 rpm
        Speed of bobbin shaft = 204.3 rpm

        Speed of bobbin = $\dfrac{204.3 \times 66}{15.24}$ = 884.7 rpm.

        Sped of bobbin = 884.7 rpm.

Drive from worm to worm wheel
        Number of teeth of worm = 1
        Speed of worm = Speed of bobbin shaft = 204.3 rpm
        Number of teeth of worm wheel = 30

        Speed of worm wheel = $\dfrac{204.3 \times 1}{30}$ = 6.81 rpm.

        ∴ Speed of cam = 6.81 rpm

Rate of winding = $\sqrt{(3.14 \times 7.6 \times 884.7)^2 + (2 \times 18 \times 6.8)^2}$

= 210.01 rpm.

Drive from initial stepped pulley to final stepped pulley
        Final stepped pulley circumference = 22 cm
        Initial stepped pulley circumference = 35 cm
        Speed of initial stepped pulley = 204.3 rpm.

        Speed of final stepped pulley = $\dfrac{204.3 \times 35}{22}$ = 325.02 rpm.

        Speed of bobbin stepped pulley = $\dfrac{325.02 \times 66}{15.24}$ = 1407.5 rpm.

Drive from worm to worm wheel

Number of teeth to the worm = 1

Speed of worm = Speed of bobbin shaft = 325.22 rpm.

Number of teeth to worm wheel = 30

$$\text{Speed of worn wheel} = \frac{325.22 \times 1}{30} = 10.83 rpm.$$

∴ Speed of cam = Speed of worn wheel

∴ Speed of cam = 10.83 rpm

∴ Traverse speed = Cam speed = 10.83 rpm

$$\text{Rate of winding} = \sqrt{(3.14 \times 7.6 \times 1407.5)^2 + (2 \times 18 \times 10.8)^2} =$$

335.9 rpm

Let us consider now different diameter of spindle:

For

D_0 = 7.6 cms	R_2 = 210 rpm
D_1 = 10 cms	R_2^1 = 277.6 rpm
D_2 = 15 cms	$R_2^{''}$ = 416.4 rpm
D_3 = 18 cms	$R_2^{'''}$ = 500 rpm

Note: Rate of winding α diameter of spindle.

5.5.2   Relation between package diameter, coil angle and number of coils per double traverse in winding machines

Case 1: Spindle winder

Here the package is driven by a separate arrangement and the traverse is also obtained separately Therefore, the n / N remains constant (the traverse speed to package speed) and thus the number of coils laid per double traverse remains constant throughout the package build.

In precision winding machines, the bobbins prepared may be either parallel or barrel shaped depending on the type of traverse mechanism employed. For parallel shaped bobbins as shown in Figure 5.11. The driving mechanism is explained in Figure 5.12.

Tin roller drives cam shaft wheel which accommodates two heart shaped cams mounted in offset condition i.e., opposite to each other. Cams push anti-friction bowl mounted on treadle levers which are fulcrumed at one end and attached to pinion by chain. Let us say a cam pushes the anti-friction bowl on the treadle "A" and it moves in downward direction as shown by arrow. This causes the chain unwind from pinion and is rotated clockwise. As pinion is connected to rack, the later moves up i.e., the traverse is towards upwards. The cycle is repeated in opposite direction as and when treadle "B" is pushed down. However if motion is not given to B, due to spring action (not shown in figure) as treadle is negative, the chain is coiled, pinion moves clock wise and thus the traverse bar moves down.

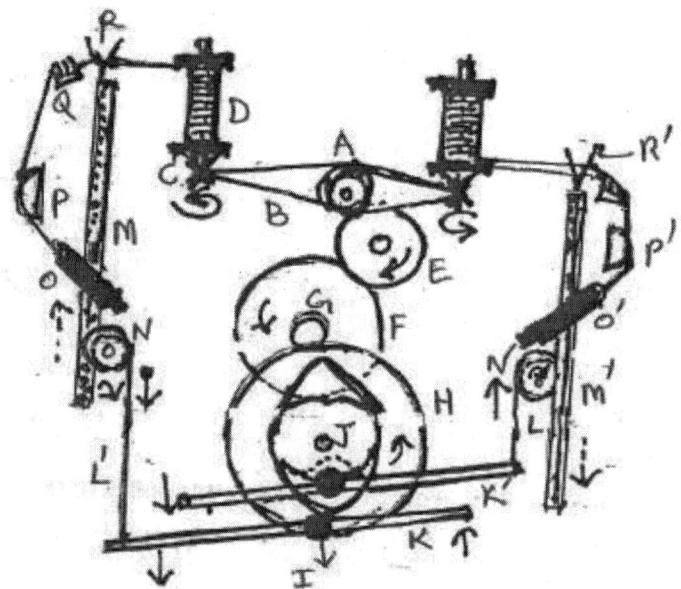

Figure 5.11   Upright spindle winder – Traverse mechanism for parallel bobbins

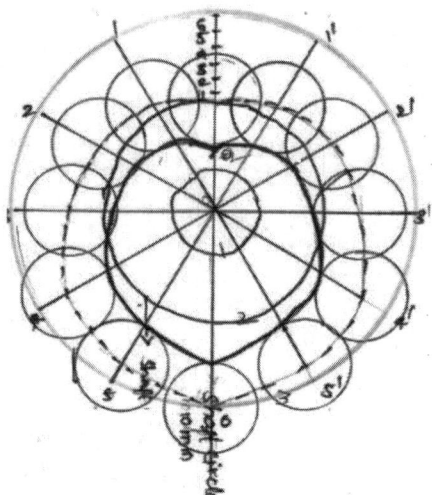

Figure 5.12   Heart shaped cam

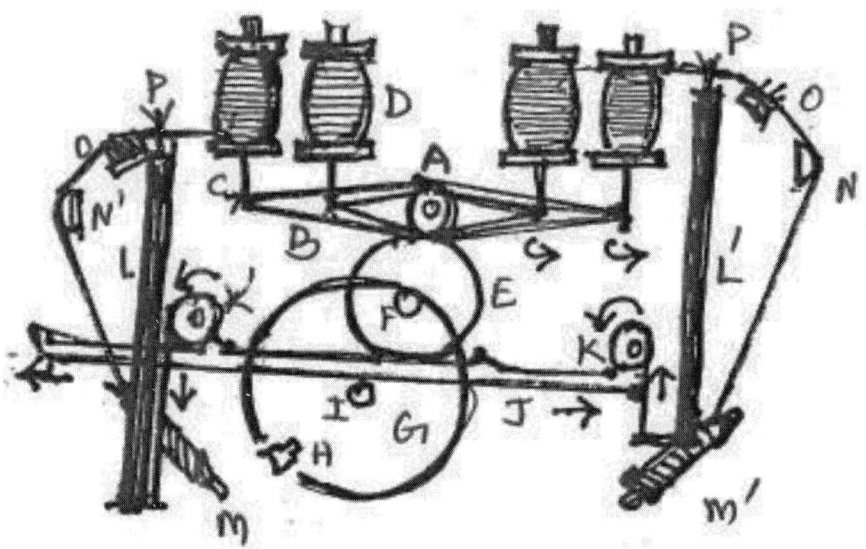

Figure 5.13　Upright spindle winder – Traverse mechanism for barrel bobbins

In precision winder a special type of Double Flanged (DF) bobbins known as barrel shaped bobbins are produced using a mangle wheel traverse. One such arrangement is shown in the driving mechanism is as follows: Tin roller drives mangle wheel through pinion "O" which is on carrier wheel "W". The pinion "O" will be able to contact the inner and outer part of wheel M. A pinion P is centered on M will be moving the rack N which carry at the end pulley with a chain or belt (flat). The poker bar or traverse bar is moved up or down as and when the belt is coiled or uncoiled over the pulley's depending upon the direction of movement of rack "N". Thus say the pinion "O" contacts the wheel M through inside the surface, the pinion "P" centered on M, rotates clockwise and hence the rack "N" moves from left to right which results in coiling of the belt on the respective end pulleys and thus the pocker bar is lifted or moves from bottom to top. Similarly the motion occurs for traverse bar from top to bottom (Figure 5.13).

Important note: It is to be noted here that as and when the pinion "O" contact M inside/outside, will face an opening "W", which cause the pinion to slide from inside to outside or vice-versa. It is during these periods there exists a small amount of "PAUSE" or "DWELL" (stationary period), at which the poker bar doesn't move in either of the directions, causing winding of more yarn at the centre than edges. Thus barrel shaped bobbins are resulted.

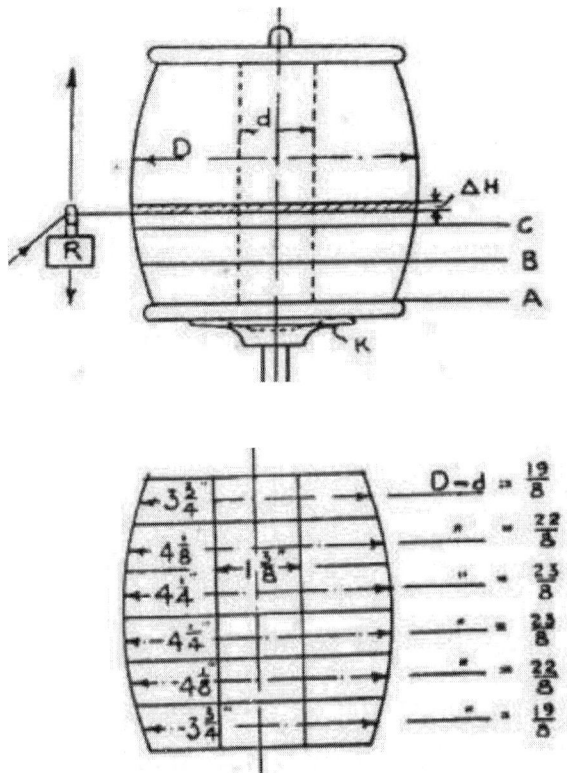

Figure 5.14   Some aspects of barrel shaped bobbin (Source: Hanton and Sengupta)

The quantity of yarn wound on any thin lamina ΔH depends on the number of coils there between the bare and full bobbin diameters. This number of coils will be in direct proportion to the difference $(D - d)$ since each coil may be assumed to increase the diameter by a constant amount. Also, since the spindle speed is constant, equal numbers of coils are put on in equal units of time; consequently the time spent in putting the yarn on any thin lamina such as ΔH will depend on the number of coils, i.e., on $(D - d)$. In designing the cam, it is therefore necessary to make it such that for equal distances moved by the rail R, such as AB, BC, etc., the times taken will be proportional to the average values of $(D - d)$ for AB, BC, etc. Figure 5.14 shows the construction for such a cam, the lift line AB being divided into equal parts corresponding to the vertical movements of the guide rail, and the angle turned through by the cam divided into parts proportional to the values of $(D - d)$. The bobbin in the upright spindle winder are driven frictional contact between the bobbin flange and a driving disc K on spindle. The friction between disc and flange depends on the weight of the bobbin, which varies approximately as the square of the outside diameter of the bobbin at any time; consequently the driving effort which the disc is able to apply to the bobbin before slipping takes place varies as (bobbin diameter)^2. The tendency for the bobbin to slip depends on the value of the retarding movement of the tension of the yarn being wound. The tension varies somewhat as the bobbin increases in diameter, due to the increased winding speed which increases the ballooning of the yarn as it leaves the cop from which it is unwound; if unwound from the side of a ring bobbin, tension increase somewhat as the wrappers bobbin fills up, due to the more rapid speeding up of the ring bobbin and spindle as unwinding changes form the nose to the other end of the chase. Also, the radius at which the yarn acts on the wrappers bobbin increases directly as the bobbin diameter increases. If the yarn tension increases directly as the bobbin diameter, then the retarding moment acting on the bobbin, equal to yarn tension × bobbin radius, varies as the square of the bobbin diameter or in other words, it varies in the same way as the driving effort which can be applied by the driving disc. Slipping, therefore, ought to be roughly independent of the bobbin diameter; if no slipping takes place at the bare bobbin, there should be none at full bobbin.

Case 2: Precision winder with drum drive for package and a separate traverse mechanism

Here we consider those machines in which the package is frictionally driven and the traverse is achieved by a separate mechanism like wing guide or grooved drum.

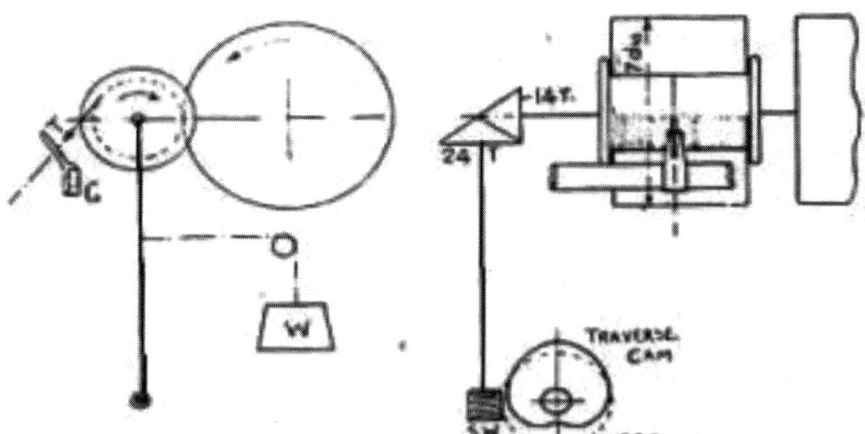

Figure 5.15   Principle of drum winding machine with separate traverse

Here the bobbins are driven by surface contact with drums rotating at constant speed. The surface speed of winding is therefore constant, and the rate of rotation of the bobbin decrease as it increases in diameter. The yarn guide bar G is (Figure 5.15) moved at a uniform rate by means of a cam, and the pitch of the coils gets greater as the bobbin diameter increases. A bobbin slightly barrel-shaped is sometimes would by reducing the rate of the guide in the centre of its movement, but the nature of the variation in speed is not quite the same as in the upright spindle machine, where the spindle speed is constant.

In this type of winder, the length of yarn would in unit time is independent of the bobbin diameter, in equal units of time, equal lengths of yarn x and y (Figure 5.16) are added. The quantity of yarn that has to be put on to form a thin lamina ΔH, say equal in thickness to the yarn diameter, is proportional to the area of the lamina $\frac{\pi}{4}\left(D^2 - d^2\right)$, or simply, proportional to (D^2-d^2). Therefore, the time taken to wind the yarn for the lamina is proportional to (D^2-d^2). Therefore, the times for equal horizontal distances moved by the guide must be proportional to the average values of D^2-d^2 for these distances.

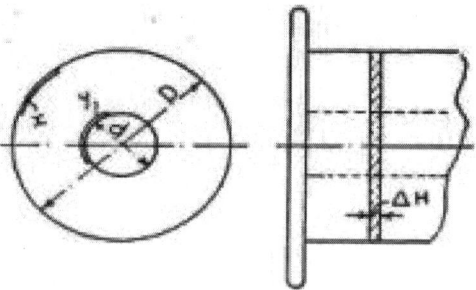

Figure 5.16   Precision winding

The frictional driving force between the drum and the bobbin is approximately constant, being caused by the weight W (Figure 5.15). As the bobbin increases in diameter the driving effort which can be applied, before slipping takes place, increases in proportion to the increasing diameter.

The yarn tension T is constant, since the rate of winding is constant, and the moment of T increases directly as the bobbin diameter. In addition to the resisting moment due to T, there is the moment due to friction at the bobbin spindle, also tending to cause slipping. This moment will remain approximately constant throughout, since it depends chiefly on the force, due to W, with which the bobbin presses against the drum, though to a small extent it will increase with the increasing weight of the filling bobbin. This means that the resisting moment will be greatest, proportionally, at the empty bobbin, and most slipping will take place then.

Case 3: Rotary traverse winder

In this type of machines as the package diameter increase the angle of wind remains the constant but the number of coils per double traverse will reduce (Figure 5.17).

Figure 5.17   A commercial cone winder

Figure 5.18   Assembly winder or rewinder

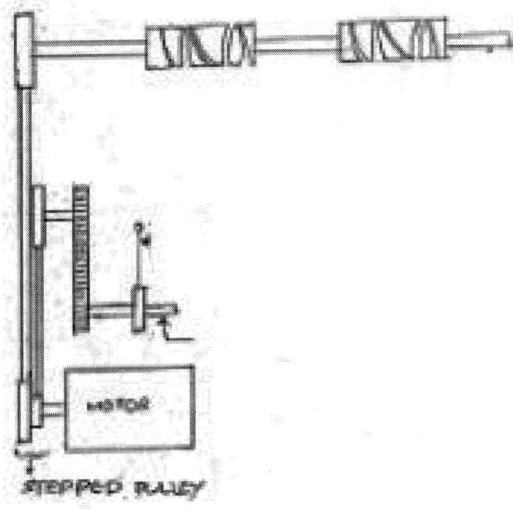

Figure 5.19   Driving arrangement of a cone winder

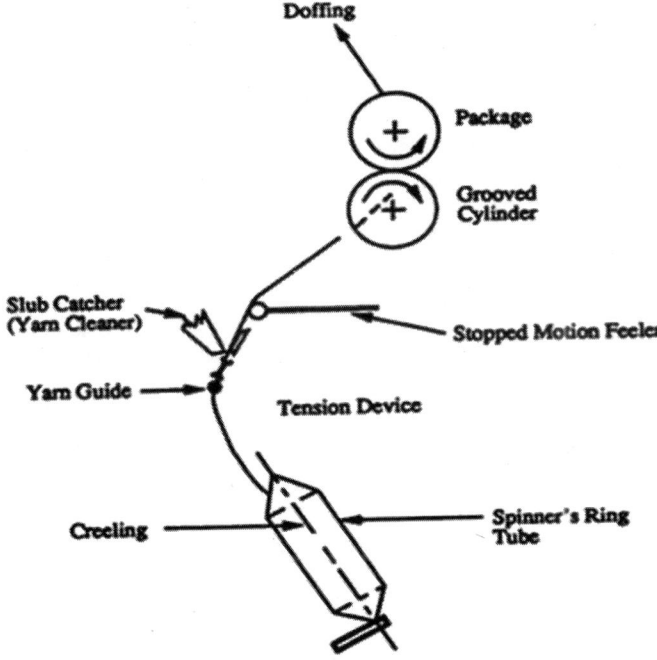

Figure 5.20   Passage of material through cone winder

In this type of machine the package is driven by the frictional contact with the drum which is rotated at a constant speed. The rate of winding in this type of machine is already discussed in the beginning of this chapter (Figure 5.18).

5.5.3   Relation between angle of wind and package diameter in winders

In drum winding, the yarn is traversed by the drum itself and thus drum is double acting in nature the type of package to be produced will depend on the requirement in further processing like warping (Figure 5.19). The factors normally considered are cone angle, cone weight, unwinding rate in the next process, etc. However, the yarn is traversed onto the package depending on the type of grooves of the drum. It is found that when yarn is wound on to the package it subtends an angle with respect to the axes of the package. This angle is called either "Coil angle" or "Angle of Wind". This angle is formed when yarn laid onto the package follow a helical path (Figure 5.20).

Consider the package to be cut into a rectangular segment for better understanding of the theory Let "θ" be the coil angel and "$\emptyset$" be the angel of wind.

Consider a package like cheese:

Let "d" be the diameter of the package at any instant "t" and let "h" be the package height (or package traverse) then the total surface area at any instant is (πdh). As we have considered the angel we can say the length of the coil will be $\pi d/\sin\theta$ and the number of such coils within one traverse will be $[h / (\pi d/\tan\theta)]$.

Then total length of yarn wound within a traverse at the diameter "d" would be

The length of yarn wound per unit surface area is given by

$$x = \frac{\text{Total length of yarn wound at the diameter "}d\text{"}}{\text{Total surface area of package at the diameter "}d\text{"}}$$

$$= \left(\frac{h}{\cos\theta}\right) \Big/ \pi d.l. = \frac{h}{\pi d.\cos\theta} \tag{1}$$

In considering the situation of a uniform stable package, we can say that the product of package diameter and cosine of the wind angle to be maintained constant. This means that as package diameter increase, the angle of wind will also increase and the coils will be progressively more parallel to the package base. If we consider the path traced out by an element of yarn on the package a vector diagram relating winding speed, traverse speed and surface speed can be constructed accordingly. We know that in a cone or precision winding the formulae for coil angle is given by

$tan\theta = ($surface speed/traverse speed$)$

$$= \frac{\pi d.n.}{v} = \frac{sin\theta}{cos\theta} = \frac{\pi dn}{v}$$

where "v" is the traverse speed, "n" is the package speed in rpm and "d" is the package diameter.

Applying the conditions as mentioned in equation (1), it can be written that

$v\,sin\theta = \pi(d\,cos\theta).n$

$= \pi($constant$).n$

It is known fact that in precision or spindle driven machines the speed of the cam to speed of the package remains constant and hence the coils laid per double traverse will also remain constant. Hence for such winders, keeping a constant value of $v\,sin\theta$ would maintain the desired relation between "d" and "θ". For surface driven winders however there exists hyperbolical relation between "n" and "d" such that

$dn = $ constant $= k$

$\therefore v\,sin\,q = p$ (constant) k/d (3)

From the above it is clear that uniform building of a cheese requires varying the traverse speed with change in package diameter in a pre-determined manner.

Consider now the case of drum winders with open cross-wind machines and assuming that at any instant, the angle of wind is "α", diameter of base is "d_1", the traverse speed is "v_1" and the traverse is h, such that

$h = v_1 t$

then by considering the package as a two dimensional structure, the surface area of the annular sector can be written as k (πd, v, t) for $\frac{1}{2} < k < h$. If the winding speed is expressed as b_1, then the length wound in time "t" is $b_1 t$. Hence

$$\frac{\text{length wound}}{\text{Area of surface covered}} = \frac{b_2 t}{k(\pi d_1 v_1 t)}$$

Let "d_2" be the second diameter with corresponding value of angle of wind β, winding speed be "b_2" and traverse speed be "v_2", then the ratio

$$\frac{\text{length wound}}{\text{Area of surface covered}} = \frac{b_2 t}{k(\pi d_2 v_2 t)}$$

The boundary condition for keeping this ratio constant can be written as

$$\frac{b_1}{d_1 v_1} = \frac{b_2}{d_2 v_2} \quad or \quad \frac{b_1}{b_2} = \frac{d_1 v_1}{d_2 v_2}$$

We know that

Tan (angle of wind)$_{1,2}$ = (surface speed)$_{1,2}$/(traverse speed)$_{1,2}$

and (winding speed)$^2_{1,2}$ = (surface speed)$^2_{1,2}$ + (traverse speed)$^2_{1,2}$

with subscript 1,2 as for the first and second diameters, one can, abbreviate surface speed as s, so that

$$b_1^2 = s_1^2 + v_1^2 \text{ and }$$

$$b_2^2 = s_2^2 + v_2^2$$

Or

$$\frac{b_1^2}{b_2^2} = \frac{s_1^2 + v_1^2}{s_2^2 + v_2^2} = \frac{v_1^2\left[(s_1/v_1)^2 + 1\right]}{v_2^2\left[(s_2/v_2)^2 + 1\right]} = \frac{v_1^2(tan^2\alpha + 1)}{v_2^2(tan^2\beta + 1)}$$

$$= \frac{v_1^2 sec^2\alpha}{v_2^2 sec^2\beta} = \frac{v_1^2 cos^2\beta}{v_2^2 cos^2\alpha}$$

but $\dfrac{b_1^2}{b_2^2} = \dfrac{d_1^2 v_1^2}{d_2^2 v_2^2}$ and hence $\dfrac{d_1^2 v_1^2}{d_2^2 v_2^2} = \dfrac{v_1^2 cos^2\beta}{v_2^2 cos^2\alpha}$

or $\dfrac{d_1}{d_2} = \dfrac{cos\beta}{cos\alpha}$

or $d \cos\alpha$ = constant

5.6   Yarn clearing and clearing efficiency

5.6.1   Optimum yarn clearing

Optimum yarn clearing is defined as the best compromise between the faults which can be allowed to remain in the yarn and the number of knots which have to replace disturbing yarn faults. Readers are instructed to refer the text book "Modern yarn preparation and Weaving Machinery" Ormerod for further details (Tables 5.1, 5.2).

Table 5.1   Norms for yarn clearer settings

Type of slub catcher	Yarn	Setting
Fixed blade	Combed (C) Karded (K)	1.5–2 D 2–2.5 D
Oscillating blade	C/K	25% more than fixed blade

Smooth edge		
Serrated heavy	C K	3–3.5 times diameter 3.5–4 D
Serrated light	C K P/C	4–5 D 4.5–5.5 D 5.5–6.5 D
Electronic yarn clearer	All	3 × 3 D

Table 5.2   Clearer efficiencies

Type of slub catchers	Clearing efficiency (%)
Fixed blade Serrated edge	2–10
Heavy weight	15–25
Light weight	25–35
Electronic yarn clearer (EYC)	70–80

1. 40's combed yarn is formed on rotoconer with fixed blade. How do you alter the setting if the yarn is to be processed on serrated, heavy slub catcher?

Solution :

$$\Rightarrow d = \frac{1}{28\sqrt{Ne}} \quad \text{Where} \quad Ne = \text{English count}$$

$$= \frac{1}{28\sqrt{40}} = 5.46 \times 0^8$$

Setting, $1.75 \times D = 1.75 \times 5.46$
$= 9.5 \times 10^{-3}$"(reader is instructed to convert to SI system).
For serrated heavy $= 3 - 3.5$
$= 3.25 \ D$
$= 3.25 \times (5.65 \times 10^{-3})$
$= 17.74 \times 10^{-3}$" (reader is instructed to convert to SI system)

2. A double-flanged bobbin is wound on a vertical spindle rotating at 900 rev/min. The centre has a diameter of 4 cm and the maximum bobbin diameter is 12 cm. The yarn is wound with a slow traverse. How does the winding speed vary from an empty to a full bobbin?

Solution :

In precision winding we know that winding rate at the empty bobbin:

$$\text{Winding rate} = \frac{900 \times \pi \times 4}{100} \cong 113 \text{ m/min.}$$

At the full bobbin:

$$\text{Winding rate} = \frac{900 \times \pi \times 12}{100} = 340 \text{ m/min.}$$

or: Winding rate at 4-cm diameter $\times$ 12 / 4 = 113 $\times$ 3 = 339 m/min.
The mean winding rate will be

$$\frac{900 \times \pi}{100} \times \frac{(4+12)}{2} = 226 \text{ m/min.}$$

Figure 5.21 illustrates the winding conditions in this example. If a constant winding rate were required, some mechanism would be necessary to reduce the spindle speed as the bobbin increases in diameter.

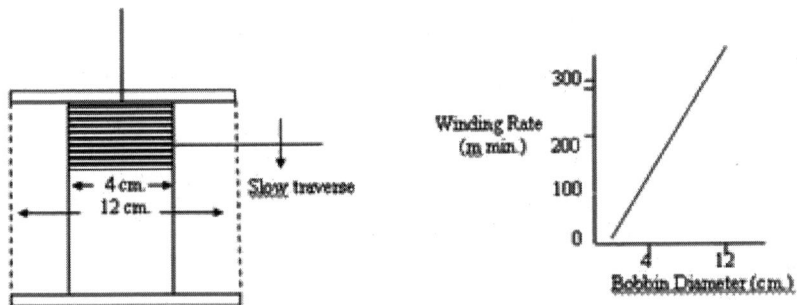

Figure 5.21   Winding on a double – flanged bobbin: relation between winding rate and diameter bobbin.

3. Referring to the previous example find the constant of the equation if it is desired to have a constant winding rate of 200 m / min.

Solution :

In precision winding, we know that, winding rev/min $\times$ bobbin diameter = constant.

Let us find the spindle speeds at different diameters like 4, 6, 8, 10 and 12 cm. The results are shown in graph.

The required spindle speed at 200 m/min on a 4-cm diameter bobbin is

$$\pi DN = \frac{200 \times 100}{\pi D} \quad \frac{200 \times 100}{\pi \times 4} = \frac{\text{winding rate in cm / min.}}{\text{bobbin circumference in cm.}}$$

= 5000/ π or 1592 rev / min.

As the product of winding rate in rev/min $\times$ bobbin diameter is a constant. We can write as

Constant = 1592 $\times$ 4 = 6368.

Similarly it can be written as:

Spindle speed at 4 cm = 6368/4 = 1592,

Spindle speed at 6 cm = 6368/6 = 1061,

Spindle speed at 8 cm = 6368/8 = 796,

Spindle speed at 10 cm = 6368/10 = 637,

Spindle speed at 12 cm = 6368 / 12 = 531.

By plotting these values we get the graph as seen in Figure 5.22.

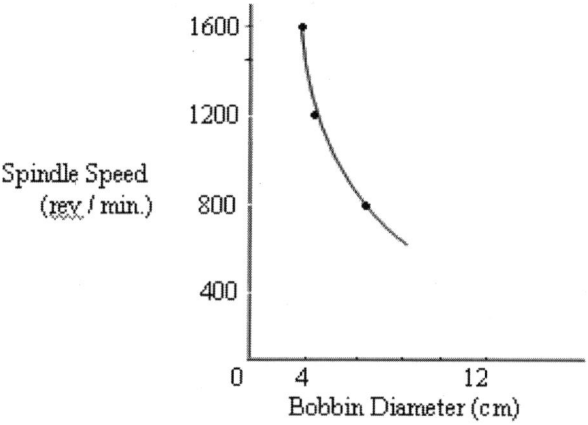

Figure 5.22   Spindle rev/min vs bobbin diameter

4. A cylindrical package is wound on a centre of 5-cm diameter. The spindle speed is constant at 3200 rev/min. If the traverse velocity is 205 m/min, determine

  (i) the net winding rate at the start of winding

  (ii) the net winding rate at a package diameter of 16 cm

  (iii) the angles of wind at the start and at a diameter of 16 cm.

Solution :

(i) The surface velocity is given by:

$$V_s = \frac{5 \times \pi \times 3200}{100} \cong 503 \text{ m/min}$$

The traverse velocity, V_t, is given as 205 m/min and thus,

tan θ = V_t / V_s = 205 / 503 = 0.3992, the angle of wind= 21°46′.

The net winding rate at 5 cm diameter = $V_{t/}$ sin θ

= 205/0.3708

= 543 m/min.

(ii) The surface velocity at 16 cm diameter = surface velocity at 5 cm × 16 / 5

= 503 × 16 / 5

$$= 1610 \text{ m/min.}$$

And the coil angle and angle of wind are given by

Coil angle, $\tan \theta$ at 16-cm diameter $= \dfrac{205}{1610} = 0.1274$, Angle of wind $= 7°16'$.

The net winding rate at 16-cm diameter $= V_t / \sin \theta$

$$= 205 / \sin 7° \ 16'$$
$$= 205 / 0.1265$$
$$= 1621 \text{ m/min.}$$

(iii) It is that the angle of wind decreases with an increase in package diameter and that the effect of the traverse velocity on the net winding rate decreases as the angle decreases. At the start, the surface velocity is 503 m/min and the net winding rate 553 m/min, a percentage increase of:

$$\frac{553 - 503}{503} \times 100 = \text{nearly } 10\%,$$

At 16-cm diameter, the percentage increase is:

$$\frac{1621 - 1610}{1610} \times 100,$$

i.e., only about 0.68% increase.

5. 40's combed yarn is formed on rotoconer with fixed blade. How do you alter the setting if the yarn is to be processed on serrated, heavy slub catcher?

Solution :

$$\Rightarrow \quad d = \frac{1}{28\sqrt{Ne}} \quad \text{where } Ne = \text{English count}$$

$$= \frac{1}{28\sqrt{40}} = 5.46 \times 10^8$$

Setting, $1.75 \times D = 1.75 \times 5.46$

$$= 9.5 \times 10^{-3}'' \text{ (reader is instructed to convert to SI system)}.$$

For serrated heavy $= 3 - 3.5$

$$= 3.25 \ D$$
$$= 3.25 \times (5.65 \times 10^{-3})$$
$$= 17.74 \times 10^{-3}'' \text{ (reader is instructed to convert to SI system)}.$$

6. Given that a 30-tex yarn requires a gap-setting of 0.3 mm, determine the setting for a 20-tex yarn.

Solution :

The required setting is given by:

$$0.3 \times \sqrt{\frac{20}{30}} = 0.3 \times \sqrt{0.67} \cong 0.25 \text{ mm.}$$

7. A mill has 3 different types of cone winders for producing cheeses of 150 mm traverse length. For all the package produce we wind the double traverse is 6. When the cheese diameter is 52.5 mm, (i) On machine "A" – the package is driven by contact with a rotary traverse drum. Workout the number of crossing in drum if drum diameter is 63 mm. (ii) On machine B – the package is driven by surface contact with a drum of 41 mm dia rotating at 2520 rpm. What is the speed of shaft? (iii) Machine "C" – is a precision winder in which the package is mounted on spindle rotating at a constant speed of 1500 rpm. Find the cam shaft speed.

Solution :

(i)   On machine A
Wind / double traverse

$$= \frac{\text{Number of revolutions of drum / double traverses}}{\text{Package diameter}}$$

$$6 = \frac{X \times 63}{52.5} = 5$$

⇒ number of revolutions drum / double traverse = 5
⇒ number of revolutions of drum / single traverse = 2.5
∴ Drum has 2½ crossings.

(ii)   For separate traverse machine like B

$$\text{Winds / D.T.} = \frac{\text{rpm of drum shaft}}{\text{rpm of cam shaft}}$$

$$6 = \frac{2520}{x} \text{ (Reader is instructed to complete the solution)}$$

(iii)   On precision winder "C"

$$\text{Winds / double traverse} = \frac{\text{Spindle speed}}{\text{cam speed}}$$

$$6 = \frac{2520}{x}$$

$x = 250$ traverses per minute.

8. In a cone winder, the drums of 95 mm ϕ with 150 mm traverse length for 2 term configurations.

Solution :

$$\alpha = 2\,tan^{-1}\left(\frac{150}{3.14(95)(2)}\right)$$

$$\therefore \alpha = 28.226$$

$$\therefore \text{ Length} = 2\,t\,/\,8\text{ m }\alpha/2 \quad \text{(back, formula)}$$

$$= 2(150/\sin(28.226)/2$$

$$= 1231.45 \text{ mm}.$$

The length of yarn wound on the drums on one double traverse given by

$$\frac{2t}{sin\,\alpha\,/\,2}$$

The number of diamonds formed on the periphery and per traverse are related with package diameter as follows:

$$\text{Package diameter} = \frac{\dfrac{\text{drum groove}}{\text{traverse}} \times \text{drum dia} \times \text{diamonds at periphery}}{\text{diamonds/traverse}}$$

9. Calculate the package diameter for 2 term configuration and with 95 mm ϕ with equal number of diamonds on periphery and per traverse.

$$\text{Package diameter} = \frac{2 \times 95 \times x}{x}\,mm = 190\ mm.$$

10. The cone winder is provided with a drum shaft running at 3500 rpm with 2½ crossings and 10 cm traverse length. The drum are 5 cm in diameter. Calculate the rate of winding and comment on the result. What is coil angle in this case?

Solution :

$\Rightarrow$ Diameter, $d = 5$ cm =   number of crossings = 2½

            S.S. = 3500 rpm = N

            T.L. = 10 cm $\therefore$ always take S = number of double traverse

Rate of winding = $\sqrt{(SS)^2 + (TS)^2}$

$$= \sqrt{(\Pi DN)^2 + \left(\frac{2LN}{S}\right)^2}$$

$$= \sqrt{(5 \times 3500 \times 3.14)^2 + \left(\frac{2(10)(3500)}{5}\right)^2}$$

= 56,705.40098 cm = 567.054 m/min.

∴ Tan θ = SS / TS

$$= \frac{S.\pi \ DN}{2LN} = \frac{5(3.14)(5)}{2(10)}$$

θ = 75.70°.

11. A mill wind chesses of 150 mm traverse length on cone winders which are rotated by surface conduct with 3 crossings. Drum diameter is 7.5 cm what is the coil angle. If identical cheeses are to be wound on machine with 7 cm drum diameter and a can operated traverse motion. The can shaft speed is 350 rpm. What is the drum shaft speed?

Solution :

Coil angle = SS/TS

= π DN/2LN

= 3.14 × 7.5 × 6 / 2 × 15

$$tan \ \alpha = \frac{SS}{TS} = \frac{\pi DN}{2xLxN \ / \ S}$$

∴ Coil angle, tan α = 4.71

= tan⁻¹(4.71)

α = 78.0°.

Because ideal package are made on both machine coil angle and traverse length remains constant.

$$\therefore \ \text{Coil angle} = \frac{\pi \times \text{Drum diameter} \times \text{drum shaft} \times \text{speed}}{2 \times L \times \text{Can shaft speed}}$$

$$4.71 = \frac{3.14 \times 7x \times x}{2 \times 15 \times 350}$$

⇒ x = 2250 rpm.

12. Consider a package produced in diameter of 30–250 mm. The drums are 75 mm in diameter and are with 2½ crossing examine the change in diameter and the wind per double traverse.

Solution :

$$\text{Wind per double travels} = \frac{\text{number of revolutions of drum} \times t_2 \ \text{drum diameter}}{\text{diameter of package}}$$

$$\text{Winds/double traverse at maximum diameter} = \frac{5 \times 75}{30} = 12.5 \ \text{coils}$$

$$\text{Winds/double traverse at maximum diameter} = \frac{5 \times 75}{250} = 1.5 \ \text{coils}$$

Wind per D.T.	Package diameter
12	31.25
11	34.09
10	37.5
9	41.67
8	46.87
7	53.57
6	62.5
5	75
4	93.75
3	125
2	187.5

D.T. = Doubt Traverse

It is clear that, as package diameter increases the number of winds per double traverse reduces maintaining a constant coil angle or angle of wind.

(*Note*: The reader is advised to plot the graph and find the values for any arbitory data.)

13. In a winding machine the sleeve is 5 cm in diameter and the package was wound-up to diameter of 25 cm. Find the stages of ribboning formation.

Solution :

The positions of formation of ribboning is 5 / 25, 10 / 25, 15 / 25, 20 / 25, and 25 / 25

14. A ring cop weighs 25 g made from 30's yarn. Estimate the length content if it is necessary to produce a cone weighing 2.85 kg. Calculate the number of cops required what is the running time of a cop if the rate of unwinding is 1000 rpm. Consider a loss of 10 s for doffing and donning. Estimate the efficiency of the process.

Solution :

$$\text{Count} = \frac{Length}{Weight}$$

$$30 = \frac{l}{25\ gms}$$

$$l = \frac{1.65 \times 840}{1.1} = 1262.6\ m$$

$$\text{Running time} = \frac{Length}{Speed}$$

$$= \frac{1262.6}{1000} = 1.26\ \text{min}$$

Number of cops required $= 2.85\ \text{kg} = \dfrac{2850}{25} = 114\ cops$

Weight of yarn = (Weight of cop + Yarn) – Weight of cop

We are assuming that the loss of waste is zero in practical. We have 0.05% waste.

$$\eta = \frac{RT}{RT + Stoppages} \times 100$$

$$\eta = \frac{1.262}{1.262 + 0.16} \times 100$$

$$\eta = 88.33\%.$$

15. A cone weighs 2.85 kg 30's and it is used to prepare beams 10,000 m length. Calculate the number of beams that can be produced from cone.

Solution :

$$2.85\ \text{kg} = 2850\ \text{g}$$
$$= 6.2816\ \text{lbs}$$
$$l = \frac{6.2816 \times 30 \times 840}{1.1}$$
$$l = 1,43,869.09\ \text{m}$$

Number of beams drafted $= \dfrac{1,43,869.09}{10,000} = 14.38$

$$\approx 14\ \text{beams.}$$

16. 20's and 30's yarns are converted to cones. On proto cone winders with speeds 580 m/min and 580 m/min. If the single thread strengths are 120 g and 125 g, what is the tension level you suggest.

Solution :

For 580 m/min speed = 1/10 × 120 = 12 g and for 850 m/min = ½ × 125 = 15.625 = 16 g approximately. The total tension range varies from 5–60 g.

17. Using the following data calculate winding speed, production /drum/ shift and production / operative / shift.

Diameter of the drum (D) = 3 1/8″, Number of slots = 2.5 (S), Speed (N) = 2000 rpm.

Count = 70's Length of the drum = 7″ (L), Shift = 7.5 h, Efficiency (η) = 80%

Solution :

$$\text{Winding speed} = \sqrt{(\pi\ DN)^2 + \frac{(2LN)^2}{25}}$$

$$= \left[(\pi\ \times 3.125 \times 2000\ )^2 + \left(\frac{4 \times 49 \times 2000 \times 2000}{25} \right) \right]^{1/2}$$

$$= 20417.92''/\text{min}.$$
$$= 567.16 \text{ rpm}.$$

Production / drum / shift at 100% η

= (Winding speed in rpm × 60 × 7.5) ÷ (840 × 2.2 × 0.5 count)

= (567.16 × 60 ×7.5) ÷ (840 × 2.2 × 70)

At 80% η

Production / drum / shift = 1.6 kg

30 drums are provided per operative.

∴. Production / operative / shift at 80% efficiency = 1.6 × 30 = 48 kg.

18. In a horizontal spindle winder the cam traverse speed is 200 m/min and the 20's yarn is wound on a double flanged bobbin with bare bobbin diameter D_o of 50 mm. The gearing arrangement in the machine run by 2 hp motor using the stepped pulleys gives the bobbin wheel speed as 2000 rpm. Find the rate of winding in the beginning and at double package diameter. Examine the effect of bobbin diameter on the coil angle and rate of winding.

Solution :

Bare bobbin dia is 50 mm / 5 cm

Speed of spindle – 2000 rpm

Traverse speed is 200 m / min.

We know that rate of winding = $\sqrt{(SS)^2 + (TS)^2}$

Case (i) when package dia is 5 cm

Surface speed $= \pi \times D \times N$
$$= 3.14 \times 5 \times 2000$$
$$= 31,400 \text{ cm/min}$$
$$= 31.4 \text{ m/min}$$

Rate of winding $= \sqrt{(31.4)^2 + (200)^2}$
$$= \sqrt{985.96 + 40,000}$$
$$= 202 \text{ m/min.}$$

And thus it proves that the winding belongs to show speed category

Coil angle $= SS/TS = 31.4/200$
$$= 83.60°.$$

Angle of wind $= 6.39°.$

Case (ii) Package dia is 100 mm.

(Note: Reader is directed to work out the above parameters and should plot rate of winding vs package dia and coil angle vs package dia.)

19. A Kamitsu winder is provided with 20 drums on a side. The drums are of bakelite type with 30 cm length and 10 cm diameter. The direct motor driving arrangement give the speed of 4000 rev/min. The drums are with three scrolls. Estimate the percentage of slippage if the speed is 1500 m/min.

Solution :

Traverse length $= 30 \text{ cm} = 0.3 \text{ m}$

Drum dia $= 10 \text{ cm} = 0.1 \text{ m}$

Speed of drum $= 4000$

Number of grooves $= 3$

Rate of winding $= 1000 \text{ m/min}$

Given scrolls $= 3$

i.e., $S = 6$

Let Y be the ratio of package surface speed to drum surface speed.

We know that, rate of wind $= \sqrt{(SS)^2 + (TS)^2}$

$$= \sqrt{(\pi D N Y)^2 + \left(\frac{2LN}{S}\right)^2}$$

$$= \sqrt{(3.14 \times 0.1 \times 4000 \times Y)^2 + \left(\frac{2 \times 0.3 \times 4000}{6}\right)^2}$$

$$= \sqrt{(1256Y)^2 + 160000}$$

$$= \sqrt{1577536Y^2 + 400}$$

$$= 1318.15Y + 400$$

Given speed is 1500 m/min.

Therefore,

$$1318.15Y + 400 = 1500$$
$$1318.15Y = 1100$$
$$Y = 0.875$$

or $1 - 0.875 = 0.125$

or slippage is 12.5%.

20. In a winding machine the rotating drum is 5 cm in diameter with a speed of 3000 rpm. On this machine a cone package is wound to a maximum diameter of 20 cm. The yarn traverse is by a cam arrangement with 320 rev/min. It is necessary to estimate the situations of ribbon formation when the speed of the package is 960 rpm with package diameter of 15 cm, making suitable assumptions, find the percentage of slippage.

Solution :

Given – Core speed 960 rpm

Core dia 15 cm

Generally it is known that somewhere across the length of the drum, a slip point exists.

$$\text{Slippage percentage} = \left(1 - \frac{\pi \times \text{Package dia} \times \text{speed}}{\pi \times Drum \text{ dia} \times \text{speed}}\right) \times 100$$

$$= \left(1 - \frac{\pi \times 15 \times 960}{\pi \times 5 \times 3000}\right) \times 100$$

$$= 4\%.$$

Drum of core = 20 cm (average value)

We know that patterning will occur when n / N is a whole number.

Given cam speed as 320 rpm.

$$r = 320 \times \text{"}w\text{"},$$

where w is a set of natural numbers.

Assume that the slippage between package and drum is remaining constant, we can write

$$320 \times w \times \text{package dia} = \text{Constant}$$
$$320 \times w \times d = 960 \times 15 = 14,400 = \text{Constant}$$

Let us substitute $w = 1$ in the above equation.

Then $w = 45$ cm which cannot be acceptable because 45 cm is greater than 20 cm (maximum diameter).

Therefore, for occurrence of patterning the package diameter from minimum to maximum dia of package can be found out by substituting value $w = 2, 3\ 4, 5$ .... and checking whether the answer is less than 20 cm.

For example if $w = 2$

$$320 \times 2 \times d = 14,400$$
$$d = 22.5 \text{ cm which is also not correct.}$$

If $w = 3$,

$$320 \times 3 \times d = 14,400$$
$$d = 15$$
$$d = \frac{14,400}{320 \times 4}$$
$$= 11.25$$

If $w = 5$ then $d = 7$

If $w = 6$ then $d = 7.5$

If $w = 7$ then $d = 6.42$ etc.

21. A lessona horizontal spindle winding machine winds 2/20's K yarn the coil angle at package diameter of 100 mm was found to be 30°. Determine the coil angle at package size of s 150 mm in diameter.

Solution :

We know that in precision winders

$$\text{Tan } \theta = \frac{SS}{TS} = \frac{SS}{\text{constant}} = \frac{\pi d N}{\text{constant}}$$

(where "θ" is coil angle)

We know that traverse speed = $2 \times L \times n$,

where "n" is number of DT/min,

N = Spindle speed

For a given machine, the values of "n",

"N" and traverse length remain constant.

$$\text{Tan } \theta = \frac{d}{\text{constant}},$$

Let θ_1, θ_2 be the coil angles at two different diameters then

$$Tan\ \theta_2 = \frac{d_2}{10 \times Tan\ 70}$$

$$= \frac{15}{10 \times 0.58}$$

$$\theta_2 = 68.86°.$$

22. A Cimmco – Muller cone winder has a bakelite drum of nearly 50 mm in diameter and produces 20's HK yarn with a coil angle of 60°. The wrap of yarn wound on the package for a single traverse is 2.5. Estimate the traverse length of the drum.

Solution :

Cimmco-Muller is a reverse traverse (RT) winder and

$$\text{Tan } \theta, \text{ Coil angle} = \frac{SS}{TS}$$

$$Tan\ 60 = \frac{\pi \times D \times N}{2 \times L \times \frac{N}{S}}$$

$$1.7320 = \frac{\pi \times 5 \times N}{2 \times L \times \frac{N}{5}}$$

$$1.7320 = \frac{\pi \times 5 \times 5}{2\ L}$$

$$2\ L = \frac{\pi \times 25}{1.7320}$$

$$L = \frac{\pi \times 25}{2 \times 1.7320}$$

$$L = 22 \text{ cm. (approximately)}$$

23. A horizontal spindle winder in a laboratory winds 10's single yarn with parallel close winds. The traverse bar moves through a distance of 200 mm in each traverse. The gearing arrangement gives a constant winding speed of 1000 m/min. The package size is nearly 100 mm and the spindle rotates at 3000 rpm. Work out the spindle speed when the package attains a diameter of 200 mm.

Solution :

Case (i): In precision winders it is known that as the package dia increases, the rate of winding also increase as the yarn will follow larger path. Let d_1 be the package diameter, "L" be the traverse length, "n" be the spindle speed, so that R/n is a constant, where "R" is the traverse speed.

$$R/n = \text{constant} = k \tag{1}$$
$$\text{Or } R = nk \tag{2}$$

We know that in precision winder

$$\text{Rate of winding} = \sqrt{(SS)^2 + (TS)^2}$$

Or (winding w speed)^2 = $(SS)^2 + (TS)^2$

$$= (\pi d_1 n_1)^2 + (2\ LR)^2 \qquad (3)$$

Now substitute for "R" from equation (2) in equation (3)

$$= (\pi \times d_1 \times n_1)^2 + (2 \times L \times n_1 K)^2$$

$$w^2 = n_1^2 \left\{ \left(\pi^2 d_1^2 + 4\ L^2\ K^2 \right) \right\}$$

Given winding speed as 1000 m/min.

$1000^2 = 3000^2 \{ (\pi^2 \times 0.1^2) + (4 \times 0.2^2 \times K^2) \}$

$K^2 = 0.0776$.

Case (ii): When package size is 20 cm or 0.2 m

Winding speed^2 $= n_2^2 \left\{ \left(\pi^2 d_2^2 + 4\ L^2\ K^2 \right) \right\}$

Substituting all values

$n_2 = 1567$ rpm. (Ans)

24. 20's cotton yarn was prepared on cheese of 15 cm traverse length and 3 as wind per single traverse. Work out the package maximum diameter if the coil angle is 82°.

Solution :

We know that in precision winder, as the package dia increases, the coil angle will increase or angle of wind decreases. Let "α" be the angle of wind and "θ" be the coil angle.

Given wind / dt = 6, L = 15 cm = 150 mm

$$\text{Tan } \alpha \text{ min} = \frac{TS}{SS} \quad (\alpha_{min} \text{ occurs at } d_{max} \text{ as explained above})$$

$$= \frac{2\ LR}{\pi \times d_{max} \times n}$$

$$= \frac{2 \times 15 \times R}{3.14 \times d_{max} \times n}$$

Given coil angle as 82° or angle of wind as 8°.

$$\text{Tan } 8 = \frac{30 \times 1}{\pi \times d_{max} \times 6} = \frac{5}{\pi\ d_{max}}$$

Or $d_{max} = \dfrac{5}{\pi \times Tan\,8} = \dfrac{5}{3.14 \times 0.1405}$

= 11.34 cm.

25. In winding department of a weaving mill, two machines are used for producing a cheese of 40's cotton K yarn. The package is with 15 cm traverse length. The minimum diameter and maximum diameter is 5 cm and 12.5

cm, respectively. One of the two machines is RT winder with 2.5 scrolls of a stainless steel drum of 7.5 cm. The other machine is precision winder with constant spindle speed and spindle move three times per single traverse. Estimate the traverse ratio for package diameters of 5 and 10 cm for the first machine. What will be the coil angle at 5 and 10 cm of package diameter if second machine is considered?

Solution :

Given $d_0 = 50$ cm; $d_{max} = 12.5$ cm

   Case (i): Drum size = 7.5 cm in diameter

   Number of winds per double traverses is 5.

$$\text{We know that wind/dt = Number of Winds/dt} \times \frac{\text{Drum dia}}{\text{Package dia}}$$

$$= 5\,p \times \frac{7.5}{5} = 7.5$$

When dia is 5 cm,

and $5 \times \dfrac{7.5}{10} = 3.75$ when package dia is 10 cm.

5.7   Gain mechanism

Gain is a mechanism employed to overcome ribbon form. Generally the machine is adjusted to the drum diameter which is not a whole number (Figure 5.23a and b). In precision winding with double traverse, when parallel close coils are laid, ribbons can be never prevented because n / N always remains constant i.e., number of coils laid /double traverse remains constant but angle of wind increases as package diameter increases lent in drum wound packages or cone winding situation is just reverse, as package decreases, its rpm decreases.

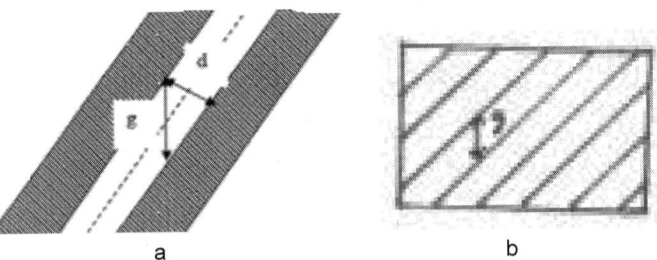

a   b

Figure 5.23 a, b   Concept of gain in cone winding

   Therefore package speed to drum speed is altered and thus number of winds/double traverse decreases, by keeping coil angle constant in cone winding gain machine is used through ribbon breakers or package breakers.

"*g*" is known as linear gain.

$$g = \frac{d}{sin\theta}$$

d = yarn diameter

θ = angle of wind or (90 – coil angle)

$$\text{revolutions gain} = \frac{g}{\text{circumference}}.$$

1. A precision winder produces a coarse count of 4's cotton with traverse length of 200 mm with a traverse ratio of 2.5. Estimate the stage of ribboning if the package attains the size of 100 mm in diameter. Note that the packing fraction for the material is 0.6.

Solution :

Given wind/*dt* $=\frac{5}{2}=2.5$ and single traverse = 1.25, the traverse length of 200

mm or 20 cm will have 1.25 warps of yarn of 4's and thus a space of 16 cm is occupied by the coil or wrap.

$$\text{Coil angle} = \frac{\pi \times 10}{16}$$

$$\theta = 63°.$$

Now it is necessary to calculate gain which is given by

$$\text{Linear } g = \frac{d}{sin\,\theta}$$

$$d = \frac{1}{28\sqrt{Ne}} \underset{=}{\frac{1}{28\sqrt{4}}} \times 2.54 = 0.04535\,cm$$

$$\therefore g = \frac{0.04535}{sin\,63} = \frac{0.04535}{0.89100}$$

$$= 0.0508.$$

$$\text{Revolutionary gain} = \frac{g}{\pi d} = \frac{0.0508}{\pi \times 10}$$

Therefore patterning can be avoided by using traverse ratio of $\frac{5}{2} \pm 0.00162$

$$= 2.501 \text{ or } 2.498.$$

2. In a warp winding section a bakelite drum of 150 mm traverse length produces cheeses of 20's HC yarn are produced with 50 mm diameter. Workout the approximate value of traverse ratio to 3 to prevent ribboning effect.

Solution :

Using Pierce formula

$$\text{Yarn dia (inches)} = \frac{1}{28\sqrt{Ne}}$$

$$\text{Yarn dia (cm)} = \frac{1}{28\sqrt{Ne}} \times 2.54$$

$$= \frac{1}{28\sqrt{20}} \times 2.54 = 0.02 \text{ cm}$$

There is equal to linear gain and revolutionary gain $= \dfrac{LG}{\pi\, d} = \dfrac{0.02}{\pi \times 5}$

$$= 0.0013$$

Therefore the nearest value to present patterning of is 3 ± 0.0013.

3. We know that in precision winder, as the package dia increases, the coil angle will increase or angle of wind decreases. Let "α" be the angle of wind and "θ" be the coil angle.

Assume wind $/ dt = 6$, $L = 15$ cm $= 150$ mm

$$\text{Tan } \alpha \text{ min} = \frac{TS}{SS}$$

$$= \frac{2\, LR}{\pi \times d_{max} \times n}$$

$$= \frac{2 \times 15 \times R}{3.14 \times d_{max} \times n}$$

Assume coil angle as $82°$ or angle of wind as $8°$.

$$Tan\, 8 = \frac{30 \times 1}{\pi \times d_{max} \times 6} = \frac{5}{\pi\, d_{max}}$$

Or $$d_{max} = \frac{5}{\pi \times Tan\, 8} = \frac{5}{3.14 \times 0.1405}$$

$$= 11.34 \text{ cm.}$$

4. Given bare bobbin diameter $d_0 = 50$ mm; $d_{max} = 12.5$ cm, drum diameter 7.5 cm

Case (i): Drum size $= 7.5$ cm in diameter

Number of winds per double traverses is 5.

We know that wind/dt = Number of winds/dt $\times \dfrac{\text{Drum dia}}{\text{Package dia}}$

$$= 5 \times \frac{7.5}{5} = 7.5$$

When dia is 5 cm. and $5 \times \dfrac{7.5}{10} = 3.75$ when package dia is 10 cm.

Case (ii): Angle of wind Tan $\alpha = \dfrac{TS}{SS}$ or

Coil angle Tan $\theta = \dfrac{SS}{TS}$

Now $Tan\ \alpha_1 = \dfrac{2\,L\,R}{\pi\,d\,n}$ and $\dfrac{n}{R} = 6$

$$= \frac{2 \times 15 \times 1}{3.14 \times 15 \times 6} = 0.3184$$

$$\alpha_1 = 17.67°$$

$$Tan\ \alpha_2 = \frac{2 \times 15 \times 1}{\pi \times 10 \times 6} = \frac{2 \times 15 \times 1}{3.14 \times 10 \times 6}$$

$$\alpha_2 = 9.05°.$$

5.8   Theory of yarn unwinding

In high speed winding, yarn is invariably withdrawn over the end from the ring-tubes before passing through a guide located at a suitable distance above the tube. As the yarn is withdrawn from the ring tube at a certain linear velocity, the portion of the yearn between the guide and the winding-off position at the new also rotates around the tube at a certain angular velocity and this causes the yarn to fly out to form a characteristic balloon as shown diagrammatically in Figure 5.24. The size and shape of the balloon change with the continuous movement of the winding-off position and this directly affects the yarn unwinding tension. The theory of year ballooning obviously applies not only to unwinding from packages as discussed above but also applicable to the winding yarn in spinning. In general, theoretically work has been carried out firstly with reference to spinning and then extended to other ballooning aspects. In 1910, Linder derived the following simple equation for the yarn tension in a narrow balloon considering only the centrifugal forces in the portion of yarn forming the balloon.

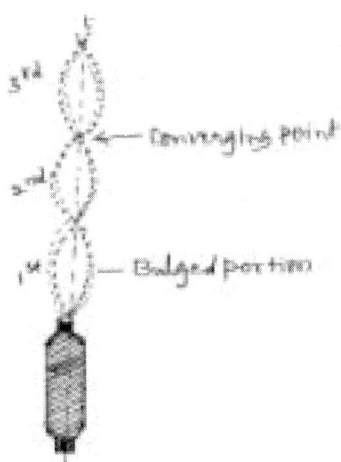

Figure 5.24 Yarn unwinding

$$T_x = \frac{0.112 \, n^2 \, NH^2}{10^3 \, (C)^2} \tag{1}$$

T_X = Vertical component of the tension in g, N = Balloon rotational speed in thousands,

H = Balloon height in cm, N = Yarn tex., C = Constant of centrifugal force.

In 1934 Greshin published work on the same problem taking into account both air drag and coroilis forces due to the linear yarn velocity with reference to balloons formed in ring spinning. Equation (1) is the basis for both Linder's and Grishin's work, but they differ in the value to be given to C. Greshin later extended his work to the case of ballooning during over-end winding from a conical shaped stationary packages at other than low speeds, and derived the following equation for a vertical component of the yearn tension.

$$T_x = mv^2 \left[2 + K \, (H)^2 \, Sin^2 \, \beta \right] \tag{2}$$

Where m = Mass per unit length, v = Linear velocity, β = Coil angle, H = Balloon height, r = winding-off radius and the co-efficient K depends upon the conditions of un-winding, mainly, on the drag of the yarn on the package (its value in practice varies between 0.83 and 0.22).

Padfield investigated the fluctuations in tension and thread shape occurring when unwinding from a package over end. Initially her results were obtained for cheeses and were then extended to cover unwinding from conical packages. The relation between tension at the guide eye and the balloon height for constant air-resistance (which depends on the yarn diameter and mass per unit length, the radius of the packages at the wind-ing-off point and the yarn speed) and unwinding angle was represented by the formula.

$$T_o = \frac{mv^2}{g}\left[A + B\left(\frac{H^2}{r}\right)\right] \qquad (3)$$

Where m = Mass per unit length in g/cm, T_o = Tension in g.wt, v = Linear velocity, g = 981 A = Constant depending on the value of air-resistance, B = Constant depending upon the unwinding angle, H = Balloon height, r = Un-winding radius.

The winding angle is assumed to be positive when the balloon is increas-ing and negative when it is decreasing. Equation (2) and (3) reveal the fol-lowing fundamental relationships.

a) Tension varies directly with yarn mass per unit length that means it varies directly with the direct or tex count or inversely with the indi-rect count. So as the yarn becomes coarse, the tension increases.

b) Tension varies as the square of the unwinding speed.

   The experimental work carried out by Brunschwailer and Moham-modin confirms the theoretical relationship.

c) The yarn tension during withdrawal at guide increases only when the balloon height becomes large in comparison with the winding-off radius. Both Greshin's and Padfield's equations are suitable when a single loop exists. But these equations do not indicate regarding the tension for multi-loop balloon.

Assuming that for n number of loops, the loop height h is a given by:

$$h = \frac{H}{n} \qquad (4)$$

A straight substitution of h and H in the Greshin's and Padfield's unwind-ing formula for tension gives an approximate value for the unwinding ten-sion for a multi-loop balloon. Other factors which affect yarn tension to some extent are coil-angle and conditions of unwinding such as drag of the yarn with the air or the package and these conditions are difficult to define quan-titatively. A number of papers have been published dealing with the tension variations during winding. Brunscheweiler and Mohammodin carried out a thorough investigation to demonstrate the unwinding behaviour and balloon characteristics during unwinding over and from ring-tubes. The major part of the following discussion has been abstracted from the work done by Brun-schwiler and Mohammodin and deals with short-term variation in cop and roving and combination build packages, long-term variations of cop, roving and combination build ring-tubes, methods of determining the optimum position of the package in relation to the guide.

5.8.1   Short-term variations

Short-term variation is the variation in unwinding yarn tension during the movement of the winding-off point from nose to base and vice-versa. With cop build packages, it has been observed that the maximum tension occurs when winding-off at the nose and minimum tension when unwinding at the shoulder or base. At both these positions, same number of loop exists. The balloon height is less at the nose than at the shoulder but tension is higher at the nose because of smaller winding-off radius. As the winding-off point moves from nose to shoulder, the increase in height is off-set by larger unwinding radius and according to the equations 2 and 3

$$\left(\frac{H^2}{r^2}\right) nose \rangle \left(\frac{H^2}{r^2}\right) base \tag{5}$$

The short interval between base to nose than between the nose to the base is because during spinning cop build ring tube, usually the lifter rail is made to descend quickly and raise slowly which helps mainly to overcome sloughing-off. With the roving build, the winding-off radius remains same during one traverse and as the yarn unwinding position moves almost the full length of the traverse, the variation is over a larger period of time at the same unwinding speed. As the winding-off point descends from top to bottom the frictional drag of yarn on yarn prevents increase in the number of loops though the height of the balloon increases and thus the yarn tension is maximum while unwinding from the bottom of the traverse. The short-term variation for combination build was not recorded by the authors, but is expected that the results will be a mixture of cop build and roving build.

5.8.2   Long-term variations

Long-term variation is the variation in yarn tension during the unwinding of a complete package. The experiments were made on three different builds used on ring-tubes (i.e., cop, roving and combination) at short-, medium- and long-guide distances of 3.8, 15.3 and 45.9 cm, respectively. The results reveal some interesting features not only within and between the packages but also regarding the guide distance settings. In general, it has been found that the tension is low at the start and as the unwinding continues, the average tension as well as the range of short-term tension variations increase and are maximum when unwinding last few layers from the package. The depth of the trace indicates the altitude of the range of variation. When the guide distance is only 0.8 cm, it has been found that, for all the builds, a single loop exists throughout the winding of the packages and average peak tension is lowest when unwinding at the start and rising slowly and steadily until it reaches its maximum value when the tube is just exhausted. For example, a yarn of 29.5 tex fibre, 17.8 cm lift and winding speed of 618 m/min the maximum tension is about 30 g. This indicates that a far greater number of tension variations per unit time occurs for the cop build than for the other builds. With a medium guide distance of 15.3 cm for a cop build package, at the start of unwinding of four loop balloon is formed and winding tension is very small with practically no variations. As the balloon height is increased with unwinding, first four loop balloon changes into a three loop balloon and further unwinding will reduce the number of loops of two and finally one loop, is formed. Associated with each change in the number of loops, there is an increase in the yarn tension proportionately to the same extent. For example, the change from a three loop to two loop balloon means there is an increase in loop height to 50 approximately with a corresponding increase in tension. According to Bruncheweiler and Mohammodin, theoretically an increase in the number of loops as given by the equation.

$$n \geq \frac{EH}{\pi} + T_X \tag{6}$$

where n is the number of loops, H is the balloon height and E is a parameter which includes the effects of centrifugal force, air-drag, angular acceleration and coriolis force. The approximate value of E is determined by Greshin in the simplified form.

$$E \geq \frac{1}{r} \sqrt{\frac{mu^2}{T_x - mu^2} + \left(\frac{mu^2}{T_x - mu^2}\right)^2} \tag{7}$$

where r = Winding-off radius, m = Yarn mass per unit length, u = Linear velocity of yearn, T_x = Vertical component of yearn tension.

5.8.3   Unwinding tension

a. Variation in tension during unwinding of a bobbin

During winding, there will be formation of balloon as the yarn unwinds from ring cop. This however depends on several factors like guide distance, type of guide, angle of inclination of cop, lift of cop, count of yarn, etc. A formulae can be expressed as

$$T\,\alpha\,\frac{(\text{Winding speed})^2}{N_e}\left[\frac{A}{\sqrt{n}}\div\frac{B\,h^2}{r^2}\right].$$

Where "n" is the number of loops of height "h", H – balloon height $\left(=\dfrac{H}{n}\right)$, r – bobbin radius, $A\,\mu\,B$ = constants.

The number of loops "n" will increase with coarser count. Generally the larger the loop, the more is unwinding tension and vice-versa. The values of $A\propto B$ depends on air drag and coil angle, tapering angle of bobbin, etc. It is generally found that higher the speed higher the unwinding tension and harder the package. This is due to the fact that the unwinding point moves away from the fixed guide. It is found that when the neck of the balloon touches the tip of bobbin, the tension will further increase and leads to unstability of bobbin. Further increase in speed, the balloons will become larger in diameter and loop height and thus the number of balloon loops are reduced and this will continue till the bobbin is exhausted. The tension variations during unwinding can be understood through short-, medium- and long-term variation. The short-term variation is caused by continuous movement of winding-off position from the nose to shoulder, with increasing winding diameter and balloon height. The former reduces the level of tension while the later increases the tension level in yarn. Of the two effects of change of the unwinding-off diameter is much more than that of the bobbin height with maximum tension within a chase, therefore occurs at maximum unwinding from the nose and the minimum when unwinding from the shoulder. In the final stages of unwinding the diameter of the shoulder goes on decreasing from one chase to the next lower. Thus there will be a rapid increase in the tension from chase to the next unless suitable care is taken. In one of the study, the path of tension was traced for different guide distances. It was observed that the number of balloons formed and their size (width) was dependent on guide distance.

b. Selection of suitable guide distance

The position of the ring cop, distance between the bobbin tip and guide etc., will be controlling the yarn tension. Therefore it is necessary to select a proper guide distance so as to maintain low level of increasing tension with balloon parameters and guide distance (GD), and also to minimise the tension variation from start to finish of ring cop. The guide distance known as optimum guide distance is obtained as follows:

$$n=\frac{G+L}{L+25}$$

Where "n" is the number of balloons,

G – Guide distance selected, L – lift of bobbin.

c. Unwinding accelerator

The tension variation will mainly depend on the size of balloon and number of bobbin. Both these can be controlled by using unwinding accelerator. This device will be placed usually 30–35 mm above the bobbin. This device breaks the balloon and thus reduces the unwinding tension.

5.9   Warping

5.9.1   Beam warping

The preparation of warp yarn is more demanding and complicated than that of the filling yarn. Each spot in a warp yarn must undergo several thousand cycles of various tensions applied by the weaving machine. In general terms, warping is transferring many yarns from a creel of single-end packages forming a parallel sheet of yarns wound onto a beam or a section beam. Warping aims to produce a quality beams and todays modern weaving demand for quality beans which is possible only when the parameters of warping are understood thoroughly. The defects occurred in the warping process will be detrimental to sizing and weaving process. Therefore the quality of warping up to a great extent influences the productivity of the weaving machine and the quality of fabric produced. The warping process is the last weaving preparation process in which tension of individual warp yarn can be controlled. The variation in tension of individual or group of yarns in warping cannot be eliminated in subsequent processes. For instance, total tension of the warp sheet should be increased in order to increase the tension of warp yarns which have low tension. However, if gaiting tension of some warp ends or sections in the weaving machine is different from the required tension, then warp ends already having adequate tension are changed. This in turn will result in woven fabric with variations in pick density, appearance, cover factor and impaired in quality due to high warp breakages.

5.9.2   Sectional warping

The sectional warping process is intended to produce wrapper's beam for stripe and check fabrics, high quality fabrics with short runs and warp yarns for which sizing is not required. Therefore sectional warping process is a vital technological process for the formation of high quality fabric. However warp yarn tension of a wrapper's beam produced by sectional warping is having more tension variations than a back beam produced by direct warping. In sectional warping certain number of ends available on the creel as a section is first wound onto an intermediate drum. Each yarn from supply packages placed at different heights and different distances from the drum is gaited through a balloon controller device, two zone tension device, various guiding elements, warp stop motion device, yarn distributer plate and reed. This number of ends is only a fraction of the total number of ends necessary on the wrapper's beam. After winding the required length in section, it is cut and tied up onto the upper surface of the section. Ends coming from the creel are used to wind the second section adjacent to the first section. The process is continued until the required number of yarns for the wrapper's beam is wound on to the drum. Then all sections on the drum are transferred on to an empty wiper's beam.

Yarn tension in sectional warping – To wind yarns onto the drum a compact package with some tension in yarn winding, is required. However excessive tension when winding impairs physical and mechanical properties of warp yarns, numerous number of breakages in subsequent processes can occur. Low tension in winding of the warp can cause unstable bulky package and it causes to disturb the proper process of fabric formation. Uneven warp yarn tension on the wrapper's beam leads to create uneven warp winding density and uneven lengths of warp yarns which can generate high waste and breakages in sub sequence processes. In addition, cone or cheese packages are placed on a creel at different geometrical positions according to the axis of the warping drum. As a result, yarn length from the packages to the warping drum varies. Therefore, tension of yarns can be different from package to package and the appearance of stripe and checked fabrics made out of such a wrapper's beam would be impaired and to have slight variations in colour for dyed fabrics. Further, uneven warp yarn tension could lead to uneven pulling force to rotate back beams, wrapper's beam or warping drum. When the warp package rotates, the maximum loads are imposed on warp yarns which are already under tension for unwinding of warp yarns from back beams in the sizing creel, weaver's beam in weaving and sectional warping drum in sectional warping. In the meantime low loads are imposed on less tensioned warp yarns and those yarns could even be stress relaxed.

5.9.3   Calculations in sectional warping machine

Find the speeds of all members of sectional warping machine
   Drive to intermediate pulley
         Speed of motor = 940 rpm.
         Diameter of motor = 7
         Diameter of intermediate pulley = 49
         Speed of intermediate pulley = $\dfrac{940 \times 7}{49}$

                        = 134 rpm.
         Speed of 21 teeth pulley = 134 rpm.

Speed of drum $= \dfrac{134 \times 106}{50} = 42.8\, rpm.$

Surface speed of drum $= 42.8 \times 3.14 \times 2 = 268.7$ rpm.

Speed of 48 teeth wheel $= \dfrac{134 \times 21}{48} = 58.6\, rpm.$

Speed of 21 teeth wheel = 58.6 rpm.

Speed of 47 teeth wheel $= \dfrac{58.6 \times 21}{47} = 26.19\, rpm.$

Speed of beam = 26.19 rpm.

Surface speed of beam = $3.14 \times 30 \times 26.18$ $(d_o = 30$ cm$)$
$$= 2466.15 \text{cm/min.}$$
$$= 4110.2 \text{cm/min.} \ (d_m = 50 \text{cm})$$

Drum speed = 42.88

Worm shaft speed $= 42.88 \times \dfrac{25}{26} = 41.23\, rpm.$

Sped of length measuring $= 41.23 \times \dfrac{1}{1.02} = 0.40\, rpm.$

Traverse speed

Speed of drum = 42.88

Speed of 20 teeth wheel $= 42.88 \, \dfrac{21}{20} = 45.024\, rpm.$

Speed of traverse $= \dfrac{45.024 \times 40}{20} = 90.048\, rpm.$

Speed of 24 teeth wheel $= \dfrac{90.048 \times 1}{24} = 3.752\, rpm.$

Pitch = 0.15″

Traverse speed = $3.752 \times 0.15 = 0.501$ rpm.

Without pattern

Width of section $= \dfrac{Creel\ Capacity}{Re\,ed\ Count}$

Width of warp = 80″

For 2″ number of dents = 40

Width $= \dfrac{100}{40} = 2.5″$

Number of sections $= \dfrac{80}{2.5} = 32$ section

Maximum number of sections = 32

Tan θ = 1.5 / 0.5 = 71.56°

Depending on θ = 71.56° type of material depends.

1. A rectangular creel with a capacity of 500 is used with drum driven wrapper for warping 19.68's or nearly 30 tex yarn for a warp width of 1400 mm. If D_0 and D_{max} of the wrapper beam is 0.3 and 075 m work out the set length and weight of warp wound on beam.

Note: beam density is 0.4 g/cm³. The tare weight of the beam is 250 kg.

Solution :

Width of warp on beam = 1.4 m, = 140 cm, Number of end used in creel = 500, Count = 30 tex, Empty beam diameter (d) = 30 cm, Full beam diameter (D) = 75 cm,

volume of yarn on the beam = $\dfrac{\pi}{4}(D^2 - d^2) \times L$

$$= \dfrac{\pi}{4}(75^2 - 30^2) \times 140 \text{ cm}^3$$

$$= 5,19,541 \text{ cm}^3$$

And thus mass of yarn = volume × density

$$= 5,19,541 \times 0.4 = 2,07,816 \text{ g}$$

$$= 207.82 \text{ kg}$$

Total weight of the beam = 207.82 + 250 = 457.82 kg

Therefore, mass of single yarn = $\dfrac{207816}{500}$ = 415.63 g

And thus length of warp = $\dfrac{415.63}{30}$ km = 13.85 km.

2. A weaving preparatory section has a beam warping machine running at 1250 rpm. The set length is fixed by full beam stop motion. The tear weight of beam is 180 kg. Width of each wrapper beam is 150 cm. V-creel is used. Yarn processed is 40's K. Beam ruffle diameter 25 cm. D_{max} is 65 cm. Beam density 0.38 g/cm³, workout necessary statistics. Time to mend break is 30 s cone used is 2.35 kg. Assume a total loss of electrical failure loss time as 30 s, total mechanical failure loss as 28 s. Assume one yarn break / 30,000 m. time lost in beam doffing and donning 30 s.

Solution :

Total yarn content on each cone.

$$C = L/W \qquad 40 = \dfrac{L}{2.35 \times 2.205}$$

$$L = \dfrac{207.27 \times 840}{1.1} = 1,58,279 \text{ ms.}$$

Volume of yarn on beam = $\dfrac{\pi}{4}(65^2 - 25^2) \times 150$

$$= 4,23,900 \text{ cm}^3$$

Yarn weight in kg = volume × density

$$= 4,23,900 \times 0.38$$
$$= 1,61,082 = 161.082 \text{ kg.}$$

Weight of single yarn (max. creel capacity) $= \dfrac{1,61,082}{540} = 298.3\,g.$

Therefore set length of each beam $= \dfrac{298.38}{14.76\ tex} = 20.215$ km.

Number of beams doffed from a cone = 20,215 m.

assuming zero wastage $= \dfrac{1,58,279}{20,215} \cong 8$ beams (approximately)

Running time of beam $= \dfrac{20,215}{1250}$

$$= 16.17 \text{ min.}$$

Time lost in mending breaks = Total number of breaks × Time to mend a break.

Total number of breaks / beam = 0.67

∴ Time lost in mending breaks = 0.67 × 30   (For 30,000 – 1

$$20,215 - ?$$
$$= 0.67)$$
$$= 20.1 \text{ s}$$

$$20.1 + 30 + 30/8 + 28/8 = 20.1 + 30 + 3.75 + 3.5$$
$$= 57.35 \text{ s or } 0.96 \text{ min.}$$

Running η / beam $= \dfrac{\text{Running Time}}{\text{Running Time + stoppages}}$

$$= \dfrac{16.17}{16.17 + 0.96} = \dfrac{16.17}{17.13} = 94\%$$

3. A pattern consists of 26 red, 18 white, 26 green, 18 white, 26 yellow, 18 white of 2/20ˢ *Ne* yarn. The sectional warping drum is 110 cm in diameter and has in line at 20°. Width of drum is 83″ and its customary to use all the 32 hooks across the drum for warping. The density is measured as 600 kg/m³. Maximum depth of yarn on beam is 15 cm. Work out necessary statistics.

Solution :

(i) First calculate the section width

$$= 26 + 18 + 26 + 18 + 26 + 18 \text{ (as all the yarns of 2/20's yarn)}$$
$$= 132 \text{ threads/section}$$

$$\text{Width occupied} = 132 \times \frac{1}{28\sqrt{10}} = 1.49''$$

(ii) Width of warp on drum $= 1.49 \times 32$

$$= 47.70'' \ (1.192 \ m)$$

(iii) Reed count = Creel capacity/width of each section

$$= 132 \, / \, 1.49 + \text{Ans} / \, 2''$$

$$= 44\text{'s Reed}$$

(iv) Total number of ends $= 132 \times 32 = 4224$

(v) $D_0 = 1.5 \ m,\ D_{max} = d + 2h = 1.1 + 2 \times \dfrac{15}{100} = 1.4 \ m.$

$$\text{Volume of yarn} = \frac{\pi(D_{max}^2 - D_0^2) \times \text{ Width of warp}}{4}$$

$$= \frac{\pi(1.4^2 - 1.1^2) \times 1.192}{4}$$

$$= \ 0.7006 \ m^3$$

$$\text{Mass of yarn} = 0.7006 \times 600$$

$$= \ 420.36$$

$$\text{Mass of single yarn} = \frac{420.36 \times 1000}{4224}$$

$$= 99.51 \ g$$

$$\text{Length} = \frac{99.5}{59.05} = 1.685 \ km.$$

$$= \ 1685.33 \ m.$$

$$\text{Tan} \ \theta = \frac{h}{H}$$

$$\text{Tan} \ 20 = \frac{15}{x}$$

$$x = \frac{15}{Tan \, 20} = \frac{15}{0.3639}$$

$x = 41.22 \ cm$

5.9.4   Sectional warping

Formulae

1. Total number of ends.   ends/inch $\times$ body width $+$ Number of selvedge

$$\text{Yarn diameter} = \frac{1}{28\sqrt{Ne}}$$

$N_e \rightarrow$ English count

$$\text{Cotton count} = \frac{590.5}{tex\ count}$$

2. Reed count $= \dfrac{end\ /\ inch}{1 + weft\ crimp}$

3. Creel capacity = Reed count × width of section

4. Total number of sections $= \dfrac{\text{width of warp on beam}}{\text{width of one section}}$

   $= \dfrac{\text{width in reed} + 1''}{\text{width of one section}}$

5. Weight of warp $= \dfrac{\text{Total number of ends} \times \text{warp length} \times 1.1}{840 \times \text{count} \times 2.205}$

6. Warp length = C.L. [H warp crimp]

7. Width in reed = C.W.[HC$_2$]

8. Weight of weft $= \dfrac{PPI \times \text{Reed width} \times \text{warp length} \times 1.1}{840 \times \text{count} \times 2.2}$.

Problems

1. Calculate the weight of warp and weft for a sheeting sort with the following data: warp and weft count 2/20's. Width in reed 9″ ends/inch = 44, weft contraction excepted 10%. Number of selvedge ends 32 on each side. Warp length 600 m. Also calculate the number of sections, width of section and total number of ends in each section.

Solution :

$$\text{Reed count} = \frac{\text{ends/inch}}{1 + \text{weft crimp}} = \frac{44}{1 + 0.1} = 40^s$$

$$\text{Width of each section} = \frac{\text{creel capacity}}{\text{Reed count}} = \frac{144}{40} = 3.6''\ \text{[Assuming creel}$$

capacity 144]

Width of warp on beam = width in reed + 1″ = 59″ + 1″ = 60″

$$\text{Total number of sections} = \frac{\text{width of warp on beam}}{\text{width of each section}} = \frac{60}{36} = 16.66''$$

Total number of ends = end/inch × width in reed + selvedge ends
$$= 44 \times 59 + 64 = 2660.$$

Weight of warp

$$= \frac{\text{Total number of ends} \times \text{Length of warp} \times 1.1}{840 \times \text{count of yarn} \times 2.205}$$

$$= \frac{2660 \times 600 \times 1.1}{840 \times 10 \times 2.205} = 94.78 = 95 kg$$

$$\text{Weight of weft} = \frac{2 \times 60 \times 600 \times 1.1}{840 \times 2.2 \times 10} = 111.17.$$

2. A sectional warping pattern consists of 26 threads white, 18 threads red, 26 threads white, 18 threads green, 26 white, 18 blue. If count selected for all these colour is 2/20 and the width selected is 36^x m beam, 40 ends/inch are expected in grey state with weft contraction around 10%. Calculate the weight of each colour required to produce 250 towels of 1.5 m length in finished state with warp contraction about 6%.

Solution :

Total length of 250 towels = 250 × 1.5 = 375 m.

Taking 6% warp contraction into consideration warp length = 375 (1 + 0.06) = 397.5 m

= 400 m

Width of each section

26 white	= 0.587
18 red	= 0.406
26 white	= 0.587
18 green	= 0.406
26 white	= 0.587
18 blue	= 0.406
132	2.98″ ≈ 3″

Number of sections = 36/3 = 12 sections

$$\text{Diameter or space} = \frac{1}{28\sqrt{Ne}} = \frac{1}{28\sqrt{10}} = 0.011''$$

Space occupied by each thread = 0.011 × 12 = 0.0225″

Weight of each colour

1. White $= \dfrac{78 \times 12 \times 400 \times 1.1}{840 \times 2.2 \times 10} = 22.28 kg$

2. $Red = \dfrac{18 \times 12 \times 400 \times 1.1}{840 \times 2.2 \times 10} = 5.1 kg$

3. Green = 5.1 kg
4. Blue = 5.1 kg

3. A sectional warping pattern consists of 64 ends/section and 19 such sections are there. Later it was found that 65 threads were in excess. If creel capacity is found to be 500 calculate the width of section and total number of ends. Assume that 62 selvedge on both sides.

Solution :

Total number of ends = 64 × 32 + 64 + 62 = 2175.

4. Find the quantity of each colour of warp and weft required to produce 880 m of fabric. If contraction is around 10 in warp and width in reed is 31″, 68^c reed is selected, 16 picks/quarter inch of 24^s wefts are used and 5% of waste was found in each hank of warp.

Solution :

Pattern	width
80 blue 24^s -	1.166
5 red 24^s -	0.073
10 blue 24^s -	0.145
5 red 24^s -	0.073
20 orange = 2/40^s –	0.319
20 blue 24^s -	0.291
10 white 2/40^s -	0.159
20 blue 24^s -	0.291
5. 5 white 2/40^s -	0.073
175	2.54″

Note : The remaining calculations to be completed by reader

5. Calculate the weight of weft and warp for a pattern towel from 250 m of warp. The reed used is 40^s and the pattern consists of 126 threads as follows:

Width of warp in reed is 32″, count of warp and weft is 2/20^s. Cost per kg of 2/20^s yarn is Rs 42. Find the cost of dozen towels allowing 5% waste both in warp and weft.

Solution :

Cost of single towel
Assuming length of each towel = 1.5 m

Weight of one towel = weight of (warp + weft) required for one towel.

$$\text{Weight of warp} = \frac{126 \times 11.6 \times 1.5(1 + 0.05) \times 1.1}{840 \times 2.2 \times 10} = 0.143 \, kg$$

$$\text{Weight of weft} = \frac{30 \times 33 \times 1.55 \times 1.1}{840 \times 2.2 \times 10} = 0.09 \, kg$$

Total weight of towel = 0.143 + 0.09 = 0.24 ≈ 0.25

Cost of towel = 43 × 0.25 = 10.5 kg.

6. A pattern has different coloured warp of equivalent count of 20 tex yarn warped onto a 1 m beam. The machine has the anglers with 15° to the axis for warp winding on to the drum. The width of the drum or maximum warp width prepared is 1380 mm from 6480 ends with density is 576 kg/m^3. The maximum height of each section on the drum is 100 mm. Work out the set length of warp and traverse speed per section.

Solution :

Yarn count = 20 tex

Warp width (L) = 1.38 m

Empty beam diameter $(d) = \dfrac{5}{\pi} m = 1m$

Full diameter of beam $(D) = d + 2h$,

where h = maximum depth of yarn the beam = 10 cm

$$= \left[1 + 2 \times \frac{10}{100} \right] m = 1.2 \, m$$

$$\text{Volume of the yarn} = \frac{\pi \times (D^2 - d^2) \times L}{4} \, m^3$$

$$= \frac{\pi \times (1.2^2 - 1^2) \times 1.38}{4} \, m^3$$

$$= 0.477 \, m^3$$

Mass of the yarn   = 0.477 × 576 kg.

= 274.75 kg.

$$\text{Mass of a single yarn} = \frac{274.75 \times 1000}{6480} = 42.4 \, g.$$

So, length of the warp $= \dfrac{42.4}{Tex} km$

$$= \frac{42.4}{20} km = 2.12 \, km$$

$$= 2120 \, m.$$

7. A warp of 20 tex blended PC yarn is to be wound on a sectional warping machine with inclines fixed at 15°C to the axis of 5 m circumference. The warp is 100 cm wide and having 40 ends/cm and at a given tension, its density is 0.575 gm/c.c. If the depth of yarn on the swift is 6.5 cm, calculate the length of warp and the traverse of reed per section.

Solution :

Radius of swift (r cm) is such that $2 \pi r = 500$ cm or $r = 79.55$ cm

Volume of the warp having a thickness h and width of warp ×

$$= \pi \times [(r + h)^2 - r^2]$$
$$= 3.14.\ 100\ [(79.55 + 6.5)^2 - 79.55^2]$$
$$= 371788.56 \text{ c.c.}$$

So the weight of yarn = volume of warp × density

$$= 371788.56 \times 0.575 \text{ g}$$
$$= 213.78 \text{ kg}$$

The weight of yarn can be written $= \dfrac{L \times 40 \times 100 \times 20}{1000} = 0.080\ L$ kg

Where L = Length of warp

Equating the two expressions,

$$0.08\ L = 213.78$$

Or $L = 2672.25$ m

The traverse of the reed $= \dfrac{\text{Thickness of warp}}{\text{Tangent of inclined angle}}$

$$\dfrac{6.5}{tan15^o} = 24.24 \text{ cm.}$$

8. Multi-coloured warps of a 20-tex spun yarn are wound on a horizontal section warping mill of 1.5 m diameter, on which the inclines are fixed at 15° to the axis. Each warp is 3000 m long and 2 m wide and contains 6500 ends. When tensioned correctly, the warp density on the mill is 0-6 g/cm². Determine the depth of yarn on the mill when the warp is completed and the corresponding reed traverse per section.

Solution :

We have

Mass of the completed warp $= 6500 \times \dfrac{3000}{1000} \times 20$

$$= 3,90,000 \text{ g.}$$

Now

Volume of yarn = mass/density $= \dfrac{3,90,000}{0.6}$

$$= 6,50,000 \text{ cm}^2.$$

The volume is also given by

$$200 \; \frac{\pi}{4}\left(d_2^2 - d_1^2\right),$$

From which

$$200 \; \frac{\pi}{4}\left(d_2^2 - d_1^2\right) = 6,50,000,$$

i.e.;

$$d_2^2 - 150^2 = \frac{650000 \times 4}{200\pi}$$

$$= 4140$$

And thus $d_2^2 = 150^2 + 4140 = 22,500 + 4140$

$$= 26,640$$

i.e., $d_2 = \sqrt{26,640}$

$$= 163.2 \text{ cm.}$$

Hence

$$d_2 - d_2 = 163.2 - 150$$
$$= 13.2 \text{ cm}$$

And $h = 13.2 / 2 = 6.6$ cm.

The depth of yarn on the mill, h, is 6-6 cm, and from let 'h' be the height of the section and let 't' be the traverse per section then $h/t = \tan 15°$. Hence the traverse per section, t, is given by

$$t = \frac{6.6}{\tan 15} = \frac{6.6}{0.2679}$$

$$= 24.6 \text{ cm}$$

The traverse per section is thus 24.6 cm.

9. A warping shed consists of Jupiter high speed machine with creel capacity of 400. The set length as full beam stop motion is 500 m. It takes nearly 45 min/set 30 s for mending a break, 1 min is lost in doffing and donning/beam, 30 s/beam is lost due to miscellaneous factors. Find η of warping shed and number of beams doffed. The rate of warping 500 m/min. Note that the norm for number of breaks/1000 m for 500 ends is 1.

Solution :

$$\eta = \frac{\text{Running time}}{\text{Running time} + \text{stoppages}}$$

Stoppages = Time lost in mending + doffing + donning
Number of breaks = 1000×500 = break (1)
$$1000 \times 500 = 1 \text{ break}$$

$$500 \times 400 \quad = ?$$
$$= 0.4 \text{ break.}$$

Total time for mending = Number of breaks × Time for mending
$$= 0.4 \times 0.5 \text{ min.}$$
$$= 0.2 \text{ min.}$$

Total time lost / beam = 0.2 +1 min. + 0.5 = 1.7

$$\eta = \frac{1}{1+1.7} = 0.37$$

Assume each cone weighing 1.8 kg with 30's yarn.

Length content = 198 beams

Length content in a cone =

length/beam   = 500 m

Expected number of beams = 198 beams
$$= 198 + 336.6$$
$$= 534.6.$$

Creel is done once in a shift. As creeling will be done once in a shift approximate number of beams doffed = 198

Weight of a beam = Total number of ends × set length metres × $\dfrac{\text{tex count}}{(1000)^2}$

$$= \frac{500 \times 500 \times \dfrac{590.5}{30}}{(1000)^2} = 4.9$$

Yarn content on a beam = 4.9.

10. The preparatory section of a weaving unit has warping section with high speed machines running at 500 m/min using 29.52 *Ne* yarn. The cones used is having yarn content of 2100 g. Loss of time is as follows

   i)   Mending time for each break 1.1 min and total breaks per beam is 30.

   ii)   Doffing time per beam 5 min.

   iii)   Time lost in recreeling a single end is 0.3 min.

   iv)   The weight of each beam is beams containing 250 kg. Compare the performance of Magazine and rectangular creel.

Solution :

Case I: Magazine creel

Length of warp yarn wound on each beam = $\dfrac{250 \times 1000}{20 \times 500} = 25km$ or 25,000 m

Running time of a beam
$$= \frac{\text{Length of warp on each beam}}{\text{Warping speed}} = \frac{25,000}{500} = 50 \text{ min}$$

Total time lost in repairing = 30 × 1.1 =33 min.

Doffing time = 5 min.

Hence total time required to produce a beam = 50 + 33+5 = 88 min.

Efficiency of wrapper = Running time / (Running time + Stoppages)

$$\times 100 \quad \frac{50 \times 100}{88} = 56.82\%$$

Production of wrapper on magazine creel = Speed × 60 × efficiency

$$= \frac{500 \times 60 \times 56.82}{100} = 17,046 \text{ m/h.}$$

Case II: Single end creel

Length of yarn on each cone = $\dfrac{2.10 \times 1000 \times 1000}{20} = 1,05,000$

So number of beams from a cone = $\dfrac{1,05,000}{25,000} = 4.2$ that is 4

Time required to creel 500 packages = 500 × 18 s = 9,000 s

= 150 min.

So for each beam time required for creeling = 150/4 = 37.5 min

Total time lost in repairing= 30 × 1.1 =33 min.

Doffing time = 5 min.

Total loss of time = 37.5 + 33 + 5 = 75.5

Running efficiency: 50 × 100 / 75.5 = 38.84%

Therefore it is better to have the Magazine creel.

11. An order for 220 pieces of woven fabric, each 110 m in length and 1.2 m wide, is received. In the finished fabric, the warp crimp is 9% and there are 38 ends / cm. The creel of the direct-warping process has a maximum capacity of 500 cones. How many back beams are required, and what length of warp is on each?

Solution :

The total number of ends, with selvedge yarn neglected, is:

ends / cm × finished width in cm.

Hence

Total number of ends = 38 × 20 = 4560.

Since the creel can hold 500 cones, the number of ends per back beam could conveniently be 456, and 10 back beams should therefore be used. The total length of warp is given by:

Number of pieces × piece length × $\dfrac{100 + \text{crimp}}{100}$

Hence,

Total length = 220 × 110 × 109 / 100 = 26,378.

12. A high speed warping machine produce beams with 420 ends for a set length of 24,000 m warps when the yarn count is 19.68 *Ne* (30 tex) at a speed of 600 rpm. The end-breakage rate is 0.4 per 1000 ends per 100 m warped, and it takes on an average 54 s to mend the breakage. Doffing and donning take 300 s and recreeling take 900 s. The cone used has sufficient yarn to produce 3 beams. Work out the running efficiency, the weight of full beam if the tare weight is 180 kg and the supply package weight.

Solution :

Running time for a beam = 24,000/ 600 = 40 min and for three beams per three beams is given by 120 min.

Loss of time due to various issues can be estimated as:

(i)   Time to mend the breaks = time per break × number of breaks

$$= 0.9 \times 0.4 \times \frac{420}{1000} \times \frac{24,000}{100} \times 3 = 109 \text{ min (Approximately)}$$

(ii) Doffing + Donning = 3 × 5 = 15 min;

(iii) Recreeling time = 15 min.

Total time lost = 109 + 15 + 15 = 139 min.

Total time = running time + stopped time = 120 +134 = 254 min

$$\text{Running efficiency} = \frac{\text{Running time}}{\text{Running time} + \text{stopages}} \times 100$$

$$= \frac{120}{254} \times 100 = 47\%.$$

Weight of yarn on beam = length in km × linear density in tex.

$$= \frac{24,000}{1000} \times 420 \times 30 \times 1/1000$$

$$= 3,02,400 \text{ g or } 302.4 \text{ kg}$$

$$\text{Weight of cone} = 3 \times \frac{24,000}{1000} \times 30 = 2160 \text{ g}$$

$$= 2.16 \text{ kg.}$$

13. A beam is prepared on sectional wrapper with 3500 ends for a set length of 3000 m. The packages in the creel are with a yarn content so that they can deliver one full beam and the warping speed is 500 m/min. It takes 0.3 min to creel a package and mend the end and 2.5 min per section for leasing and moving the traverse, etc. Breaks during warping average 20 per beam, and the average repair time is 1.1 min. The beaming speed is 60 m/min, and the time lost in preparing the beam is 8 min. Work out necessary statistics given the capacity of creel is 540 but the only 520 ends can be used due to some issues.

Solution :

It is mentioned that the workable capacity of the creel is 520 and for making the calculations simple and for the problem less operation assume a total of 500 ends on each beam so that we require 7 sections.

Then running time of a section is = 3000/500 × 7 = 42 min;

Beaming time = 3000/60 = 50 min;

The various times lost is: Mending of breaks = 20 × 1.1 = 22 min;

Creeling = 500 × 0.3 = 150 min;

Lease band insertion = 2.5 × 7 = 17.5 min;

Doffing of beam = 8 × 1 = 8 min;

Total time lost = 197.5 min.

Total time per beam = running time + beaming time + loss of time = 42 + 50 + 197.5 = 289.5 Min. running efficiency = $\dfrac{42}{289.5}$ × 100 = 14.5%.

5.10   Mechanics of weft winding

Weft winding – Conventional pirn winders – Overview

Introduction to weft preparation

Objects of rewinding of weft

- To get a sufficient length of yarn on the pirn so that the loom does not stop frequently.
- To give some tension to get a compact pirn which can be unwound easily during weaving.
- To control the full pirn diameter so that it grips properly in the shuttle.
- To facilitate automatic pirn changing.
- To prepare different types of weft packages for different looms.

Systems of weft preparation

- Direct system
- Indirect system
- Loom winders
- Unconventional system

5.10.1   Indirect system (or) rewinding of weft

Principles of pirn winding

1. Spindle principle – used in handloom type

2. Spindle less principle
3. Drum principle – This is not widely used in staple yarns and some filaments yarns and specially used for propylene yarn because of the grooved drum which put up of the coils in a cross-wound fashion, which in turn becomes difficult unwound the pirn.

5.10.2   Classification of pirn winding

It is most widely used system of weft preparation in which packages like cons, cheese, spool, etc., are unwound and rewound on to pirns which can be placed in shuttle and used in loom. Different pirn winders are classified on the following basis

I   Basis on the speed

i)   Slow speed pirn winders – 5,000 rpm
ii)   High speed pirn winders – 9,000 rpm
iii)   Super speed pirn winders – 12,000 rpm

II   Based upon the type of automation

i)   Non-automatic
ii)   Semi-automatic
iii)   Fully automatic

Effect of direction of twist in the yarn with respect to the direction of twist in pirn winder machine.

i)   If the package of yarn is of Z twist and if we provide the direction of rotation of a pirn winding machine in the same direction (Z) which will lead in the increasing of twist by 5%.

ii)   If the direction is opposite to each other which will lead in de-twisting of points to the extent of 10%.
        S = clock wise   Z = Anti-clock wise

Sunrise pirn winder

Speed calculations

Speed of motor = 1425 rpm.

$$\text{Speed of worm shaft} = 1425 \times \frac{14.5}{43} \times \frac{43}{27} = 765.3 rpm.$$

$$\text{Speed of pirn holder} = 765.3 \times \frac{27}{12} = 1721.9 rpm.$$

$$\text{Speed of cam / worm wheel} = 765.3 \times \frac{1}{32} = 24 rpm.$$

∴ Number of double traversers
(or to and fro motions) of the layer locking device = 24 per minute.

Semi-auto pirn winder

Calculations

Speed of motor = 930 rpm, πD of motor = 11″, πD of intermediate pulley = 49″

Then speed of intermediate = $\dfrac{930 \times 11}{49}$ = 208.77 rpm.

Speed of spindle = $\dfrac{208.77 \times 49}{5}$ = 2045.9 rpm.

Speed of 16th teeth = 208.77 rpm.

Speed of 58th teeth = $\dfrac{16 \times 208.77}{58}$ = 57.56 rpm.

Time taken for 1 pirn = 1.22 min.
Weight of empty pirn = 26.1 g
Weight of yarn + pirn = 40.9
Weight of yarn = 40.9 – 26.1 = 14.8 g
Number of traverses = 185
Weight of lea = 5.7

Count $Ne = \dfrac{64.8}{5.7}$

$$= 11.85^S = \dfrac{2^S}{22}$$

Count = $\dfrac{1}{W}$ \qquad $1 \Rightarrow C \times W$

$11 = \dfrac{X}{840} \times \dfrac{453.6}{5.7}$ \qquad\qquad $\Rightarrow \dfrac{11 \times 5.7 \times 840}{453.6}$

x = 116 yarns \qquad\qquad l = 116 yards
x = 106 m \qquad\qquad\quad = 106 m

Rotary speed

Speed of rachet = $\dfrac{208.77 \times 8}{77}$ = 21.6 rpm.

Speed of rotary = $\dfrac{21.6 \times 34}{18}$ = 40.8 rpm.

Calculations of automatic pirn winding machine

Calculations

Given motor speed = 1440 rpm.

$$\text{Speed of intermediate} = \frac{1440 \times 13}{11}$$

$$= 505.94 \text{ rpm.}$$

$$\therefore \text{ Speed of fast and loose wheel} = \frac{505.94 \times 37}{10}$$

$$= 1871.978 \text{ rpm.}$$

$$\text{Speed of spindle} = \frac{1871.978 \times 32}{16} = 3743.956 \, rpm.$$

$$\text{Speed of eccentric wheel} = \frac{1871.978 \times 1}{27} \left(worm \; wheel \right)$$

$$= 69.33 \text{ rpm.}$$

$$\therefore \text{ Speed of traversing motion of the pirn} = 69.33 \text{ rpm.}$$

$$\text{Speed of projection wheel} = \frac{3743.956 \times 1}{60}$$

$$= 62.39 \text{ rpm.}$$

For 35 count

Number of teeth dragged by prawl = 5 teeth

$$\text{Speed of rotary motion of pirn} = \frac{62.39 \times 5 \times 26}{82 \times 26}$$

$$= 3.8048 \text{ rpm.}$$

For 10 count:

Number of teeth dragged by prawl = 13 teeth

$$\text{Speed of rotary motion of pirn} = \frac{62.39 \times 13 \times 26}{82 \times 26}$$

$$= 9.8925 \text{ rpm.}$$

Speed calculations in Hacoba automatic pirn winder

Speed calculations

Speed of motor = 950 rpm

$$\text{Speed of fast and loose pulleys} = 950 \times \frac{6\,3}{19\,75} \times \frac{37}{9} = 1246 \, rpm$$

$$\text{Speed of pirn holders} = 1246 \times \frac{31}{18} = 2146 \, rpm$$

Number of to and fro motions of the traversing guide = 150 / min.

5.10.3   A review on passage of material through pirn winder of conventional type

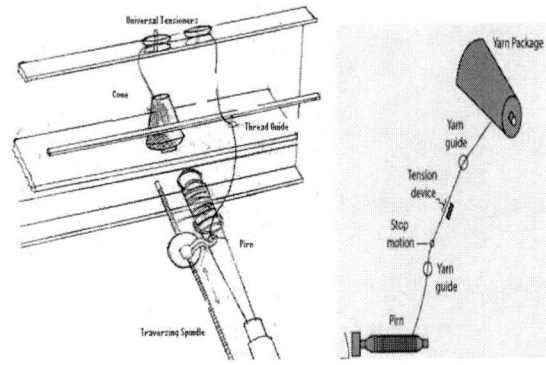

Figure 5.25   Passage of material through semi-auto pirn winder

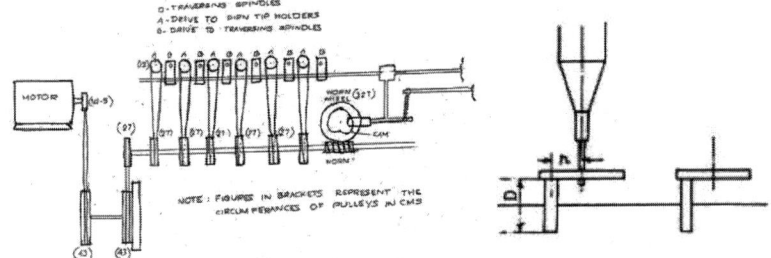

Figure 5.26   Driving arrangement of semi-auto pirn winder

To withstand stresses and strains imparted by picking and checking mechanisms. The winding of weft yarn on pirns at constant tension necessitates a varying pirn speed. This is achieved with constant surface speed of piris; frequently spring tensioning devices are employed to help to minimise variations in tension. Variation in tension, when winding at uniform pirn speed, depends largely on the way in which the yarn is unwound before winding on the pirn. Generally a balloon will result as the yarn unwinds from the supply package. With this method, the yarn balloons (Figure 5.25) to a certain extent between the cone and the guide eye above it, and centrifugal force acting on this ballooning yarn causes a tension which increases the winding tension. The centrifugal force of the ballooning yarn increase when the winding rate increases, and this causes extra tension on the yarn. To get sufficient winding on tension to make a firm pirn, it is generally necessary to apply some resistance to the yarn before it reaches the guide. This resistance generally consists of tension units, over which the yarn is passed.

Various forms of drag washers are also used for tensioning. In all of these methods tension is applied by friction, and since friction is nearly independent of speed the change in speed, as the rate of winding changes, should not materially affect the tension caused by this friction. Various automatic devices are used on constant speed machines for minimising tension variation.

A common method of driving pirn winder spindles at a varying speed, in order to keep the rate of winding on constant, is shown in Figure 5.26. Each spindle is driven by a pulley running at a constant speed and acting at a varying radius r on a disc which drives the spindle. The bowl shaft is reciprocated in conjunction with the guide, so that as the winding on diameter changes the radius r and the rate of rotation of the pirn is changed to keep constant rate of winding.

Pirn speed must vary as $\dfrac{1}{\text{Pirn diameter}}$; also pirn speed $\alpha \dfrac{D}{r}$; i.e., as $\dfrac{1}{r}$

since D = constant.

$$\therefore \frac{1}{r} \alpha \frac{1}{\text{Pirn diameter}} \text{ or } r \, \mu \text{ pirn diameter.}$$

The yarn guide and the driving bowl must be reciprocate in unison; as the guide approaches the nose of the pirn the bowl must move nearer to the centre of the disc. If a uniform cam is used for reciprocating the guide, the same kind of cam must be used for moving the bowl shaft. Such an arrangement gives a varying pitch to the coils, which get closer together towards the nose of the chase, due to the high speed of rotation there; this results in too much yarn accumulating at the nose, and generally the guide is moved so as to keep the pitch of the coils approximately uniform i.e., the guide is speeded up as the pirn speed increases. In such circumstances the movement given to the bowl shaft must be of a similar nature, to keep the correct relation between the winding on diameter and the driving disc radius.

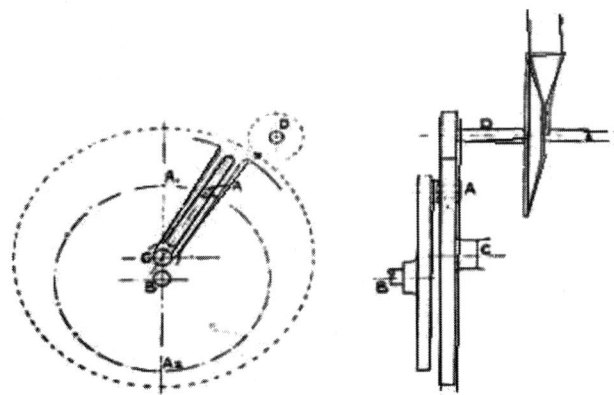

Figure 5.27   Eccentric method of dringing the pirn

Figure 5.27 shows an eccentric wheel drive for giving a varying pirn speed. A driving pin A is driven at a constant speed about the shaft centre B. A engages with the slotted arm of a gear wheel which is centred at C; this wheel rotates at a varying speed and drives a disc shaft D which rotates the spindles by frictional contact. The fastest speed of the driven wheel will evidently be reached when the pin A is at A_1, nearest to the wheel centre C and the slowest speed when A is in position A_2, farthest from centre C. The guide must therefore move from the nose to the largest pirn diameter while the pin A is moving from A_1 to A_2.

In an actual machine, the radius of A's path is 6″ the distance BC 1.8″, and the pirn diameter varied from 1″ to 3/8″. The guide moves as shown on the pirn chase in Figure 5.28 while the pin A moves through three equal angles from A to A_3. The actual speed of the spindle is always proportional (ignoring slip) to the angular velocity of the variable speed wheel driven by A. Assume that the linear velocity of A is 420 inch per minute; this number has been taken for convenience to give a maximum speed of 100 rad/min. Then when the crank pin is at A the angular velocity ω of the variable speed wheel is $\omega = \dfrac{V}{r} = \dfrac{420}{4.2} = 100$ rad/min, since a point on the arm of the wheel 4.2″ from the wheel centre is moving at the same speed as the driving crank pin. Similarly at A_3 $\omega = \dfrac{420}{7.8} = 54$ rad/min.

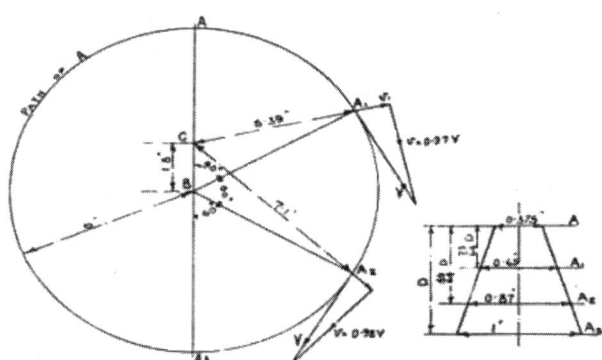

Figure 5.28   Analysis of pirn winding

But at A_1 and A_2 the crank pin is not moving at right angles to the slotted arm, and since it is only the velocity of the crank pin at right angles to the arm that gives the latter its angular velocity, it is necessary to resolve V into its components v and v_1 at right angles to and along the slot, respectively.

Then at A_1 $\omega = \dfrac{v}{A_1 C} = \dfrac{0.97 \times 420}{5.39} = 75.5$ rad/min

And at A_2 $\omega = \dfrac{v}{A_2 C} = \dfrac{0.98 \times 420}{7.1} = 58$ rad/min.

To give uniform rate of winding the angular velocity should vary inversely as the pirn diameter; thus if $\omega = 100$ rad/min is correct for post position A when the diameter is 0.375".

Then ω for A_1 should be $\dfrac{0.98 \times 420}{7.1} = 58$ rad/min.

The relation between speed and pirn diameter is shown in Figure 5.29.

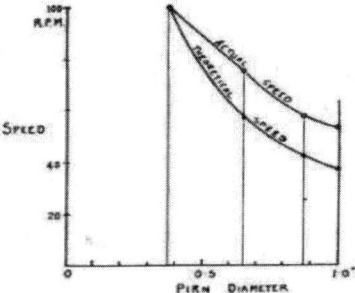

Figure 5.29   Relation between pirn speed and pirn diameter

Elements of a pirn
Following are the elements of an ideal pirn.

a. Butt or Head or Base; b. Back chase; c. Chase; d. Front chase; e. Shoulder angle.

Referring to the Figure 5.30, consider the full diameter of a pirn wound from cotton yarn is 32 mm, and the bare-pirn diameter at the nose of the chase is 14 mm. Determine the chase angle when the traverse is 34 mm.

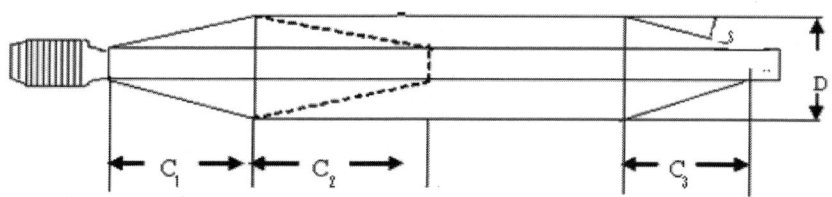

Figure 5.30   Elements of a pirn

It will be seen that

$$\text{Tan } \alpha = \frac{(D-d)}{2} \times \frac{1}{C_n}.$$

Hence,

$$\text{Tan } \alpha = \frac{(32-14)}{2} \times \frac{1}{34} = 0.2646,$$

from which the chase angle, $\alpha = 14°\,49'$.

1. A 7-wind pirn is wound on a spindle that rotates in an anti-clockwise direction. The yarn is an S-twist yarn with 8 turns/cm. If the length of yarn wound per traverse is 44 cm, calculate the turns / cm in the pick.

Solution :

Since the spindle will generate S twist, 7 turns of twist will be added to the twist in 44 cm of yarn. Hence the total turns in 44 cm of yarn in pick form will be:

$$(44 \times 8) + 7 = 359.$$

From which: Turns/cm. $= 359 / 44 == 8.16$.

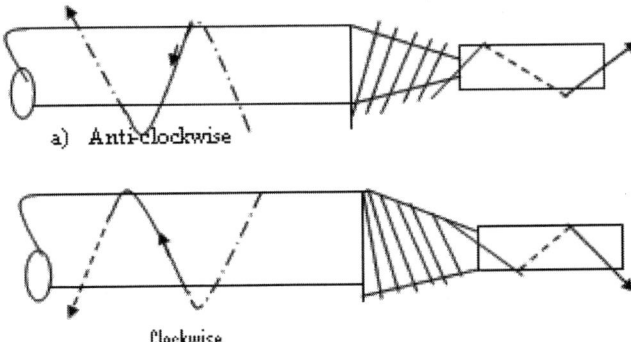

a) Antirclockwise

Clockwise

Figure 5.31 Pirn rotation: the direction of spindle rotation

The difference in this case is only about 0.16 turn/cm and would not alter the yarn behaviour significantly. However, for yarns with very few turns/cm, such as continuous-filament types, it could be good practice to make sure that what little twist is already present is not reduced by a wrong choice of the winding-spindle direction of rotation. Where the yarn is already highly twisted, e.g., as in crepe yarn, it would be good practice to abstract a little twist and reduce the snarling behaviour of such a yarn. A Z-twist crepe yarn would therefore be wound on an anti-clockwise spindle and an S-twist crepe yarn on a clockwise spindle (Figure 5.31).

5.11   Mechanics of sizing

5.11.1   Beam winding by PIV gear

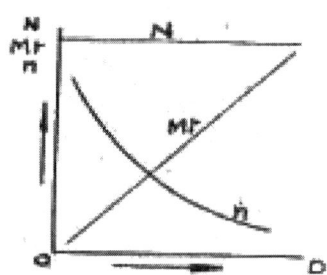

Figure 5.32   Mechanics of PIV gear in sizing

Figure 5.33   Mechanics of PIV gear in Joseph Hibbert sizing

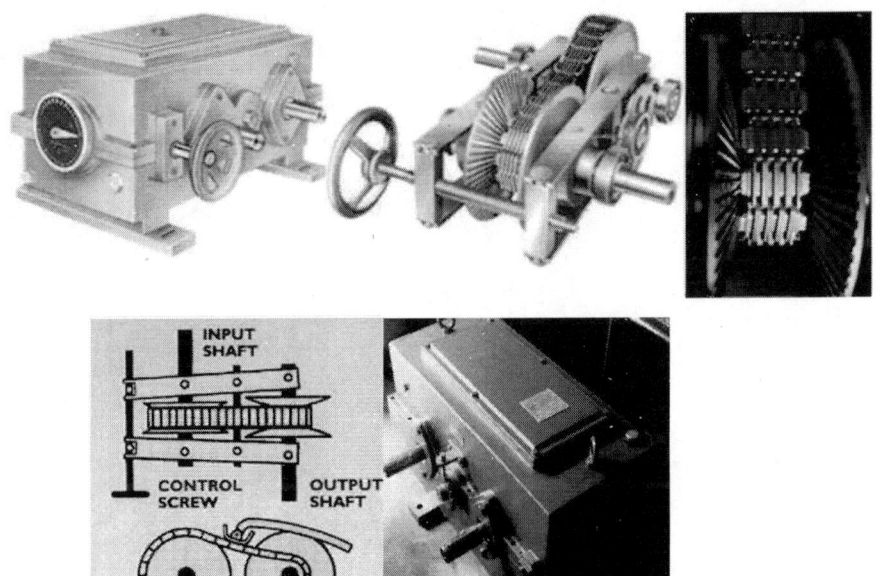

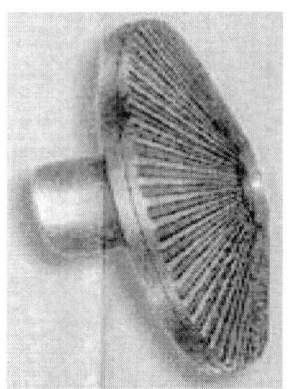

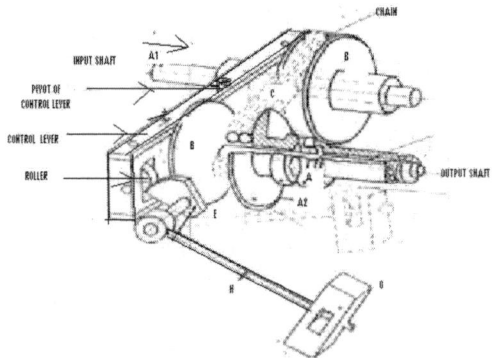

A1.A2: Thrusts, A - Pressure Sleeve, B - Conical Disc, C - Chain, D - Control lever,
E - Control Cam, F - Concial disc 2, G - Weigh, H - Weight lever, I - Central of Motor,
K - frame, X - Driver shaft, Y - Driven shaft

Figure 5.34   Details of PIV gear

For winding procedures the law holds good in theory that speed and tension and consequently also efficiency must remain constant over the whole winding range.

$$P \times V = NW = \text{Constant}$$

This requirement is complied with when the speed of the winding shaft is changed inversely proportional to the winding diameter.

$$n = \frac{v}{x \quad D}$$

The torque of the winding shaft changes in proportion to the winding diameter.

$$Mt = \frac{D}{Z} \times P$$

The theoretical course of speed and torque is shown in Figure 5.32.

Practical operation as a rule requires a characteristic winding feature that deviates from these theoretical values. The PIV winder meets this requirement of practical work. Through the form of its control curve the required course of tension can be determined.

5.11.2   Regulation with PIV winders

The PIV winder is designed for winding (WI) and unwinding (brake winder) (BR) operations. For regulating the speed of the winding shaft it needs no accessories, such as relays, contactors and servomotor.

A pre-requisite for this kind of speed regulation is that the speed of the goods to be wound is determined by a producing machine or a pair of driven rollers.

The PIV winder can be inserted in the flux of a machine or a winding device or can be driven directly by a motor, as illustrated in Figure 5.33.

Among the infinitely variable PIV genre, the PIV winder holds a special position as it exercises the function of a regulator as well as transmitting power at the same time. The winder tends to change its ratio automatically if the load changes.

If, for instance, during a winding operation the torque at the winding shaft increases, the winder decreases its output speed automatically until the product from speed and torque has reached its initial value again. The PIV winder is, therefore, capable of solving winding problems of all kinds. The tension desired for the goods strip is adjusted by shifting the sliding weight on the weight lever and can be changed in the proportion of up to a maximum of 1:7.5. During the winding operation the weight need not be moved.

The winding range $R = \dfrac{\text{Outer diameter}}{\text{Inner diameter}}$ is 7.5

With the types available, capacities of $N = 0.005$–10 hp can be transmitted.

5.11.3   Mode of operation of the PIV winder

Figure 5.34 shows the essential structural elements of the winder. The position of the chain and that of the weight lever for the greatest speed ratio of a winder are illustrated here. The tension is regulated by shifting the sliding weight G on the weight lever h, via control came, and control lever d, the lever movement gives rise to a force A1 at the axially movable conical disc f of input shaft I.

A1, A2 – Axial thrusts

I   Input shaft

II   Output shaft (to winding shaft)
    a.   Pressing sleeves
    b.   Conical disc on output shaft
    c.   Chain
    d.   Control lever
    e.   Control can
    f.   Conical disc on input shaft
    g.   Sliding weight
    h.   Weight lever
    i.   P Pivot of control lever
    j.   r Roller

Output shaft II is now connected to the winding shaft. Through the balls tying between the pressing sleeves a, the torque introduced exerts an axial thrust A2. On their helical faces which acts upon the movable conical disc b. If thrust A2 becomes greater than thrust A1, the conical disc b and f are moved towards slow motion in the sense of a variation of speed. As soon as thrusts A1 and A2 are balanced the variation of speed will end.

5.11.4   The winding procedure

At the beginning of the winding procedure the weight lever is in its lowest position. Chain c, runs between the pair of conical discs of shaft II on its smallest pitch circle. Shaft II is connected with the winding shaft, input shaft I with the machine or an individual motor. In this way the smallest winding diameter is combined with the greatest gear ratio into quick motion. In accordance with the position of the sliding weight there results a thrust A1 at the conical disc f. The growing winding diameter causes a rise in torque on shaft II and through the pressing sleeves s, an increase of thrust A2 at the conical disc b. This increase of thrust is followed by an axial displacement of the conical discs b and f and a rotation of control cam e, via control lever d. The weight lever is lifted in direction s. The effective lever arm of the control cam is lessened and at the same time the back pressure against control lever d and conical disc f is increased until throats A1 and A2 are balanced.

The rotation of the control cam is followed by a variation of speed towards slow in the winder. The control cam is shaped in such a way that the product of speed and torque at the output shaft II is constant or follows a given law. At the end of the winding procedure and after severing the strip of goods, the weight lever swings back down into its initial position, or can be pushed down by hand. The winding procedure may begin a new.

5.11.5   Mode of operation of control head on PIV winder

Weight lever and control cam are attached to the shaft of the control head. The weight lever has a calibration in cm for setting and re-adjusting the correct position of the weight and the correct tension. Sliding weight at the end of the lever results in maximum tension, weight at the hub of the lever results in minimum tension. The shape of the control cam determines the course of tension in dependence on the winding diameter. Two shock absorbers are provided to lessen vibrations and shocks which may occur during the winding procedure.

To adapt the speed corresponding to the initial diameter, the control head is provided with a square head to be used only when the speed of the strip of goods is not in harmony with the circumferential speed of the winding core. In order to ensure that the ratio reached at in the winter does not change when the strip of goods breaks, a locking device can be inserted. It keeps the weight lever from lowering.

1. Calculate the solids concentration percentage of 50 kg of size mixture which contains 10 kg of size ingredients if the moisture contents of its ingredients are: starch 13%, lubricant emulsion 60%, antiseptic 0.5% and the proportion of these in the size mixture are as 10:2:0.5.

Solution :

$$100 \text{ kg of the size mixture contain } \frac{10 \times 100}{50} = 20\,kg.$$

size ingredients.

The proportions of the individual ingredients are 10:2:0.5 i.e., 12.5 of the total ingredients contain 10 starch, 2 lubricant emulsion and 0.5 antiseptic.

∴ 100 kg of the size mixture contains:

$$\frac{10 \times 20}{12.5} = 16\,kg \text{ of starch.}$$

$$\frac{2 \times 20}{12.5} = 3.2\,kg \text{ of lubricant.}$$

$$\frac{0.05 \times 20}{12.5} = 0.8\,kg \text{ of antiseptic.}$$

Since starch contains 13% moisture, it contains 87% solid dry starch.

∴ 16 kg of starch contain 13.92 kg solid, dry starch.

Lubricant contains 60% moisture, i.e., 40% solid.

∴ 3.2 kg lubricant contains 1.28 kg solid.

Antiseptic is all solid, without moisture.

∴ 100 kg of size mixture contain 13.92 + 1.28 + 0.8 kg solids = 16% solids.

Ans: 16%.

2. Calculate the weight of size ingredients in kg and the volume of water in litres required to prepare 240 litres (l) of 10% concentration (on the volume) size mixture, if the overall moisture content of the size ingredients used is 20% and steam condensate is 15% of the volume of cold water with the size ingredients. The specific volume of the sizing agents in the mixture is 0.7 dm³/kg.

Solution :

Weight of size ingredients in kg.

$$= \frac{\text{Volume of size mixture in } 1 \times concentration\%}{(100 \text{ - moisture content }\%)}$$

$$= \frac{240 \times 10}{100 - 20} = 30$$

Water requirement in litres

$$= \left(\frac{\text{Volume of size mixture} \times 100}{(100 \text{ - condensate }\%)}\right) - \left(\text{Weight of size ingredients in kg} \times \text{sp. volume}\right)$$

$$= \frac{240 \times 100}{100 + 15} - (30 \times 07)$$

$$= 187.7\,l$$

Ans.: 30 kg and 187.7 l.

3. The consumption of the size paste is measured by measuring the depth of size mixture in the storage beck (also called reserve beck or reservoir) at the commencement and at the end of the experiment by means of a dip stick immersed perpendicularly in the reserve beck. The only necessary conditions in this are that there should be no addition to the reserve beck during the experiment and the level of the size in the size box should be constant.

The size consumption then $= l \times b \times h$, where

l = length of the reservoir in cm,

b = width of the reservoir in cm, and

h = difference in size level in cm.

(The size may circulate between the reservoir and the size box, via the overflow in the size box.)

4. When 1000 m of a warp sheet having 4000 ends of 50 Nm cotton yarn is passed through the size paste, the size level in the reservoir is reduced by 20 cm. The concentration of the size paste is 10% and I cm level difference in the reservoir corresponds to 6.5 kg of size paste pick-up in the yarn. Find the size pick-up and add-on percentages.

Solution :

$$\text{Size mixture consumption} = 20 \times 6.5$$
$$= 130 \text{ kg.}$$
$$\text{Weight of warp in kg.} = \frac{4000 \times 1000}{50 \times 1000}$$
$$= 80$$
$$\text{Size pick-up\%} = \frac{\text{Size mixture consumption} \times 100}{\text{Weight of the warp}}$$
$$= \frac{130 \times 100}{80}$$
$$= 162.5\%$$
$$\text{Size add-on\%} = \frac{\text{Size pick-up\%} \times \text{Size concentration}}{100}$$
$$= \frac{162.5 \times 10}{100}$$
$$= 16.25$$

Ans.: 162.5% and 16.25%.

5. A cotton warp having 4000 ends of 20 tex yarn is sized on a sizing machine running at a speed of 60 m/min.

The size mixture has a concentration of 11% and the pick-up is 140%. Calculate –

i)   The warp capacity in kg of warp processed in one hour (kg/h), at 100% efficiency.

ii)   Effective warp capacity if the efficiency of the machine is 60%.

iii)   The water vaporising capacity of the dryer, in kg/h, if the moisture content before and after sizing is 8% and 6.5%, respectively.

Solution :

$$\text{Warp weight in g/running m} = \frac{\text{Number of threads} \times tex}{1000}$$

$$= \frac{4000 \times 20}{1000} = 80$$

Warp capacity in kg/h

$$= \frac{\text{Warp weight in g/running metre} \times \text{machine speed} \times 60}{1000}$$

$$= \frac{80 \times 60 \times 60}{1000}$$

$$= 288$$

Effective warp capacity in kg/h = Warp capacity × efficiency %

$$= \frac{288 \times 60}{100}$$

$$= 172.8 \text{ W}$$

(iii) Water vaporising capacity in kg/h

$$= \frac{\text{Warp Capacity} \times \text{Size pick} - up\% \times (100 - \text{Size conc.}\%)}{100 \times 100} + \frac{\text{Warp Capacity} \times \text{Moisture \% difference}}{100}$$

Moisture % difference = Moisture content % before the sizing
                                      − Moisture content % after the sizing.

Water vaporising capacity in kg/h

$$= \frac{288 \times 140 \times (100 - 11)}{10,000} + \frac{288 \times 1.5}{100}$$

$$= 358.85 + 4.32$$

$$= 363.17$$

Ans.: 363.17 kg/h.

6. Find (i) the quantity of size paste pick-up, (ii) the machine running speed, (iii) effective warp capacity, (iv) the running time for a sett to be sized, from the following particulars:

Material	= Polyester – Cotton
Count of warp	= 40 Nm

Total number of threads $= 4000$

Warp weight/running m $= 100$ g

Sett length $= 10,000$ m

Size concentration $= 10\%$

Expected size pick-up $= 150\%$

Expected evaporating capacity of dryer/h $= 600$ kg

Efficiency of dryer, (without setting time) $= 90\%$

Moisture difference between unsized and sized warp $= 0\%$

Solution :

(i)   Total size pick-up by warp $= \dfrac{\text{Weight of warp} \times \text{pick up\%}}{1000}$

$$= \frac{10,000 \times 4000 \times 150}{40 \times 1000 \times 100}$$

$$= 1500 \text{ kg.}$$

(ii)   Machine speed: This can be found among the data given, from the length of warp, evaporating capacity of dryer, the concentration of size paste and the total size paste picked-up by warp.

Total size paste: 1500 kg. Water in this size paste

$$= \frac{(100 - 10) \times 1500}{100} = 1350 \, kg.$$

Time required to evaporate this $= \dfrac{1350 \times 60}{600} = 135 \, min.$

Length of warp passing through: 10,000 m

$\therefore$ Machine speed $= \dfrac{10,000}{135} = 74.07 \, m \,/\, min.$

(iii)   Effective warp capacity in kg/h

$$= \frac{\text{Warp weight in g/running m} \times \text{machine speed} \times 60 \times \text{efficiency}}{1000 \times 100}$$

$$= \frac{100 \times 74.07 \times 60 \times 90}{1000 \times 100}$$

$$= 400$$

(iv)   Running time per sett in minutes

$$= \frac{\text{Weight of warp per sett} \times 60}{\text{Effective warp capacity}}$$

$$= \frac{10,000 \times 4000 \times 60}{40 \times 1000 \times 400} = 150$$

Ans.: (i) 1500 kg, (ii) 74.07 m/min, (iii) 400 kg/h, (iv) 150 min.

7. A 25 tex cotton warp containing 3600 ends has 9% size add-on. The maximum drying capacity is 500 kg/h. The size concentration is 7%. Calculate the optimum sizing speed which ensures that the moisture regain of the sized yarn is 8. The moisture regain of the unsized warp is 8.5%.

Solution :

Oven-dry weight of warp per running m

$$= \frac{3600 \times 25}{1000} \times \frac{100}{108.5}$$

$$= 82.95$$

Moisture in unsized yarn per running m

$$= \frac{82.95 \times 8.5}{100} = 7.05g.$$

Size added-on 1 m of yarn $= \frac{82.95 \times 9}{100}$

$$= 7.46 \text{ g.}$$

Weight of sized yarn per running metre = 82.95 + 7.46 = 90.41 g.

Size pick-up per running metre of warp

$$= \frac{\text{Sixe add-on per m} \times 100}{\text{Conc. of size paste \%}}$$

$$= \frac{7.46 \times 100}{7} = 106.57g.$$

Water in the size paste per m of warp

$$= 106.57 - 7.46 = 99.11 \text{ g.}$$

Total weight of water per m, before drying

$$= \text{natural moisture of warp} + \text{Water in the size paste}$$

$$= 7.05 + 99.11 = 106.16$$

Amount of moisture per running m of warp on the weaver's beam

$$= \frac{90.41 \times 8}{100} = 7.23g.$$

∴ Total quantity of water to be evaporated per m.

$$= 106.16 - 7.23 = 98.936.$$

Time required to evaporate this much water

$$= \frac{98.93 \times 60}{500 \times 1000} = 0.0118719 \, min.$$

This is therefore the time required to run 1 m.

$$\therefore \text{ Running speed } = \frac{1}{0.0118716} = 85.47 \, m \, / \, min.$$

Ans.: 85.47 m/min.

8. Calculate the efficiency of a sizing machine from the following particulars:-

Yarn count	= 20 Nm
Length of yarn on wrapper's beam	= 12,000 m
Total number of ends	= 3800
Speed of the sizing machine	= 40 m/min
Number of lappers / 3000 ends / 1000 m	= 2.5
Average time to cut a lapper	= 1.5 min
Length of yarn on waver's beam	= 1200 m
Time to doff a beam and insert a fresh lease	= 10 min
Time to creel wrapper's beam and change a sett	= 100 min
Miscellaneous loss of time	= 10 min.

Solution :

$$\text{Running time of one beam } = \frac{1200}{40} = 30 \, min.$$

$$\text{Number of lappers/beam } = \frac{2.5 \times 3800 \times 1200}{3000 \times 1000} = 30 \, min.$$

$$= 3.8$$

Time required for cutting lappers/beam = 3.8 × 1.5

$$= 5.7 \text{ min.} \qquad (a)$$

Number of weaver's beams per set of wrapper's beams

$$= \frac{12,000}{1200} = 10$$

$$(b)$$

Time required for creeling etc.

$$\text{Weaver's beam } = \frac{100}{10} = 10 \, min. \qquad (c)$$

Time required for doffing a beam etc. = 10 min.

Total time lost per weaver's beam

$$= (a) + (b) + (c) + \text{miscellaneous loss of time}$$
$$= 5.7 + 10 + 10 + 10 = 35.7 \text{ min.}$$

$$\text{Efficiency of the sizing m/c } = \frac{\text{Running time} \times 100}{}$$

$$= \frac{30 \times 100}{30 + 35.7} = \frac{3000}{65.7} = 45.66\%$$

Ans.: 45.66%.

9. Other particulars remaining the same in the above example, if the speed of the sizing m/c is increased to 60 m/min., find the efficiency.

Solution :

$$\text{Running time } = \frac{1200}{60} = 20 \ min.$$

Since the total stoppages are the same,

$$\text{Efficiency } = \frac{20 \times 1000}{20 + 35.7} = 35.91\%$$

Ans.: 35.91%.

10. In a sizing machine the stretch percentages at the creel zone, wet and drying zone and winding zone are 0.5%, 1.75% and 0.5%, respectively. Calculate the total stretch percentage in the machine for 300 m of warp sheet unwound from the wrapper's beam.

Solution :

Stretched length of yarn in the creel zone:

$$= 300\left(1 + \frac{0.5}{100}\right) = 301.5 \ m$$

Stretched length of yarn in the wet and drying zone

$$= 301.5\left(1 + \frac{1.75}{100}\right) = 306.78 \ m$$

Stretched length of yarn in the winding zone

$$= 306.78\left(1 + \frac{0.5}{100}\right) = 308.31 \ m$$

Total increase in length $= 308.31 - 300 = 8.31$ m.

$$\text{Overall stretch percentage} = \frac{8.3 \times 100}{300} = 2.77.$$

11. Calculate the dead loss percentage when 25,000 kg of size materials had been issued for sizing 1,40,000 kg of unsized warp and 1,59,600 kg of sized warp had been obtained. The moisture content of the sized warp was 8%, that of unsized warp 7.5% and the overall moisture content of the sizing materials 15%. During the same time the unsized waste was 475 kg and sized waste 950 kg.

Weight of sizing materials issued(Z)	= 25,000 kg
Weight of unsized warp (X)	= 1,40,000 kg
Weight of sized warp (Y)	= 1,59,600 kg
Moisture content in unsized warp(P)	= 8%
Moisture content in sized warp(Q)	= 7.5%
Moisture content in sizing materials (R)	= 15%
Unsized waste (U)	= 475 kg
Sized waste (W)	= 950 kg

Solution :

Using ATIRA's equation:

Dead loss =

$$\frac{Z\left(1-\dfrac{R}{100}\right)-\left[(Y+W)\left(1-\dfrac{Q}{100}\right)-(X-U)\left(1-\dfrac{P}{100}\right)\right]}{Z\left(1-\dfrac{R}{100}\right)}\times 100$$

$$=\frac{25000\left(1-\dfrac{15}{100}\right)-\left[(159600+950)\left(1-\dfrac{7.5}{100}\right)-(14000-475)\left(1-\dfrac{8}{100}\right)\right]}{25000\left(1-\dfrac{15}{100}\right)}\times 100$$

$$=\frac{21250-\left[148509-128363\right]}{21250}\times 100 = 5.20\%$$

Ans.: 5.2%.

Using a simple but crude (and erroneous) equation,

$$\text{Dead Loss } =\frac{Z-(Y-X)}{Z}\times 100$$

$$= 21.6\%.$$

12. Find (i) the cost per kg of the sizing ingredients and (ii) the cost of sizing materials per 1 kg of the warp yarn from the following data regarding the recipe of the size mixture. Also calculate (iii) the cost of size ingredients per litre of the size paste, if the concentration of the paste is 12% (over-dry).

Ingredient	Weight in kg	Price/kg
Starch	40	4.30
Polyvinyl alcohol	10	62.00
Carboxy Methyl Cellulose (CMC)	5	11.30
Gum	2	8.30
Softener	2	14.20
Antistatic agent	0.5	24.00

Size add-on 16%   Dead loss 7.41%

Moisture content of both unsized and sized yarn is 6.25%.

Overall moisture content of sizing materials is 19% and there is no waste of yarn at all.

Solution :

(i)   Total cost $= (40 \times 4.30) + (10 \times 62.00) + (5 \times 11.30) + (2 \times 8.30) + (2 \times 14.20) + (0.5 \times 24)$

$= $ Rs. 905.50

Total weight of the sizing ingredients $= 59.5$ kg.

Overall cost of the ingredients $= \dfrac{905.50}{59.5} = Rs.15.22 / kg.$

(ii) As the add-on is 16% and the dead loss 7.41% the weight of sizing ingredients issued for sizing 100 kg. of warp would be Z and

Dead loss $= 7.41 = \dfrac{Z(1-0.19) - \left[116(1-0.0625) - 100(1-0.0625)\right]}{Z(1-0.19)} \times 100$

$\therefore 0.81\, Z \times 7.41 = (0.81\, Z - [108.75 - 93.75]) \times 100$

$= 81\, Z - 1500$

$\therefore 6.0\, Z = 81\, Z - 1500$

$\therefore 1500 = 75\, Z$

$\therefore Z = 20.$

$\therefore$ Cost of sizing ingredients issued for 100 kg of warp in rupees would be $= 20 \times 15.22 = 304.4.$

$\therefore$ Cost per kg of warp would be Rs. 3.04.

(iii) A litre of the paste contains 0.12 kg of oven-dry size ingredients. Therefore the weight of size ingredients with their moisture, per litre of size paste would be:

$$= \dfrac{0.12 \times 100}{81} = \dfrac{4}{27}\, kg$$

The cost of this would be $\dfrac{4}{27} \times \dfrac{15.22}{1} = \dfrac{60.88}{27}$

$= 2.25$ approximately.

Ans.: Rs. 15.22, Rs. 3.04 and Rs. 2.25, respectively.

13. In a weaving unit 14s yarn of 100 kg (bone dry) warp yarns was sized with an add-on of 8% and dried on conduction system with to an overall (yarn and dry size) moisture content of 10%. Workout necessary statistics.

Solution :

Bone dry warp = 100 kg, Add-on = 8%

So, $\dfrac{\text{Mass of dry size}}{\text{Mass of bone dry weight of warp}} \times 100 = 8$

or, $\dfrac{\text{Mass of dry size}}{100 \text{Kg}} \times 100 = 8$

or, Mass of dry size without moisture = 8 kg.

Total mass of bone dry (size + yarn) = 108 kg, moisture content = 10%

Therefore $10 = \dfrac{W}{W + 108} \times 100$, where W is mass of water.

Or, $W = 12$ kg

So, final weight of sized yarn is = (108 + 12) kg = 120 kg.

14. A medium Poplin fort for a decentralized sector has 20 kg cotton warp, with moisture regain of 8.5%, and is sized with a paste containing 15% con-centration of solid ingredients. The sow box is arranged to give a 10% add on the bone dry weight of the yarn, find the wet yarn pick-up? Workout the amount of water to be evaporated to give the moisture content of the sized yarn as 8%.

Solution :

Moisture regain (MR) of cotton is 8.5%

$\dfrac{W}{D} \times 100 = 8.5$, where, W = mass of water and D = oven dry mass of yarn

$W = 0.085\ D$, Total mass of yarn = oven dry mass of yarn + mass of water

$D + 0.085\ D = 20$ kg $D = 18.43$ kg

Dry size added

$$= D \times \frac{add\ on\%}{100}$$

$$= 18.43 \times \frac{10}{100} = 1.843\ \text{kg}$$

mass of water $(W) = (20 - 18.43)$ kg $= 1.57$ kg

$$\text{Wet pick up (WPU)} = \frac{\text{Add on \%}}{\text{Concentration \%}} = 10/15 = 0.667.$$

Total mass of size paste picked up by yarn $= 0.667 \times 18.43$ kg $= 12.29$ kg.

Mass of water within size paste picked up by yarn =

$$12.29 \times \left(\frac{100 - \text{concentration}\%}{100} \right)$$

$$= 12.29 \times \left(\frac{100 - 15}{100} \right) = 10.45 kg$$

Total mass of water in the yarn after pick-up
= Mass of water in the size paste + Mass of water originally present in yarn
= (10.45 + 1.57) kg = 12.02 kg
Water to be retained (for 7=8% moisture content in yarn and size film)

$$= \frac{8}{92} \times 18.43 + \frac{1.843 \times 8}{92}$$

$$= 1.76\ \text{kg}$$

The mass of water to be evaporated $= (12.02 - 1.76)$ kg $= 10.26$ kg.

15. Find the oven dry mass of warp yarn made of 40 tex cotton yarn with add on of 8% and the moisture regain of the yarn is 10%.

Solution :

Size add on = 8%, Moisture regain = 10%, So, $\frac{W}{D} \times 100 = 10$, where $W =$

mass of water and $D =$ oven dry mass of yearn

$W = 0.1D$, total mass of unsized yarn or = 1 kg

So, $D + W = 1$ kg

$D + 0.1D = 1$ kg $D = 0.909$ kg and add on is 8%.

$$\frac{\text{Mass of dry size}}{\text{Mass of oven dry yarn } (D)} \times 100 = 8 \qquad \text{Dry size} = D \times 0.08$$

$$= 0.073\ \text{kg} = 73\ \text{g}.$$

16. A stretch meter in a sizing machine records 3%, 5% and 2% stretch in the creel zone, sizing zone and drying zone, respectively. If it is expected that the fabric has warp crimp of 10%, determine the length of fabric that could be produced from 1000 m length warp sheet.

Solution :

Total stretch in sizing can be calculated using the following multiplicative expression.

$$\text{Total stretch} = \left(1+\frac{3}{100}\right) \times \left(1+\frac{5}{100}\right) \times \left(1+\frac{2}{100}\right)$$

$$= 1.103$$

Length of sized yarn $= (L_y) = 1.103 \times 1000 = 1103$ m

Crimp % = 10%

$$10 = \frac{\text{Length of yarn} \left(L_y\right) \text{ - Length of fabric} \left(L_f\right)}{\text{Length of fabric} \left(L_f\right)} \times 100$$

$$L_f = \frac{L_y}{1.1} = \frac{1103\,m}{1.1} = 1002.7\,m$$

Length of fabric is 1002.7 m.

17. A sizing machine is running at 150 m/min with 6000 ends. The add-on requirement is 12% and concentration of the size paste is 18%. If yarn count is 20 tex (without moisture) and residual moisture content in the sized yarn and film after drying is 10%, calculate the number of drying cylinder required, if a drying cylinder can evaporate 4 kg water per min.

Solution :

$$\text{Oven dry mass of yarn} = \frac{150 \times 6000 \times 20}{1000 \times 1000}\,kg = 18\,kg$$

$$= \frac{\text{Add on \%}}{\text{Concentration \%}} = \frac{12}{18} = 0.667$$

Total mass of size paste $= 0.667 \times$ Oven dry mass of yarn.

$$= \frac{12}{18} \times 18\,kg = 12\,kg$$

Mass of dry size = Mass of size paste × Concentration
Total mass of yearn and dry size = (18 + 12 × 0.18) kg = 20.16 kg for moisture content of 10%.

Mass of water to be retained in yearn and size film $= \dfrac{1}{9} \times 20.16\ kg$

$$= 2.24\ kg.$$

Mass of water in size paste of 12 kg $= 12 \left(\dfrac{100 - \text{Concentration \%}}{100} \right) = 9.84\ kg.$

Water to be evaporated $= (9.84 - 2.24)\ kg/min = 7.6\ kg/min.$

Drying capacity of one cylinder $= 4\ kg/min$ and 2 cylinders are required.

Construction of displacement, velocity and acceleration diagrams

6.1   Introduction – Significance

In any textile production system we find at least any one member will be moving with different stages and times to perform specific functions. Therefore a member when moves over a period of time, it is possible to know the displacement at any time during the movement and from this one can estimate the maximum displacement covered by the member. For example if the shuttle is travelling across the reed space, one can find the maximum distance travelled and the corresponding time so that the velocity of the shuttle can be estimated. Thus it is possible to estimate or determine the maximum displacement, maximum velocity or minimum velocity which in turn will be helpful in understanding the kinetics of shuttle propulsion or mechanics of picking, etc.

In knitting, it is important to know the displacement of needle to understand the type of the structure formed like Knit or Miss or Tuck. Similarly the distance travelled by the sword pin for one complete revolution of crank shaft in the loom will be helpful in understanding the beat-up motion. Following are the examples from the textile field to understand the displacement of the member for further analysis of velocity and acceleration.

1. Movement of the feed lattice and motion of cotton
2. Movement of the cotton in the long trunk of ERM cleaner so that the feed can be regulated by photocells
3. Motion of the flats in carding with respect to the motion of the cylinder
4. Motion of the Nippers, top comb and back detaching rollers in combing
5. Motion of the bobbin rail with bobbin building
6. Movement of the ring rail in ring frame
7. Motion of the traverse bar or guide on precision winder
8. Motion of the sheeting rollers in warping and sizing to understand the kinetics involved in comb mechanism.
9. Motion of the shuttle or Picker or Heald in loom

10. Motion of the traverse reed or mechanism in sectional warping.

11. Motion of the fabric in wet processing machineries

12. Displacement of the needle for one complete revolution of the cylinder

13. Motion of the cross-lapper in non-woven.

6.2   Construction of displacement, velocity, and acceleration diagrams

The movement of machine parts may be studied conveniently by means of displacement, velocity and acceleration diagrams. The general method is to construct first a displacement diagram, from which a velocity diagram can be obtained, and finally the velocity diagrams is used for the construction of the acceleration diagram.

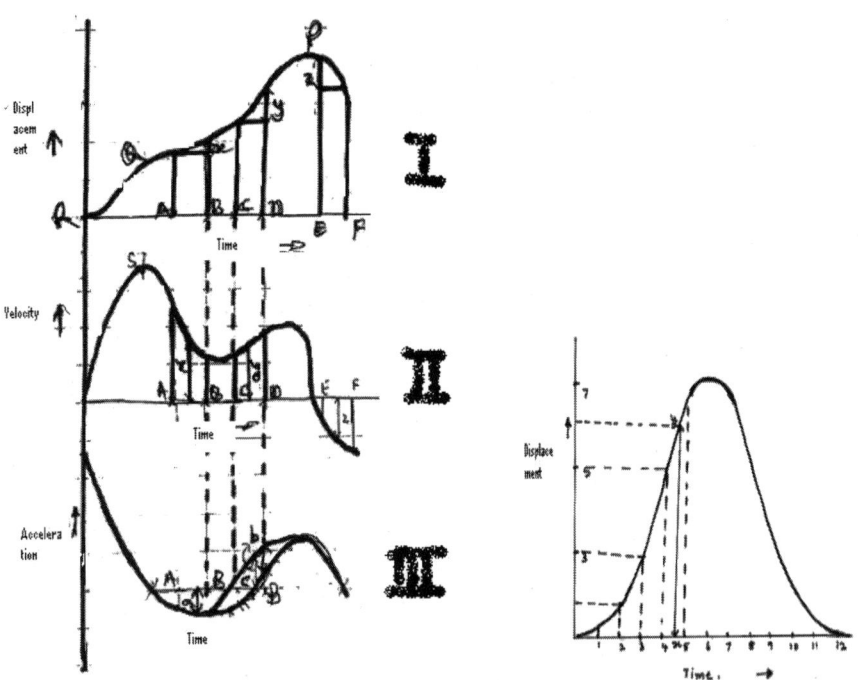

Figure 6.1   Construction of displacement, velocity and acceleration diagrams by graphical method (I: Displacement diagram, II: Velocity diagram, III: Acceleration diagram)

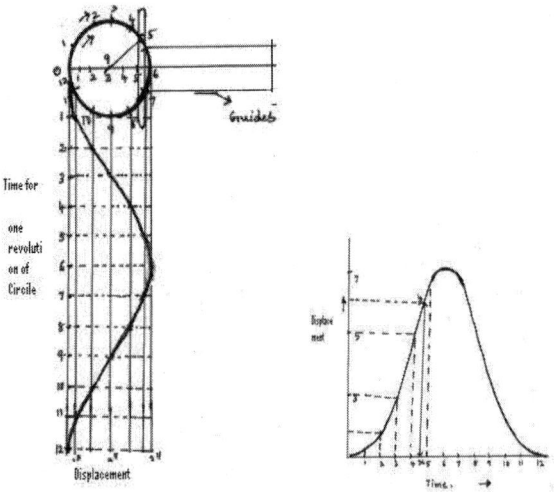

Figure 6.2 Driving arrangements for a thread guide of reeling machine

The movement of the machine part may be conveniently studied by means of the displacement, velocity and acceleration diagrams. The usual method is first to construct the displacement diagram from which a velocity diagram can be obtained and finally velocity diagram is used to construct the acceleration diagram. The displacement diagram shows where the part is at any instant. The velocity diagram show how fast is moving and the acceleration diagram the rate at which the speed is changing.

The relationship between the various diagrams and a simple method of getting the velocity and acceleration diagrams from the displacement diagram is illustrated in Figure 6.1. The displacement diagram in first part show the displacement and distance from the starting point on the basis of time. In the start time interval from A to B the displacement increases by "x". The average speed during that time is

x / time represented by AB

Similarly during the time CD the average speed is

y / time CD

The average speeds may be calculated and plotted on a new base as in diagram B and setting of the average speed during AB midway between A & B as the average speed will be reached approximately half-way through the time or if it is preferred to do the whole work graphically if all the time intervals AB, CD, EF etc., are equal, the distances x, y, z, etc., will represent the average speed during AB, CD etc., and can be set-up from the velocity diagram time base against in the centres of time interval. The distance "z" has been set-off below the base time because during time interval EF the displacement form the starting position is decreasing. At the point "*p*" the greatest displacement is reached and obviously the direction of movement is reversed there. If the velocities during AB and CD are positive then the displacement is increasing. The velocity during EF must be considered negative. Note that when the displacement reaches its maximum at "*p*", the velocity is zero and the velocity curve (stage II) crosses the base line. Note that maximum velocity occur when the displacement curve is steepest at Q and R.

The acceleration diagram (stage III) from the velocity diagram is deduced in exactly the same way as the latter is got from the displacement diagram. Thus in the time interval AB the velocity changes by the amount represented by the height "*a*". The average acceleration is therefore change of velocity represented by "*a*" divided by the time represented by AB. If all the time intervals are equal, the heights *a*, *b*, *c* etc., in diagram "*c*" represents the acceleration approximately half way through the time interval AB, CD etc., and the heights *a*, *b*, *c* etc., can be set-up front he base line of an acceleration diagram as shown in c of Figure 6.1.

It will be noticed that the acceleration "*a*" during AB, has been set-off below the base line. This is because as is obvious from the velocity diagram, the velocity is falling during AB. Hence if acceleration is considered positive when velocity increasing as during CD it should be negative when the velocity is decreasing. Note here that when the slope of the velocity curve is steepest, the acceleration has its maximum value and as the velocity changes direction as at "5" the acceleration is "zero".

Explanation for Figure 6.2

Figure 6.2 shows the method of obtaining a displacement diagram for the reciprocating movement of the thread guide bar used in reeling machine. Divide the crank circle into any number of equal parts say 12. Number these consequently in the direction of rotation of the crank from a zero position preferable one of the extreme position reached by the bar. The crank pin will occupy these position after equal intervals of time. Draw the vertical lines 1-1', 2-2' through equal positions of the crank. The displacement diagram can be drawn directly below the crank circle, base line 0-0 is drawn vertically below the zero position of the crank pin. This base line may be of any convenient length and it should be divided and numbered similarly to the crank circle into 12 equal parts. It represents the time for one revolution of the crank. If for example, speed *pf* the crank is 120 rpm or 2 rev/s, the time represented by the base line will be 0.5 s and one division will represent 1/24 of a second. By projection vertically downwards as shown from the points 1-1', 2-2' etc., to horizontal line drawn through the corresponding points 1, 2 etc., on the base line points are obtained to which the displacement diagram curve must pass. Draw a smooth curve to pass through these points.

From this curve positions of the bar can be determined at any time during the revolution and not only at the time when the crank is in the marked position 1, 2, etc. For example in the position in which the crank is drawn midway between position 4 and 5, its displacement from its zero or extreme left hand position is the distance xy on the displacement diagram.

6.3   Displacement diagram of Moscrop thread testing machine

Figure 6.3 shows the driving mechanism for a Moscrop thread testing machine. The crank AB rotates at uniform speed and rocks the slotted lever CD. C is a slider which is constrained to move in a horizontal path as shown, by attaching it to a horizontally moving carriage which it drives.

To draw a displacement diagram for pin C, trace out the path of C and mark its position for a number of different positions of the driving crank. As B moves at uniform speed, if the points 0, 1, 2 etc., on the crank circle are equidistant, the positions 0', 1', 2', etc., of C are the positions it occupies after equal intervals of time. The distances of C from its zero position (taken as its extreme left hand position) are now set-up from a base line representing the time for one revolution of the crank (Figure 6.3), 0' -1' being set up at crank position 1, 0' - 2' position 2, and so on. The extremities are joined by a smooth curve which gives, at any time during the crank's revolution, the displacement of C from its zero position.

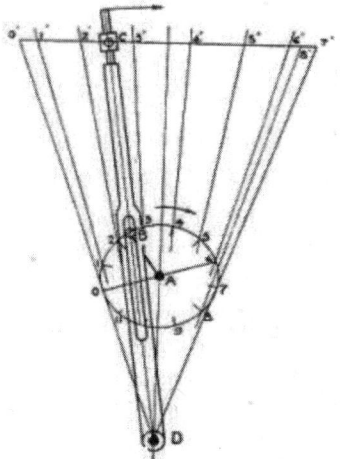

Figure 6.3 Crank movement (Source: Hanotn)

Velocity diagram – From the displacement diagram a diagram showing the velocity of C at any time can be constructed. The average velocity of a moving body during any interval of time is the distance it moves divided by the time; thus the average velocity of C during the interval of time 0-1 is 0'- 1' divided by the time taken for the crank pin to move from 0-1 which can easily be found if the speed of the crank and the diameter of the crank circle are known. This average velocity may be taken to be the velocity of C when the crank pin is midway between 0 and 1. If the velocities are calculated in this way between 0-1; 1-2, etc., and set up to any scale from a base line representing time, as before, the velocity diagram for C can be drawn by joining the points so found (Figure 6.4). To distinguish velocity from left to right from that from right to left, velocity from left to right has been plotted above the base line, and the other below the base line.

The diagram maybe constructed graphically as in Figure 6.4 which reproduces part of the displacement diagram (Figure 6.5) to a larger scale.

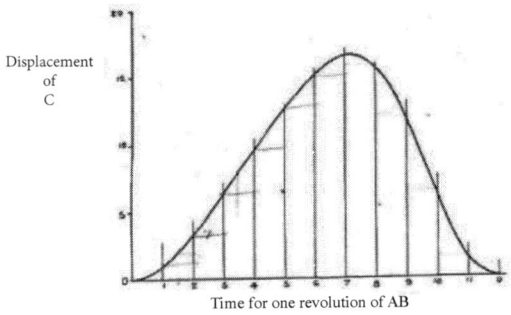

Displacement of C

Time for one revolution of AB

Figure 6.4 Displacement Diagram (Source: Hanotn)

Since the intervals of time 0-1, 1-2 etc., are all equal, the average velocities will be in proportion to the distances AB = 0' 1'; CD =1' 2' etc., and may be represented, as shown, by AB setting up midway between 0-1; CD midway between 1-2 and so on. The scale of the diagram can be found as follows: Suppose the time interval 0-1, 1-2 etc., to be 0.1 s. Then if the scale of the displacement diagram is full size, if AB measures say 0.5", the average velocity between 0-1 is $\frac{0.5"}{0.1} = 5\,\text{inch}$ per second. This is represented by a height = AB, or 0.5" on the velocity diagram, whose scale therefore is 1", representing 10 inch per second. If the scale of the displacement diagram had been, say, ¼ full size, then AB would have measured 1/8", and that height would have represented a velocity of 5 inch per second, corresponding to a scale of 1 inch to 40 inch per second.

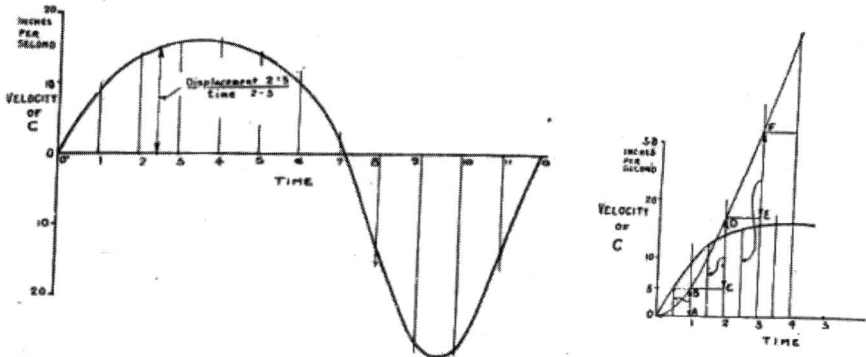

Figure 6.5   Displacement Diagram (Source: Hanotn)

From the above it will be seen that the velocity scale is obtained by dividing the displacement represented by a height of 1″ on the displacement diagram by the time interval; thus in the first example, the scale is 1″ ÷ 1/10 s =10 inch per second per inch, and in the other 4″ ÷ 1/10 s = 40 inch per second per inch. It may be convenient to double or otherwise change the heights AB, CO, etc. in making the velocity diagram, in order to get a better diagram, care being taken, of course, to alter the scale accordingly. It will be seen that these methods are only the approximate, since the average velocity may not be exactly in the middle of the time interval. The shorter the time interval, the more accurate will the velocity diagram be. The velocity diagram may be obtained in another way (Figure 6.6).

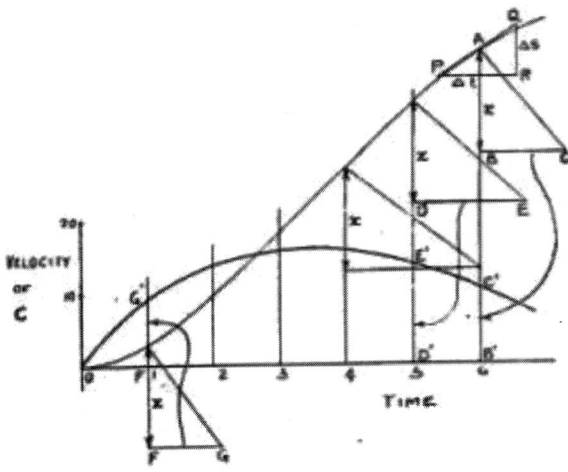

Figure 6.6   Velocity Diagram (Source: Hanotn)

The velocity at any instant is in proportion to the slope of the displacement curve. Thus at No. 6 crank position, the velocity is in proportion to $\dfrac{\Delta_s}{\Delta_t} = \dfrac{BC}{AB}$ where PQ is a tangent to the curve at A, whilst AC is a normal (at right angles to the tangent). If this construction is repeated as shown and AB = x kept constant, then the horizontal distances BC, DE, etc., represent the velocities at 6, 5, etc., and can be set up as ordinates of the velocity diagram at these points. Then if 1″ on the displacement scale represents s inch and t is the time represented by AB (to the scale of the time base line), the scale of the velocity diagram is s/t inch per second per inch.

Acceleration diagram – The acceleration diagram can be obtained from the velocity diagram in exactly the same way as the velocity diagram is got from the displacement diagram, and any of the methods given above can be used. Thus the average acceleration during any interval of time is the change in velocity during that time divided by the time. The average acceleration can be assumed to be correct for the middle of the time interval, and the calculated average accelerations can be set up as the ordinates of the acceleration diagram in the middle of the time intervals. Figure 6.7 shows the application of the second method, the changes of velocity AB, CD etc., being set up as the accelerations in the middle of the time intervals 0-1, 0-2, etc. Where the positive velocity is increasing the acceleration is positive and is set above the base line where the velocity is decreasing, the acceleration is negative; but where the velocity changes in direction, as represented by the part of the velocity diagram below the base line, the acceleration is negative for an increasing velocity and positive for a decreasing velocity. The reason for this can be seen well if the force required for the acceleration is considered. The force required is always in direct proportion to the acceleration, and it is evident that the same direction of force is needed to bring a body to rest as is required to start it up in the opposite direction. The scale of the acceleration diagram (Figure 6.7) is obtained by dividing the velocity represented by a height of 1″ on the velocity diagram by the time interval, the result being the acceleration represented by 1″. Of height on the acceleration diagram, Figure 6.8 shows the tangent method, the scale being the velocity represented by a height of 1″, divided by the time represented by the length AB, to the base line scale of time. When the tangent to the velocity curve becomes parallel to the base line, as at P, the acceleration is zero.

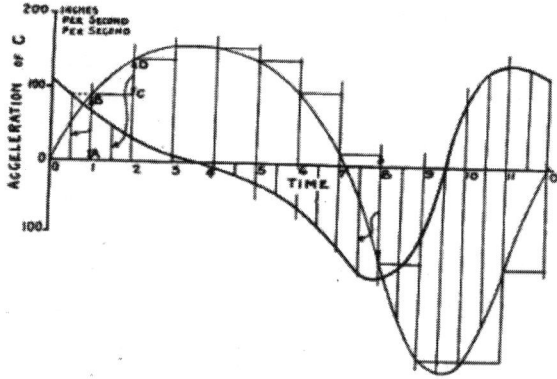

Figure 6.7   Acceleration Diagram (Source: Hanotn)

But at R, although the velocity is zero, there is an acceleration. Although the body stops, when at R, and changes the direction of its motion, an infinitely short time before R it is moving positively, an infinitely short time after R it is moving negatively. It has had negative acceleration and has required the continuous application of accelerating force to stop and start it.

Accelerating force diagram – The force required to accelerate a body is in direct proportion to the acceleration and is equal to the mass of the body in engineer's units

$$\left(\frac{lbs.}{g = 32.2}\right) \text{ multiplied by the acceleration.}$$

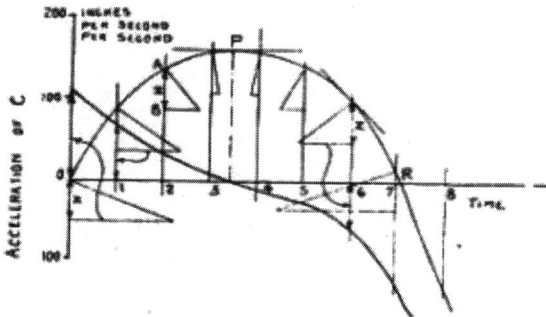

Figure 6.8   Acceleration Diagram (Source: Hanotn)

The accelerating force diagram will therefore be exactly similar to the acceleration diagram, with the scale corrected. If the same diagram is used for acceleration and accelerating force, and if the scale for acceleration is 1″ representing f feet per second per second, the force scale will be 1″ representing

$$f \, X \frac{mass\ of\ body\,(lb.)}{32.2} lb.$$

Numerical examples

1. The thread guide of cone winding machine is displaced as follows, as it moves from larger to the smaller end of the cone in the time of "3" revolutions of roller driving the cone.

Revolution of roller	0	0.5	1.0	1.5	2.0	2.5	3
Guide displacement (in inches)	0	0.75	1.6	2.6	3.65	4.8	6.0

Solution :

Plot the curve of displacement on a base representing time. From the displacement curve draw the speed curve. Estimate the speeds if the guide after 0.5 and 2.5 revolutions of the roller is the speed of the roller is 900 rpm.

∴ Roller speed is 15 rpm.

Time of 1 revolution $= \dfrac{1}{15} s$

∴ Time of 3 revolutions of roller $= \dfrac{3}{15} = \dfrac{1}{5} s = 0.2s$

Time of 0.5 revolution of roller $= \dfrac{0.5}{15} = 0.033s$

Time of 1.0 revolution of roller $= \dfrac{1.0}{15} = 0.066s$

Time of 1.5 revolution of roller $= \dfrac{1.5}{15} = 0.1s$

Time of 2.0 revolution of roller $= \dfrac{2.0}{15} = 0.133s$

Time of 2.5 revolution of roller $= \dfrac{2.5}{15} = 0.166s$

Time of 3.0 revolution of roller $= \dfrac{3.0}{15} = 0.2s$

Velocity diagram $V = \dfrac{\text{Change in displacement}}{\text{Time interval}}$

Roller revolution	Velocity
0–0.5	$\dfrac{0.75 - 0}{0.03} = 25 \text{ in/s} = 2.08 \text{ ft/s}$
0.5–1.0	$\dfrac{1.6 - 0.75}{0.03} = 2.36 \text{ ft/s}$
1.0–1.5	$\dfrac{2.6 - 1.6}{0.03} = 2.78 \text{ ft/s}$
1.5–2.0	$\dfrac{3.65 - 2.6}{0.03} = 2.92 \text{ ft/s}$
2.0–2.5	$\dfrac{4.8 - 3.65}{0.03} = 3.19 \text{ ft/s}$
2.5–3.0	$\dfrac{6 - 4.8}{0.03} = 3.33 \text{ ft/s}$

2. A carriage moves along a straight path as follows

Time (t), s	0	4	10	15	22	38	45	50
Distance in meters, m	4	7	30	73	150	393	543	710

Solution :

Plot the displacement and velocity diagrams

$$\text{Velocity} = \frac{\text{change in displacement}}{\text{time}}$$

1. $V = \dfrac{7 - 4}{4} = \dfrac{3}{4} = 0.75 \text{ m/s}$

2. $V = \dfrac{30 - 7}{10 - 4} = \dfrac{23}{6} = 3.83 \text{ m/s}$

3. $V = \dfrac{73 - 30}{15 - 10} = \dfrac{43}{5} = 8.6 \text{ m/s}$

4. $V = \dfrac{150 - 73}{22 - 15} = \dfrac{77}{7} = 11 \text{ m/s}$

5. $V = \dfrac{393 - 150}{38 - 22} = \dfrac{243}{16} = 15.2 \text{ m/s}$

6. $V = \dfrac{543 - 393}{45 - 38} = \dfrac{150}{7} = 21.43 \text{ m/s}$

7. $V = \dfrac{710 - 543}{50 - 45} = \dfrac{167}{5} = 33.4$ m/s.

3. The distance moved by the run of a pulley on a loom driving shaft driving 14 equal time intervals between one beater and the next were found to be as follows.

Time interval number	1	2	3	4	5	6	7	8	9	10	11	12	13	14
Rim movement in inches	2.6	2.3	2.15	2.1	2.2	2.05	2.3	2.5	2.35	2.2	2.10	2.15	2.4	2.6

Plots displacement and velocity graph. Plot a displacement curve showing approximate variation in speed through speed of driving shaft is 162 rpm. What are approximate highest and lowest speed drives the revolution.

Solution :

Time interval number	1	2	3	4	5	6	7	8	9	10	11	12	13	14
Rim movement in inches	2.6	2.3	2.15	2.1	2.2	2.05	2.3	2.5	2.35	2.2	2.10	2.15	2.4	2.6
Cumulative displacement speeds	2.6	4.9	7.05	9.15	11.35	13.4	15.7	18.2	20.55	22.75	24.85	27	29.4	32

$$\text{Velocity} = \frac{\text{change in displacement}}{\text{Time}}$$

$$\text{Velocity} = \frac{4.9 - 2.6}{1}$$

In 60 → 162 new

? ← 1 revolution.

$$\frac{1 \times 60}{162} = 0.37s$$

∴ Each time interval $= \dfrac{0.37}{14} = 0.026s$

Total displacement = 32

∴ Any displacement $= \dfrac{32}{14} = 2.286.$

$$= 2.29''.$$

Time interval number	1	2	3	4	5	6	7	8	9	10	11	12	13	14
Displacement	2.6	2.3	2.15	2.1	2.2	2.05	2.3	2.5	2.35	2.2	2.10	2.15	2.4	2.6
***	183.93	162.7	152.09	148.5	155.63	145.02	162.7	176.85	166.84	155.63	148.55	152.09	169.78	183.93

Average displacement $n = 2.29''$
Average speed = 162 rpm.
2.29″ $\rightarrow$ speed 162

$\therefore$ 2.6″ $\rightarrow$ speed $= \dfrac{162 \times 2.6}{2.29} = 183.93$

For 2.3 speed $= \dfrac{162}{2.29} \times 2.3 = 162.7$ rpm

For 2.15 speed = 152.09 rpm
For 2.1 speed = 148.56 rpm
For 2.2 speed = 155.63 rpm
For 2.05 speed = 145.02 rpm
For 2.3 speed = 162.7 rpm
For 2.5 speed = 176.85 rpm
For 2.35 speed = 166.24 rpm
For 2.2 speed = 155.63 rpm
For 2.10 speed = 148.55 rpm
For 2.15 speed = 152.09 rpm
For 2.4 speed = 169.78 rpm
For 2.6 speed = 183.93 rpm.

4. A loom shuttle has a displacement from its starting position at the end ten equal time intervals are as follows.

Time interval number	1	2	3	4	5	6	7	8	9	10
Displacement in inches	0.02	0.01	0.3	0.65	1.2	1.95	2.9	4.0	5.2	6.5

Total time for acceleration is 0.03 s. Draw the displacement and velocity diagrams and also acceleration diagram.

Solution :

Total time interval = 0.03

One time interval $= \dfrac{0.03}{10} = 0.003$ s

1. Velocity $= \dfrac{\text{displacement}}{\text{Time}} = \dfrac{0.02 - 0}{0.003} = 6.67$

2. Velocity $= \dfrac{0.1 - 0.02}{0.003} = 26.67 \, \text{in. /s}$

3. Velocity $= \dfrac{0.3 - 0.1}{0.003} = 66.67 \, \text{in. /s}$

4. Velocity $= \dfrac{0.65 - 0.3}{0.003} = 116.67 \, \text{in. /s}$

5. Velocity $= \dfrac{1.2 - 0.65}{0.003} = 183.33 \, \text{in. /s}$

6. Velocity $= \dfrac{1.95 - 1.2}{0.003} = 250 \, \text{in. /s}$

7. Velocity $= \dfrac{2.9 - 1.95}{0.003} = 316.67 \, \text{in. /s}$

8. Velocity $= \dfrac{4.0 - 2.9}{0.003} = 366.67 \, \text{in. /s}$

9. Velocity $= \dfrac{5.2 - 4.0}{0.003} = 400 \, \text{in. /s}$

10. Velocity $= \dfrac{6.5 - 5.2}{0.003} = 433.33 \, \text{in. /s}$

Acceleration $= \dfrac{\text{Change in velocity}}{\text{Time}}$

1. Acceleration $= \dfrac{667.0}{0.003} = 2223.33 \, \text{in. /s}^2 = 185.28 \, \text{ft/s}^2$

2. Acceleration $= \dfrac{26.67 - 6.67}{0.003} = 6666.67 \, \text{in. /s}^2 = 555.56 \, \text{ft/s}^2$

3. Acceleration $= \dfrac{66.67 - 26.67}{0.003} = 13{,}333.33 \, \text{in. /s}^2 = 1338.89 \, \text{ft/s}^2$

4. Acceleration $= \dfrac{116.67 - 66.67}{0.003} = 16{,}666.67 \, \text{in. /s}^2 = 1851.67 \, \text{ft/s}^2$

5. Acceleration $= \dfrac{183.33 - 116.67}{0.003} = 22{,}220 \, \text{in. /s}^2 = 1851.94 \, \text{ft/s}^2$

6. Acceleration $= \dfrac{250\text{-}183.33}{0.003} = 22,223.33$ in. $/s^2 = 1851.94$ ft/s^2

7. Acceleration $= \dfrac{316.67\text{-}250}{0.003} = 22,223.33$ in. $/s^2 = 1388.88$ ft/s^2

8. Acceleration $= \dfrac{366.67\text{-}316.67}{0.003} = 16,666.67$ in. $/s^2 = 925.83$ ft/s^2

9. Acceleration $= \dfrac{400\text{-}366.67}{0.003} = 11,110$ in. $/s^2 = 925.83$ ft/s^2

10. Acceleration $= \dfrac{433.33\text{-}400}{0.003} = 11,110$ in. $/s^2 = 1111.11$ ft/s^2.

5. A sword pin on a reciprocating loom sley has displacement as follows: From its forward or beater position. During M one revolution of crank which drives it. The crank shaft speed being 180 rpm. Draw displacement, velocity and acceleration diagram for the sword pin.

Displacement from beater	0	30	60	90	120	150	180	210	240	270	300	330	360
Pin displacement its starting line	0	0.45	1.55	3.1	4.5	5.5	6.0	5.8	4.95	3.45	1.8	0.5	0

Solution :

Rotation per minute (RPM) $= 180$

In $(180 \times 360)°$ it takes 60 s,

$\therefore$ For $30°$ it takes $\dfrac{30 \times 60}{280 \times 360}$

$= \dfrac{1}{36}$ s

$= 0.0277$ s.

Total timing $= 0.027 \times 12$ (time intervals)

$= 0.333$ s.

Velocity $= \dfrac{\text{Change in displacement}}{\text{time}}$

1. $\dfrac{0.45 - 0}{0.028} = 16.07$ unit $/s$

2. $\dfrac{1.55 - 0.45}{0.028} = 39.29 \text{ unit / s}$

3. $\dfrac{3.1 - 1.55}{0.028} = 55.38 \text{ unit / s}$

4. $\dfrac{4.5 - 3.1}{0.028} = 50 \text{ unit / s}$

5. $\dfrac{5.5 - 4.5}{0.028} = 35.71 \text{ unit / s}$

6. $\dfrac{6.0 - 5.5}{0.028} = 17.85 \text{ unit / s}$

7. $\dfrac{5.8 - 6}{0.028} = -7.14 \text{ unit / s}$

8. $\dfrac{4.95 - 5.8}{0.028} = -30.36 \text{ unit / s}$

9. $\dfrac{3.45 - 4.95}{0.028} = -53.57 \text{ unit / s}$

10. $\dfrac{1.8 - 3.45}{0.028} = -58.93 \text{ unit / s}$

11. $\dfrac{0.5 - 1.8}{0.028} = -46.43 \text{ unit / s}$

12. $\dfrac{0.0 - 0.5}{0.028} = -17.86 \text{ unit / s}$

$$\text{Acceleration} = \dfrac{\text{Change in velocity}}{\text{time}}$$

1. $A = \dfrac{16.7 - 0}{0.028} = 573.92 \text{ unit / s}^2$

2. $A = \dfrac{39.29 - 16.07}{0.028} = 829.28 \text{ unit / s}^2$

3. $A = \dfrac{55.36 - 39.29}{0.028} = 573.93 \text{ unit / s}^2$

4. $A = \dfrac{50 - 55.38}{0.028} = -192.14 \text{ unit / s}^2$

5. $A = \dfrac{35.71 - 50}{0.028} = -510.36 \text{ unit / s}^2$

6. $A = \dfrac{17.85 - 35.71}{0.028} = -637.89 \text{ unit / s}^2$

7. $A = \dfrac{-7.14 - 17.85}{0.028} = -892.5 \text{ unit / s}^2$

8. $A = \dfrac{-30.36 + 7.14}{0.028} = -829.27 \text{ unit / s}^2$

9. $A = \dfrac{-53.57 + 30.36}{0.028} = -828.93 \text{ unit / s}^2$

10. $A = \dfrac{-58.03 + 53.57}{0.028} = -191.43 \text{ unit / s}^2$

11. $A = \dfrac{-46.43 + 58.93}{0.028} = 446.43 \text{ unit / s}^2$

12. $A = \dfrac{-17.86 + 46.43}{0.028} = 1020.36 \text{ unit / s}^2$

6. A drop box on a multiple shuttle box one positively controlled C is lifted 6″ during the time taken by the loom crank shaft to move 144° at a speed of 120 rpm. Distance moved by the box in successive time intervals of 0.1 of the total are as follows.

Time intervals number	1	2	3	4	5	6	7	8	9	10
Displacement inches	0.15	0.45	0.65	0.8	0.95	0.95	0.8	0.65	0.45	0.15

Note the displacement and velocity diameter from the diagram find out how fast the box is moving when it has been lifted 1″ and 5″. What are the approximate accelerations then?

Solution :

Cumulative displacement	0.02	0.04	0.06	0.08	0.10	0.12	0.14	0.16	0.18	0.2
Time intervals number	1	2	3	4	5	6	7	8	9	10
Displacement inches	0.15	0.45	0.65	0.8	0.95	0.95	0.8	0.65	0.45	0.15
Cumulative displacement in inches	0.15	0.60	1.25	2.05	3.0	3.95	4.75	5.4	5.85	6

$$t = \frac{144 \times 60}{360 \times 120} = 0.2 \ s \text{ for 10 intervals}$$

$$\therefore t = 0.02 \text{ s for 1 interval.}$$

$$\text{Velocity} = \frac{\text{Change in displacement}}{\text{Time}}$$

1. $\text{Velocity} = \dfrac{0.15 - 0}{0.02} = \dfrac{0.15}{0.02} = 7.5 \text{ in /s}$

2. $\text{Velocity} = \dfrac{0.6 - 1.5}{0.02} = \dfrac{0.45}{0.02} = 22.5 \text{ in /s}$

3. $\text{Velocity} = \dfrac{1.25 - 0.6}{0.02} = \dfrac{0.65}{0.02} = 32.5 \text{ in /s}$

4. $\text{Velocity} = \dfrac{2.05 - 1.25}{0.020} = \dfrac{0.8}{0.02} = 40 \text{ in /s}$

5. $\text{Velocity} = \dfrac{3.0 - 2.05}{0.02} = \dfrac{0.95}{0.02} = 47.5 \text{ in /s}$

6. $\text{Velocity} = \dfrac{3.95 - 3.0}{0.02} = \dfrac{0.95}{0.02} = 47.5 \text{ in /s}$

7. $\text{Velocity} = \dfrac{4.75 - 3.95}{0.02} = \dfrac{0.80}{0.02} = 40 \text{ in /s}$

8. $\text{Velocity} = \dfrac{5.4 - 4.75}{0.02} = \dfrac{0.65}{0.02} = 32.5 \text{ in /s}$

9. $\text{Velocity} = \dfrac{5.85 - 5.4}{0.02} = \dfrac{0.45}{0.02} = 22.5 \text{ in /s}$

10. $\text{Velocity} = \dfrac{6 - 5.85}{0.02} = \dfrac{0.15}{0.02} = 7.5 \text{ in /s}$

A) $\text{Acceleration} = \dfrac{\text{Change in velocity}}{\text{Time}} \text{ in/s}^2$

1) $\dfrac{7.5 - 0}{0.02} = 375 \text{ in/s}^2$

2) $\dfrac{22.5 - 7.5}{0.02} = 750 \text{ in/s}^2$

3) $\dfrac{32.5 - 22.5}{0.02} = 500$ in/s^2

4) $\dfrac{40 - 32.5}{0.02} = 375$ in/s^2

5) $\dfrac{47.5 - 40}{0.02} = 375$ in/s^2

6) $\dfrac{47.5 - 47.5}{0.02} = 0$ in/s^2

7) $\dfrac{40 - 47.5}{0.02} = -375$ in/s^2

8) $\dfrac{32.5 - 40}{0.02} = -375$ in/s^2

9) $\dfrac{22.5 - 32.5}{0.02} = -500$ in/s^2

10) $\dfrac{7.5 - 22.5}{0.02} = -750$ in/s^2

Role of clutches and breaks in textile production

7.1 What are clutches?

A clutch is a mechanical device that provides for the transmission of power (and therefore usually motion) from one component (the driving member) to another (the driven member) when engaged, but can be disengaged. Clutches allow a high inertia load to be stated with a small power. Clutches are used whenever the transmission of power or motion needs to be controlled either in amount or over time (e.g., electric screwdrivers limit how much torque is transmitted through use of a clutch; clutches control whether automobiles transmit engine power to the wheels). In the simplest application, clutches are employed in devices which have two rotating shafts (drive shaft or line shaft). In these devices, one shaft is typically attached to a motor or other power unit (the driving member) while the other shaft (the driven member) provides output power for work to be done. In a torque-controlled drill, for instance, one shaft is driven by a motor and the other drives a drill chuck. The clutch connects the two shafts so that they may be locked together and spin at the same speed (engaged), locked together but spinning at different speeds (slipping), or unlocked and spinning at different speeds (disengaged). Therefore, a clutch is a form of connection between a driving and a driven member on the same axis. It is so designed that the two members may be engaged or disengage at will either by hand operated device or automatically by the action of some power device.

7.2 Types of clutches

Clutches are mainly classified as follows

- Pneumatic,
- Mechanical,
    - Positive contact clutches,
        - Jaw clutch,
        - Toothed clutch

- Friction clutches,
  - Disc clutch,
  - Cone clutch, and,
  - Centrifugal clutch,
  - Hydraulic, and,
- Electromagnetic.

7.3   Materials of construction

Various materials have been used for the disc friction facings, including asbestos in the past. Modern clutches typically use a compound organic resin with copper wire facing or a ceramic material. A typical coefficient of friction used on a friction disc surface is 0.35 for organic and 0.25 for ceramic. Ceramic materials are typically used in heavy applications such as trucks carrying large loads or racing, though the harder ceramic materials increase flywheel and pressure plate wear.

A simple manually operated friction clutch is shown in Figure 7.1. Basically an operating lever is moved either left or right. As the lever is moved to the right the rotating pulley operates against the friction disc. If sufficient pressure is applied to the operating lever, it will cause the friction between the driving pulley and the friction disc to rotate the driven member. During the time necessary for sufficient friction to be created between the members, slippage will occur which is an important point to be understood with this type of arrangement.

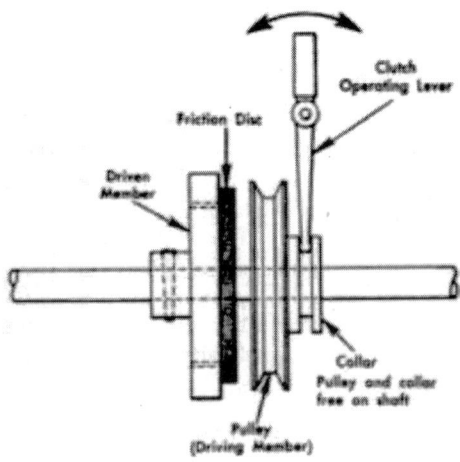

Figure 7.1   Elements of a friction clutch

7.4   Simple plate clutch

7.4.1   Need for clutch

The looms can be fitted with a direct gear drive for working of the loom and the motor is switched on and off as and when loom stops due to stop motions. This calls for high starting torque and such arrangements may not work suitable with loom banging off set up. Because the momentum of the motor will cause it to continue running and thus necessitates the use of clutch. Such a clutch will absorb the motor energy in such sudden stoppages. It is also to be noted that the side movement should not result in poor life of motor assembly. Further it can be said that loom stoppage to crash stop is most essential to avoid the starting marks.

In the simple plate clutch system is illustrated in Figure 7.2, the distance r can be assumed to be the average distance of the driving point from the centre of the shaft, and the friction force Fr will then be equal to μPr, where μ is the coefficient of friction and P the sideways thrust required to bring the clutch plate into effective contact with the driving wheel. In Figure 7.2, however, the normal force thrust he related to the sideways thrust required and thus, since

$$F = \mu R$$

and

$$\text{Sin } \theta = \frac{P}{R}$$

then:

$$R = \frac{P}{\sin \theta}$$

and hence

$$F = \frac{\mu P}{\sin \theta}$$

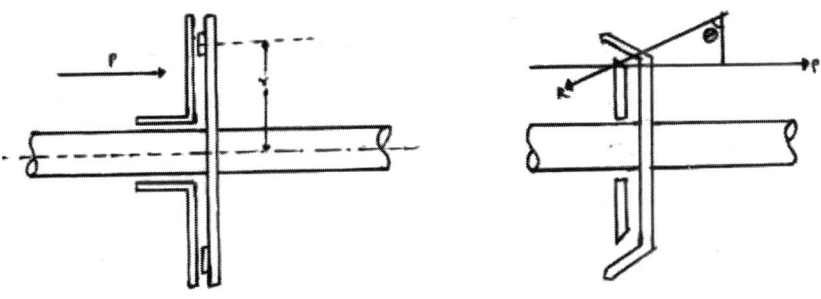

Figure 7.2 Simple plate clutch

$$F = \frac{\mu P}{\sin \theta}.$$

For example if a frictional force of 400 N is required to drive a loom and μ is 0.50, then a sideways thrust of

$$P = \frac{F}{\mu} = \frac{400}{0.5} = 800 \ N$$

will be required with a plate clutch, but, with a conical clutch having an angle of 30°, the required thrust will be:

$$P = \frac{F \sin \theta}{\mu} = \frac{400 \times 0.5}{0.5} = 400 \ N$$

and, if the cone angle is reduced to 10°, the required thrust will be:

$$P = \frac{F \sin \theta}{\mu} = \frac{400 \times 0.1736}{0.5} = 138.88 \ N$$

Hence the conical clutch needs a lower sideways force, and furthermore, the smaller the cone angle, then the lower will be the sideways thrust required to drive the loom. There will, however, be a greater amount of wear and an increased risk of jamming in the engaged position.

Sideways displacement involving forces of these magnitudes is generally un-desirable in spite of the popularity of these systems because wear will occur in the main shaft bearings unless adjustable thrust bearings are used. Expanding clutches overcome this problem. The clutch is usually mounted inside the driving wheel, and it is made to expand or contract by the starting handle of the loom, which gives a partial rotation to a rod. Each end of the clutch band carries an internally screwed sleeve, but the two sleeves are screwed in opposite directions. The ends of the rods are similarly screwed in opposite directions to pair with the sleeves. The linkage between the rod and the starting handle is so arranged that the clutch expands and the loom starts when the handles is moved to the ON position. Movement of the handle in the other direction will cause the clutch to contract and thus cease to drive the loom. In a more recent arrangement, the clutch band in this type of clutch is loaded with a series of springs, which force the band onto the driving wheel when the starting handle is moved to the ON position.

7.5   Friction clutch

A friction clutch has its principal application in the transmission of power of shafts and machines which must be started and stopped frequently. Its application is also found in cases in which power is to be delivered to machines partially or fully loaded. The force of friction is used to start the driven shaft from rest and gradually brings it up to the proper speed without excessive slipping of the friction surfaces. In automobiles, friction clutch is used to connect the engine to the driven shaft. In operating such a clutch, care should be taken so that the friction surfaces engage easily and gradually brings the driven shaft up to proper speed. The proper alignment of the bearings must be maintained and it should be located as close to the clutch as possible. It may be noted that

1. The contact surfaces should develop a frictional force that may pick up and hold the load with reasonably low-pressure between the contact surfaces.

2. The heat of friction should be rapidly dissipated and tendency to grab should be at a minimum.

3. The surfaces should be backed by a material stiff enough to ensure a reasonably uniform distribution of pressure.

The friction clutches of the following types are important from the subject point of view:

1. Disc or plate clutches (single disc or multiple disc clutch),

2. Cone clutches, and,

3. Centrifugal clutches.

The clutches are discussed with necessary inputs. It may be noted that the disc and cone clutches are based on the same theory as the pivot and collar bearings.

7.5.1   Single disc or plate clutch

A single plate friction clutch consisting of a clutch disk between the flywheel and a pressure plate. Both the pressure plate and the flywheel rotates with the engine crankshaft or the driving shaft. And both sides of clutch disc are faced with friction material (usually of ferrodo). The clutch disc is mounted on the hub which is free to move axially along the splines of the driven shaft but not turn able towards the transmission input shaft. The pressure plate pushes the clutch plate towards the flywheel by a set of strong springs which are arranged radially inside the body. The three levers (also known as release levers or fingers) are carried on pivots suspended from the case of the body. These are arranged in such a manner so that the pressure plate moves away from the flywheel by the inward movement of a thrust bearing. The bearing is mounted upon a forked shaft and moves forward when the clutch pedal is pressed. By pressing the clutch pedal down, the thrust bearing moves towards the flywheel by means of linkage force, and press the longer end of the lever inwards. Due to this, the lever turns on their suspended pivot and forces the pressure plate to move away from the flywheel this action compresses the clutch springs which in turn moves the pressure plate away from the clutch plate and remove the pressure from the clutch plate. This enables the clutch plate to move back from the flywheel and thus, the driven shaft becomes stationary by moving the foot back from the clutch pedal, the thrust bearing moves back and allows the spring to extend which pushes the clutch plate backwards the flywheel. This engages the flywheel and the clutch plate which starts the motion of the driven shaft (Figure 7.3).

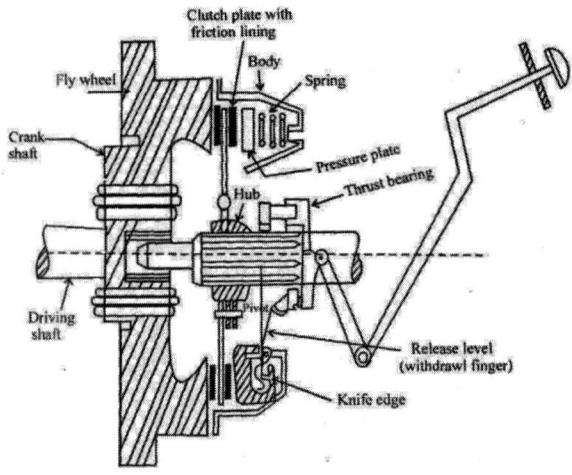

Figure 7.3   Plate clutch

Formula

Friction torque between each pair of contact surfaces for each surface of clutch:

$$T = \frac{\mu \, W \, (r_1 + r_2)}{2}$$

Where W = axial pressure
r_1 and r_2 = external,
x = internal radii

For single plate clutch having two pair of contact surfaces then,

$$T = \frac{\mu W (r_1 + r_2)x^2}{2}$$

$$= \mu \, W \, (r_1 + r_2).$$

7.5.2   Cone clutches

A simple cone clutch is shown in Figure 7.4. Cone clutch consists of an inner cup attached to driving shaft and follower cone. Follower cone is movable; it can axially slide over the driven member. The inner side of the driver cup exactly fits the outer surface of the cone. The slope of the cone is made small, that help to give higher normal forces. The recommended angle of slope is between 8–15°. According to the allowable normal pressure and coefficient of friction required the contact face of the driven member is lined with material like leather, asbestos, wood, etc. The clutch engaged by bringing two cone surface together in contact. A spring is provided on the driven shaft to hold the face of clutch in touch by producing required axial force. A forked lever is used to disengagement of the clutch.

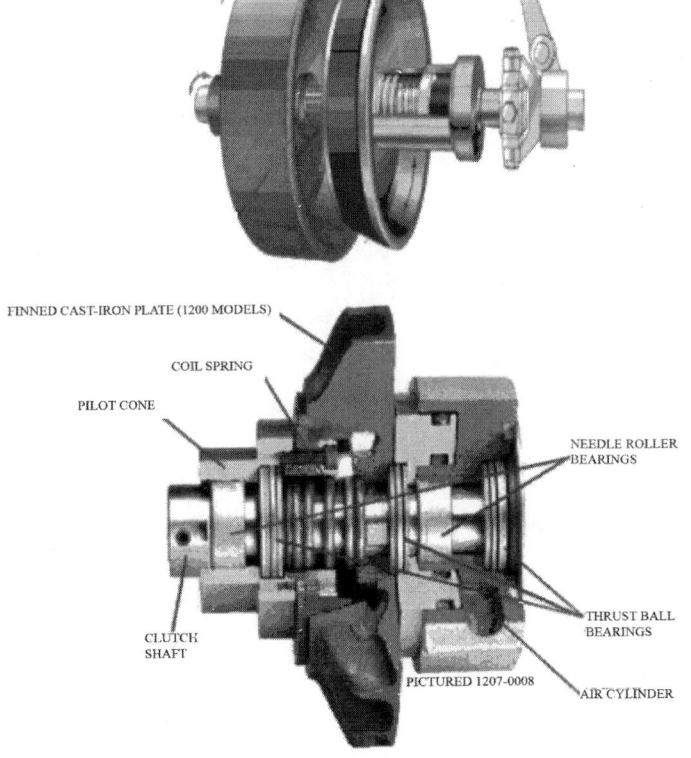

Figure 7.4 Cone clutches

Advantages

- Small axial force is required to keep the clutch engaged.
- Simple design.
- For a given dimension, the torque transmitted by cone clutch is higher than that of a single plate clutch.

Disadvantages

- One pair of friction surface only.
- The small cone angle causes some reluctance in disengagement.

7.5.3   Centrifugal clutch

These are used when it is required to engage a load after the driving member attained a particular speed. This clutch is helpful with an engine that has trouble starting under load. They are usually used in motor pulleys. The centrifugal clutch consists of a number of shoes or friction pads arranged radially symmetrical position inside the rim. It can slide along the guide's integral with the boss on the driving shaft. The shoes are held against boss by using a spring that exerts a radially inward force. As the inner hub rotates, the weight of the shoe causes a radially outward force known as centrifugal force. This force depends on the weight of the shoe and the speed at which it rotates. At low speed, the centrifugal force also low, the shoes remain in the same position. As speed increases, the centrifugal force also increases, when centrifugal force becomes equal to spring force the shoes start floating. When the driver rotates fast enough the centrifugal force exceeds the spring force the shoes moves outward. At a certain speed, it gets contact with the inner surface of the drum and torque is transmitted. As the load increases, speed decreases; the shoes return to their original position and clutch gets disengaged (Figure 7.5).

Advantages

- Simple and inexpensive and need little maintenance.
- The centrifugal clutch is automatic any kind of control mechanism is not necessary.
- They help to prevent the engine from stalling.
- The engagement speed can precisely control by selecting spring.

Disadvantages

- Loss of power due to friction and slipping.
- This type of clutch not appropriate for the high amount of torque, the shoes will slip at the heavy loaded condition.

- They engage at full or near-full power, shoes get heated very quickly may cause overheating.

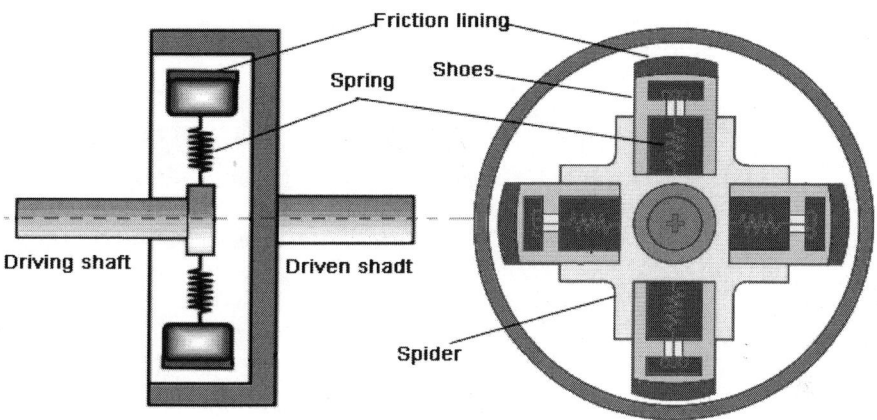

Figure 7.5   Centrifugal clutch

7.6   Commercial clutches

Morse mechanical CAM clutches use a cam or wedging element to lock the inner and outer races of the clutch. This transmits torque in only one direction. CAM clutches are often referred to as freewheels, sprag, overrunning, backstop or one-way clutches. These clutches are used for three basic modes of operation: Overrunning, Indexing and Backstopping. Morse CAM clutches set the standard for higher overrunning speeds, greater torque capacities and longer service life. Morse clutches offer one of the most complete offerings in the industry.

Morse B200-B500 clutches
The B200-B500 series clutches share the same design. A full complement of cams is retained in an outer race. This requires a bearing supported hardened shaft as the inner race. These are primarily used for back stopping applications (Figure 7.6).

Morse B200-B500 clutches   Morse BR series clutches   Morse HT clutches

Figure 7.6   Commercial Morse clutches

Morse BR series clutches

The BR series clutch has cams that are designed to lift-off. They have no contact with either the inner or outer race when it overruns. This provides extended overrunning wear life.

Figure 7.7   Morse KK clutches, Morse M clutches, and accessories Morse MZEU clutches

Morse HT clutches

The HT series clutch was designed primarily for indexing applications. These cam clutches are not symmetrical and therefore must be ordered according to the rotation required in application.

Morse KK clutches

The KK series clutch is a compact cam clutch with built-in bearing support. This unique construction combines a cam clutch with a 6200 series metric ball bearing for use in applications ranging from exercise equipment to industrial machinery (Figure 7.7).

Morse M clutches and accessories

The M series clutch has a dual ball bearing supported design. It contains high quality steel cams in a cage. They are hardened precision formed and finished.

Morse MZEU clutches

The MZEU series clutch has a dual metric ball bearing supported design with metric bores. It is suitable for overrunning and middle speed backstopping.

Figure 7.8   Morse NFS clutches, Morse NSS clutches, Morse PB clutches

Morse NFS clutches

The NFS Series clutch is available in 13 sizes. It is designed with the same overall dimensions as a light metric series ball bearing. The bore of the clutch is metric and must be installed with ball bearing support (Figure 7.8).

Morse NSS clutches

The NSS series clutch is available in 12 sizes. It is designed with the same overall dimensions as a light metric series ball bearing.

Morse PB clutches

The PB series clutch features precision formed cams providing high torque capacities size-for-size combined with excellent wear life. They have a bronze bearing to maintain race concentricity.

7.7   Electromagnetic clutch

The operating principle of the actuator of electromagnetic (EM) clutch is an electromagnetic effect, but torque transmission is mechanical. The difference between electromagnetic clutch and the regular clutch is in how they control the movement of pressure plates. In the normal clutch, a spring used to engage the clutch whereas in EM clutch an electromagnetic field is used for engagement. The electromagnetic clutch comes various forms, including magnetic particle clutch and multi-disc clutch. There are even no contact clutches such as hysteresis clutch and eddy current clutch. However, most widely used form is single face friction clutch.

7.7.1   Working principle

The main components of EM clutch are a coil shell, an armature, rotor, and hub. The armature plate is lined with friction coating. The coil is placed behind the rotor. When the clutch activated the electric circuit energises the coil, it generates a magnetic field. The rotor portion of clutch gets magnetised. When the magnetic field exceeds the air gap between rotor and armature and then it pulls the armature toward the rotor. The frictional force generated at the contact surface transfer the torque. Engagement time depends on the strength of magnetic fields, inertia, and air gap. When voltage is removed from the coil, the contact is gone. In most design a spring is used to hold back the armature to provide an air gap when current is removed (Figure 7.9).

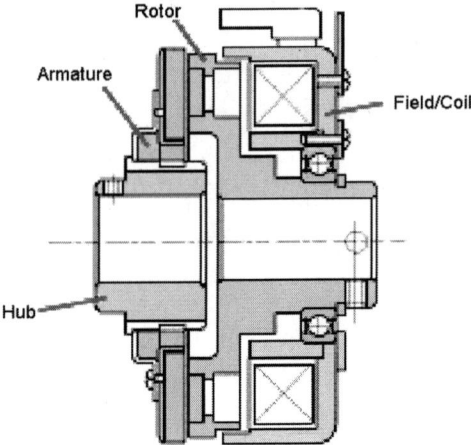

Figure 7.9 Electromagnetic clutch – general

Advantage

• The complicated linkage is not required to control clutch.
*Disadvantage*s

• High initial cost.

• Operating temperature is limited because at high temperature insulation of the electromagnet gets damaged.

• The risk of overheating during the engagement.

• The brushes used to energise coils are needed a periodic check.

7.7.2   Hysteresis clutch

Hysteresis clutch is the type of electromagnetic clutch have an extremely high torque range. It has no mechanical contact between the rotating parts. An electrically powered hysteresis clutch is illustrated in Figure 7.10. It consists of a field (coil), rotor and hysteresis plate. The hysteresis plate passes through the rotor, without touching it. When current applied to the coil, it creates a magnetic flux. When flux is high enough, it magnetises hysteresis disc. The magnetic drag during the rotor rotation cause rotation of the hysteresis disc, eventually it matches the input speed. The torque created depending on the current that energises the coil. In a hysteresis clutch, there is no contact, they do not depend on a friction or shear forces for torque transfer, so no wear occurred during the operation. As a result, it provides an extremely long life and gives superior torque repeatability. They ensure a smooth operation with almost no maintenance.

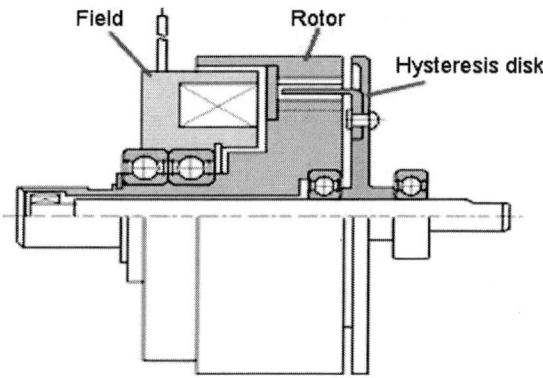

Figure 7.10   Hysteresis clutch

7.8   Diehl motor clutch

All the Toyoda type automatic looms are provided with a clutch attached with the motor – Crank-shaft drive. Whenever the starting handle is moved either left or right, it can be seen that the clutch on the crank shaft will act as the plates of the clutch will engage and thus the drive will be transferred.

7.8.1   Construction of the Diehl motor clutch

The main part is Stator and Rotor fixed on the central shaft which carry the driving pinion at one end and at the other end is thrust bearing which in turn is connected to starting handle. A diaphragm connects to motor casing and which in turn connected to Stator and Rotor through plate as shown in the Figure 7.11.

7.8.2   Working

Movement of the starting handle to the ON position will cause the clutch plate to be pushed in contact with the rotor because the centre shaft the motor, to which the plate is connected will receive a very slight lateral movement. This arrangement is shown in Figure 7.11. The machine is directly driven by gear arrangement. When the starting handle is put to OFF position, pressure on the clutch plate is released, and the plate is moved off the rotor and onto the casing of the motor, which acts as a brake plate. This system has proved very effective in high speed weaving machines since it allows the required quick start and the cyclic variation in the angular velocity of the crank shaft is no problem. However, there is a safety problem, because the rotor continues to revolve freely for about half an hour after the isolator has been switched off. Under these circumstances, if an operative on seeing the isolator in OFF position pulls the starting handle ON to perform some repair work, the machine will turn over for two or three picks before gradually slowing down to a stop. Before any repair work is undertaken, the shuttle should be removed and the machine should be in protection and then slowly pull the starting handle ON to stop the machine. The advice of removing the shuttle is given to prevent any damage to the weft fork and also prevention of shuttle flying out in such cases. For the same reason the machine must not be stopped when the sley is moving away from the cloth fell. The brake handle should be engaged only while the sley is moving forward to beat-up.

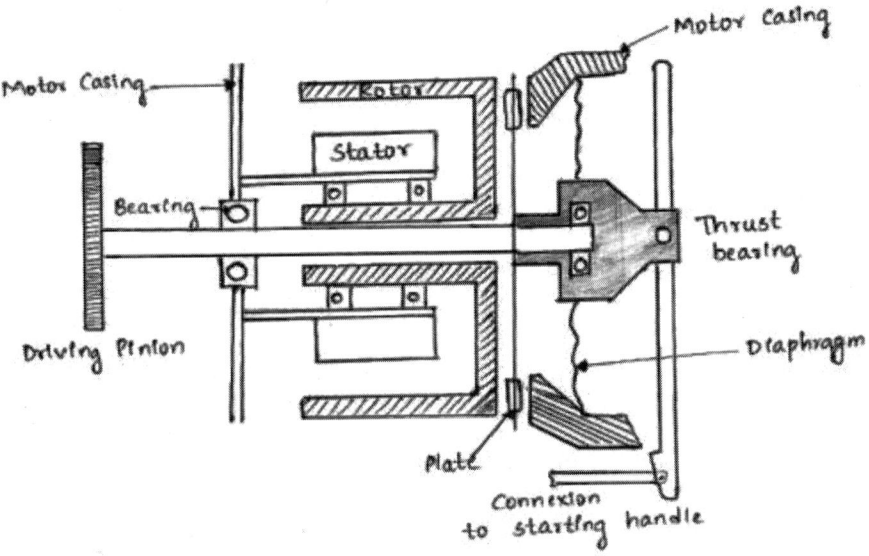

Figure 7.11   The Diehl clutch motor

7.9   Multiple plate clutch

7.9.1   Construction

A multiple plate clutch has more number of clutch plates. A typical clutch consist of the following components: Clutch basket or cover, clutch hub, drive (friction) plates, driven (steel) plates, pressure plate and the clutch springs. These can be seen in the Figure 7.12.

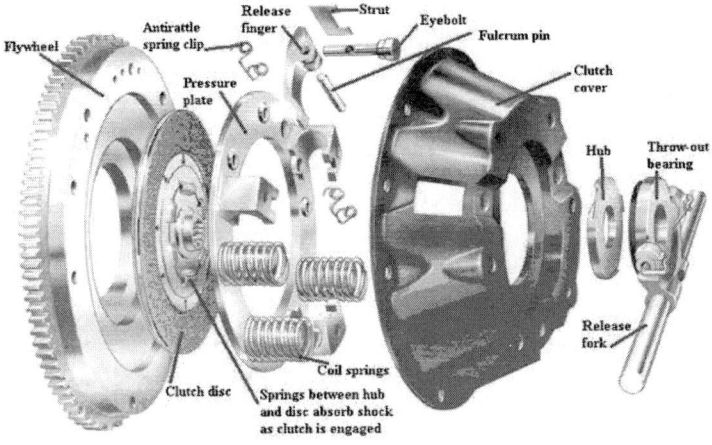

Figure 7.12   Multiple plate clutch

The clutch housing is attached to the engine crank shaft flywheel. The pressure plate is fixed on the flywheel through the clutch springs. The engine flywheel turns the clutch housing. The inner circumference of the clutch basket is splined to carry the thin metal plates. The clutch basket splines engage the tabs on the friction drive plates. This sources the clutch housing and the drive plates to rotate together. Additionally they are free to slide axially within the clutch basket. Interleaved with the drive plates, there are many number of driven plates. These driven friction plates have inner splines. These splines engage with the outer splines on the clutch hub. As such, the driven friction plates can slide on the clutch hub. The clutch hub is linked to the input shaft of the transmission gear box.

Figure 7.13 shows various types of multiple plate clutch. The drive plates and the driven plates are firmly pressed together by the pressure plate due to the clutch springs. The drive plates, driven plates and the strong clutch coil springs are assembled within the clutch basket.

Figure 7.13 Types of multiple plate clutches

7.9.2   Working

During clutch engagement, spring pressure forces the pressure plate towards engine flywheel. This causes the friction plates and the steel driven plates to be held together. Friction locks them together tightly. Then the clutch basket, drive plates, driven plates, clutch hub and the gearbox input shaft all spin together as one unit. Now power flows from the clutch basket through the plates to the inner clutch hub and into the main shaft of the transmission (Figure 7.14).

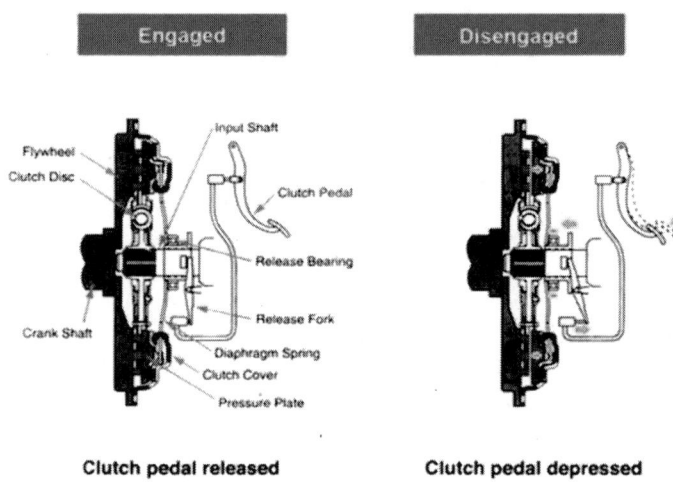

Figure 7.14   Working of multiple plate clutch

The clutch gets released or disengaged when the clutch pedal is pressed. This causes the clutch pressure plate to be moved away from the drive and driven plates, overcoming the clutch spring force. This movement of the pressure plate, relieves the spring pressure holding the drive and driven plates together. Then the plates float away from each other and slip axially. Thus, the clutch shaft speed reduces slowly. Finally, the clutch shaft stops rotating. Power is no longer transferred into the transmission gearbox. Multiple plate clutches are used widely in motor cycles and scooters. The multiple plate clutch is used in some types of epicycle gearboxes. A multiple plate clutch provides a very compact, yet a high friction coupling between the engine and the gearbox. With multiple plates, the friction surface area, strength and friction in the engaged clutch are increased. The increased friction surfaces, increases the torque transmission capacity of the clutch. Depending upon the power output of the engine, and the weight of the vehicle, four to eight sets of plates (four to eight drive plates and four to eight driven plates) may be housed in the clutch basket. The multiple plate clutch may be of dry type or wet type. When the clutch functions in atmosphere, it is called a dry clutch. When the clutch operates in an oil bath, it is named as a wet clutch. Some multiple plate clutches work dry. Then the driving plates are lined on each side with a friction fabric. A dry clutch can withstand high temperatures and permits frequent gear shifts without much loss of power initiated by heat. A dry clutch should never come in contact with oil. In dry clutches, seals are used to inhibit entry of oil into the clutch basket. Oil will source clutch slippage and will ruin the clutch friction discs. Most multiple clutches run in an oil bath. Wet clutches are used for several reasons: Debris resulting from clutch wear can be drained with the oil and trapped by the oil filter. Oil helps the clutch to run cool. In the case of two wheelers, since the primary drive requires lubrication, it is less costly to use a wet clutch. A wet clutch operates smoother and uses more plates. Some clutches have alternative steel and bronze plates running in an oil bath and are quire smaller in diameter for minimal inertia.

7.10   Applications of clutch in textile production

7.10.1   Application of clutch in warping

In warping – specially drum driven wrapper, the beam is driven with the frictional contact of drum. The latter is driven directly by main motor. The speed of warping is nearly 700–2000 rpm and it is necessary to stop the beam from running in the event of an end break. In order to stop the beam instantly the assembly of clutch and break is required. It is known fact that, as the speed of warping is in the above mentioned range, following an end break if the machine does not stop, a long length of warp would be wound without the end that is broken and may create problems at weaving. Therefore it is very much essential that, as soon as an end breaks the machine should be brought to a crash stop which is possible only with the help of clutch and break (Figure 7.15).

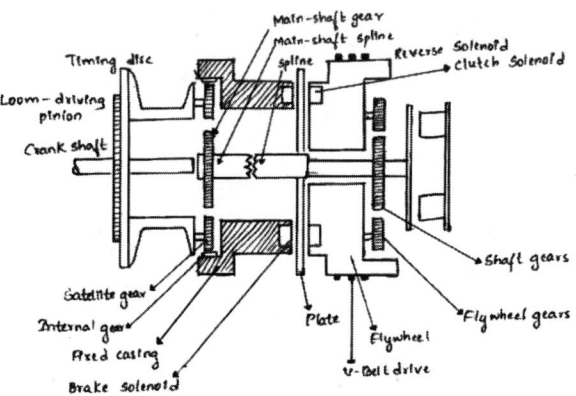

Figure 7.15   Clutch and Brake in warping

Theory of beam winding in sizing

Headstock forms the last part in sizing like in warping. The beam winding mechanism in sizing is complex and involves the thorough understanding of the process. If one considers the driving arrangements, the main motor first gives the drives to the beam via the apparatus which may consist of slip friction clutch or Reeves gear or PIV. In any of these cases we will find that initially at D_0 the force required to drive the weaver's beam is very low as the force acting from the motor end will be very high and hence the beam is driven at higher speeds. This results in yarn build up within no time and slowly the force now shifts from centre to rim and finally when the beam is D_{max}. Hence the force required to rotate the beam will be very high and this calls for sophisticated tools like clutch or PIV. In this section an attempt has been made to explain the role of such friction clutch.

Friction clutch – Slipping friction drives were used on sizing machines for several years to drive the weaver's beam. It is negative in character and is unsound in principle and practice because of heat generation during its operation. Nevertheless it enables the angular velocity of the beam to decrease with increase in diameter. This form of drive comprises a pair of friction plates held against each other. One plate is driven by the main driving element and is pressed against the other plate. Between the two plates a disc is provided which drives a hollow shaft into which one end of the loom beam is placed. A driving arrangement fastened onto this shaft is set with its projection against one of the radial ribs on the head of the loom beam, thus turning the beam with the shaft. All the slippage takes place between the two plates of the clutch only; at the start of the beam slippage is about 10%–15%, whereas when the beam is practically full it is likely to be several hundred per cent. As the beam diameter increases due to winding of the warp on it, the point at which the pull of the yarn is being exerted shifts further and further from the axis of the beam (Figure 7.16). Therefore a greater force will be needed to turn the beam to maintain the tension constant on the warp. With slipping friction motion this can be achieved by increasing the pressure between the plates. Adjustment is made either with a weighted lever, by moving the weight away from the fulcrum or by the use of a hand wheel. This adjustment is made from time to time by the sizer. On certain machines there is also provision to adjust the weight automatically using a linkage or a chain mechanism. In practice this system is only satisfactory for light sorts. For heavy set, double friction drives are often used. Fig.7.14 Effectweavers beambeams

attthereThe following figure shows the working principle of a common type of such a mechanism much seen in India.

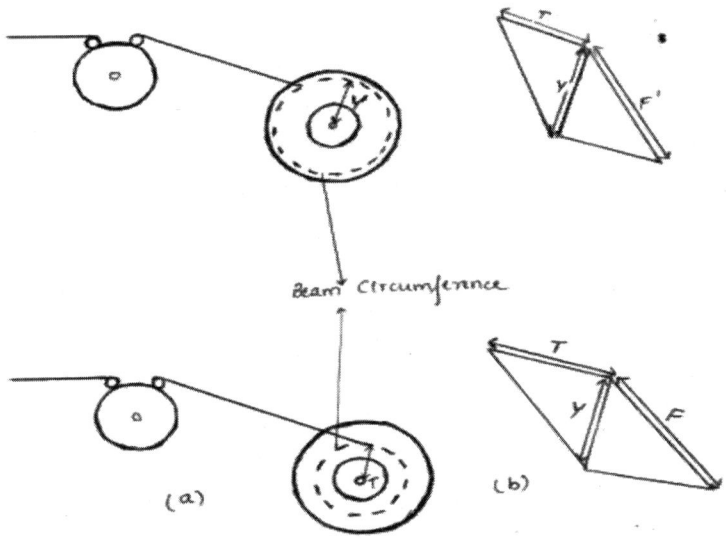

Beam Circumference

(a)   (b)

Figure 7.16 Effect of increase in yarn diameter of weavers beam on the force required to rotate it. (a) The beam start at D_o and end at D_{max} and beam diameter at end, there are varieties of slipping friction motion. (b) The resulting parallelogram of forces.

The friction wheel B, the prime mover of the slipping friction motion, mounted loosely on the beam shaft G is driven by the drag roller wheel A. The wheel has two inner flanges C which are bent at right angles at their ends to form a rim. The outer flanges D fit inside the rims of inner flanges. Each of the outer flanges carries a small lug W which fits inside a recess in the inner flange and locks them together. Rotation of the friction wheel B and its flanges does not impart motion directly to the beam shaft G since the outer flanges are fitted loosely on this shaft. Whereas friction wheel and inner flanges fit loosely on the bosses F and R. The bosses are fitted on the beam shaft by means of a feather key. A channel is cut on the beam shaft for fitting the feather key so that the shaft can be adjusted laterally to suit different widths of weaver's beam. The motion of B is transmitted to the beam shaft G by means of thin steel plates, X, riveted to the bosses and each plate is encased between a pair of flannel washers Q, saturated with oil.

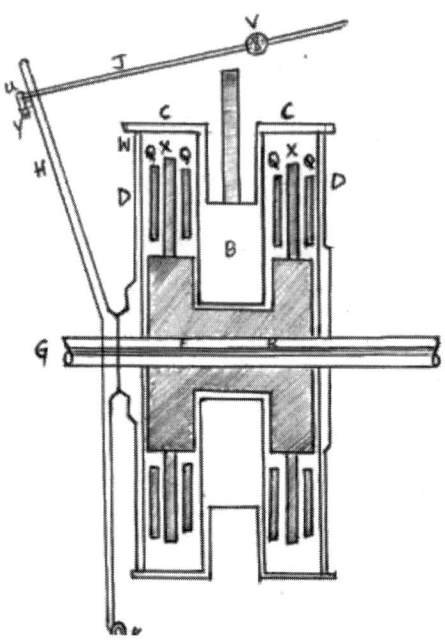

Figure 7.17　Slipping friction clutch found on sSizing.

A: Drag roller wheel, B: Friction wheel, C: Inner flange, D: Outer flange, FL Boss,
GL Shaft, H: Lever, J: Weight lever, K: Fulcrum for H, R: Boss, Q: Flannel washer,
U: Pin, V: Weight, W: Lug, X: Steel plate, Y: Fulcrum for J.

When pressure is applied between the inner and outer flanges the discs X along with the flannel washers will be pressed causing them to grip and thus transmit the motion to the shaft G. Any opposing force on the shaft G which is greater than the pressure applied between the inner and outer flanges causes the discs and the bosses to slip and in this way constant yarn speed is obtained. Greater the force applied between flanges, more is the fore required to slip and therefore greater will be the tension on the warp sheet. This means that tension on the warp can be adjusted by adjusting the pressure between the outer flanges. Pressure is applied to the friction drive by means of an upright lever H, fulcrumed at K which fits round the shaft G and bears against the outer plates of the friction clutch A. A weight lever J, fulcrumed at Y, carries a pin U, which bears against lever H. Thus any downward pressure on the lever J forces the lever H against the outer flanges of the friction drive and applies the pressure between the flanges. The amount of pressure is controlled by the position of weight V on its lever, more when it is away from the fulcrum. It may be remembered that at the start the weight V is close to the fulcrum. With increase in the diameter of the beam, from time to time, it is shifted away from the fulcrum to maintain the tension uniform. This is the main drawback of this type of machine that the tension is subjected to manual error. The drawback of this type of drive is that considerable amount of heat generated and this has a marked effect in reducing the warp tension towards the top of the beam. As the beam increases in diameter the rate of slippage and the pressure between the plates increase and this increases the heat liberated which goes in the wasted work. This results in increased form of temperature and decrease of co-efficient of friction due to decrease in the viscosity of the oil with which the felt linings are provided saturated (Figure 7.17).

7.10.2   Application of clutch in spinning machines

Flange mounted clutch type – UFC is used in various textile machinery to transmit power, whenever required (Figure 7.18). The list of some important machines are given below:

LR High Production Carding Machine C 1/3 (UFC-50 / D402 / 24VDC)
LR Bale Opener MBO igh (UFC-250 / D402 / 24VD)

Figure 7.18   UFC clutch Ring

7.10.3   Clutch systems in loom (weaving machine drives)

In almost all the composite textile mills it is found that a large number of machines are housed under one roof. The number varies anywhere between 200–1000 or even more. These machines used to get their power from one main shaft driving a number of line shafts through transmission belts. In some cases these shafts are suspended from the roof, and in others shafts are housed in underground alleys. These line shafts have large diameter pulleys from which, by means of another transmission belt, the weaving machines are driven. The machines are fitted with fast and loose pulleys, so that when the machine has to run the belt on the loose pulley is shifted, by means of a fork connected to the starting handle, onto the fast pulley. This would mean that the drum or pulley on the line shaft would have to be of width greater than twice the width of the loose or fast pulley. In this type of drive considerable skill on their part of the weaver is called for so that the crankshaft, on which the fast pulley is mounted, get enough momentum to ensure successful and complete traverse of the shuttle on the first pick. Under this condition, it is very often seen that the weavers pull the sley towards the cloth no sooner the starting handle is shifted to on position. This would obviously call for more power from the weavers themselves. Therefore it is preferred to have a single driving point which is economical in driving a large number of machines. But this type of drive involves two issues, one of which is the loss of transmission power due to belt slippage, and secondly, in case of any breakdown in the main shaft a huge loss in production has to be incurred. The system was gradually and eventually replaced by a number of large motors driving a group of machines. This also, later, gave way to the present day system of driving of the weaving machines by individual motors. This system has several points to its favour – minimum power loss, they are almost built-in taking no extra space, spaces taken by several shafts are no longer there, accident potentialities due to the transmission belts are absent, no more obscuration of light due to shafts being suspended from the roof top, with underground shafting maintenance problems were very acute, huge loss of production in case of any breakdown of a single shaft has been eliminated. However with individual drive the risk of fire hazard has also multiplied. It requires considerable skill on the part of the weaver to "inch" forward even with an individual motor although V-belt and gear transmissions are the order of the day, unless the drive has incorporated in it the various functions that a weaver has to do, e.g., inching, reversing, single pick. The weaving machines that are not provided with electromagnetic drives, the weaver has to depend on his or her skill and starting places would be still a headache for all cloth productions.

The use of direct gear drive between the motor and the loom means that the motor has to be switched on and off whenever the loom is started or stopped. This system thus requires a motor having a high starting torque to ensure that the first pick can be successfully inserted. With a loom using this type of drive, there is a problem when the loom is stopped suddenly, in the event of a bang off, for example. The momentum of the motor will cause it to continue running and it is thus necessary to have a slipping clutch arrangement in the driving wheel on the main shaft of the loom. Such a clutch will slip and absorb the motor energy in the event of a sudden stoppage of the loom. The clutch should be adjustable in the event of wear occurring, because slipping must be avoided during normal starting, stopping, and running of the loom if the optimum loom speed is to be achieved. Furthermore, it is desirable that the loom should be brought to rest at the appropriate position for each type of loom stoppage (i.e., warp break or weft break) in order to minimize the work of the weaver. Direct drives are not confined to gear transmission, and a similar arrangement incorporating a slipping clutch has been used extensively with V-belt transmission. The main advantage of direct drive is that there is a saving of power whenever the loom is stopped because the motor is switched-off. A further advantage is that no sideways movement is required in the driving arrangement on the main shaft of the loom. Sideways movement of either the driving wheel of the clutch (usually the latter), although minimal, is necessary if a friction clutch arrangement is used between the motor and the loom. High initial torque is then quite unnecessary because the freely rotating clutch (or may be driving wheel) is already rotating at the normal running speed. Provided that a highly efficient clutch system is used, the first pick will be projected at its full velocity if the loom is placed in the back-centre (180°) position before starting. The main types of driving clutch used in indirect driving systems are the plate clutch, the conical clutch, and the expanding clutch. The plate clutch has been used extensively because it is capable of spreading the area of wear over a larger area, but the conical clutch is capable of transmitting a greater torque (i.e., more power) for a given pressure applied to frictional faces. Both direct and indirect methods of drive require some type of flywheel on the main drive shaft of the loom to reduce the effect of cyclic variation in the speed of the loom on the frictional properties of the clutch. An additional wheel may be used, but the driving wheel or clutch wheel, whichever is fitted to the main shaft, is usually quite effective for this purpose. Such a wheel must not be too large or too heavy in order to prevent slow starting and over running on stopping.

Plate clutch drive of Crompton and Knowles loom

The loom is fitted with a plate clutch drive and works on the principle as described under the section plate clutch above. The loom is started or stopped by a set of starting handles and a brake handle provided at the end of breast beam. Each mechanism is actuated by a separate handle through linkages. The brake handles are connected to the brake band and the shipper (starting handle) operate the clutch. The arrangement is set such that first the clutch is operated followed by brake application as we observe even in beam warping machines (Figure 7.19).

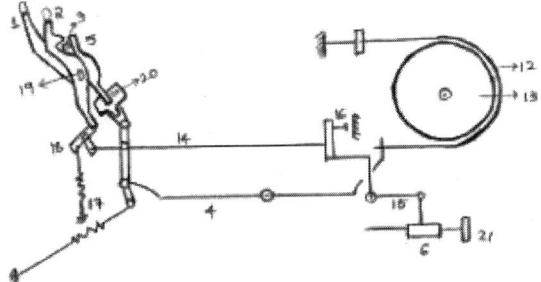

Figure 7.19 Crompton and Knowles drive system (Side view)

In this clutch the devise exerts pressure against the drive gear of the machine and forces it against a friction plate located besides it. The drive gear is meshed with the motor pinion and is continually turning when the motor is running. The clutch plates are mounted on crank shaft like Toyoda loom system with several staggered rows of cork inserts. These are brought in contact with the drive gear and provide enough friction surface to drive the machine without slippage even under extreme loaded conditions.

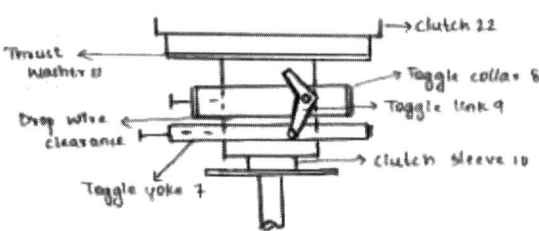

Figure 7.20   Plate clutch system of Crompton and Knowles system (top view)

Figure 7.19 show the shipper handle 2 if OFF position and the brake handle 1 in ON position that is the machine is in stopped situation when the brake band 12 will be pressing the brake 13. The shipper handle is pivoted on shaft 19, and is connected to the spring loaded T-lever 18 which is also fulcrumed. At one end of the T-lever is connected a spring and the other end is linked with the shipper connector 14. The other end of the connector carries a bumper 16 which limits the movements of the shipper handle. The bumper has knuckle connection with the shipper – elbow lever 15 which is pivoted with the machine framing. The right-hand end of the elbow lever is connected to the clutch lever 21 through shipper lever 6. The top of the clutch is pivoted with framing bracket of the drive system and it actuates the clutch through the clutch sleeve 107.16b. A toggle yoke is mounted at the end of the clutch sleeve which makes it fast on the crankshaft by means of an adjustable screw to increase or decrease pressure on the friction plate. On the toggle yoke mounted a toggle link which is pivoted on the toggle collar 8. There should be a clearance between the clutch sleeve and toggle collar to the extent of a drop wire for easy movement of the clutch sleeve. It is also necessary to have some clearance (say 2 mm) between the toggle collar and the thrust washer 11. The brake band has a lap of 180° round the brake drum 13. The upper end of the brake band is connected to the framing and can be adjusted by means of screw and nut (Figure 7.20).

The working of the assembly is very simple. For instance if any one of the shipper handles are pushed to OFF position, the link levers and connections will move such that there exists a lateral pressure on the brake drum and thus halting the speedy loom instantly.

7.10.4   Cone clutch drive in Northrop automatic looms

Figure 7.21a shows a typical arrangement of cone clutch drive is used in Northrop automatic weaving machine for heavy fabrics. Similar drives may be obtained with certain simple modifications for driving other weaving machines. The clutch rod (A) is connected with the starting handle) which when moved to ON position gives a clockwise twist of the clutch rod. This action will produce a lateral movement to the clutch fork (B) through clutch link (G) so that the V-pulley clutch (D) moves towards the left and is engaged with the friction cone (E). The V-pulley being loose on the pinion shaft (F), is capable of lateral movement. The friction cone is fast on the pinion shaft driving the crankshaft through the bottom shaft. The bottom shaft is driven by the pinion on the left end the friction cone. As soon as the starting handle is moved to its OFF position. The V-pulley clutch moves laterally to the right and with the application of brake the machine is brought to a stop instantaneously, buy the V-pulley being loose on the pinion shaft continues to rotate with the motor. The V-pulley assembly works as a flywheel, and the insertion of the first pick with full shuttle velocity poses no problem when the machine is restarted.

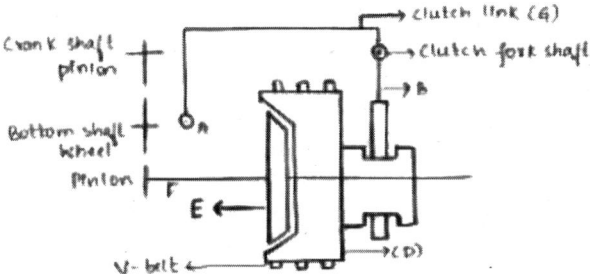

Figure 7.21   Clutch drive system of Northrop auto loom

Loom brake and drives

The brake arrangement of cone clutch drive is shown in Figure 7.22. The brake band (A) circumscribes almost the whole periphery of the brake drum (B). The lower end of the band is fixed to a pin (C) on the machine drive bracket, and the other end terminates in brake band lifting catch (D). This catch is connected to a strong spring (E) via a hook (F). Through this hook passes the brake band lifter (G) which is connected to the lower end of the brake band lever (H). Latter is pivoted at (I) and above it the brake rod (J) is connected with (H) The length of the brake rod is adjustable by means of turn buckle (K). The forward end of the brake rod is joined with the lower end of the brake lever (1). The starting handle (H) is fulcrumed on the clutch rod (N) and carries a cam face which is capable of pushing and lifting a small bowl (O) which in turn gives a twist to clutch rod lever bracket (P). There is a small brake easing cam (Q) on the brake easing lifter (R) to lift the brake lever and release the brake band when the machine requires any manual operation. It will be seen that the rear end of clutch rod is connected to the clutch drive system through the link (S).

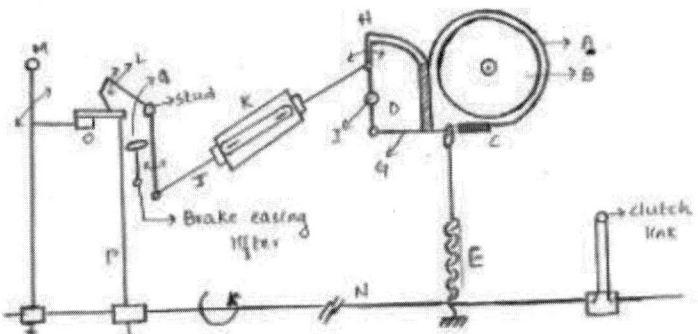

Figure 7.22   Clutch drive system of Northrop auto loom

When the starting handle is shifted to the right, the cam face on the handle pushes the small bowl so that the clutch rod lever bracket is lifted slightly and brake rod lever moves upward which will allow the brake band lever to release its control on the lifting catch and at the same time the hook lifts the brake band, and the brake drum is now free. This action simultaneously engages the clutch with the friction cone to run the machine. In the event of the starting handle, which is spring loaded, is thrown to OFF position, the face cam releases the small bowl which in turn will cause the brake rod lever to move down. This will allow the brake rod to be pushed to the rear of the machine and will allow the brake band lever to bite on the lifting catch due to the action of the spring and the brake band now presses on the brake drum stopping the machine immediately. In the process, the clutch rod is given a counter-clockwise turn and the machine is at once declutched.

Inching and reversing

In older types of shuttle weaving machines there are no provision of inching or reversing, or inserting a single pick if required. In those types of machines inching is generally done by the weaver by short spells of pulls and pushes of the starting handle which requires considerable skill on the part of the weaver together with his considerable loss in energy. Reversing has to be done manually taking much of his time and energy. But the present day shuttle weaving machines are provided with all these aspects of a weaver's job, that is reversing, inching and single pick insertion together with the normal run of course. Thus a weaver's load is reduced. The number of such control stations in the present day machines are such that a machine can be stopped controlled easily at any moment in case of an emergency from any one station. In some cases, the machine can be restarted without the weaver having to move to the main starting station or the handle. This is only possible when the position of the machine is not critical on starting. This condition is only possible when the weft insertion is done positively, and rapier machine is a typical example in this case. It is desirable that the weaving machine should come to stopped position always after the monitoring devices have detected any fault so that the faults could be repaired under desirable conditions every time. But with conventional mechanical drive this is probably too much for the asking. It is frequently seen that the machines are positioned by the weaver by fiddling with the machine handle, and at the last stage the correct position is obtained by turning the flywheel. This manipulation requires considerable skill and no guarantee could be assumed that for a particular purpose the machine could be brought to the same position every time.

It is desirable that during mending of the broken warp or weft yarns the machine should be reversed from the top and the bottom centre positions. In the first case, the healds normally are level at the top centre position when mending broken warp yarns, and in the second case, the broken weft yarn has to be found out before commencement of production. In Saurer and Ruti C-1001 weaving machines a small motor, rotating in the opposite direction, could be switched on for reversal of the machine slowly, and the reversal will continue till the press-button controlling the motor is released. In some of the machines a slowly reversing motor V-pulley engages with a segment on the main shaft of the machine the reversal stops as soon as the segmented part loses contact with the V-pulley. In this case the machine is brought to the back centre position for restarting. In certain cases it is desirable that the machine should be capable of inserting a single pick and then come to a stop. This can be performed by use of push-button controls, and for as many number of picks as is required. Crompton & Knowles C-5 machines are provided with foot pedals at each end for single pick insertion and also for taking out the shuttle from the box when necessary.

Push-button controls

The present day high speed weaving machines are provided with push-button controls to provide reasonable opportunity to the weavers to produce good fabrics with minimum walking time so that a larger allocation of machines could be made. The weaving machine should come to a stop between the back and top centre positions, and when this is done in single shuttle ordinary weaving machine with either simple clutch or flat belt drive, it becomes a rather difficult proposition. Push-button controls gained wide popularity due to the facts that had already been stated, and for better appreciation it would be better if we repeat the advantages of such controls which are:

- Most reliable way of ensuring a quick and accurate start and a complete successful first pick.
- It brings the sley always at the same position with the heald staves in level conditions approximately.
- The weaving machine can be run forward at normal speed, or at a speed of reduced magnitude to the selected position.
- The weaving machine can be run in the reverse direction slowly to the back centre position so that restarting problem at full speed does not arise.
- For the purpose of pick finding, reversing can be continued for a longer period.

Weaver's control stations are mounted at each end of the breast beam and contain start, stop and reverse push-buttons. Mounted at that sides of the control stations are the brake release and single pick toggle switches. The push-button control enclosure is gas kitted against lint or oil and also serves as a junction box for different units and switches.

The clutch and brake magnets are mounted on the transmitter left located at the drive end of the machine. The brake magnet is located immediately adjacent to the machine side while the forward and reverse clutch magnets are located on either side of the large V-belt pulley. A three-section slipping assembly provides power to the rotating clutch magnets. The electromagnetic drive is used in some of the advanced models of Crompton & Knowles C type shuttle weaving machines. A rotary timing switch is attached to the drive end machine side and is driven by a small chain from the crankshaft. This is really the "heart" of the control system. It has several functions to do, for example, it determines whether the shuttle is picked properly or not, stops and machine after the shuttle is boxed in insertion of a single pick, prevents reversing beyond back centre position and cancels the weft stop motion before the first pick.

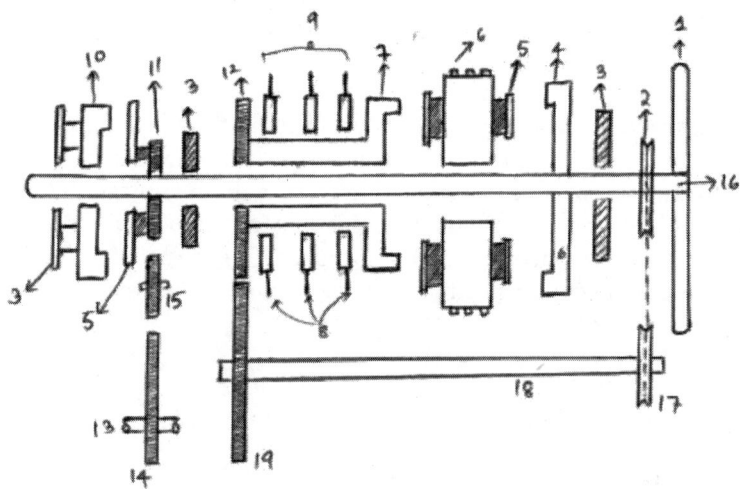

Figure 7.23   Electromagnetic drive

Figure 7.23 shows the working principle of the electromagnetic drive to a shuttle weaving machine. The transmitter shaft housed in three bearings (3) at the extreme fight end is the rotating handle (1). There are three electromagnets (4) (7) and (10) in which (4) moves the machine in the forward direction being fast on the transmitter shaft (16) (7) is the reversing electromagnet loose on the transmission shaft on which the reversing wheel (12) is fixed, and the brake electromagnet (10) is fast on the machine framing. Two friction plates (5) are mounted on either side of the main V-pulley. The V-pulley (6) being loose on the transmission shaft is capable of slight lateral movement to become fast with either of the electromagnets (4) or (7) The loose reversing magnet has at its end a small wheel (12) which drives the wheel (19) mounted on the reversing shaft (18). At the right end of the shaft (18) is a small reversing pulley (17) driving another pulley (2) fast on the transmission shaft.? The crankshaft wheel (14) mounted on the crankshaft (13) is driven from the driver wheel (11) through a carrier wheel (15). The wheel (11) is fast on the transmission shaft, and also carries the friction plate (5). There are three slip rings (8) which have individual brush contacts (9) which magnetises any one of the electromagnets (4) (7) or (10) at any one time. In order to obtain maximum electrical efficiency from the transmitter, the brushes and slip rings should be cleaned periodically with carbon tetrachloride, or a similar non-flammable cleaning fluid. In some extreme conditions very fine sand paper may be used to clean the rings.

When the push-button for starting the machine is pressed the forward drive magnet is energized and will allow the friction plate on the right-hand side of the main V-pulley to clutch. And the former being fast with the transmission shaft the machine runs with normal speed and direction. But with the depression of the stop button, the forward magnet is de-energized and the brake magnet becomes energized causing the friction plate near to it to be attracted towards it. As the brake magnet is fast to the framing the friction plate is prevented from rotating and so also the transmission to the crankshaft. In this case the stop button must be kept depressing until the machine comes to a stop because the machine may not stop instantaneously due to the reason that has been timed to stop only during specific periods of its cycle. In case of an emergency the machine can be stopped by any one of the following methods;

A)   Depress stop button

B)   Depress push button

C)   Depress a drop wire of the warp stop motion.

Reversing the machine under power requires certain precautions. The machine must be stopped and the sley must not be between top and back centre position to reverse under power. It is recommended by the maker that the machine be reversed under power ONLY from the front centre to back centre. As soon as the reverse button, is depressed the reverse magnet is energized and it will be drawn towards the friction plate in its front and in turn will draw the main drive V-pulley. The friction plates being fast with V-pulley by slip key on which they can also slide, will actuate the reversing wheel to drive the pulley in the opposite direction, and as soon as the bottom centre reached it stops automatically due to the action of the rotary timing switch driven from the crankshaft. In order to obtain single pick at a time, the machine must be in the position of starting. The single pick switch is downed and depressing the start button the machine will make one pick and stop automatically just past the back centre. Additional single picks may be made by depressing the start button each time a pick is desired. Before starting the machine in normal way it must be seen that the single pick switches are in UP position. Before starting the machine one must be certain that the cabinet door is closed and handle is in ON position. The brake safe switch and single pick switches and brake release switches are in ON or UP position. Now by depressing the start button will allow the machine to run normally. It is always a better practice to start the machine when the sley is at the back centre position so that enough force is available for insertion of the first pick.

Brakes for looms with weft stop motion, warp protector and warp stop motion

It is an additional mechanism for a power loom fitted with Weft stop motion, Warp stop motion and Warp protector motion to bring the loom to a crash stop to prevent the defects from occurring. One such set-up is to have a brake band, brake shoe and brake lever. As and when these stop motions come in to action, the shift in the starting handle from ON to OFF position, the brake lever is dragged and thus a lateral force is exerted on to the brake drum. The brake liner will be a leather piece acting as brake shoe. The brake shoe will be acting against the brake wheel and thus the loom is brought to crash stop. The clutch shoe brake is shown in Figure 7.24.

Figure 7.24   Clutch shoe with brake

Problems

1. A car engine rated at 12 hp gives a maxi torque of 65 ft lb. The clutch is of single plate type, both sides of plate being effective. If $\mu = 0.3$ mean axial pressure = 12 lb/in². External radius of friction surface is 1.25 lines the internal diameter. Find the dimensions of the clutch plate and total axial pressure which must be emitted by spring?

Solution :

Let r_1 and r_2 be external and internal radii of friction surfaces.

Total axial strength

$$W = 2\pi \, (p(r_1 - r_2)) \qquad P = \text{Axial pressure} \times r_2$$
$$= 2 \times 3.142 \times (12 \times r_2) \, (1.25 \, r_2 - r_2)$$
$r_1 = 1.25 \, r_2 \qquad = 2 \times 3.142 \times 12 \times 0.25 \, r_2^2$
$$= 18.849 \, r_2^{\,2}$$

$$T = \mu \, w \, (r_1 + r_2)$$
$$= 0.3 \, (18.849 \, r_2^2) \, (1.25 \, r_2 + r_2)$$
$$= 0.3 \times 18.849 \, r_2^2 \times 2.25 \, r_2$$
$$65 = 0.3 \times 18.849 \times 2.25 \times r_2^{\,3}$$

$$r_2^{\,3} = \frac{65}{0.3 \times 18.849 \times 2.25}$$

$r_2 = 1.72$
$r_1 = 1.25 \times r_2$

$$= 1.25 \times 1.72$$
$$= 2.15 \text{ ft}$$
$$= 2 \pi P (r_1 - r_2)$$
$$W = 2 \times 3.142 \times 12 \times 1.72 (2.15 - 1.72)$$
$$= 55.76 \text{ ft lb.}$$

2. A friction clutch is to transmit 15 hp at 3000 rpm is to be of the single plate type with both the plate defective. The axial pressure limited to 0.09 kg/cm². If the external radii of friction line is 1.4 times of the internal radii. Find required dimension of the lining if $\mu = 0.3$.

Solution :

$$W = 2 \pi P (r_1 - r_2)$$
$$= 2 \pi \times 0.9 \times 10^4 \times r_2 (1.4 \, r_2 - r_2)$$
$$= 2 \pi \times 0.9 \times 10^4 \times 0.4 \, r_2^2 \text{ Newtons}$$
$$= 22690.467 \, r_2^2$$

$$\text{Torque} = \mu \, w \, (r_1 + r_2)$$
$$= 0.3 \times 22619.46 \, r_2^2 \times (1.4 \, r_2 + r_2)$$
$$= 0.3 \times 22619.46 \times 2.4 \, r_2^3$$

$$r_2^3 = \frac{35.8}{0.3 \times 22,619.46 \times 2.4}$$

$$= 0.13 \text{ m}$$
$$r_1 = 1.4 \, r_2 = 0.18 \text{ m}$$
$$15 \text{ hp}, \ 15 \times 0.75 = 11.25 \text{ watts}$$

$$P = \frac{2 \pi N T}{60} = \frac{2 \times \pi \times 3000 \times T}{60}$$

$$\frac{11.25 \times 1000 \times 60}{2 \times \pi \times 3000} = T$$

$$T = 35.8 \text{ Newtons}$$

A friction clutch is required to transmit 10 hp at 2500 rpm. It is single type disc with the both sides of disc being effective. The axial pressure is limited to 0.75.

3. Determine the maximum, minimum and average pressure in a plate clutch when the axial force is 4 kN. The inside radius of the contact surface is 50mm and the outside radius is 100 mm. Assume uniform wear.

Solution :

Given: $W = 4 \text{ kN} = 4 \times 10^3 \text{ N}; \ r_2 = 50 \text{ mm}; \ r_1 = 100 \text{ mm}.$

Maximum pressure

Let P_{max} = Maximum pressure.

Since the intensity of pressure is maximum at the inner radius (r_2), therefore,

$$P_{max} \times r_2 = C \quad \text{or} \quad C = 50 \, P_{max}$$

We know that the total force on the contact surface (W),

$$4 \times 10^3 = 2 \, \pi \, C \, (r_1 - r_2) = 2 \, \pi \times 50 \, P_{max} \, (100 - 50) = 15710 \, P_{max}$$

$$\therefore \qquad P_{max} = 4 \times 10^3 / 15710 = 0.2546 \text{ N/mm}^2 \qquad \text{Ans.}$$

Minimum pressure

Let P_{min} = Minimum pressure,

Since the intensity of pressure is minimum at the outer radius (r_1), therefore,

$$P_{min} \times r_1 = C \quad \text{or} \quad C = 100 \, P_{min}$$

We know that the total force on the contact surface (W),

$$4 \times 10^3 = 2 \, \pi \times (r_1 - r_2) = 2 \, \pi \times 100 \, P_{min} \, (100 - 50) = 31{,}420 \, P_{min}$$

$$\therefore \qquad P_{min} = 4 \times 10^3 / 31{,}420 = 0.1273 \text{ N/mm}^2 \text{ Ans.}$$

Average pressure

We know that average pressure,

$$P_{av} = \frac{\text{Total normal force on contact surfaces}}{\text{Cross -sectional area of contact surfaces}}$$

$$= \frac{W}{\pi[(r_1)^2 - (r_2)^2]} = \frac{4 \times 10^3}{\pi[(100)^2 - (50)^2]} = 0.17 \text{ N/mm}^2 \qquad \text{Ans.}$$

4. A single plate clutch, with both sides effective, has outer and inner diameters 300 mm and 200 mm, respectively. The maximum intensity of pressure at any point in the contact surface is not to exceed 0.1 N /mm². If the coefficient of friction is 0.3, determine the power transmitted by a clutch at a speed 2500 rpm.

Solution :

Given $d_1 = 300$ mm   or $r_1 = 150$ mm; $d_2 = 200$ mm   or $r_2 = 100$ mm; $p = 0.1$ N/mm²; $\mu = 0.3$; $N = 2500$ rpm   or $\omega = 2 \, \pi \times 2500 / 60 = 261.8$ rad/s.

Since the intensity of pressure (p) is maximum at the inner radius (r_2), therefore for uniform wear,

$$p r_2 = C \quad \text{or} \quad C = 0.1 \times 100 = 10 \text{ N/mm.}$$

We know that the axial thrust,

$$W = 2 \, \pi \, C \, (r_1 - r_2) = 2 \, \pi \times 10 \, (150 - 100) = 3142 \text{ N}$$

and mean radius of the friction surfaces for uniform wear,

$$R = \frac{r_1 + r_2}{2} = \frac{150 + 100}{2} = 125 mm = 0.125 m$$

We know that torque transmitted,

$$T = n\mu WR = 2 \times 0.3 \times 3142 \times 0.125 = 235.65 \text{ N-m}$$

.... ($\because n = 2$, for both sides of plate effective)

$\therefore$ Power transmitted by a clutch,

$$P = T\omega = 235.65 \times 261.8 = 61,693 \text{ W} = 61.693 \text{ kW}\quad \text{Ans.}$$

5. A single plate clutch, effective on both sides, is required to transmit 25 kW at 3000 rpm. Determine the outer and inner radii of frictional surface if the coefficient of friction is 0.255 the ratio of radii is 1.25 and the maximum pressure is not to exceed 0.1 N/mm². Also determine the axial thrust to be provided by springs. Assume the theory of uniform wear.

Solution :

Given $n = 2$; $P = 25 \text{ kW} = 25 \times 10^3 \text{ W}$; $N = 3000$ rpm,   or

$\omega = 2\pi \times 3000/60 = 314.2 \text{ rad/s}$; $\mu = 0.255$; $r_1/r_2 = 125$; $p = 0.1$ N/mm².

Outer and inner radii of frictional surface

Let r_1 and r_2 = Outer and inner radii of frictional surfaces, and

T = Torque transmitted

Since the ratio of radii (r_1/r_2) is 1.25, therefore,

$$r_1 = 1.25\, r_2$$

We know that the power transmitted (P),

$$25 \times 10^3 = T\omega = T \times 314.2$$

$\therefore$ $T = 25 \times 10^3/314.2 = 79.6 \text{ N-m} = 79.6 \times 10^3 \text{ N-mm}$

Since the intensity of pressure is maximum at the inner radius (r_2), therefore,

$$pr_2 = C \quad \text{or} \quad C = 0.1\, r_2 \text{ N/mm}$$

and the axial thrust transmitted to the frictional surface,

$$W = 2\pi C (r_1 - r_2) = 2\pi \times 0.1\, r_2\, (1.25\, r_2 - r_2) = 0.157\, (r_2)^2 \quad \text{(i)}$$

We know that mean radius of the frictional surface for uniform wear,

$$R = \frac{r_1 + r_2}{2} = \frac{1.25 + r_2 + r_2}{2} = 1.125\, r_2$$

We know that torque transmitted (T),

$$79.6 \times 10^2 = n\mu WR = 2 \times 0.255 \times 0.157\, (r_2)^2\, 1.125\, r_2 = 0.09\, (r_2)^3$$

$\therefore$ $(r_2)^3 = 79.6 \times 10^3/0.09 = 884 \times 10^3$ or $r_2 = 96$ mm   Ans.

and $r_1 = 1.25\ r_2 = 1.25 \times 96 = 120$ mm   Ans.

Axial thrust to be provided by springs

We know that axial thrust to be provided by springs

$$W = 2 \pi C (r_1 - r_2) = 0.157 (r_2)^2 \quad \text{[From equation (i)]}$$
$$= 0.157 (96)^2 = 1447 \text{ N} \quad \text{Ans.}$$

6. A single dry plate clutch transmits 7.5 kW at 900 rpm. The axial pressure is limited to 0.07 N/mm². If the coefficient of friction is 0.25, find (1) Mean radius and face width of the friction lining assuming the ratio of the mean radius to the face width as 4 and (2) Outer and inner radii of the clutch plate.

Solution :

Given: $P = 7.5$ kW $= 7.5 \times 10^3$ W; $N = 900$ rpm or
$\omega = 2 \pi \times 900/60 = 94.26$ rad/s; $p = 0.07$ N/mm²; $\mu = 0.25$.

(1) Mean radius and face width of the friction lining

Let R = Mean radius of the friction lining in mm, and
              w = Face width of the friction lining in mm.

Ratio of mean radius to the face width,
$$R/w = 4 \qquad \text{(Given)}$$

We know that the area of friction faces,
$$A = 2 \pi R w$$

∴ Normal or the axial force acting on the friction faces,
$$W = A \times p = 2 \pi R w p$$

We know that torque transmitted (considering uniform wear),
$$T = n \mu W R = n \mu (2 \pi R w p) R$$

$$= n \mu \left[2 \pi R \times \frac{R}{4} \times p \right] R = \frac{\pi}{2} \times n \mu p R^3 \qquad (\because w = R/4)$$

$$= \frac{\pi}{2} \times 2 \times 0.25 \times 0.07 R^3 = 0.055 \, R^3 \, N-mm \qquad (i)$$

$$(\because n = 2 \text{ for single plate clutch})$$

We also know that power transmitted (P),
$$7.5 \times 10^3 = T \omega = T \times 94.26$$
∴ $T = 7.5 \times 10^3 /94.26 = 79.56$ N-m $= 79.56 \times 10^3$ N-mm   (ii)

From equations (i) and (ii)
$R^3 = 79.56 \times 10^3 /0.055 = 1446.5 \times 10^3$ or $R = 113$ mm   Ans.
And $w = R/4 = 113/4 = 28.25$ mm   Ans.

(2) Outer and inner radii of the clutch plate

Let r_1 and r_2 = Outer and inner radii of the clutch plate, respectively.

Since the width of the clutch plate is equal to the difference of the outer and inner radii, therefore,
$$w = r_1 - r_2 = 28.25 \text{ mm} \qquad (iii)$$

Also for uniform wear, the mean radius of the clutch plate,

$$R = \frac{r_1 + r_2}{2} \text{ or } r_1 + r_2 = 2\,R = 2 \times 113 = 226 \text{ mm.} \qquad \text{(iv)}$$

From equations (iii) and (iv)

$$r_1 = -127.125 \text{ mm; and} \qquad\qquad r_2 = 98.875 \text{ mm} \qquad \text{Ans.}$$

7. A multiple disc clutch has five plates having four pairs of active friction surfaces. If the intensity of pressure is not to exceed 0.127 N/mm², find the power transmitted at 500 rpm. The outer and inner radii of friction surfaces are 125 mm and 75 mm, respectively. Assume uniform wear and take coefficient of friction = 0.3.

Solution :

Given: $n_1 + n_2 = 5$; $n = 4$; $p = 0.127 \text{ N/mm}^2$; $N = 500$ rpm   or
$\omega = 2\pi \times 500/60 = 52.4$ rad/s; $r_1 = 125$ mm; $r_2 = 75$ mm; $\mu = 0.3$.

Since the intensity of pressure is maximum at the inner radius r_2, therefore,

$$p\,r_2 = C \quad \text{or} \quad C = 0.127 \times 75 = 9.525 \text{ N/mm}$$

We know that axial force required to engage the clutch,

$$W = 2\,\pi\,C\,(r_1 - r_2) = 2\pi \times 9.525\,(125 - 75) = 2990 \text{ N}$$

And mean radius of the friction surfaces,

$$R = \frac{r_1 + r_2}{2} = \frac{125 + 75}{2} = 100 mm = 0.1 m$$

We know that torque transmitted

$$T = n\mu W R = 4 \times 0.3 \times 2990 \times 0.1 = 358.8 \text{ N-m}$$

∴ Power transmitted,

$$P = T\omega = 358.8 \times 52.4 = 18800 \text{ W} = 18.8 \text{ kW} \qquad \text{Ans.}$$

8. A multi-disc clutch has three discs on the driving shaft and two on the driven shaft. The outside diameter of the contact surfaces is 240 mm and inside diameter 120 mm. Assuming uniform wear and coefficient of friction as 0.3, find the maximum axial intensity of pressure between the discs for transmitting 25 kW at 1575 r.p.m.

Solution :

Given: $n_1 = 3$; $n_2 = 2$; $d_1 = 240$ mm or $r_1 = 120$ mm ; $d_2 = 120$ mm
or $r_2 = 60$mm;
$\mu = 0.3$; $P = 25$ kW $= 25 \times 10^3$ W; $N = 1575$ rpm or $\omega = 2\,\pi \times 1575/60$
$= 165$ rad/s.

Let $T =$ Torque transmitted in N-m and
$W =$ Axial force on each friction surface.

We know that the power transmitted (P),
$$25 \times 10^3 = T\omega = T \times 165 \quad \text{or} \quad T = 25 \times 10^3/165 = 151.5 \text{ N-m}$$
Number of pairs of friction surfaces.
$$n = n_1 + n_2 - 1 = 3 + 2 - 1 = 4$$
and mean radius of friction surfaces for uniform wear
$$R = \frac{r_1 + r_2}{2} = \frac{120 + 60}{2} = 90mm = 0.09m$$
We know that torque transmitted (T),
$$151.5 = n\mu WR = 4 \times 0.3 \times W \times 0.09 = 0.108 \ W$$
$$\therefore W = 151.5 /0.108 = 1403 \text{ N}$$
Let p = Maximum axial intensity of pressure.
Since the intensity of pressure (p) is maximum at the inner radius (r_2) therefore for uniform wear
$$pr_2 = C \quad \text{or} \quad C = p \times 60 = 60 \ p \text{ N/mm}$$
We know that the axial force on each friction surface (W),
$$1403 = 2 \pi C (r_1 - r_2) = 2 \pi \times 60p (120 - 60) = 22622 \ p$$
$$p = 1403 /22622 = 0.062 \text{ N/mm}^2 \quad \text{Ans.}$$

9. A conical friction clutch is used to transmit 90 kW at 1500 rpm. The semi-cone angle is 20° and the coefficient of friction is 0.2. If the mean diameter of the bearing surface is 375 mm and the intensity of normal pressure is (not the exceed) 0.25 N/mm², find the dimensions of the conical bearing surface and the axial load required.

Solution :

Given: $P = 90 \text{ kW} = 90 \times 10^3 \text{ W}$; $N = 1500 \text{ rpm or } \omega = 2 \pi \times 1500/60$
= 156 rad/s;

$\alpha = 20°$; $\mu = 0.2$; $D = 375$ mm or $R = 187.5$ mm; $P_n = 0.25$ N/mm²

Dimensions of the conical bearing surface

Let r_1 and r_2 = External and internal radii of the bearing surface, respectively.

b = Width of the baring surface in mm, and

T = Torque transmitted.

We know that power transmitted (P),
$$90 \times 10^3 = T\omega = T \times 156$$
$$\therefore \quad T = 90 \times 10^3/156 = 577 \text{ N-m} = 577 \times 10^3 \text{ N-mm}$$
And the torque transmitted (T),
$$577 \times 10^3 = 2 \pi \mu p_n R^2 b = 2\pi \times 0.2 \times 0.25 (187.5)^2 b = 11046 \ b$$
$$\therefore \quad b = 577 \times 10^3/11046 = 52.2 \text{ mm} \quad \text{Ans.}$$
We know that $r_1 + r_2 = 2R = 2 \times 187.5 = 375 \text{ mm}$ \hfill (i)

and $r_1 - r_2 = b \sin \alpha = 52.2 \sin 20° = 18$ mm   (ii)

From equations (i) and (ii)

$r_1 = 196.5$ mm, and $r_2 = 178.5$ mm   Ans.

Axial load required

Since in case of friction clutch, uniform wear is considered and the intensity of pressure is maximum at the minimum contact surface radius (r_2), therefore,

$P_n r_2 = C$ (a constant)   or $C = 0.25 \times 178.5 = 44.6$ N/mm

We know that the axial load required,

$W = 2\pi C (r_1 - r_2) = 2\pi \times 44.6 (196.5 - 178.5) = 5045$ N   Ans.

10. An engine developing 45 kW at 1000 rpm is fitted with a cone clutch built inside the flywheel. The cone has a face angle of 12.5° and a maximum mean diameter of 500 mm. The coefficient of friction is 0.2. The normal pressure on the clutch face is not to exceed 0.1 N/mm².

Determine: 1. The axial spring force necessary to engage to clutch, and
            2. The face width required.

Solution :

Given: $P = 45$ kW $= 45 \times 10^3$ W; $N = 1000$ rpm or $\omega = 2\pi \times 1000/60 = 104.7$ rad/s;

        $\alpha = 12.5°$; $D = 500$ mm or $R = 250$ mm $= 0.25$ m; $\mu = 0.2$; $p_n = 0.1$ N/mm².

1. Axial spring force necessary to engage the clutch.

First of all, let us find the torque (T) developed by the clutch and the normal lad (W_n) acting on the friction surface.

We know that power developed by the clutch (P),

$45 \times 10^3 = T\omega = T \times 104.7$ or $T = 45 \times 10^3/104.7 = 430$ N-m

We also know that the torque developed by the clutch (T),

$430 = \mu W_n R = 0.2 \times W_n \times 0.25 = 0.05 W_n$

$\therefore$ $W_n = 430/0.05 = 8600$ N

And axial spring force necessary to engage the clutch,

$W_e = W_n (\sin \alpha + \mu \cos \alpha)$
    $= 8600 (\sin 12.5° + 0.2 \cos 12.5°) = 3540$ N   Ans.

2. Face width required.

Let b = Face width required.

We know that normal load acting on the friction surface (W_n),

$8600 = p_n \times 2\pi Rb = 0.1 \times 2\pi \times 250 \times b = 157 b$

$b = 8600/157 = 54.7$ mm   Ans.

11. A leather faced conical clutch has a cone angle of 30°. If the intensity of pressure between the contact surfaces is limited to 0.35 N/mm² and the breadth of the conical surface is not to exceed one-third of the mean radius, find the dimensions of the contact surfaces to transmit 22.5 kW at 2000 rpm. Assume uniform rate of wear and take coefficient of friction as 0.15.

Solution :

Given: $2\alpha = 30°$ or $\alpha = 15°$; $p_n = 0.35$ N/mm²; $b = R/3$; $P = 22.5$ kW $= 22.5 \times 10^3$ W; $N = 2000$ rpm or $\omega = 2\pi \times 2000/60 = 209.5$ rad/s; $\mu = 0.15$.

Let $r_1 = $ Outer radius of the contact surface in mm,

$r_2 = $ Inner radius of the contact surface in mm,

$R = $ Mean radius of the contact surface in mm,

$b = $ Face width of the contact surface in mm $= R/3$, and

$T = $ Torque transmitted by the clutch in N-m.

We know that power transmitted (P),

$$22.5 \times 10^3 = T\omega = T \times 209.5$$

$$T = 22.5 \times 10^3/209.5 = 107.4 \text{ N-m} = 107.4 \times 10^3 \text{ N-mm}$$

We also know that torque transmitted (T),

$$107.4 \times 10^3 = 2\pi\mu p_n R^2 b = 2\pi \times 0.15 \times 0.35 \times R^2 \times R/3 = 0.11 R^3$$

$$R^3 = 107.4 \times 10^3/0.11 = 976.4 \times 10^3 \text{ or } R = 99 \text{ mm} \quad \text{Ans.}$$

The dimensions of the contact surface are shown in the following Figure.

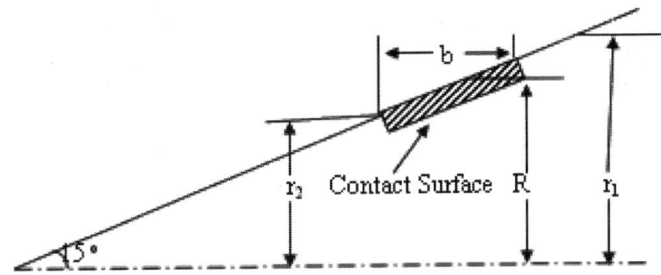

From figure, we find that

$$r_1 - r_2 = b\sin\alpha = \frac{R}{3} \times \sin\alpha = \frac{99}{3}\sin15° = 8.54 mm \qquad \text{(i)}$$

and $\qquad r_1 + r_2 = 2R = 2 \times 99 = 198$ mm $\qquad\qquad$ (ii)

From equations (i) and (ii),

$$r_1 = 103.27 \text{ mm, and } r_2 = 94.73 mm \qquad \text{Ans.}$$

7.11   Brakes

A brake is a mechanical device which produces frictional resistance against moving machine member, in order to slow down the motion of machine. In the process of performing this function, the brake absorbs kinetic energy of moving member and the brake absorbs potential energy of lowering member. The energy absorbed by brakes is released to surrounding in form of heat there are two main functions of brakes: (i) To slow down or stop the machine in the shortest possible time at the time of need, (ii) To control the speed of machine so that the material will not move further.

Electromagnetic brakes slow an object through electromagnetic induction, which creates resistance and in turn either heat or electricity. Friction brakes apply pressure on two separate objects to slow the vehicle in a controlled manner.

7.11.1   Machine brakes

The principle of operating machine brakes is very similar in wide range of weaving machines. The movement of the starting handle from ON to OFF position causes a brake band to bite on the surface of the flat part of the fly-wheel. It is obvious that the fly-wheels are mounted on the main shaft, excepting the brake arrangement in the electromagnetic drives already described earlier. The efficiency of a brake is dependent upon the angle of lap over which the brake band makes contact with the fly-wheel, the co-efficient of friction the condition of the two contacting surfaces. If some foreign matters are present between the two surfaces, the braking efficiency will be reduced. It is quite likely that due to careless manual lubrication oil may find a place between the two surfaces, and brake is likely to fail to perform its work properly. It is desirable that the brake system should bring the sley to rest at the desired position in order to allow a broken yarn to be mended without any other assistance.

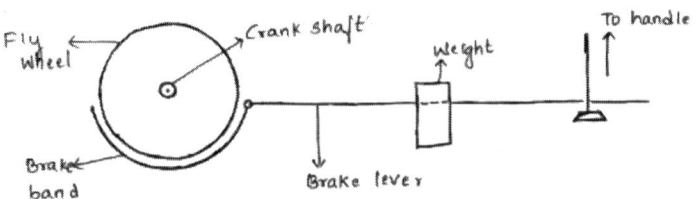

Figure 7.25 Orthodox loom brake system

The machine should come to stop, in case of warp break, when the healds are level and, when a weft thread breaks the sley should be around the back centre after side weft fork monitoring device has detected such a break; in case of central weft monitoring detecting a weft break the sley should stop prior to beat up otherwise in pick finding setting on marks cannot be avoided. In the older types of orthodox machines, the angle of lap of the brake shoe is very small, as can be seen in Figure 7.25. This system takes recourse to the gravitational pull of a weight supported by weight lever. The braking efficiency of such a machine is not very high. The type of brake system used in conjunction with cone clutch drive would be very effective since the brake band encompass almost the whole periphery of the brake well. Similarly, the braking efficiency would also be high.

7.11.2   Characteristics of brakes

Brakes are often described according to several characteristics including:

1. Peak force – The peak force is the maximum decelerating effect that can be obtained.

2. Continuous power dissipation – Brakes typically get hot in use, and fail when the temperature gets too high. The greatest amount of power (energy per unit time) that can be dissipated through the brake without failure is the continuous power dissipation. Continuous power dissipation often depends on e.g., the temperature and speed of ambient cooling air.

3. Fade – As a brake heats, it may become less effective, called brake fade. Some designs are inherently prone to fade, while other designs are relatively immune. Further, use considerations, such as cooling, often have a big effect on fade.

4. Smoothness – A brake that is grabby, pulses, has chatter, or otherwise exerts varying brake force may lead to skids.

5. Power – Brakes are often described as "powerful" when a small human application force leads to a braking force that is higher than typical for other brakes in the same class.

6. Pedal feel – Brake pedal feel encompasses subjective perception of brake power output as a function of pedal travel. Pedal travel is influenced by the fluid displacement of the brake and other factors.

7. Drag – Brakes have varied amount of drag in the off-brake condition depending on design of the system to accommodate total system compliance and deformation that exists under braking with ability

to retract friction material from the rubbing surface in the off-brake condition.

8. Durability – Friction brakes have wear surfaces that must be renewed periodically. Wear surfaces include the brake shoes or pads, and also the brake disc or drum. There may be trade-offs, for example, a wear surface that generates high peak force may also wear quickly.

9. Weight – Brakes are often "added weight" in that they serve no other function. Further, brakes are often mounted on wheels, and un-sprung weight can significantly hurt traction in some circumstances. "Weight" may mean the brake itself, or may include additional support structure.

10. Noise – Brakes usually create some minor noise when applied, but often create squeal or grinding noises that are quite loud.

7.11.3　Classification of brakes – Case I

On the basis of method of actuation

(a) Foot brake (also called service brake) operated by foot pedal In sectional warping the machine is controlled by a brake operated by applying pressure to a foot pedal then it is called foot brake.

(b) Hand brake – it is also called parking brake operated by hand

Parking brakes or emergency brakes are essentially mechanical brakes operated by hand.

On the basis of power

(a) Mechanical brakes

Internal expanding shoe brakes are most commonly used in automobiles. In a beam warping machine the band is fitted on a drum. The brake shoes come in contact with inner surface of this drum to apply brakes.

(b) Hydraulic brakes

The brakes which are actuated by the hydraulic pressure (pressure of a fluid) are called hydraulic brakes. Hydraulic brakes are commonly used in modern looms or textile machines.

(c) Air brakes

Air brakes are applied by the pressure of compressed air. Air pressure applies force on brakes shoes through suitable linkages to operate brakes.

An air compressor is used to compress air. This compressor is run by engine power.

(d) Vacuum brakes

Vacuum brakes are a piston or a diaphragm operating in a cylinder. For application of brakes one side of piston is subjected to atmospheric pressure while the other is applied vacuum by exhausting air from this side. A force acts on the piston due to difference of pressure. This force is used to operate brake through suitable linkages.

(e) Electric brakes

In electrical brakes an electromagnet is used to actuate a cam to expand the brake shoes. The electromagnet is energized by the current flowing from the battery. When flow of current is stopped the cam and brake shoes return to their original position and brakes are disengaged. Electric brakes are used in modern sizing machine for stoppage of machine instantly.

On the basis of method of application of braking contact

(a) Internally – expanding brakes
(b) Externally – contracting brakes

On the basis of application

(a) Service or foot brakes
(b) Parking or hand brakes

On the basis of operation

(a) Manual
(b) Servo
(c) Power operation

On the basis of applying brake force

(a) Single acting brakes
(b) Double acting brakes

7.11.4   Classification of brakes – Case II

In broad the brakes are classified into

- Mechanical
  - Block brake
  - Band brake
  - Internal or external expanding shoe brake
- Electrical
- Hydraulic.

Brakes are one of the most important safety features on any machine. The brakes used to stop a machine while running are known as the service brakes, which are either a disc and drum brake. Machines are also today equipped with other braking systems, including anti-lock and emergency brakes.

7.11.5   Mechanical brake

The brake is shown in Figure 7.26.

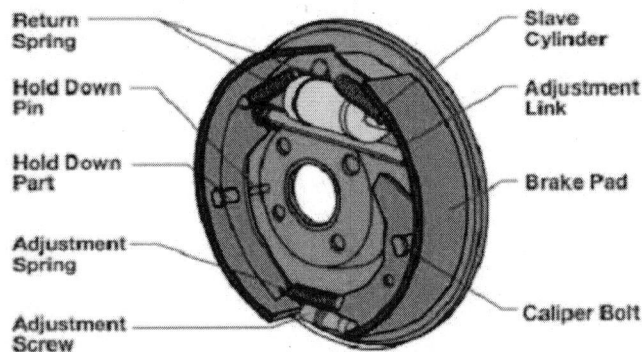

Figure 7.26   A simple mechanical brake

7.11.6   Block brake

Single block or shoe brake: This consists of a block which is pressed against the rim of revolving brake wheel drum. The proxy brake is example of block brake (Figure 7.27).

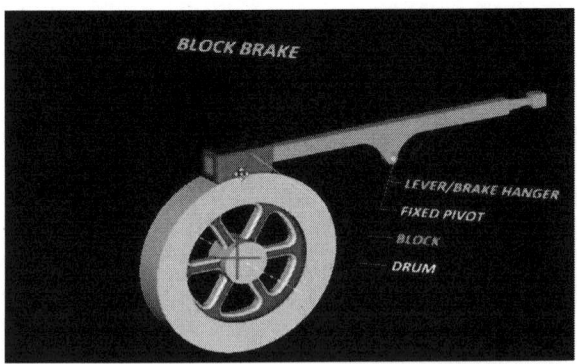

Figure 7.27   Single block or shoe brake

Double block or shoe brake:

- It consist of two brake blocks at the opposite ends of the wheel.
- These shoes apply force to both the sides of wheel and reduces the unbalanced force on the shaft.
- This spring pull the upper end of the brake arms together and brake is applied.
- When a force is applied to the bell crank level , the spring is compressed and the brake is released

7.11.7   Band brake

A band brake consists of a flexible steel band lined with friction material, which wrap to the brake drum. When an upward force is applied to the lever end, the lever turns about the fulcrum pin and tightens the band on the drum and hence the brakes are applied. The friction between the band and the drum provides the braking force. This type of brake is used in civil construction equipment and also in automobiles as hand brake. The rope brake is example of band brake (Figure 7.28).

Figure 7.28   Band brake

7.11.8   Internal expanding shoe brake

An internal expanding shoe brake consists of two shoes. The outer surface of the shoes are covered with friction material. Each shoe is pivoted at one end about a fixed fulcrum and other end rests against cam. When the cam is operated, the shoe are pushed outwards against brake drum. The friction between the shoes and the drum produces the braking torque (Figure 7.29).

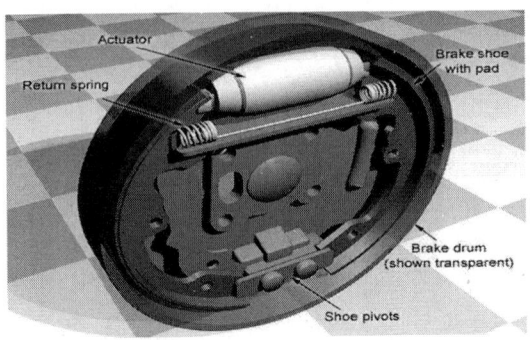

Figure 7.29   Internal expanding brake

7.11.9   Disc brakes

Disc brakes are different from drum brakes in that the drum is replaced by a circular metal disc and the brake shoe are replaced by a calliper which supports a pair of friction pads, one on each side of the disc. These pads are forced inward by the operating force and also retard the disc. Disc brakes consist of a disc brake rotor – which is attached to the wheel – and a calliper, which holds the disc brake pads. Hydraulic pressure from the master cylinder causes the calliper piston to clamp the disc brake rotor between the disc brake pads. This creates friction between the pads and rotor, causing your car to slow down or stop. (Figures 7.30 and 7.31).

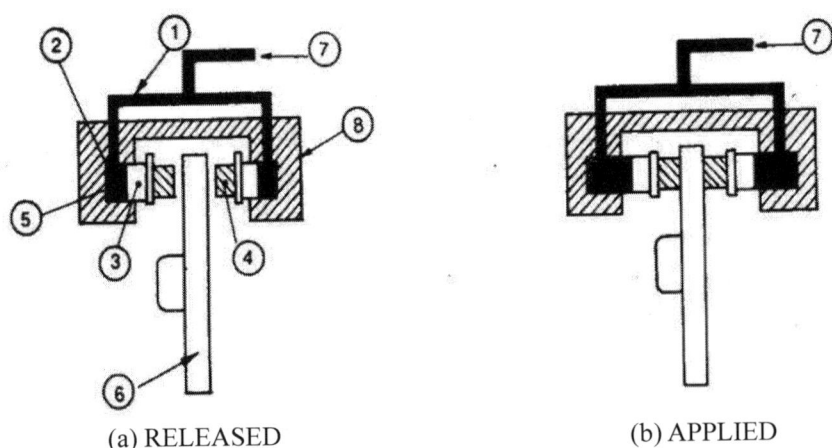

(a) RELEASED   (b) APPLIED

Figure 7.30   Working principle of disc brake

Details of parts: 1. Connecting tube, 2. Cylinder, 3. Piston, 4. Friction pad, 5. Hydraulic fluid, 6. Brake disc, 7. From master cylinder, 8. Calliper.

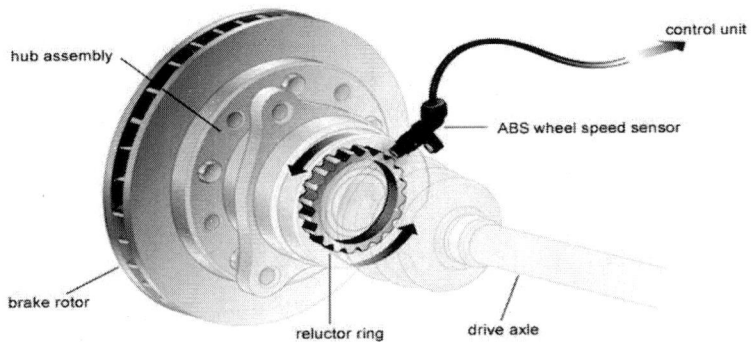

hub assembly

control unit

ABS wheel speed sensor

brake rotor

reluctor ring

drive axle

Figure 7.31   Disc brake

Working of disc brake

When the driver applies pressure on the brake pedal, hydraulic pressure pushes the pistons out from their housing. The pistons, in turn, press the brake pads against the moving disc faces, causing friction and hence slowing it down. Hydraulic pressure is equally applied by the hydraulic fluid to the floating pistons on either side. When the driver takes his foot off the brake pedal, hydraulic pressure on the friction pads is released; the pistons move inwards and break their contact with the disc.

7.11.10   Drum brakes

The main components of drum brakes are

1. Brake drum.
2. Back plate.
3. Brake shoes.
4. Brake liners.
5. Retaining springs.
6. Cam.
7. Brake linkages.

Hydraulic pressure from the master cylinder causes the wheel cylinder to press the brake shoes against the brake drum. This creates friction between the shoes and drum to slow or stop machine. (Figure 7.32).

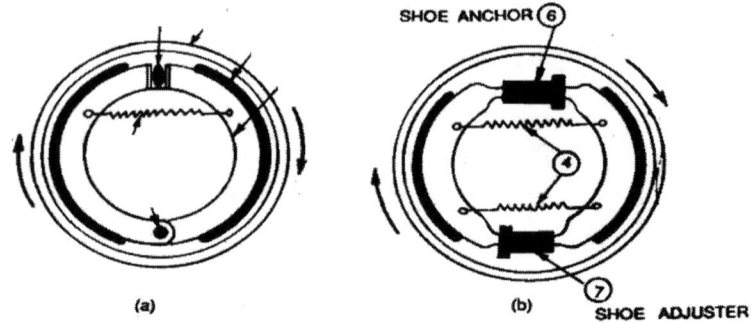

Figure 7.32   Drum brakes

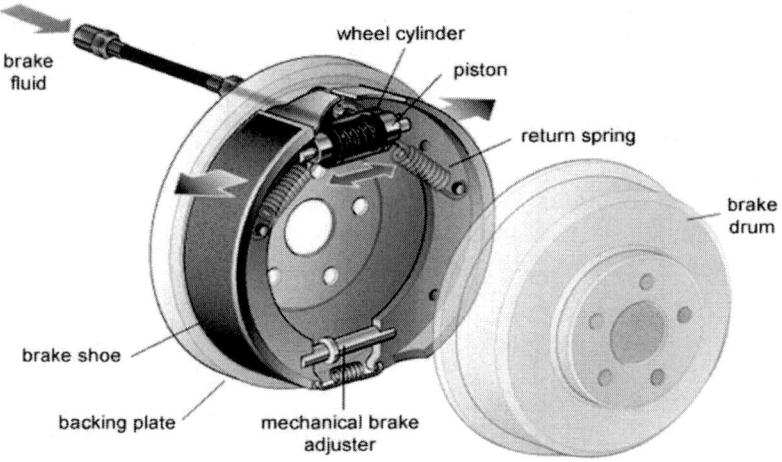

Figure 7.33   Drum brake construction

Construction

In this system the wheel is attached to drum. There are brake shoes used to contact the rotating drum for braking operation. The shoes provide lining on their outer surface. The cam is used to lift the brake shoes at one end, other end is connected by some method so as to make as the brake sleeve come into contact in the brake drum. The retaining spring is provided for bringing the brake shoes back to its original position, after releasing the brake pedal. All these parts are fitted in the back plate and enclosed with brake drum (Figure 7.33).

Working

When the pedal is pressed the cam moves the shoes outwards through linkages, there by coming in frictional contact with the rotating drum. As soon as the brake pedal is released the retaining springs help the brake shoes to brought back and release the brakes.

7.11.11   Crash stop or emergency brakes

Some of the textile machines like beam warping machine, loom, etc., needs sudden or emergency or secondary braking system or parking brakes. Emergency brakes are independent of the service brakes, and are not powered by hydraulics. Parking brakes use cables to mechanically apply the brakes (usually the rear brake). There are a few different types of emergency brakes, which include: a stick lever located between the driver and passenger seats; a pedal located to the left of the floor pedals; or a push button or handle located somewhere near the steering column. Emergency brakes are most often used as a parking brake to help keep a vehicle stationary while parked. And, yes, they are also used in emergency situations, in case the other brake system fails!

7.11.12   Hydraulic brakes

Hydraulic brakes make used of hydraulic pressure to force the brake shoes. It works based on PASCAL'S law. Hydraulic brake is an arrangement of braking mechanism which uses brake fluid, typically containing ethylene glycol, to transfer pressure from the controlling unit, which is usually near the operator of the machine, to the actual brake mechanism, which is usually at or near the driving arrangements of the machine (Figure 7.34).

The most common arrangement of hydraulic brakes consists of the following:

a)   Brake pedal or lever

b)   A pushrod (also called an actuating rod)

c)   A master cylinder assembly containing a piston assembly (made up of either one or two pistons, a return spring, a series of gaskets/ O-rings and a fluid reservoir)

d)   Reinforced hydraulic lines

e)   Brake calliper assembly usually consisting of one or two hollow aluminium or chrome-plated steel pistons (called calliper pistons), a set of thermally conductive brake pads and a rotor (also called a brake disc) or drum attached to the driving unit. When the brake pedal is pressed the piston is forced in to the master cylinder, the hydraulic pressure is applied equally to all wheel cylinders. The pistons in the wheel cylinders pushed outwards against the brake drum. When the driver release the brake pedal, the piston in the master cylinder returns back to its original position due to the return spring pressure. Thus the pistons in the wheel cylinder come back in its original inward position. Thus the brakes are released.

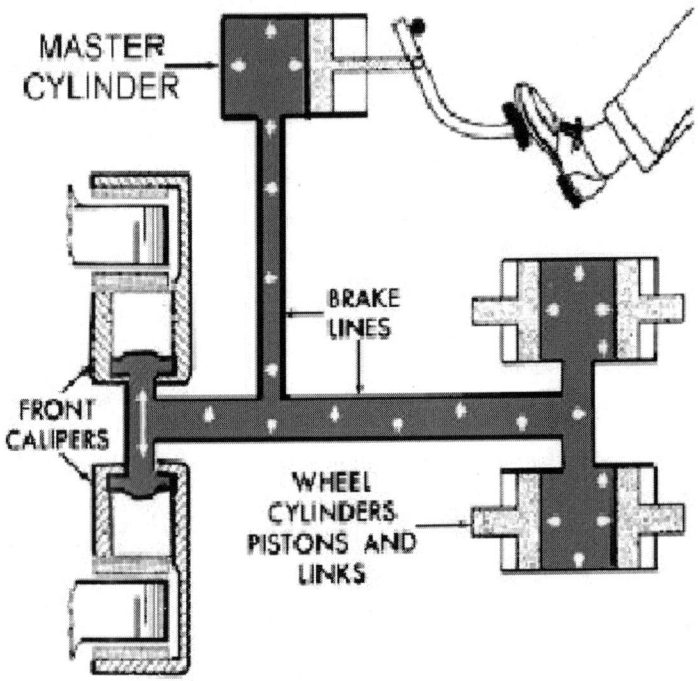

Figure 7.34 Hydraulic brake

Advantages and disadvantages of hydraulic brakes

Advantages

- Equal braking action on all wheels.
- Increased braking force.
- Simple in construction.
- Low wear rate of brake linings.
- Flexibility of brake linings.
- Increased mechanical advantage.

Disadvantages

- Whole braking system fails due to leakage of fluid from brake linings.
- Presence of air inside the tubing's ruins the whole system.

7.12   Applications of brakes in various industries

7.12.1   Anti-lock brakes

Computer-controlled anti-lock braking systems (ABS) is an important safety feature which is equipped on most newer vehicles. When brakes are applied suddenly, ABS prevents the wheels from locking up and the tires from skidding. The system monitors the speed of each wheel and automatically pulses the brake pressure on and off rapidly on any wheels where skidding is detected. This is beneficial for driving on wet and slippery roads. ABS works with the service brakes to decrease stopping distance and increase control and stability of the vehicle during hard braking.

Brake shoes (Figure 7.35).

Figure 7.35   Brake shoe of different materials

7.12.2   Brakes in textile industry

Different types of brakes in multiple voltages and can have either standard backlash or zero backlash hubs are used in various textile machines. Multiple disks can also be used to increase brake torque, without increasing brake diameter. There are two main types of holding brakes. The first is spring applied brakes. The second is permanent magnet brakes. These brakes are typically used on or near an electric motor.

Features: Standard/zero backlash, Brake torque, Multiple voltages

Technical features: Torque ranges from 50NM to 3000NM ,High operating Reliability / frequency,

Raw materials according to DIN standards
Different armature designs for various types of machines
Also with different voltage optional: 24 VDC / 90 VDC / 190 VDC
Applications in textile machines include: SULZER, PICANCOL, DORNIER, GAMMA, RIETER DRAW FRAME, ZINSER DRAW FRAME, TSDOKOMA JAPAN, AIRJET LOOMS

7.12.3   Textile machinery brake

Electromagnetic DC (direct current) flange mounted brakes are normally Off type brakes. These brakes are applied after disconnecting the supply to the drive or driven. External DC voltage will be given to the brake stator (coil) which attracts the armature and thus the brake is applied (Figure 7.36).

Figure 7.36   Textile machinery brake

Specifications:
    Torque: 8 Nm–2550 Nm (Higher Torques on Request)
    Design: 202 / 204 / 206
    Voltage: 24 / 96 / 105 / 190 / 205 VDC (Other Voltages on Request).
Applications
    Carding C 1/3, TEXTOOL - TYGA Winder, TEXTOOL WS 60 & WS 90, HOW A Draw Frame
    TOYODA Ring Frame, STATEX Wrap Reel.

7.12.4   Clutch brake unit / clutch brake combinations on textile machinery

Clutch brake combination units / clutch brake combinations are used where intermittent operations required. Instead of controlling the prime mover, intermittent operations are achieved by controlling clutch and brake by giving DC voltage. These unit is encased, low maintenance and low power consuming (Figure 7.37)

Torque: 8 Nm–500 Nm
Voltage: 24 / 96 / 105 / 190 / 205 VDC.

Figure 7.37   Clutch brake unit found on textile machines

Applications:
Packaging and printing Mmachines, indexing machinery, loom and weaving preparatory machines, cut to length machines, Aauto-Llabelling machines.

Spinning – Ring spinning
In ring spinning each spindle is provided with break system as shown in the Figure 7.38.

Figure 7.38   Spindle brakes

7.12.5   Brakes in weaving preparatory – Sizing

Sizing machine creel braking systems (fig.7.34a

Regardless of the type of creel configuration or thread-up, there are several major goals which any creel braking system must provide (Figure 7.39).

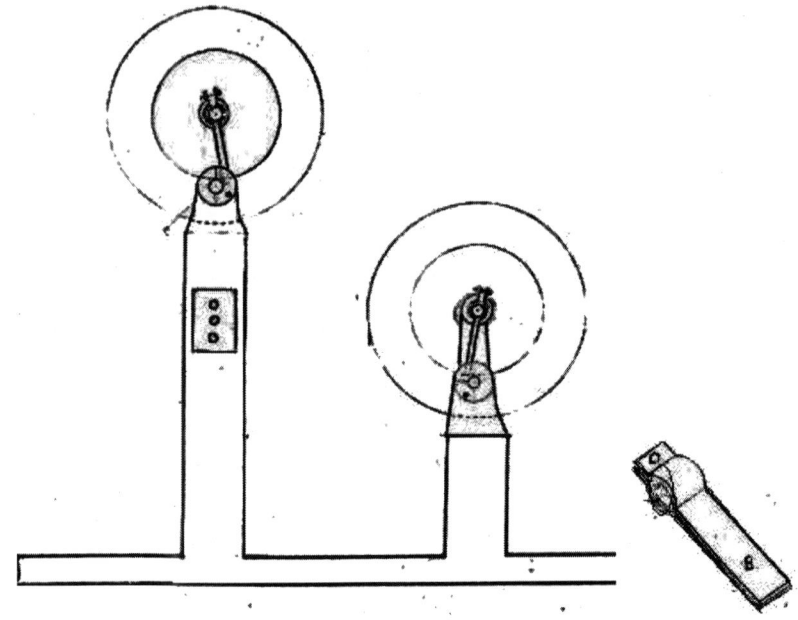

Figure 7.39   U – Clip in Negative sizing creel

These include:

1. Allow the lowest possible tension to retain as much elongation capability in the yarn as possible while insuring there is adequate tension € or proper control of the yarn.

2. Adjustment of the tension level to compensate for slack ends or other yarn problems from warping.

3. Maintain a desired tension level while accommodating the reduction of beam diameter which occurs as yarn is removed from the section beam.

The ideal effect of a braking system would be to allow a constant web tension as the beam diameter decreases from full to empty.

Loom brakes

The principle of operating loom brakes is very similar in a wide range of looms. The movement of the starting handle from the ON to the OFF position causes a brake band to close round the flat perimeter of a wheel (Figure 7.40) which is mounted on the main shaft of the loom and is acting as a brake drum. Such a system is likely to vary in efficiency, depending upon the angle of lap over which the brake band makes contact with the brake drum, the coefficient of friction, and the condition of the two surfaces. The presence of foreign matter, such as oil on the surface is quite detrimental to efficient braking. It is preferable that the braking system should allow the loom to be brought to rest at the desired position in order to allow a yarn break to be repaired without intermediate manual or mechanical adjustment of the loom position. Examples of this requirement are:

a)   With the healds level if the warp stop-motion stops the loom for an end-break.

b)   With the sley around the back-centre position on the pick after a weft break has been detected by the side-weft fork motion; and

c)   Before beat-up on the actual pick of the weft break if the centre weft-fork motion is responsible for stopping the loom in weaving a fabric in which pick-finding is almost certain to create a setting on place.

The braking system in the last of these instances needs to be highly efficient and, for this reason, brake drums on looms having centre weft forks usually have larger angles of lap, the brake band having a high coefficient of friction and covering almost the whole of the perimeter of the brake drum in order to spread the weaver over a greater area and dissipate the heat created.

Loom brake system
A brake is a device by means of which artificial frictional resistance is applied to moving body in order to stop the motion of a loom (Figure 7.41).

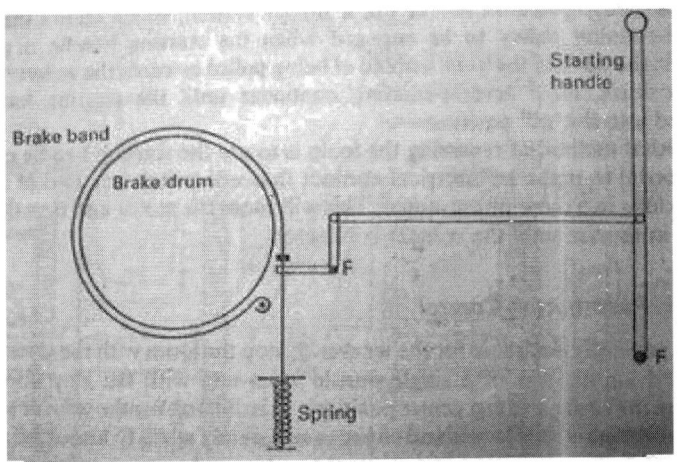

Figure 7.40   Loom brake system

Figure 7.41 Loom clutch and brake on Toyoda loom

The assumption that the loom will always come to rest in the required position is quite unrealistic, and frequently it is necessary to adjust the position of the stopped loom manually by releasing the brake and turning the hand-wheel of the loom. Alternatively, the starting handles maybe drawn towards the ON position and released as soon as the sley has moved forward to the required position. A small movement of the sley (inching) is generally achieved with each pull of the starting handle, but it is tricky manoeuvre, which lacks accuracy. Mechanical reversing of the loom from the top and front centre positions becomes desirable after the redrawing of a broken end or the detection of a weft break. It saves time and effort on the part of the weaver.

Kinetics of shedding, picking and beat-up motions

8.1 Introduction

Weaving or loom motions can be divided mainly into primary, secondary and tertiary motions. The primary motions include shedding, picking and beat-up motions. The first primary motion shedding is defined as the process of separation of warp into two or more layers. The shed depth is important from the smooth operation of shed and loom during the passage of shuttle is concerned. If the shed depth is not adjusted, the warp during separation may collide with shuttle and thus the shuttle may get trapped. The shed depth depends on many factors like type of weaving, type of shuttle, total number of ends, etc.

8.2 Shed opening diagram and an expression for shed depth

Shedding involves the opening of the warp layers into two or more layers for the passage of shuttle in loom. The shed movement follow the simple harmonic motion with dwell of one third of a pick or 120° (from 30° to 150° of the crank circle diagram). Referring to the Figure 8.1, the movement is traced from the displacement diagram of the heald shaft or warp threads. The shed opening at any time is being the distance between the top and bottom lines of warp threads. The fastest rate of closing the shed is seen when two sets of warp threads are at level with each other as at the number 9 position (Figure 8.2).

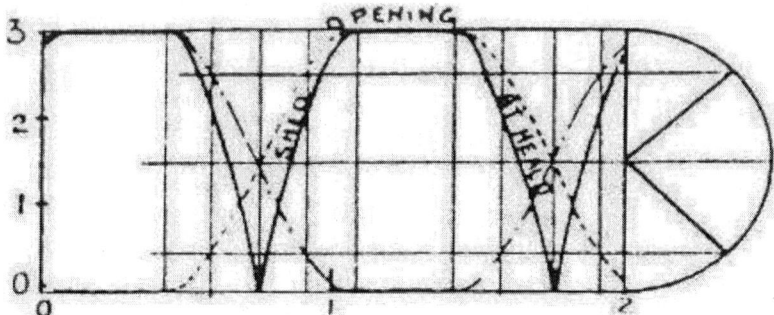

Figure 8.1 Dwell period of heald shaft and time for two revolutions of crankshaft (Source: Hanton)

The shed opening at the box front side of the shuttle box when the time available for the shuttle passage will depend on the shed opening at the heald or shed angle but also on the distance moved by shuttle front side from the fabric fell. The same is shown in the graphical diagram in Figure 8.3. The diagram of reed displacement of the reed from the fabric fell is shown in Figure 8.4. It can be observed that the diagram is similar to the sley displacement. In the figure, *AB* give the distance from the original base by a distance equal to shuttle width. The displacement of the shuttle from the fell is always less than the displacement of the reed by a distance equal to shuttle width and thus the measurement from the base *AB* to the reed displacement curve will give shuttle movement.

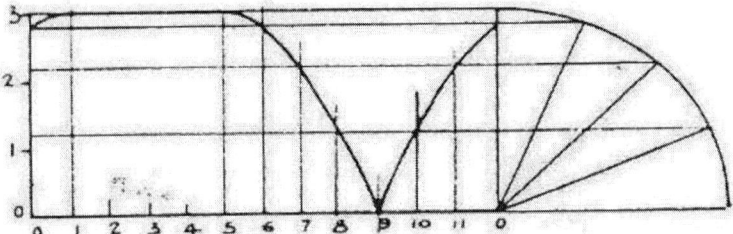

Figure 8.2 Heald movement for one revolution of crankshaft (Source: Hanton)

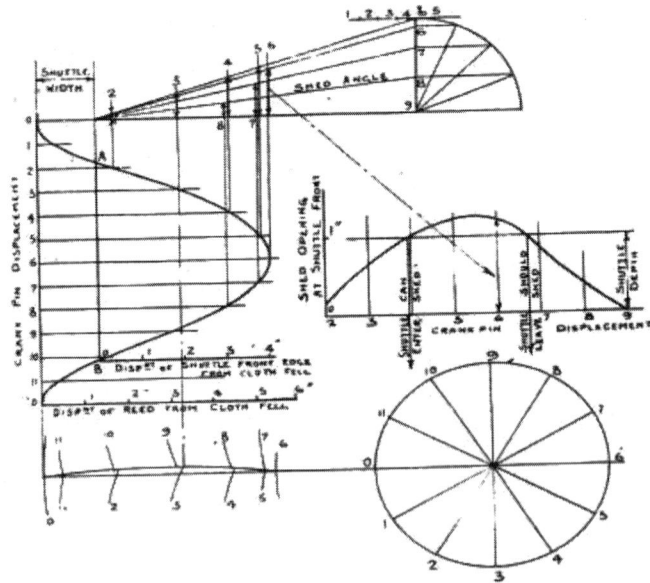

Figure 8.3 Displacement diagram of the heald analysed (Source: Hanton)

The Figure 8.4 shows the shed opening at the shuttle front with earliest and latest crankshaft positions when shuttle is in the shed. From these the time for the passage of shuttle when its front edge enters shed at one side until its rear end clears the shed at the other end can be easily found.

By referring to the Figure 8.5, the displacement of the sword pin can be understood through geometry. Figure shows the shedding set up in a shuttle loom. The figure also show the race board and crank arm along with shed in opened state.

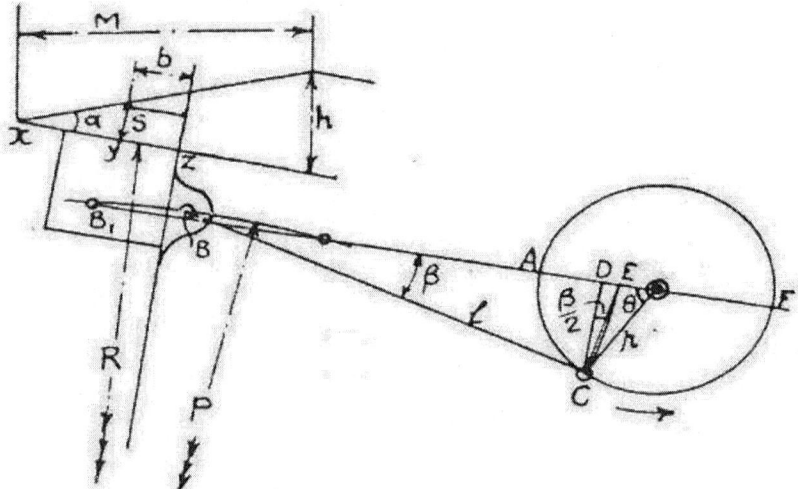

Figure 8.4 Shed opening diagram (Source: Hanton)

The displacement of the sword pin from the beating up position

$$= BB_1 = AE$$

$$= AD + DE$$

$$= (2r - DF) + DE$$

$$= 2r - (r + r + \cos \theta) + DC \tan \frac{\beta}{2}$$

$$= (r - r \cos \theta) + r \sin \theta \tan \frac{\beta}{2}$$

$$= r (1 - \cos \theta + \sin \theta \tan \frac{\beta}{2})$$

$$= r \left(1 - \cos \theta + \sin \theta \frac{\sin \beta}{1 + \cos \beta} \right)$$

$$= r\left(1 - \cos\theta + \sin\theta \, \frac{\sin\beta}{2}\right) \text{ taking } \cos\beta = 1 \text{ since } \beta \text{ is always a}$$

small angle.

But $DC = 1$ $\sin\beta = r\sin\theta$

$$\therefore \sin\beta = \frac{r}{1}\sin\theta$$

and $BB_1 = r\,(1 - \cos\theta + \sin\theta \, \dfrac{r}{2l}\sin\ \theta)$

$$= r\,(1 - \cos\theta + \frac{r}{2l}\sin^2\theta) \tag{1}$$

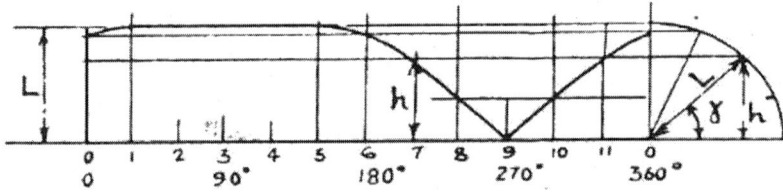

Figure 8.5 Shed opening at heald dwell of one third of a pick and
heald level at top centre (Source: Hanton)

Also, the displacement of the reed from the fell of the cloth

$$= xz = BB_1 \times \frac{R}{P}$$

and the displacement of the shuttle front from the fell of the cloth

$$= xy = \left(BB_1 \times \frac{R}{P}\right) - b$$

The shed opening at the shuttle front $S = xy \tan\alpha$

$$= \left\{\left(BB_1 \cdot \frac{R}{P}\right) - b\right\} \tan\alpha$$

$$= \left[\left\{r\left(1 - \cos\theta + \frac{r}{2l}\sin^2\theta\right) \cdot \frac{R}{P}\right\} - b\right] \tan\ \alpha \ . \tag{2}$$

Again $\dfrac{h}{2} = M \tan\dfrac{\alpha}{2}$ *and* $\dfrac{\alpha}{2} = \tan^{-1}\left(\dfrac{h}{2M}\right)$

$$\therefore \alpha = 2 \tan^{-1}\left(\frac{h}{2M}\right) \text{degrees} \qquad (3)$$

Between $\theta = 30°$ and $\theta = 150°$, the shed angle α is constant, and $h = L$

$$\therefore \text{Between these limits } \alpha = 2 \tan^{-1}\left(\frac{L}{2M}\right) \text{degrees} \qquad (4)$$

Between $\theta = 150°$ and $\theta = 270°$ the shed depth $h = L \sin \gamma$

Also $\gamma (202.5 - 0.75\ \theta)°$

$$\therefore h = L \sin (202.5 - 0.75\theta)$$

$$\therefore \text{From (3) } \alpha = 2 \tan^{-1}\left\{\frac{L\sin(202.5 - 0.75\theta)}{2M}\right\} \text{degrees}$$

8.2.1   Numerical example on shed angle

From the data given below find the shed angle, shed height, stroke of the tappet.

Movement of sley angle when at back centre – 12° angle between race board and reed – 87.

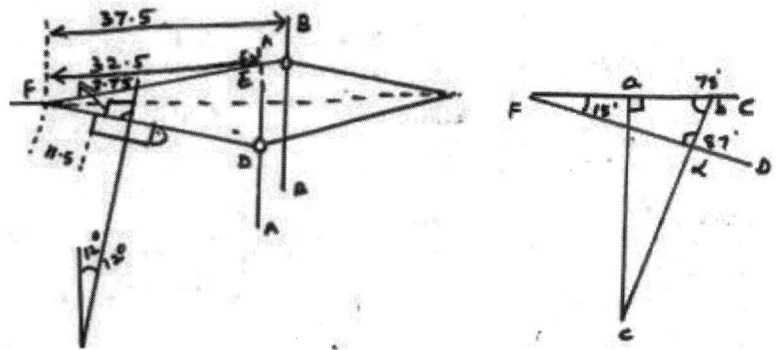

Figure 8.6   Shed geometry

Angle between bottom shed line to horizontal = 15°.

Front of shuttle from cloth fell – 11.5 cm. Shuttle height – 43.75 cm.

Clearance between top front edge of shuttle and top line of shed = 0.5 cm.

Solution :

Referring to Figure 8.6 angle *EFD*

$$\text{Tan } EFD = \frac{3.75 + 0.5}{11.5}$$

$$= \frac{4.25}{11.25}$$

Or angle $EFD = 20°17'$

Angle $EFC = 20°17' - 15° = 5°17'$

(Angle between top shed line and the horizontal)

The vertical movement of heald stares AA & BB is as follows:

Movement of heald stares $AA = CD + DE$

In triangle CDF, $CD/FC = \text{Tax } 15$

$\quad CD = 32.3 \times 0.2679 = 8.71$ cm

Again from triangle $EFC = \dfrac{CE}{FC} = \text{Tax } 5°17'$

Or $CD = 3$ cm.

Total movement of stare $= 8.71 + 3 = 11.71$ cm.

Similarly stare $BB = 13.2$ cm.

8.2.2 Geometry of shed

The size of the shed

The selection of a right type of a shuttle depends on many factors like warp count, weft count, type of loom, type of picking etc., and thus the dimensions of the shuttle like width and depth of the shuttle depends on the weft count or diameter. It is to be noted that the size of the shed for a given size of the shuttle depends on shed depth as it is necessary to allow a gap of at least half an inch to three fourth of an inch above the top wall of the shuttle and the top layer of the warp so that the shuttle will not collide with the warp.

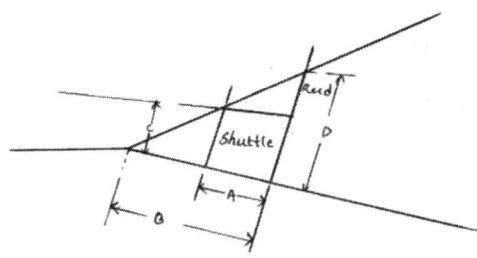

Figure 8.7 Geometry of the warp shed

In other words the size of the shed depends on the shed depth at the front wall of the shuttle and is shown in Figure 8.7 by C. The shuttle width is A, the distance of the fell from the reed or race board is B, and D gives the depth of the shed at reed in back centre of the loom i.e., 180° of the crankshaft. As the shuttle moves through the shed both B and D will vary as the sley advances and recedes during crank motion and thus D will also vary depending on the shuttle passage movement coinciding with the dwell period.

A simplified geometry of the shed is shown Figure 8.8.

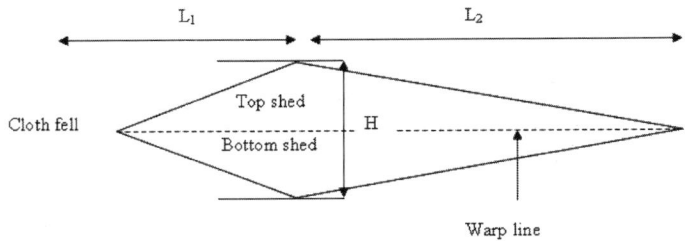

Figure 8.8   Geometry of shed (simplified) (Source: NPTL)

Let L_1: Length of the front shed, L_2: Length of the back shed, H: Shed height be the parameters of the shed considered under study. When the healds are on the warp line (healds are levelled), the path taken by the warp is the shortest. However, as the healds move away from the warp line, the warp takes a longer path. Thus, warp yarns are extended which has to be compensated either by the extensibility of the warp or by the regulation of the yarn delivery system. If the length of the back shed is increased, then yarn extension is reduced and this if preferred for weaving delicate yarns like silk. However, shorter back shed creates clearer shed and it is preferred for weaving coarser and hairy yarns. It is important to understand the factors which influence the degree of yarn extension during the shed formation. A simplified mathematical model has been presented to relate the warp strain with the shed parameters.

Calculation of warp strain during shedding

Let us consider h as half of the shed height (Figure 8.9). Therefore, $H = 2h$

Elongation in the front shed = E_1

Now, $E_1 = AD - AC$

$$= \left(L_1^2 + h^2 \right) - L_1$$

$$= L_1 \left[1 + \left(\frac{h}{L_1} \right)^2 \right]^{\frac{1}{2}} - L_1$$

$$= L_1 \left[1 + \frac{1}{2}\left(\frac{h}{L_1}\right)^2 + \frac{\frac{1}{2}\left(\frac{1}{2}-1\right)}{2}\left(\frac{h}{L_1}\right)^4 + \ldots \right] - L_1 = \frac{h^2}{2L_1}$$

(It is here by higher power of $\frac{h}{L_1}$ which is <1 is being neglected for simplicity)

The ratio of lengths of front and back shed is called shed symmetry parameter (i).

Therefore, $\frac{L_1}{L_2} = I =$ Shed symmetry parameter

Initial length of warp $= L = AB$

$$= L_2 + L_2$$

$$= L_1 + \frac{L_1}{i}$$

$$= L_1\left(\frac{1 + L_1}{i}\right), \text{Total elongation} = E = E_1 + E_2$$

$$= \frac{h^2}{2L_1} + \frac{h^2}{2L_2} = \frac{h^2}{2L_1}(1 + i)$$

$$\text{Strain} = \frac{\text{Elongation}}{\text{Initial length}} = \frac{E}{L}$$

$$= \frac{1}{L} \times \frac{h^2}{2L_1}(1 + i)$$

$$= \frac{h^2}{2L^2}\left(\frac{(1 + i)^2}{i}\right)$$

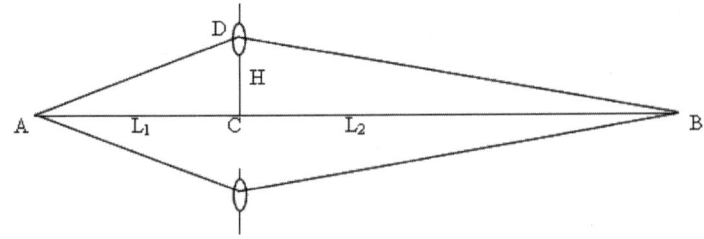

Figure 8.9 Warp strain during shedding (Source: NPTL)

From the above equation the following things can be inferred.

- Warp strain increases with the increase in shed height.
- Warp strain reduces with the increase in shed length.
- Warp strain reduces as the shed becomes symmetric (the value of i increases).

8.2.3   Bending factor and its significance in loom

The depth of the shed at the front wall of the shuttle is expressed as a fraction of the height of the shuttle front wall known as bending factor or interference factor. It indicates the extent of the warp threads deflection

Bending factor is defined as the ratio of depth of shed in front of shuttle (s) and the actual height of the shuttle (h) as shown in Figure 8.10.

So, bending factor $= \dfrac{s}{h}$

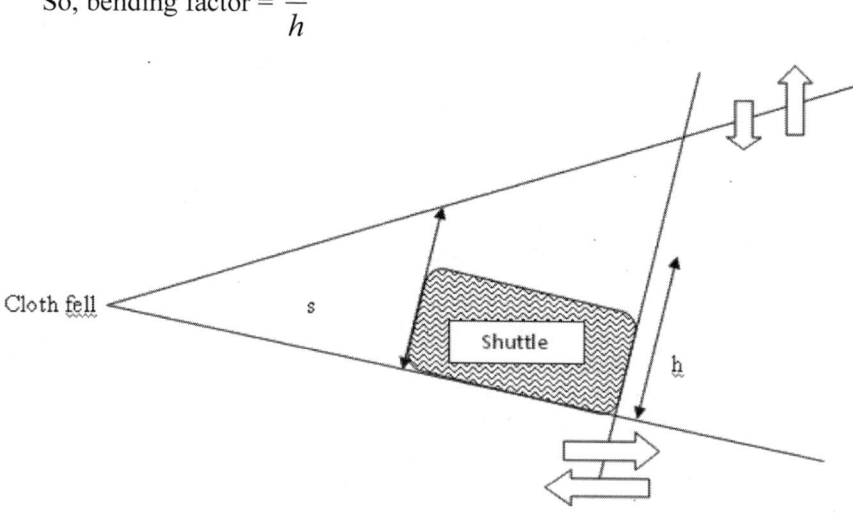

Figure 8.10 Concept of bending factor (Source: NPTL)

Bending factor less than one indicates that the warp line will deflect and there will be severe abrasion observed between warp layers and shuttle (Figure 8.11). And if it is greater than 1, then there will not be any deflection or abrasion between warp sheets and the shuttle. But bending factor less than one may lead to high warp end breakage rate and even the trapping of shuttle in the shed. The bending factor changes continuously as it is influenced by the follow two factors namely: Movement of the healds and Movement of the sley

The bending factor will reduce as the top shed line will move in the downward direction causing reduction in the value of s and vice versa. Besides, as the reed moves towards the cloth fell, the depth of shed in front of shuttle (s) reduces. Thus the bending factor reduces.

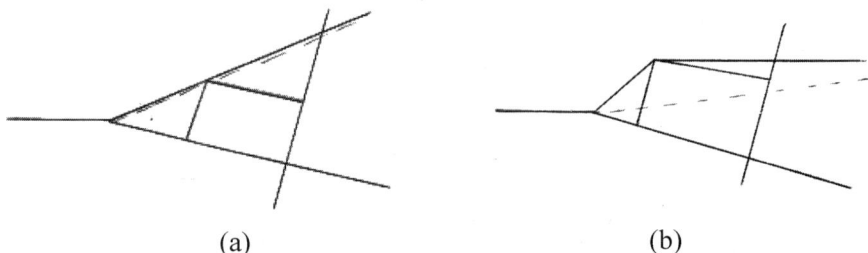

(a)   (b)

Figure 8.11   (a) and (b) shows the situation with low and high bending factor

The reed moves towards the back of the loom between 0°–180°. Then it moves towards the front of the loom between 180°–360°. For late shedding (where the shed levels at 0), dwell occurs between 120°–240°. Therefore, during this period, the healds are stationary. So, the depth of shed in front of the shuttle varies only due to the sley movement. As the sley moves to the back centre at 180°, the depth of shed becomes the maximum at the point as shown by the blue line of Figure 8.12. However, the reed moves forward after 180° and thus the depth of shed reduces. After 240°, the shed starts to close and sley is still moving forward. Both the factors synergistically reduce the depth of shed at a faster rate. Actual shuttle height (2.8 cm) has been indicated with the broken horizontal line. It is observed that the depth of the shed is very close (slightly less) than the shuttle height when the shuttle enters and leaves the shed. So, the bending factor is close to 1. For early shedding (where the shed levels at 270°), dwell occurs between 30°–150°. After 150°, shed starts to close. But, sley moves backwards till 180°. Therefore, between 150°–180°, two factors are countering each other in influencing bending factor. The maximum shed height is obtained around 160° as indicated by the red line in Figure 6.16. After 180°, sley starts to move forward and shed is still closing (till 270°). Therefore, the shed depth reduces very fact after 180°. At the time of shuttle exit (240°), the depth of shed is around 1 cm which is just about one third of the shuttle height Thus, severe abrasion between the warp sheet and shuttle is quite obvious while the latter leaves the shed. Table 8.1 shows the values of bending factors for early and late shedding in a typical case study conducted by research workers.

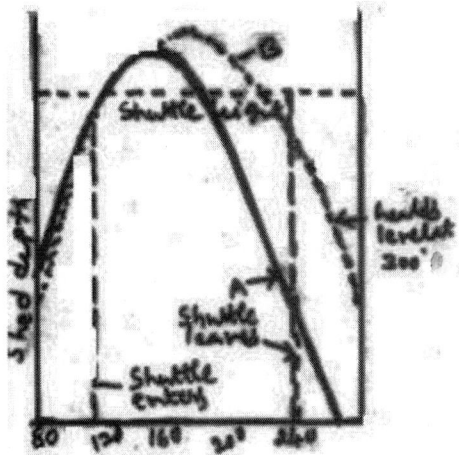

Figure 8.12   Shed depth curves

Table 8.1   Bending factors in relation to shed timing of a case study

Healds crossing time	Bending factors	
	Entering	Leaving
270° Early shedding (curve A)	0.87	0.34
0° Late shedding (curve B)	0.84	0.9

8.2.4   Method to increase the cover factor of the grey fabric in a loom

The effect of the back rest position on the cover factor of the fabric can be very well understood by the Figure 8.13. A research study was conducted in a weaving laboratory by raising the position of the back rest by half an inch above and below from its regular position and the results were very interesting. The reader is advised to refer the article mentioned at the end of the chapter.

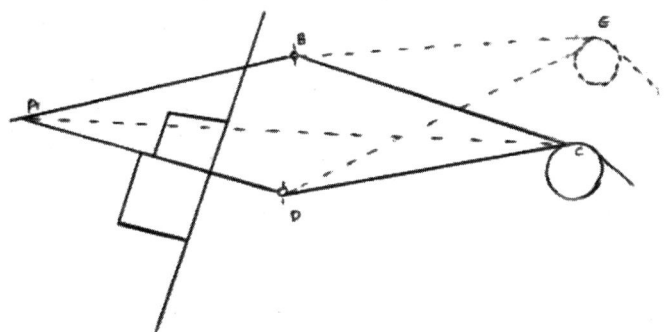

Figure 8.13   Effect of backrest position on the fabric cover

It is usual to set the heald shafts so that warp sheet *ADC* will be contact with the sley race when the crank is at back centre. The lift of the heald is adjusted to give some gap between the upper line of the shed and top of the shuttle front wall of shuttle. The normal position of the backrest would be with its top on the line *AC* which bisects the angle *BAD*. The lengths namely *ABC* and *ADC* will be equal at this moment and the tensions in warp yarn layers will also be equal. However if the backrest is raised to position *E*, with the heald crossing before beat-up and the warp lengths *ADE* and *ABE* are different. Thus the length *ABE* is shorter than *ADE*. In other words the upper line is slack and lower line is taut and the same at the beat-up position is depicted in Figure 8.14.

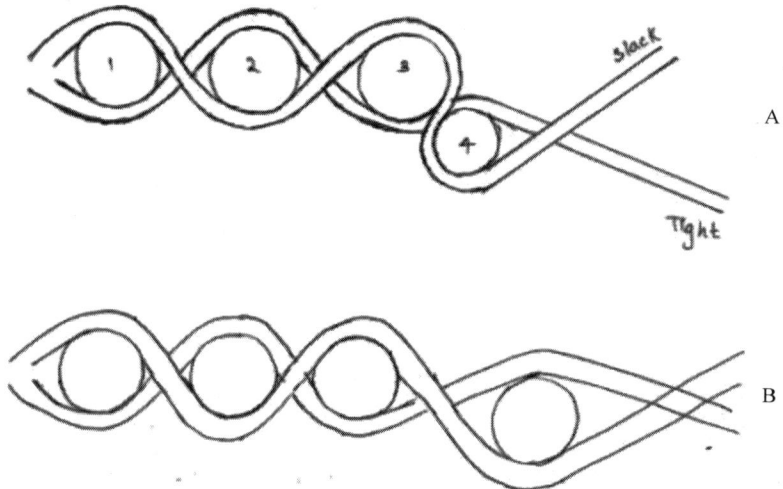

Figure 8.14   Position of the weft at beat-up with change in the backrest position

8.2.5   Numerical example on warp strain

A loom is set with plain weaving with negative tappet shedding mechanism. Two heald shafts are used namely front and back heald. Work out the ratio of strain created in the warp threads during shedding mechanism by the front heald and back heald if the total shed length (distance between the cloth fell and back rest) is 1.2 m, front shed length for the front heald is 0.2 m, distance between the front and back heald is 0.04 m, diameters of reversing rollers are 0.05 m and 0.06 m.

Solution :

The strain in warp yarns can be expressed by the following equation.

$$\text{Strain} = \frac{h^2}{2L} \times \frac{(1+i)^2}{i}$$

For the front heald

Front shed length $(L_1) = 0.2$ m or 20 cm.

Therefore, the back shed length $= L_2 = (1.2{-}0.20) = 1$ m or 100 cm.

So, shed symmetry parameter $= i_1 = \dfrac{L_1}{L_2} = \dfrac{0.20}{1.00} = 0.2$

For the back heald

Front shed length $(L_1) = (0.2 + 0.04)$ m $= 0.24$ m or 24 cm.

Therefore, the back shed length $= L_2 = (1.2{-}0.24) = 0.96$ m or 96 cm.

So, shed symmetry parameter $= i_2 = \dfrac{L_1}{L_2} = \dfrac{0.24}{0.96} = 0.25$

Now, $\dfrac{h_2}{h_1} = \dfrac{d_2}{d_1}$ and the total shed length L is same for both the healds.

So, ratio of strain $= \dfrac{h_1^2}{h_2^2} \times \dfrac{\dfrac{(1+i_1)^2}{i_1}}{\dfrac{(1+i_2)^2}{i_2}} = \dfrac{h_1^2}{h_2^2} \times \dfrac{(1+i_1)^2}{(1+i_2)^2} \times \dfrac{i_2}{i_1}$

$$= \frac{5^2}{6^2} \times \frac{(1.2)^2}{(1.25)^2} \times \frac{0.25}{0.2} = 0.8.$$

So, the ratio of strain is 0.8:1 which implies that the back heald is creating more strain in the warp yarns.

8.3   Head movement and displacement diagrams

Figure 8.15 shows the common sheds formed in loom. The displacement diagrams are drawn for a cycle of 4 picks. *Y*-axis give the heald displacement and *x*-axis give the number of picks.

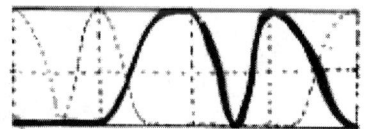
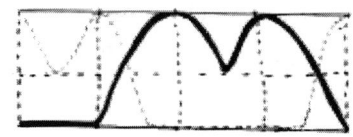

Bottom closed shed   Semi-open shed

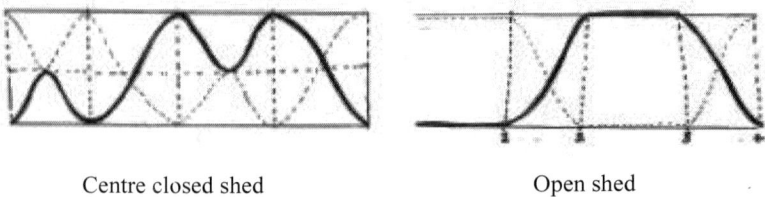

Centre closed shed   Open shed

Figure 8.15   Heald displacement diagrams

The heald movement is nearly Simple Harmonic Motion (SHM). Figure 8.16 show the displacement on Y-axis and time on X-axis for benefit of understanding three pairs.

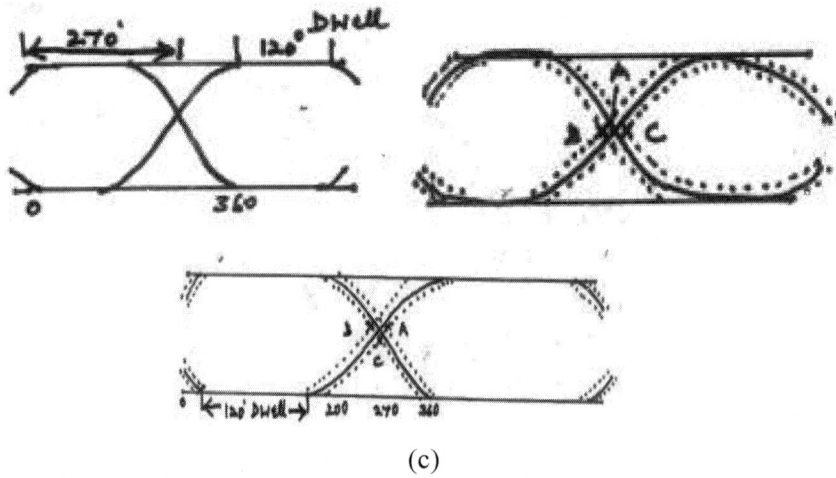

(c)

Figure 8.16   a – All threads crossing at the same time and height, b – Healds crossing at same height but at different time, c – Healds crossing at the same time but at different height

8.3.1   Staggering of healds in weaving

Let d_1, d_2 be the dia of smaller and larger reversing rollers. t_1, t_2 are treadle levers c_1, c_2 be the corresponding tappets on bottom shaft (Figure 8.17). Let b_1 and b_2 are corresponding anti-function bowl with treadles t_1 and t_2 with fulcrum f. Let l_1, l_2 be the weft or through of cams. h_1 and h_2 are shed opening or height. B is the distance two heald shaft. Let "a" be the distance between cloth fell and fun heald so that the back heald is away from fell by a distance $a + b$ for ideal shedding meets $h_1 = h_2$ $d_1 = d_2$, $a = b$. Which are hypothetical conditions. Figure 8.18 show the displacement of heald staggering.

$$h_1 = l_1 \frac{(x + y + b)}{x}$$

$$h_2 = l_2 \frac{(x + y)}{x}$$

$$\frac{l_1}{l_2} = \frac{h_2}{h_1} \left(\frac{x + y + b}{x + y} \right)$$

But $\dfrac{h_2}{h_1} = \left(\dfrac{a + b}{a} \right)$

$$\frac{l_2}{l_1} = \frac{(a + b)(x + y + b)}{a(x + y)}$$

to understand better $b = 3$ cm, $x = y = 22.5$ cm, $a = 20$ cm

$$\frac{l_2}{l_1} = \frac{(3 + 20)(22.5 + 22.5 + 3)}{20(22.5 + 22.5)}$$

$$\frac{l_2}{l_1} = \frac{1104}{900}, \quad \frac{l_2}{l_1} = 1.23 .$$

In other words the lift of 2nd tappet should be 23% greater than that of tappet one. This derivation indicates preferring the tappets of different lifts.

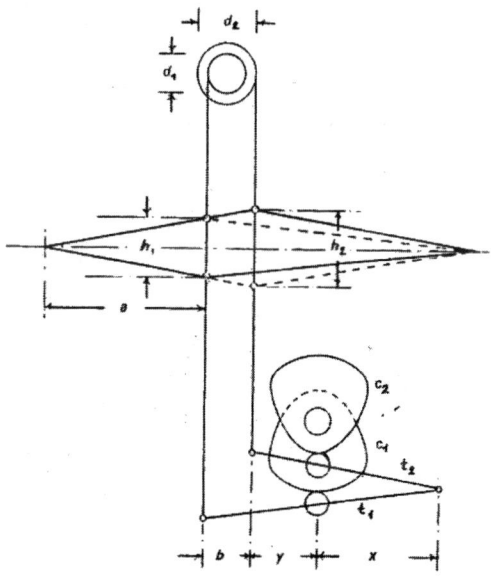

Figure 8.17   Heald staggering in plain loom

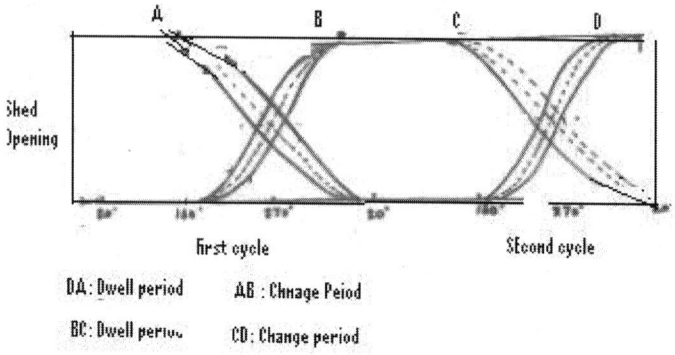

Figure 8.18   Displacement diagram of staggered healds

8.3.2   Numerical example on heald staggering in loom

Calculate the throw (lift) of the cam controlling the back heald from the following particulars:

Cam lift for the front heald = 8 cm

The distance between the front and back heald = 4 cm

The distance between the fulcrum and bowl on the treadle = 20 cm

The distance between the bowl and the fastening point of the back heald = 20 cm

Diameter of small reversing roller = 5 cm

Diameter of large reversing roller = 6 cm.

Solution :

The expression to be used for calculating the throw or lift of the cam is as follows:

$$\frac{l_2}{l_1} = \left(\frac{x+y+b}{x+y}\right) x \left(\frac{a+b}{a}\right)$$

Here

l_2 = Throw (lift) of the cam for the back heald

l_1 = Throw (lift) of the cam for the front heald = 8 cm

x = Distance between the fulcrum and bowl on the treadle = 20 cm

y = Distance between the bowl and the fastening point of the back heald = 20 cm

b = Distance between the front and back heald = 4 cm

d_2 = Diameter of large reversing roller = 6 cm
d_1 = Diameter of large reversing roller = 5 cm
a = Distance between cloth fell and front heald (not given)

using the equations we get

$$= 8 \times \frac{20 + 20 + 4}{20 + 20} \times \frac{6}{5} = 10.56$$

So, the lift of the cam controlling the back heald is 10.56 cm.

.71 + 3 = 11.71 cm.

Similarly stare BB = 13.2 cm.

8.4   Kinetics of tappet shedding, dobby shedding and Jacquard shedding mechanisms

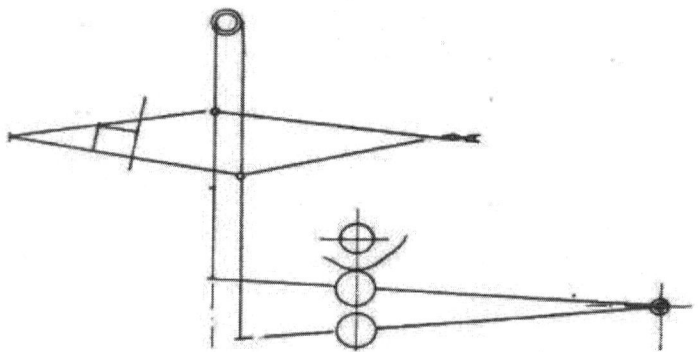

Figure 8.19   Negative tappet shedding device (Source: Hanton)

Figure 8.19 show the simple set-up of negative tappet shedding mechanism with treadle levers, antifriction bowls and top roller mountings. The movement of the warp threads is nearly always simple harmonic, the threads begin to move slowly and are moving fastest when halfway up or down and when the threads are slack. In the following examples the working is shown of two threads, one working one up two down – shown in full lines – and the other one down two up – shown dotted. This shows the behaviour of the warp threads when changing, and when needed up or down for two – or more – picks in succession. With tappets (Figure 8.20), the movement of the warp threads can be regulated in any desired way; the dwell, or pause, generally starts about 30° alter the beating up point and lasts for, say, 120°. No unnecessary movement takes place, the threads only moving when a change is required.

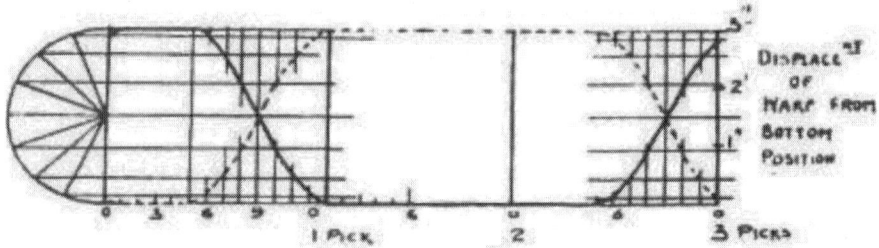

Figure 8.20   Displacement of warp from bottom position in negative tappet shedding motion (Source: Hanton)

Tappets the displacement diagram is different as shown in the Figure 8.21.

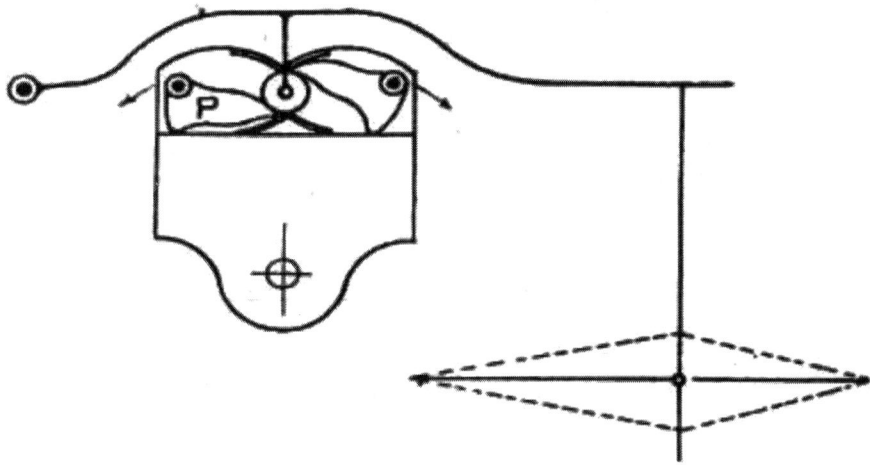

Figure 8.21   Oscillating tappet shedding (Source: Hanton)

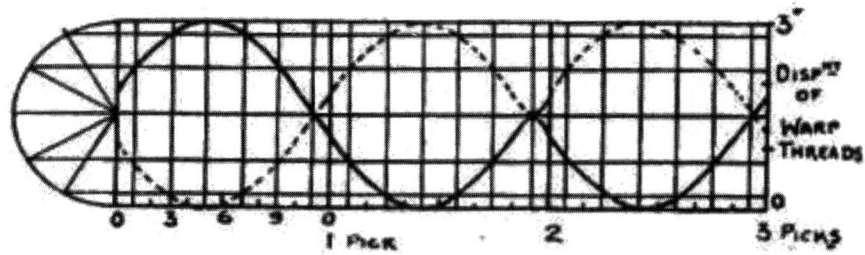

Figure 8.22   Displacement diagram for oscillating tappets (Source: Hanton)

With the oscillating tappet (Figure 8.22), the warp threads are all brought level after the shuttle has crossed and consequently there is much unnecessary movement. The movement is approximately simple harmonic, and depends on the shape of the groove and plates *P* (Figure 8.21). Dwell can be obtained by a suitable shaping of the groove, but often there is only the reduced speed of the threads as the crank drives the rocking plate passes its dead centre.

The single lift Jacquard (Figure 8.23) lifts only the threads which are needed to form the top line of the shed, but all the threads are returned to the bottom of the shed after the passage of the shuttle. There is no dwell when the shed is formed, other than the reduced speed as the driving crank passes its dead centre, but there is a pause or dwell in the movement when threads reach their lowest position, while the lifting knife clear of the hook to enable, the fresh selection to be made. This makes the movement given to the warp not quite simple harmonic, the threads stopping suddenly at the end of the downward movement and starting up again suddenly at the beginning of the lift.

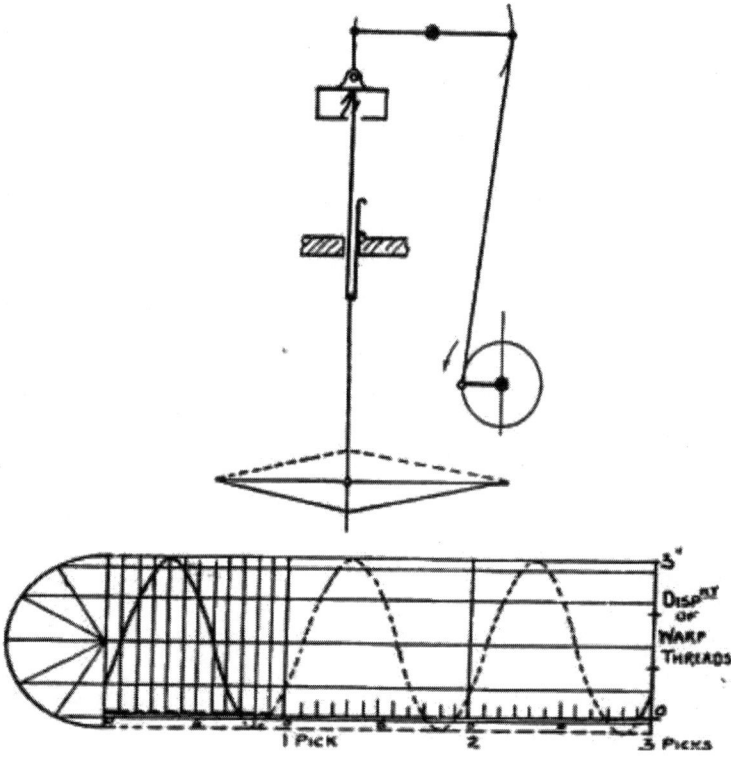

Figure 8.23   Heald displacement in single lift Jacquard (Source: Hanton)

Figure 8.24 shows the diagram from the Hattersley or Keighley Dobby. Here again there is no dwell of the top line of the shed, except such as is caused by slackness at the hearings of the driving gear, but there is a pause of the bottom line to enable the fresh selection of hooks to be made. Also, when a heald shaft is to be up for two picks in succession, although the dobby is called an open shed one, the shaft falls about ¼" along with those which are going down, before the second lifting knife comes up to the hook head and supports the shaft; the shaft rises up to its original height as the lifting knife gets to the end of it movement, because the other knife, which has been lowering the shaft now moves away from its hook head, to give the clearance necessary for the next selection.

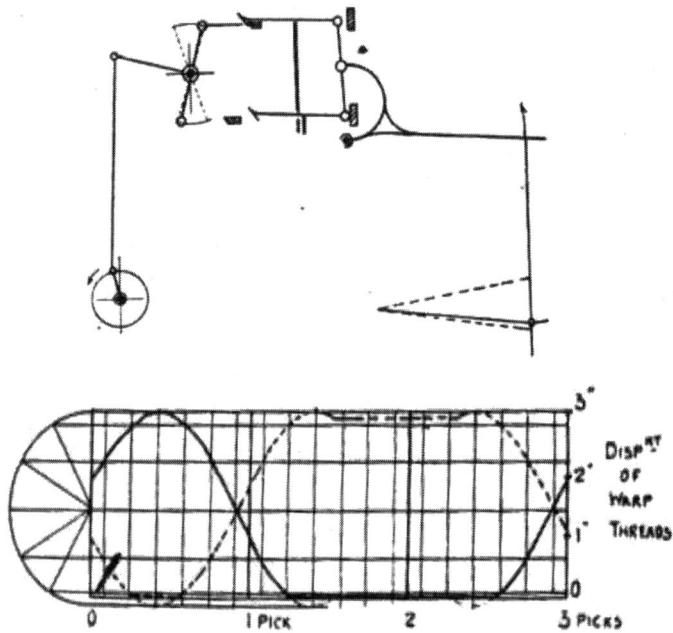

Figure 8.24   Displacement diagram of Keighly Dobby (Source: Hanton)

The Knowles Dobby (Figure 8.25) gives a movement to the warp threads very similar to that given by tappets. The movement is nearly simple harmonic, and the heald shafts have dwell in both their top and bottom positions. With the usual arrangement of the driving cylinders CC_1 the dwell given must be approximately 180° of the crankshaft's revolution. The cylinders CC_1 rotate at the same speed as the crankshaft, and since the teeth on these cylinders extend round approximately half of the circumference, movement is given to the heald shafts for about half the time of the crankshaft's revolution, and dwell during the other half. To get a dwell other than half a pick, it would be necessary to make the gears G different in diameter from the cylinders CC_1. Thus for a dwell of one fourth of a pick, C would require to be two-thirds the diameter of G and to have teeth extending approximately round three-quarters of its circumference. The rotation of G, when in contact with C, would be $\frac{3}{4} \times \frac{2}{3} = \frac{1}{2}$ revolution, as required and this half revolution, with the movement of the heald shaft, would have occupied three-fourths of the time for one revolution of C or of the crankshaft, leaving one fourth of the crankshaft's revolution for dwell. A double-lift Jacquard (Figure 8.26) forms what is known as a semi-open shed, threads which are going up or coming down doing so without unnecessary movement, but a thread which is required up for two picks in succession sinks to the centre of the shed between the two picks, after which the other hook takes it back to the top position. Again there is no dwell for the top line of warp, whilst the bottom line is stationary during the time for selecting the hooks next to be lifted.

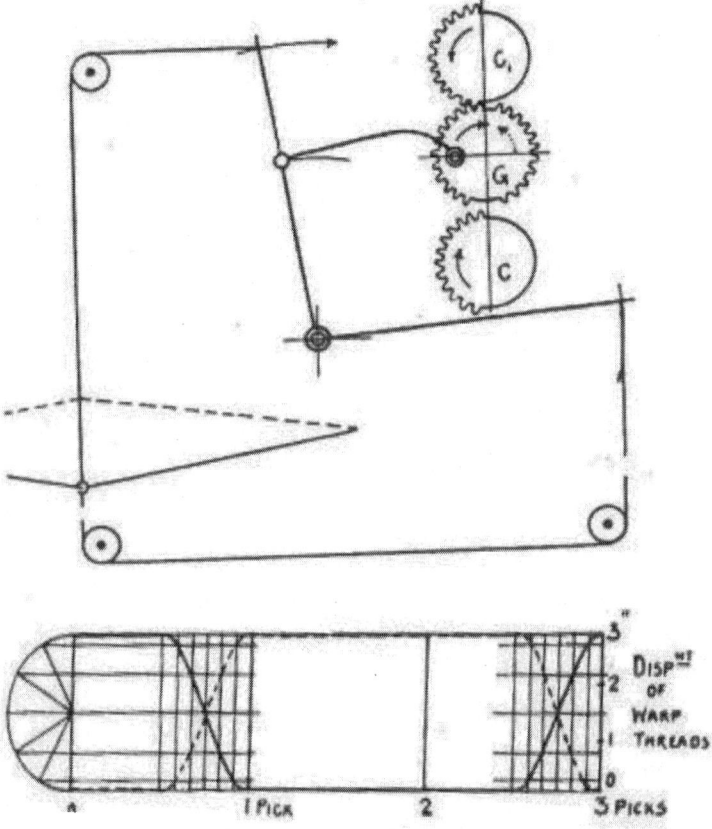

Figure 8.25   Heald displacement diagram of Knowles Dobby (Source: Hanton)

The movement given is simple harmonic, except when stopping a starting up at the bottom position.

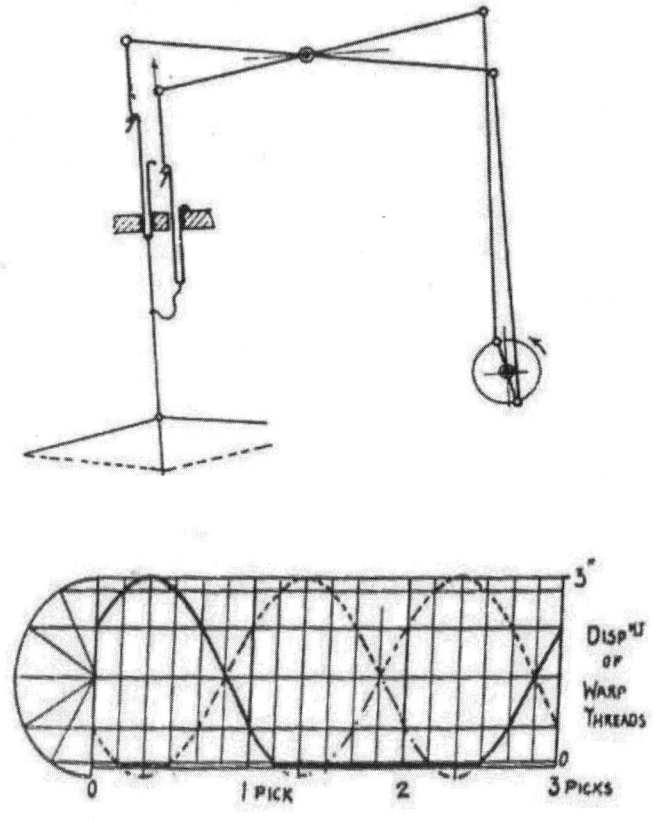

Figure 8.26   Displacement diagram of heald in double lift Jacquard (Source: Hanton)

8.4.1   Heald reversing

The tapped shedding mechanism is of two types namely positive and negative depending on the control of heald shaft direction. Present day loom are provided with positive shedding motions and therefore the problems of conventional heald reversing motion is totally absent. However, a large portion of the weavers still use the old looms and thus the discussion is made here for heald shaft reversing motions for negative shedding. The mechanism aspects are not discussed here and the reader is directed to read any book on *Fabric Formation* to understand the working concept. However, the mechanics part is dealt here (Figure 8.27).

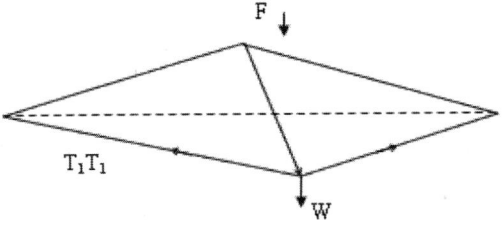

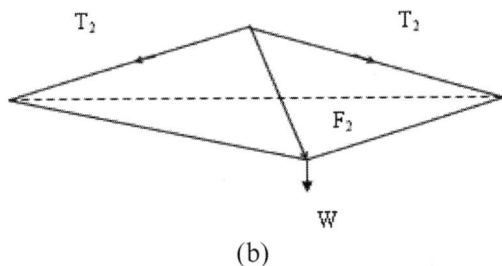

(b)

Figure 8.27 Forces on warp sheet

The healds are returned to their original position by means of spring which exerts a force "*F*". This tension is required on the stane is considerable and force may be so great that the treadle lever may slip-off the tapped surface. Referring to the Figure 8.27, it will be seen that the tension in the spring and in the connection to the stane will be greatest when stane is down and when the spring is doing no work of any use. With such an arrangement the maximum tension in the spring is much greater than is required to raise the stane. In Figure 8.27, the stane is down and the motion in warp produces a force F, acting upwards so that if W be the weight of the stane and the accessories, the force required to raise it at that point will be $W + T_1$ and similarly in the other condition, $W + T_2$ as the tension T_2 and force F_2 are acting down wards and hence this force is not sufficient. This is because the force exerted by the spring should increase as stane raises, actually it decreases owing to the decrease in the extension of the spring. The problem can be overcome by using traction springs and generally two springs heald is greater than an unstretched spring. This is observed from Figure 8.28.

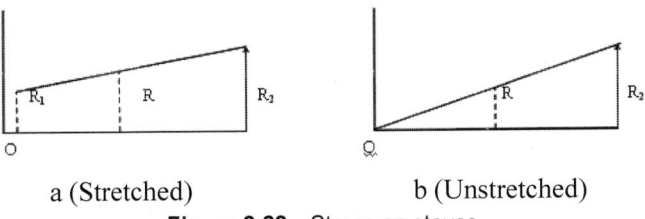

a (Stretched)   b (Unstretched)

Figure 8.28 Stress on staves

$$\frac{R_1 + R_2}{2} = R \text{ in first case}$$

and $$O + \frac{R_2}{2} = 2 \text{ in second case.}$$

8.4.2   Kenyons under motion

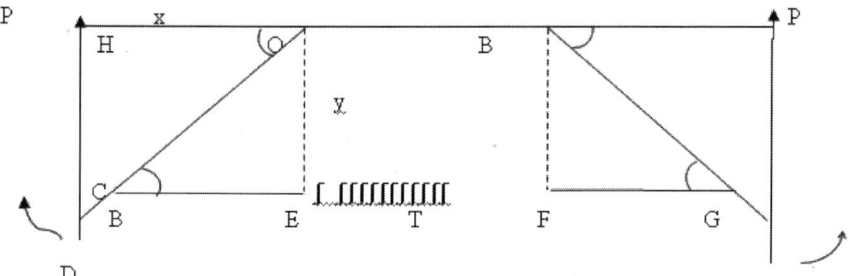

Figure 8.29   Kenyons under motion

It is summarised above that the system of reversal with single action of springs are not efficient. But in Keyons system it will be seen that traction springs are used in such way that they not only stretch less than the lift of stane, but as the springs stretch and their pull increase they act in a less effective manner on the stane and hence on the warp threads with the result that the force applied to the sane remains constant throughout the lift. Figure 8.29 shows the system with spring attached to levers in such a manner that the rate of stretching decreases as the stane is lifted. Not only the stretch falls but the line of action of the springs gets nearer to the lever fulcrum A and B and hence the spring acts less effectively on the lever and heald stane.

From Figure $CG = CE + EF + FG$

$= AB + 2\ CE$ (because $AB = EF$ and $CE = FG$)

As $CE = AC \text{ Cos } \theta$, $CG = AB + 2\ AC \text{ Cos } \theta$ (1)

Stretch of the spring

$CG -$ unstretched length

$= AB + 2\ AC \text{ Cos } \theta - n$, "$n$" be the unstretched length of the spring.

Therefore stretch of spring $= 2\ AC \text{ Cos } \theta + a$ where $a = (AB - n) =$ Constant.

We know that tension is proportional to stretch.

= K × stretch, where K = Spring constant.

Consider movements about A

$P{-}x = T{-}y$ where P = Force on stain and T = Spring tension

$$P = T - \frac{y}{x}$$

But $y = AC \sin \theta$, $x = AD \cos \theta$

$$\therefore P = T - \frac{AC}{AD} \tan \theta$$

Total tension acting on stane is = $2\,T\,\dfrac{AC}{AD}\,\tan \theta$

Or P proportional to $T \tan \theta$

As "y" decreases and "x" increases, the value of "θ" diminishes gradually so that pull will be reduced at $\theta = 0°$ and θ= 98-, Tan 90 infinity so the pull increases infinitely.

8.4.3   Quadrat spring over motion (Figure 8.30)

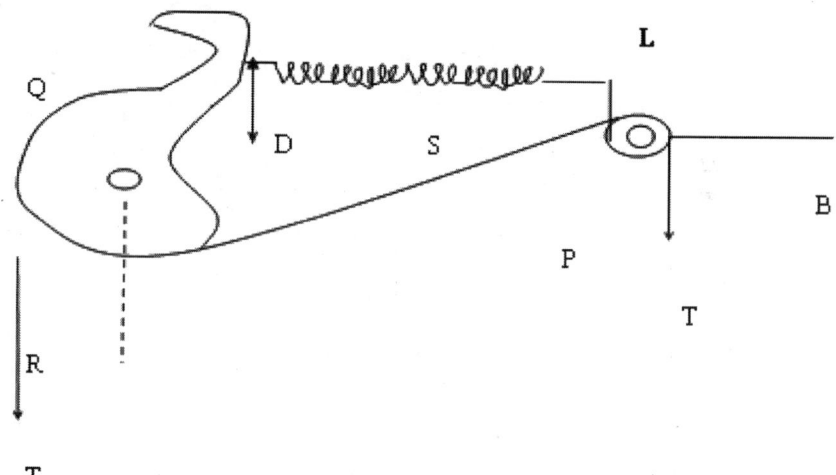

Figure 8.30   Quadrant spring over motion

$2\,TR = SD$

$$T = \frac{SD}{2\,R} \text{ (as "}R\text{" is constant)}$$

Or $T \propto SD$

8.5   Tappet drive in a loom

Generally the ratio between crankshaft and bottom shaft in a plain loom will be 1:2. But this may not be the case as the picks per repeat of the design may decide the ratio. For example let us consider the picks per repeat as two as in the case of a plain weave the ratio will be 1:2. If the picks per repeat is 4, the ratio may be 1:4; for picks 5 the ratio may be 1:5, similarly we can indicate for 6 picks ratio is 1:6, for 7 picks ratio is 1:7, for 8 picks it is 1:8, for 9 picks it is 1:9 and for 10 picks it is 1:10. But if one can observe carefully it is understood that up to 7 picks the sum of the wheels are same as the initial case of 2 picks i.e., 120; but for 8 and 9 it is 180 which is more than 120 and similarly for 10 picks it is totally different value as 186. Hence it is necessary to plan a good inventory of wheels if it is necessary to change the picks per repeat in design.

Let us consider the arrangement of wheels in loom. For 2 picks the two tappets are placed on bottom shaft, for more number of picks up to 5 we can use countershaft or auxiliary shaft as shown in Figure 8.31 in which the drive is transferred from bottom shaft and a set of gear can be used for specific requirement of picks per repeat.

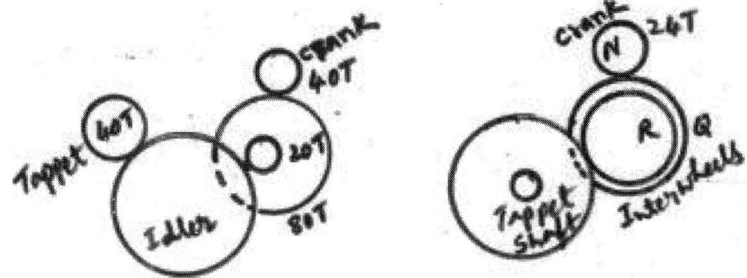

Figure 8.31   Tappet drive with idler wheel and without idler wheel

On the other hand consider a crank wheel of 40 teeth driving the 80 teeth bottom shaft wheel, a 20 teeth wheel on later driving a idler gear and finally idler gearing to 40 teeth tappet shaft then we can write for every revolution of the crankshaft the tappet shaft will rotate, say for 4 picks repeat design by

$$\frac{1 \times 40 \times 20}{80 \times 40} = \frac{1}{4} \; revs.$$

Thus the relation between the crankshaft and the tappet shaft rotation will be in the ratio of 4:1 or, it is seen that the rotation of the crankshaft will be equal to the number of picks per design repeat. It is customary for the tappet wheel to contain a number of teeth which is multiple of the weaves produced. Here the tappet wheel has 120 teeth which is divisible by any number up to and including 10. But this number is not divisible by 7 and 9. The number of teeth required in the change wheel or crank wheel, for weaves with picks per repeat as 2, 3, 4, 5 and 6 is obtained by dividing the particular number into the number of teeth contained in the tappet wheel and in that case the intermediate wheels will be replaced by a single carrier wheel. For example, if a design repeats on 6 picks, then we may require $\frac{120}{6} = 20$ teeth. Change wheel.

But for tappets intended for weaving 7, 8, 9 and 10 picks, the intermediate will be required in place of a single carrier wheel as these will not divide 120, and if 8 and 10 picks are intended wheel numbers would be respectively 15 and 12 which will not just fit in. Now, by adopting a small wheel on the crankshaft with 24 T, the intermediate wheels required would be in inverse proportion as 24 multiplied by the number of picks per repeat of the tappets to be used is to the number of teeth in the tapped wheel 120. For a 8 pick pattern the wheel ratio is found to be

$$\frac{24 \times 8}{120} = \frac{192}{120},$$

but these wheels being too large, wheels of same ratio will have to be used. Thus the intermediate wheel with 48 T and the wheel concentric on intermediate wheel with 30 T give the same ratio and can be conveniently used. In general we can write $\dfrac{p \times N \times R}{Q \times R} = r.p.m.$ of the tappet shaft, where, $p =$ number of picks per repeat, $N =$ crankshaft wheel, $Q =$ intermediate driven wheel, $R =$ intermediate driving wheel, and $T =$ tappet shaft wheel. For 8-pick pattern, the ratio Q/R is found to be $\dfrac{192}{120} = \dfrac{96}{60} = \dfrac{48}{30}$.

For 7-pick pattern the same ratio is found to be $\dfrac{168}{120} = \dfrac{84}{60} = \dfrac{42}{30}$;

for 9-pick pattern the ration is $\dfrac{216}{120} = \dfrac{108}{60} = \dfrac{54}{30}$.

Thus, from the above values it can be concluded that with given number of wheels, the generalised relation is $\dfrac{(6\,p)}{30}$.

It must also be remembered that though the ratio may be kept same, the number of teeth in the wheels should not be too small and prevent any meshing of the wheels. The wheel ratio must be kept within the limitations of the machine designer.

On the other hand if one can consider the usual arrangement of gearing between the crankshaft and the shafts on which shedding tappets are mounted and as each pick is inserted for each revolution of crank, the speed of the crankshaft divided by the speed of the tappets shaft must be equal to the number of picks per repeat of design. For 2-picks design, the tappets are placed on the bottom shaft (Figure 8.32). If the speed of the crankshaft is 180 rpm, the bottom shaft will have a speed of $\dfrac{180 \times 30}{60} = 90 \ rpm.$

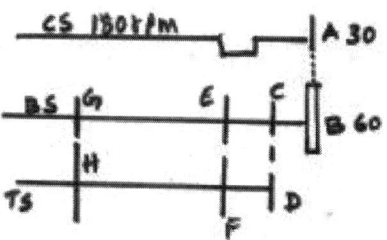

Figure 8.32 Tappet drive

provided the crank and bottom shaft wheels have 30 T and 60 T, respectively. Or, in other words, the bottom shaft will rotate at half the speed of the crankshaft. Similarly, the bottom shaft will rotate one-third of the crankshaft speed when weaving a 3-pick design. Whereas, in case of 2-picks repeat, it is possible to put shedding tappets on the bottom shaft, it is impossible to do so for a 3-pick design without changing the bottom shaft wheel, and this too will not be possible unless the pitch of both crank and bottom shaft wheels are changed, or provisions for changing the centre distance between these two shafts are incorporated in the machine. This problem leads us to the concept of mounting tappets on another shaft, known as counter or tappet shaft, to be driven from the bottom shaft. It is observed that under certain conditions the sum of the teeth in both the wheels are kept constant with a tolerance of only one tooth. Similarly, for drive to the counter shaft the sum of the counter shaft driven wheel and the bottom shaft wheel (being driver) teeth must be constant with a tolerance of one tooth only. This facilitates algebraic relations to be used in calculating the teeth in the wheel for the required purpose. Hence, for a 2-pick design, we can write

$\dfrac{A}{B} = \dfrac{1}{2},$ or $B = 2A,$ and $A + B = 90,$ where A is the crank wheel and B is the bottom shaft wheel, so that we have $2A + A = 90,$ from which we get $A = \dfrac{90}{3} = 30.$ Thus, wheel A, the crankshaft wheel, will have 30 teeth, and the bottom shaft wheel B to have 60 teeth, keeping the ratio undisturbed and the sum of the wheel teeth unchanged. For a 3-pick design, as stated earlier, the tappet will be mounted on the counter shaft. The counter shaft with additional wheels like C and D and wheel is D driven by wheel C on the bottom shaft, and whose sum is 70. So we obtain $C + D = 70$, and from the drive relation we find $\dfrac{A \times C}{B \times D} = \dfrac{1}{3}$,

so that $\dfrac{30 \times C}{60 \times D} = \dfrac{1}{3}$, $C = 2 = \dfrac{2\,D}{3}$.

substitution the value of C in the earlier relation, we have $5\,D = 210$, or $D = 42$, and $C = 28$, and the speed ratio of 2/3 is maintained. Similarly for designs with larger picks per repeat, like for 4-pick design, we have $E + F = 70$, and the drive ration $\dfrac{30 \times E}{60 \times F} = \dfrac{1}{4}$, $\dfrac{E}{F} = \dfrac{1}{2}$. and

$E = \dfrac{F}{2}$ so that as before $F = \dfrac{140}{3} = 46.67$,

and $E = 23.33$.

As fractional wheels are inconceivable, the number must either be increased or decreased to give a whole number, and at the same time the speed ratio must not be sacrificed. But, as the difference of one tooth does not make any difference in transmission of motion, we can use a 23 T wheel of E and a 46 T for wheel F, and at the same time the ration of 1/2 is maintained.

For 5-pick design the wheels are found to be 50 T for wheel H, and 20 T for wheel G. The above train of wheels are capable of weaving up to 5-pick designs, and for 6 and 7-pick designs other combination of bottom shaft and counter shaft wheels must be found out. For a 6-pick design, if the total number is taken to be $C + D = 80$, then with $D = 60$ T, and $C = 20$ T. Again for a 7-pick design with $C + D = 90$, we must have $D = 70$ T, and $C = 20$ T. Under this condition the pitch of the wheels will have to be so adjusted that the centre distance between the two shafts are not altered.

$$\frac{1 \times 40 \div 20}{80 \times 40} = \frac{1}{4}\ revs.$$

Thus the relation between the crankshaft and the tappet shaft rotation will be in the ratio of 4:1. In other words, for every rotation of the tappet shaft the crankshaft will rotate four times. So, it is seen that the rotation of the crankshaft will be equal to the number of picks per design repeat. The carrier wheel is omitted in the calculation as it works as an idler wheel to convey the rotation in the opposite direction, and if the space available is not suitable for the two main wheels to mesh.

It is customary for the tappet wheel to contain a number of teeth which is multiple of the weaves produced. Here the tappet wheel has 120 teeth which is divisible by any number up to and including 10. But this number is not divisible by 7 and 9. The number of teeth required in the change wheel N, for weaves which are complete on 2, 3, 4, 5 and 6 picks per repeat, is obtained by dividing the particular number into the number of teeth contained in the tappet wheel and in that case the intermediate wheels Q and R will be replaced by a single carrier wheel. For example, if we have a design repeating on 6 picks it will require a change wheel of $\dfrac{120}{6} = 20$ teeth.

But for tappets intended for weaving 7, 8, 9 and 10 picks, the two wheels Q and R will be required in place of a single carrier wheel. This is due to the reason that neither 7 nor 9 will divide 120 into a whole number. And, if 8 and 10 picks are intended they will render the wheels calculated to be useless as they will be too small, i.e., wheel numbers would be respectively 15 and 12 which will not just fit in. Now, by adopting a small wheel on the crankshaft with 24 T, the intermediate wheels required would be in inverse proportion as 24 multiplied by the number of picks per repeat of the tappets to be used is to the number of teeth in the tapped wheel 120. For a 8 pick pattern the wheel ratio is found to be

$$\frac{24 \times 8}{120} = \frac{192}{120},$$

but these wheels being too large, wheels of same ratio will have to be used. Thus the wheel Q with 48 T and wheel R with 30 T give the same ratio and can be conveniently used. For various pick-pattern the following wheel ratios could be used with the relation

$$\frac{p \times N \times R}{Q \times R} = rpm \text{ of the tappet shaft,}$$

where, p = number of picks per repeat,
     N = crankshaft wheel,
     Q = intermediate driven wheel,
     R = intermediate driving wheel, and
     T = tappet shaft wheel.

For 8-pick pattern, the ratio Q/R is found to be $\dfrac{192}{120} = \dfrac{96}{60} = \dfrac{48}{30}$.

For 7-pick pattern the same ratio is found to be $\dfrac{168}{120} = \dfrac{84}{60} = \dfrac{42}{30}$;

for 9-pick pattern the ration is $\dfrac{216}{120} = \dfrac{108}{60} = \dfrac{54}{30}$.

Thus, from the above figures we can conclude that with given number of wheels in the figure, the generalised relation becomes $\dfrac{Q}{R} = \dfrac{(6\,p)}{30}$.

It must also be remembered that though the ratio may be kept same, the number of teeth in the wheels should not be too small and prevent any meshing of the wheels. The wheel ratio must be kept within the limitations of the machine designer.

8.5.1   Drive without idler wheel

Hanton takes an approach which is normally found in the ordinary weaving machines. Figure shows the usual arrangement of gearing between the crankshaft and the shafts on which shedding tappets are mounted. Since one pick is inserted in the cloth for every rotation of the crankshaft, the speed of the crankshaft divided by the speed of the tappets shaft must be equal to the number of picks per repeat of design. For 2-picks design, the tappets are placed on the bottom shaft. If the speed of the crankshaft is 180 rpm, the bottom shaft will have a speed of $\frac{180 \times 30}{60} = 90\ rpm$ provided the crank and bottom shaft wheels have 30 T and 60 T, respectively. Or, in other words, the bottom shaft will rotate at half the speed of the crankshaft. Similarly, the bottom shaft will rotate one-third of the crankshaft speed when weaving a 3-pick design. Whereas, in case of 2-picks repeat, it is possible to put shedding tappets on the bottom shaft, it is impossible to do so for a 3-pick design without changing the bottom shaft wheel, and this too will not be possible unless the pitch of both crank and bottom shaft wheels are changed, or provisions for changing the centre distance between these two shafts are incorporated in the machine. This problem leads us to the concept of mounting tappets on another shaft, known as counter or tappet shaft, to be driven from the bottom shaft.

It is seen that under certain conditions the sum of the teeth in both the wheels are kept constant with a tolerance of only one tooth. Similarly, for drive to the counter shaft the sum of the counter shaft driven wheel and the bottom shaft wheel (being driver) teeth must be constant with a tolerance of one tooth only. This facilitates algebraic relations to be used in calculating the teeth in the wheel for the required purpose. Hence, for a 2-pick design, we can write

$$\frac{A}{B} = \frac{1}{2}, \quad or \quad B = 2A, \text{ and}$$
$$A + B = 90,$$

so that we have $2A + A = 90$, from which we get $A = \frac{90}{3} = 30$. Thus,

wheel A, the crankshaft wheel, will have 30 teeth, and the bottom shaft wheel B to have 60 teeth, keeping the ratio undisturbed and the sum of the wheel teeth unchanged.

For a 3-pick design, as stated earlier, the tappet will be mounted on the counter shaft. The counter shaft with D wheel is driven by wheel C on the bottom shaft, and whose sum is 70. So we obtain $C + D = 70$, and from the drive relation we find $\dfrac{A \times C}{B \times D} = \dfrac{1}{3}$,

so that $\dfrac{30 \times C}{60 \times D} = \dfrac{1}{3}$, $C = 2 = \dfrac{2D}{3}$.

Putting the value of C in the earlier relation, we have 5 D = 210, or D = 42, and C = 28, and the speed ratio of 2/3 is maintained. Again, for 4-pick design, we have E + F = 70, and the drive ration $\dfrac{30 \ x \ E}{60 \ x \ F} = \dfrac{1}{4}$, $\dfrac{E}{F} = \dfrac{1}{2}$. and

$E = \dfrac{F}{2}$ so that as before $F = \dfrac{140}{3} = 46.67$,

and $E = 23.33$.

As fractional wheels are inconceivable, the number must either be increased or decreased to give a whole number, and at the same time the speed ratio must not be sacrificed. But, as the difference of one tooth does not make any difference in transmission of motion, we can use a 23 T wheel of E and a 46 T for wheel F, and at the same time the ration of 1/2 is maintained.

For 5-pick design the wheels are found to be 50 T for wheel H, and 20 T for wheel G. The above train of wheels are capable of weaving up to 5-pick designs, and for 6 and 7-pick designs other combination of bottom shaft and counter shaft wheels must be found out. For a 6-pick design, if the total number is taken to be $C + D = 80$, then with $D = 60$ T, and $C = 20$ T. Again for a 7-pick design with $C + D = 90$, we must have $D = 70$ T, and $C = 20$ T. Under this condition the pitch of the wheels will have to be so adjusted that the centre distance between the two shafts are not altered.

8.6   Kinematics of shuttle picking

The movement of the shuttle in a conventional weaving machine is complex and it took some years before it was understood fully from the works done by Vincent and Thomas, Vincent and Cetlow. In order to get a proper understanding of the whole thing it is quite important to have a critical look from the very elementary aspects of the total system. Generally speaking, at every machine cycle, the shuttle is accelerated from rest to the maximum velocity within a distance of 20–25 cm at the commencement of picking cycle and then it travels with a slight retardation during its free flight through the shed and then again is decelerated within a short distance in the other shuttle box. Thus the whole thing can be divided into three parts which are not necessarily interconnected but each has some effect on the other. These three stages can be described as:

i) Shuttle acceleration,
ii) Retardation during its free flight, and
iii) Shuttle checking.

8.6.1   Relationship between loom speed and weft carrier velocity and picking angle

- There is a simple numerical relation between the loom speed, the distance travelled by the weft carrier through the shed, the degrees of crankshaft rotation available for its passage, and the average velocity of the shuttle during its passage.
- n is the loom speed in ppm,
- θ is the angular crankshaft rotation available for weft carrier traverse in degrees,
- v is the average shuttle velocity (m/s),
- K_b is the reed-space (m), and L_s is the weft carrier length (m),
- T is time in seconds of a weaving cycle, t is time in seconds available for weft carriage passage.
- s is the carrier path.

8.6.2   Loom speed and weft carrier velocity

- For a given width of loom and length of shuttle, the denominator in the equation is constant, in which case we may write $n \, \mu \, v \, \theta$.
- Thus, if we wish to increase the loom speed, we must increase the shuttle velocity or the fraction of the loom-cycle available for its passage or both.
- Then, if t is the time in seconds available for its passage

$$t = \frac{60}{n} \times \frac{\theta}{360} = \frac{\theta}{6n}$$

$$v = \frac{s}{t} = \frac{(K_b + L_s)6n}{\theta}$$

Thus, the practical loom speed as: $n = \frac{1}{6} \cdot \frac{\theta \cdot v}{K_b + L_s}$

1. Suppose that the maximum tolerable value for the average shuttle speed in a loom with an effective reed-space of 1.15 m is 13.75 m/s, that 135° of crankshaft rotation can be allowed for the passage of the shuttle, and that the effective length of the shuttle is 0.30 m, then the maximum permissible loom speed will be:

$$n = \frac{13.75 \times 135}{6(1.15 + 0.30)} = 213 \text{ picks/min.}$$

2. The reed space of a loom is 120 cm with shuttle length 30 cm allowing the average shuttle velocity of 12.5 m/s when crankshaft turns through 120° calculate the loom speed.

$$n = \frac{12.5 \times 120}{6(1.20 + 0.30)} = 167\, ppm$$

8.7   Shuttle acceleration

Vincent and Thomas during their experimental studies with an ordinary weaving machine were able to plot with some accuracy the shuttle displacement against angular position of the crankshaft during picking and checking. The displacement curve for the shuttle during weaving is quite different from the curve plotted from observations made while the machine is turned over by hand. The later type of displacement was defined as the "normal" displacement, and the former was called the actual displacement. The nature of this difference is shown in Figure 8.33. Curves (A) and (B) relate to a non-automatic over pick machine. Curve (A) obtained is that of the "normal" displacement, and the curve (B) is the "actual" observed during actual running of the machine. The 0° on the graph is actually the 75° position of the crankshaft rotation when the picker and the shuttle are just about to begin to move. Comparing the two curves, it can be noted that the nominal and actual displacement of the shuttle are the same at the start and again after about 30° of crankshaft rotation when the shuttle leaves the picker. Between 0° and 30° on the graph, there is a lag in the position of the shuttle and picker compared to that of the nominal position.

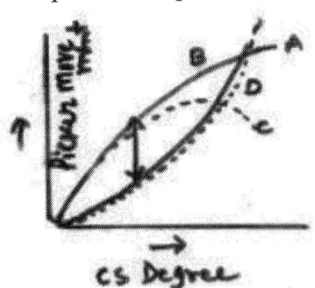

Figure 8.33   Nominal and actual displacements of Picker

The lag increases from 0° to 15°, where it reaches a maximum value as is indicated by the width of the loop. Thereafter, the lag decreases as the actual displacement gradually catches up with the nominal displacement, and finally overtakes it at about 30°, where the two curves intersect. When the machine is turned over slowly by hand, the force exerted on the shuttle by the picker is enough to overcome frictional resistance, which is due to pressure exerted by the swell on the shuttle. If, for example, the swell exerts a force of 65 N on the shuttle back wall, there will be similar and opposite force of 65 N between the box front and shuttle front wall. If the co-efficient of friction between the shuttle and the swell, and between the shuttle and box front are both 0.25, the force exerted by the picker on the shuttle in overcoming this friction will be 2 × 65 × 0.25 = 32.5 N.

A similar force will be required to overcome friction when the machine is running at speed, but, in addition, a much larger force will be necessary to accelerate the shuttle against the inertial resistance. Suppose, that the mass of the shuttle is 0.5 kg., and it is uniformly accelerated from rest to a speed of 12.5 m/s over a distance of 0.2 m.

Then since, we have $V_2 = 2\,aS$, and $\quad f = ma,$ we get, $f = \dfrac{\left(mV^2\right)}{2\,S}.$

Where, V = the final velocity (m/s),
$\quad a$ = the uniform acceleration (m/s²),
$\quad f$ = force (N),
$\quad s$ = the distance over which acceleration occurs (m),
$\quad m$ = mass of the shuttle (kg).
$\quad$ Putting the values in the above equation, we obtain

$$f = \frac{0.5 \times 12.5^2}{2 \times 0.2} = 195\ N.$$

Now, comparing the above result with the one obtained for nominal displacement to overcome friction, we find that the force required to overcome the inertial force is about six times the force required to overcome friction. It will be seen later that the acceleration is never uniform, and the peak force generated in overcoming picking may be as much as twice what would be if the acceleration were uniform. For our present discussion, the force required to overcome friction is neglected. This is justified in the sense that it is relatively small compared with the force required to overcome the inertial force. Going back to Figure 8.33, the curves (A) and (B) are due to lag produced by the force required to overcome inertia of the shuttle. The picking mechanism is not rigid, and the stresses set up in the mechanism by the resistance of shuttle to its acceleration result in strains in the picking mechanisms: the picking band or lug strap stretches, the picking stick

bends, and the picking shaft twists. In the early part of the shuttle acceleration (0°–15°), these stresses and strains are building up because the picker
is trying to overcome the shuttle resistance. In the later part (15°–30°), the
stresses and strains diminish and ultimately disappear when the shuttle
leaves the picker where the two curves intersects. The action of the picker,
the picking band and the picking stick can be compared to that of an elastic
of a catapult and the shuttle as the missile – this analogy has been for-
warded by Thomas. The maximum lag-at about 15° can be compared to the
straining of the catapult, and the second half of 15°–30° can be compared to
the release of such elastic strain. As soon as the elastic becomes slack the
missile, that is the shuttle in our case, leaves the leather, i.e., the picker, at
the intersection of the two curves. If there is a linear relation between stress
and strain during picking, the lag at any instant would be proportional to the
force exerted on the shuttle by the picker at that instant. Because of other
complicating factors, this is approximately true. Even if it is accepted as an
approximation, it is clear that the force of acceleration increases from zero
at the commencement of a pick to a maximum about halfway through the
pick and then falls to zero as soon as the shuttle leaves the picker. If it were
uniform the peak value would be larger than what would be needed. The
two additional curves (C) and (D) shown in Figure 8.33 are for cone under
pick in automatic weaving machine. In absence of picking band the type of
picking stick is used in over pick machine, the picking mechanism is more
rigid. In this case the acceleration is complete over a short distance but
over the same angular movement of the crankshaft. The width of the loop
at 15° is less than the one between the curves (A) and (B), which shows
that the lag in this case is less, but none the less the picking stick does have
a tendency to bend under the influence of shuttle's inertia force. The use
of picking tappet designed to give uniform nominal acceleration, does not
give uniform actual acceleration, but gives fluctuating acceleration with a
peak value equal to twice the designed value. Repeated application of such
a transient force has several undesirable consequence. Lateral vibration of
large amplitude is set up in the machine frame; there is excessive wear of
the teeth of the two main gear wheels in mesh during acceleration, resulting
in loss in machine speed; the checking of picking mechanism, which has to
be made sufficiently robust to withstand oscillating stresses imposed on it,
leads to wasted energy and more vibration.

The maximum force involved in picking should be reduced, and it can
be achieved in two ways: by accelerating the shuttle more uniformly and
accelerating it over a longer distance. The best possible conditions would
be obtained by accelerating the shuttle uniformly over the whole available length of the shuttle box. Improvement obtained by increasing the
distance without increasing the degree of uniformity of acceleration, was

demonstrated by Thomas and Vincent by use of double length of rubber tubing. This showed a reduction in force of about 40% compared to one when normal picking band is used. But as the rubber tubing of sufficient durability having large extension is not industrially available it became an academic interest only.

8.7.1   Shuttle acceleration in projectile loom

Compared to shuttle loom, the weft carrier in projectile loom has the gripper projectile and both are the projected bodies and when imparted initial speed move across the shed due to inertia without any possibility of influencing the course of their flight. The gripper projectile is fully controlled in the shed whereas the shuttle control is partial. The Figure 8.34 shows the shuttle acceleration of gripper projectile in Projectile loom. Three phases can be seen like Phase I as Acceleration, Phase II as free flight and Phase III as checking or retardation.

It can be written as

$\phi = \omega t$, where ϕ is degree of movement of shaft, ω is the angular speed of index shaft and "t" is the time of travel.

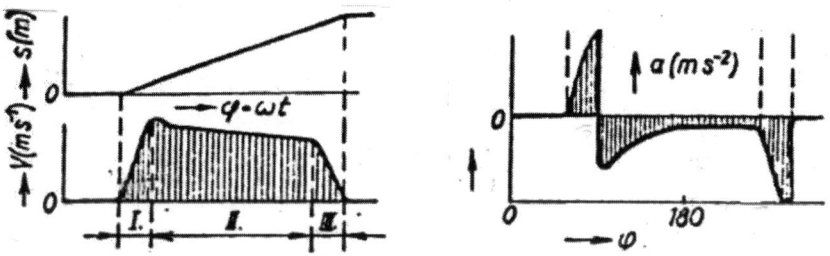

Figure 8.34   The path, speed and acceleration of the weft carrier in projectile loom

The values of the carrier acceleration at the start and is retardation during braking are not constant and are determined by the types of the mechanisms used. The retardation during the free flight (Phase II) depends on:

- The resistance of the weft to unwinding from a package or a weft accumulator.
- The quality, unevenness and oscillation of the projectile guiding segments.
- Wavy shuttle or gripper projectile motion.
- Frictional resistance of the weft carrier to the guide path etc.,

There are two mechanisms exit for imparting motion to a weft carrier like

A. Mechanisms that generate energy when it is required by using a cam to displace the picker against the inertial resistance offered by the weft carrier.

- The shuttle speed varies with the loom speed.

- The force acting on the shuttle tends to increase to a maximum about halfway through the period of acceleration and then to decrease to zero at the instant at which the shuttle loses contact with the picker.

B. Mechanism that generate and store energy in a spring or torsion rod and release it suddenly when required.

- Because the energy is released suddenly the weft carrier speed is independent of the loom speed. The force exerted by the spring or torsion rod is greatest at the instant of release and then decreases steadily to zero if the spring or torsion rod is allowed to return to its unstressed state at the end of the pick.

- The uniform acceleration is unattainable with mechanisms of type A and more uniform acceleration with mechanisms of type B.

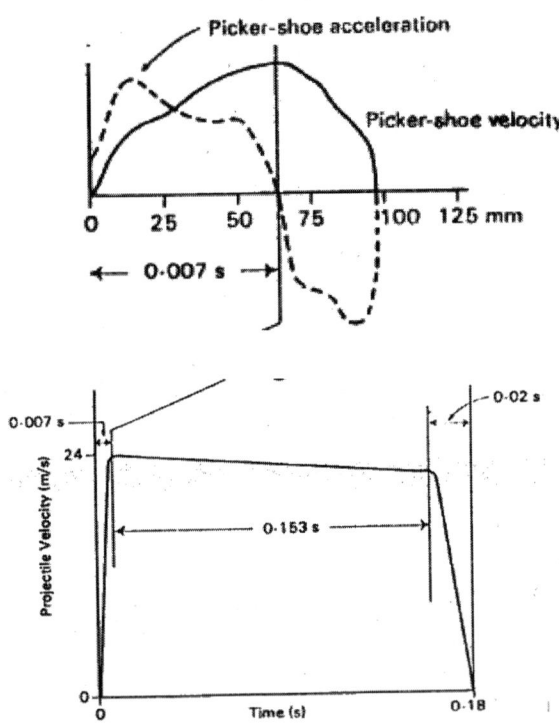

Figure 8.35   Velocity of gripper projectile weft carrier

The picker accelerates (Figure 8.35) the projectile over a distance of about 65 mm in 0.007 s. The energy utilisation of the projectile system is not much more efficient than that of the shuttle projection and all other weaving machines are less efficient. The residual energy in the picking system, some 62% of the whole is absorbed in the by hydraulic buffer the body and plunger. The picker and the rest of the lever system are then brought to rest over the next 40 mm of the picker movement by the oil brake. The speed of the projectile as it leaves the picker is normally about 24.5 m/s.

Acceleration is far from uniform reaches a peak value of 6630 m/s^2 when the projectile has travelled nearly 15 mm. The peak deceleration has a value of 9920 m/s^2 and occurs about 12.5 mm before the picker comes to rest.

Acceleration of the projectile occupies 0.007 s. The velocity of the projectile varies during its acceleration flight across the loom and retardation to its rest position at the opposite side of the loom. Retardation occurs over a substantially longer period than acceleration and is therefore relatively gentle. The peak forces acting on the projectile are of the same order as those acting on the shuttle loom because the projectile has less than 40 g. This gives a peak accelerating force of about 262.5 N and a peak retarding force of about 397 N

8.7.2   Elastic properties of picking mechanism

When the stiffness of the picking mechanism is shown, and it is easy to measure under static conditions, it can be used in conjunction with the nominal and actual displacement curves to estimate the force exerted by the picker on the shuttle during picking and to compare these with constant force that would produce the same final shuttle speed. Thomas and Vincent obtained certain figures on which the following calculations are based and are given in Table 8.2, and relates to loose-reed over pick machine.

Table 8.2   Experimental details of the set up

Mass of shuttle	0.32 kg
Stiffness of mechanism	3.65 kN/m
Shuttle speed	12.2 m/s
Stroke of picker	0.20 m
Maximum lag	0.075 m

Since a force of 3.65 kN acting on the picker parallel to the shuttle axis produces a deflection of 0.075 m, force required to produce such a lag is:

$$0.075 \times 3650 \text{ N} = 274 \text{ N}.$$

This is the peak force exerted by the picking on the shuttle during picking. Now, the uniform force that would produce the same final speed over the same distance is given by:

$$f = \frac{mV^2}{2s} = \frac{0.32 \times 12.2^2}{2 \times 0.20} = 119 \text{ N.}$$

The actual peak force of 274 N includes the force required to overcome friction. An example, gave a typical value of 32.5 N for this, so it may be reasonably assumed that, of actual peak force of 274 N, a force (274–32.5) N = 241.4 N was used to overcome inertia. This is about twice the uniform force of 119 N. Forces encountered in the modern automatic weaving machines are substantially greater than those calculated above because the shuttle tends to be heavier, the effective stroke shorter, and the speed of the shuttle greater. In the machine to which the curves (C) and (D) in Figure 8.35 relates, the shuttle speed was 13.3 m/s, the mass of the shuttle was 0.51 kg including a full pirn, and the effective length of stroke was 0.165 m. The required uniform accelerating force would be:

$$f = \frac{0.51 \times 13.3^2}{2 \times 0.165} = 273.5 \text{ N.}$$

This compares with 119 N in the preceding example, in which the actual peak force was twice as great. The actual peak force in the last example might exceed 500 N.

The above calculations assumed the following conditions:

i)   the picking mechanism as a simple elastic system obeying Hooke's law.

ii) the stiffness of the system remained constant during picking, and

iii) the force acting on the shuttle given by the product of the stiffness and lag remained constant.

But these assumptions are not completely true.

The stiffness of the picking system close to the main drive is different from that of from the side opposite to the main drive, this is attributed to the twisting of the camshaft which results in reduced stiffness during picking from the other side, and is truer for wider machines. Lord has shown that the stiffness may substantially increase from the beginning to the end of a pick. Thus, that picking mechanism is analogous to a simple elastic system is only of approximate nature. A mathematical treatment is given in Appendix A at the end of this chapter, for the picking mechanism as a simple elastic.

8.7.3   Initial and average shuttle speed during traverse

A knowledge of the relation between the two speeds is desirable in order to co-ordinate the picking with shedding. It is clear, however, that if the shuttle is pressed between the two sheets of warp during the part or all of its traverse, no definite relation between the two speeds can exist. Thomas and Vincent did some experiment to find this difference and observed that there was only slight difference and it could be easily ignored unless some critical value affecting other factors come into play. Some typical values are given from their work here in Table 8.3.

Table 8.3   Initial and average shuttle speed during traverse

Crank position (in degrees)	Initial speed (m/s)	Average speed (m/s)
60 after beat-up	10.4	10.1
75 after beat-up	10.7	10.5
120 after beat-up	11.5	11.2

They found that the initial and average shuttle speeds are approximately equal and ranged from 9.4 to 14.6 m/s. The sizes of the nose bits of over pick system are approximately directly proportional to the nominal displacement after 30° of the crankshaft rotation. The shuttle speed is approximately directly proportional to the machine speed and inversely proportional to the distance from the picking shaft. Thus we write

$$V_s \propto v/d$$

Where, v_s = shuttle speed,
v = machine speed, and
d = distance of picking tappet to tappet shaft.

They also found that the shuttle speed is independent of the shuttle mass and buffer position, shortening of picking bank, lowering of the picking bowl increases the shuttle speed. When picking time is progressively made later, the shuttle speed is seriously lowered due to the greater interference in shed, i.e., the bending factor is reduced as the shuttle leaves the shed. With greater reed width the relative shuttle speed, which is defined as the ratio between the average shuttle speed and the machine speed, is also increased.

8.7.4   Factors affecting initial shuttle speed

There are several factors which affect the initial shuttle speed in an over pick machine which were thoroughly investigated by Thomas and Vincent. The primary factors which are responsible for controlling the shuttle speed are discussed below:

Shape of picking tappet

Shape of that part of the picking tappet in contact with the picking bowl, while the shuttle is being accelerated, other things being unaltered, it will depend on the length of the nose bit of the picking tappet. Depending on the size of the machine width the nose bits are marked by punches to indicate the range of shuttle speed that would be required for a machine. Wider the machine greater the number of punches marked.

Machine speed

The average machine speed during the shuttle traverse does not matter materially, that is, during picking the machine speed does not differ much. But there is evidence at present time, using sophisticated measuring instruments that machine speed does fall during shuttle acceleration. However, it will be evident that with increasing machine speed the average shuttle speed also increases but not strictly in direct proportion.

Machine speed (ppm):	216	202	185	171
Shuttle speed (m/s)　:	13.0	12.7	11.9	11.7

Time of picking

Variation in the time of picking has a tendency to give higher shuttle speed when picking was found to be late with respect to the crankshaft position starting from 45°. The changes in speed with timing were less with under pick than with over pick system, and were absent with over pick when the sley was fixed. Higher shuttle speed is necessary as it has to force through the shed as it progressively lowers as the shuttle emerges out of shed.

Length of picking bank

Changing the length of the picking band is one of the simplest and most frequently adjusted part of the picking mechanism that is handled by the tacklers or weavers during routine weaving. It is also a change that may occur spontaneously through stretching, if the band is new, though, recovery from stretch when the machine is stopped, and through changes in the humidity. Such changes produce double effects, it alters the time of picking and also the position of the picking bowl on the nose bit at the instant the bank is taut. This also results in change of picking force. The length of picking band has a direct bearing on the shuttle speed greater the band length lesser the speed and vice-versa.

Swell resistance

The swell exerts a frictional retardation force on the shuttle and before the start of the shuttle it is constant. The shuttle will not move until the elastic displacement of the picking mechanism gives rise to a force on the shuttle sufficient to overcome this frictional force due to the swell pressure. Changes in the swell resistance may affect the shuttle speed in two ways. There is a direct effect resulting from changes in the resistance offered to the shuttle during acceleration and there may be an indirect effect if the swell pressure is great enough resulting from variation in the position at which the shuttle comes to rest after the previous pick, since variation of the starting position modifies the nominal movement during picking. With increased swell resistance the shuttle speed is increased due to straining of the picking mechanism. With the sudden release of the swell resistance an unbalanced force is available for shuttle acceleration. Some of the modern under pick weaving machines have in them incorporated the swell releasing or "easing" motion, and this precludes any unbalanced force being released during picking. The effect of swell resistance is shown against the shuttle speed below:

Maximum swell resistance (N):	2.24	8.95	20.01	27.60	32.10
Shuttle speed (m/s) :	12.4	12.6	12.7	12.75	12.00

Mass of shuttle

The amount of checking, a shuttle requires, is dependent on its mass and its speed. Reduction in momentum is the direct resulting reduction of mass and not its speed. It is often viewed that a lighter shuttle moves more slowly than a heavier one but this is not true. From experimental observations it was found that the shuttle speed is substantially independent of its mass. Timing was not affected though the heavier shuttle has a tendency to be picked later.

Height of the picking bowl

It was observed that within the limits set by the machine designers raising of the picking bowl causes reduction in shuttle speed, and the following figures substantiate the facts:

Height of picking bowl axis above the floor (cm):	40.7	41.30	41.80	42.70
Shuttle speed :	10.80	10.20	9.40	9.20

Picking tappet distance from picking shaft

For a given increment in the radius of the picking tappet at its point of contact with its picking bowl there will be a large increase in the angular movement of the picking shaft where the tappet is nearer to it than when the picking shaft is further away. For this reason an increased shuttle speed would be expected as the picking tappet was moved nearer to the picking shaft. The regulation between this parameter and the shuttle speed is given below:

Distance between the central plane of picking tappet and axis of picking shaft (cm) :	10.0	11.3	12.5	13.5
Shuttle speed (m/s) :	11.9	11.2	10.2	9.4

Position of buffer

It is known that if the picking mechanism is checked immediately after the picking tappet has ceased to exercise control, the "pick will become chocked" and the shuttle speed will be lowered. This contention has been negative by experiments. Progressively increasing the distance between the spindle stud and buffer showed that the shuttle speed remained unaltered. Thus under the circumstances the position of buffer has no bearing on the shuttle speed.

Prolonged weaving

It is desirable that there should be no changes in shuttle speed as weaving continues. Such changes may occur through the working of loose nuts and bolts, changes in the length of the picking band, and changes in the degree of checking. Whilst the first and the last named causes are not systematic in nature and hence not useful, the picking bank stretches and causes lowering of the shuttle speed. Though this stretching is not a continuous process, as bands recover when the machine is stopped resulting in increased speed on restarting. After sometime stretching ceases and changes in shuttle speed is also stopped. With well-regulated relative humidity and temperature the effects of other disturbances are small.

Initial gap between picker and shuttle

From time to time in course of routine weaving the shuttle may be obstructed in its passage through the shed and causes a gap between the shuttle and the picker. This naturally leads to a reduction in shuttle speed for the next pick causing "loom bang-off" or shuttle trap. With ordinary nose bits the fall in speed becomes serious when the initial gap exceeds 25 mm, but with constant nominal acceleration the fall is not that serious.

Shuttle checking

After the shuttle leaves the picker at the end of picking it enjoys a free flight across the shed until it reaches the other box. However, during this free flight there is some retardation due to

(i)   friction with the bottom warp shed line, and

(ii) friction between the shuttle edge and the reed.

The usual retardation of a shuttle for a slow running machine is in the vicinity of 6.1 m/s², and it is about 10.7 m/s² in case of a fast running machine.

From the equations of motions we find that the maximum velocity of shuttle as it enters the shed by:

$$v = \frac{S + 0.5 \, at^2}{t}$$

Where, S = total shuttle flight, (m),
  = $(R + L)$,
a = acceleration (m/s²)
t = flight time (s),

If the width of the machine plus the effective length of the shuttle is 1.08 m and the machine speed is 190 ppm, the shuttle retardation is 10.7 m/s², and if the shuttle takes 105° of the crankshaft rotation, the time for the shuttle flight is

$$t = \frac{105 \times 60}{360 \times 190} = 0.092\ s$$

Then the maximum shuttle velocity is found out from the above relation:

$$v = \frac{\left(1.08 + 0.5 \times 10.7 \times 0.092^2\right)}{0.092} = 12.23\ m/s.$$

It can be seen from Figure 8.36 has been found from the research studies that after about a width of 200 cm, the maximum shuttle velocity changes very slowly increase in the machine width so that it can be ignored. Velocity of the shuttle during traverse is found if we assume that at any instant, say, t_x, during the total time of traverse t, let the shuttle velocity be V_t, then

$V_t = V - a t_x$. Also therefore, the speed of the shuttle at the end of its flight or on emerging out of the shed at time t,
$V_t = V - at$,
where V_t = terminal velocity (m/s).

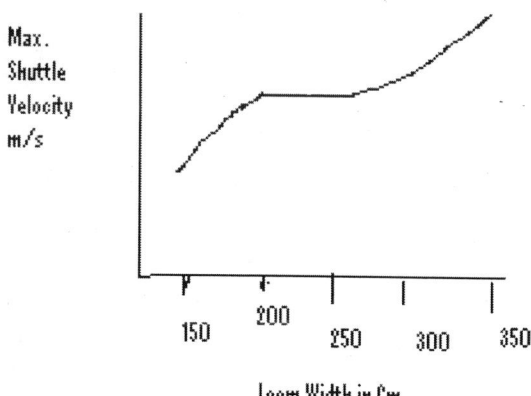

Loom Width in Cm

Figure 8.36   shows a simple arrangement of a shuttle checking device in a fast reed weaving

Therefore, if the maximum shuttle velocity be 12.23 m/s and the shuttle retardation is 10.7 m/s² and its traverse time is 0.092 sec, then the terminal velocity of shuttle at the end of its flight would be

$$V_t = 12.23 - 10.7 \times 0.092 = 11.27 \text{ m/s}.$$

So, it is found that the shuttle emerges out of the shed at a slightly lesser speed that it was projected into the shed, and it must be brought to rest at a reasonably smoother way over a distance about the same as, or rather less than, that over which it was accelerated. This becomes increasingly difficult as the speed of the machine increases, since the energy that has to be dissipated, that is the kinetic energy of the shuttle, is proportional to the square of its velocity. As because the problem of shuttle checking is one of the main factors that limit the speed of the machine, the design of high speed machine must be accompanied by a very efficient checking system. Shortly after the leading end enters the shuttle box, the shuttle strikes the swell which is usually situated at the back of the shuttle box, though in some makes, it is found in front of the box as well, the swell is displaced. The swell has two principle functions

(i)   it operates the warp protection device in a fast reed machine and monitors the shuttle movement, that is, it acts as a detecting system, and prevents any damage to warp;

(ii) it helps to reduce the shuttle speed at the end of its trajectory.

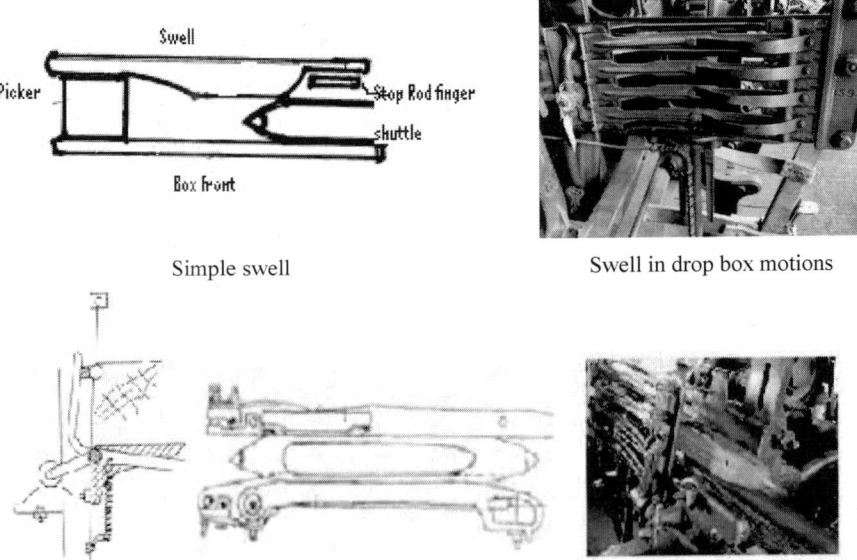

Simple swell   Swell in drop box motions

Swell in loose reed loom   Shuttle box plan view   Swell in shuttle changing loom

Figure 8.37   Swell in looms

machine. In this system the swell is pivoted at the outside of the shuttle box and is spring loaded either by means of a flat or spiral spring, depending on the machine design, and protrudes inside the box. As the shuttle strikes the swell, it is displaced, thus causing the stop-rod finger to be displaced at the same instance. The stop-rod finger is also spring loaded. Pressure due to the springs cause the shuttle to slow down as the swell assembly resists any displacement. The pressure between the swell and shuttle, and between shuttle and box front are caused by the swell spring or the stop-rod spring or both. In some cases additional pressure are called into play which is released during picking to avoid excessive swell pressure. This mechanism of easing the swell pressure is often known as the "swell releasing" mechanism. The shuttle retardation curve obtained by Thomas and Vincent for a fast reed machine is shown in Figure 8.37, where it is found that the shuttle strikes the swell at about 14.3 m/s. Here the X-axis is giving the distance moved by shuttle after striking picker and Y-axis is giving the shuttle speed in m/s.

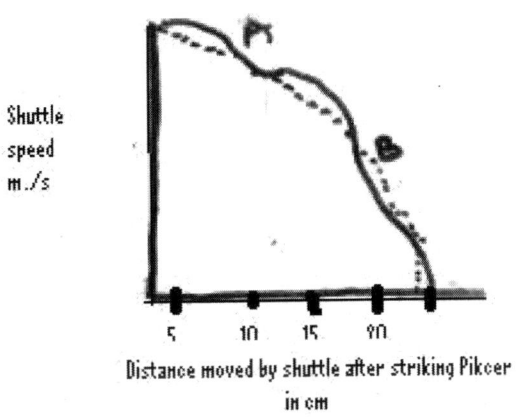

Shuttle speed m./s

Distance moved by shuttle after striking Pikcer in cm

Figure 8.38   Shuttle retardation curve with hinged swell

It is found that during the next 0.114 m of travel of the shuttle its velocity is reduced by about 32%, i.e., the velocity becomes 9.8 m/s when the shuttle strikes the picker at (B). Thereafter, the combined action of the picker and swell and the check strap reduces the shuttle velocity to zero over a further distance of 0.05 m of its travel. It is clear that the retardation is not uniform (Figure 8.38). The actual retardation between the shuttle striking the swell first and its contact with the picker at (B) is found by using the relation:

$V^2 = U^2 - 2\ as,$

and putting the values from the curve, we have

$9.8^2 = 14.3^2 - 2\ a \times 0.114$

Or, $a = 478$ m/s^2.

Again, the shuttle is brought under the action of the checking system (check strap) at the point C, and during this period an actual retardation of about 2000 m/s² takes place. The dotted line in the figure shows the uniform retardation curve over the total distance of 0.165 m. Under this condition the uniform retardation between the first contact of shuttle with the swell and the position (B) would be 616 m/s². Thus it can be said that the swell failed to reduce the shuttle speed, and this had been verified in the actual observation that the pivot type swell, on contact with the in-coming shuttle flies back, and there is a momentary loss of contact between the two, and during this period the shuttle moves at a fairly high speed inside the box. This type of contact and slapping takes place for a considerable length of time until the shuttle comes to rest. Thus the curve takes the form, of slip-stick nature. The curve further reveals that at the point of contact with the picker the shuttle is retarded heavily with considerable reduction in shuttle speed – the actual curve takes a dip under the dotted curve at (C). Thus throughout the retardation period inside the shuttle box the shuttle is never retarded uniformly, which is due to sley construction mainly. More efficient checking can be obtained by separating the two functions of the swell. In the modern weaving machines the tendency is to use two swells or a swell in two parts. Attempts were made by Roy with the use of "piano-swell" where as many as eight mini-swells, each being individually spring loaded, in the form of piano keys, were used. Though the condition of uniform retardation was quite appreciably good, it was thought that to maintain such a number would be a problem in actual practice, Roy gradually reduced the number of such swells to two through four, and the results were very encouraging. One of the arrangements of swell in a modern machine and the shuttle retardation is illustrated in Figure 8.39. In this diagram Y-axis give the shuttle speed in m/s and X-axis give the distance moved by the shuttle after striking the swell. In this type of arrangement it is found that a small swell at the box mouth is hinged, and this is connected to the stop rod system. This part thus functions as the shuttle monitoring system. This part thus functions as the shuttle monitoring system. This small swell has a very small effect on the shuttle retardation. The second swell, which floats on two spiral

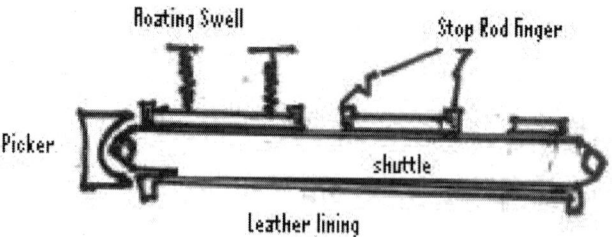

Figure 8.39   Floating swell

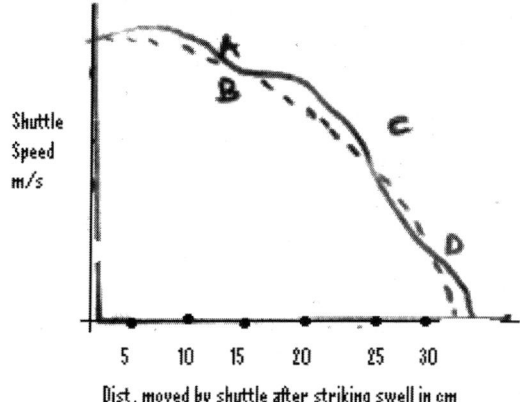

Figure 8.40   Shuttle retardation with floating swell

springs, is the main swell and it is to be noted that this one is not hinged like the former type. As the shuttle enters the box the front end of the swell is pressed back at position (*B*) of Figure 8.40. This will cause the rear end of the swell to retract a little, but never will fly back as in the pivoted case. Both the ends of the swell retract at position (*C*), Figure 8.40. From this position on-wards the shuttle speed is reduced fairly uniformly and rapidly. As the shuttle strikes the picker at (*D*), the check strap cushions the speed. In the present day machines this cushioning is done by hydraulic system instead of check strap. If we compare the two curves we find that the curves at Figure 8.40 has a lesser gradient than that of the curves, which indicates that deceleration in the second case is more uniform and takes place over a longer distance. This shows a substantial improvement in shuttle checking. In place of using wooden swells, two metal units with greater moment of inertia provides pro-gressive action on the shuttle so that it is accurately positioned in the box. Swells with greater moment of inertia have been found to absorb the kinetic energy of the shuttle in a shuttle box more efficiently.

8.7.5   Shuttle mass and checking

The mass of the shuttle gradually decreases as the pirn weaves down. How-ever, the checking force remains unaltered, and hence the effectiveness of checking the shuttle must have to be better. As the shuttle mass is gradually reduced the impact velocity of the shuttle will also be less. The checking system should be efficient to allow for variation in the shuttle speed due to reduction of its content. Furthermore, it should be able to allow for any variation in the shuttle speed due to friction and other resistances during its trajectory through the shed. Consider, for example, the mass of the shuttle including its contents to be 0.51 kg when the pirn is full, and it becomes 0.48 kg when it is nearly empty, that is during weft insertion 30 g of yarn have been consumed. Let us also assume that the impact velocity of the shuttle should not be less than 4.5 m/s at any time during correct functioning of the machine, its calculated speed is 13.75 m/s as the shuttle strikes the swell, and is uniformly retarded over a distance of 0.20 m up to impact with the picker.

Now, $v^2 = U^2 - 2\ as$,

Where, $U =$ initial velocity,

$v =$ impact velocity,

$a =$ uniform retardation,

$s =$ retardation distance.

The impact velocity is least when the pirn is nearly empty, i.e., when

$13.75^2 - 4.5^2 = 2\ a \times 0.20$,

Thus, $a = 422$ m/s^2.

Since, $f = ma$, retardation is inversely proportional to the mass being retarded, so that when the shuttle is full, we have

$$a = 422 \times \frac{0.48}{0.51} = 397\ m/s^2.$$

And, $13.75^2 - V^2 = 2 \times 397 \times 0.20$,

From which, $V = 5.5$ m/s.

In this example, the impact velocity is about 22% greater than the pirn when it is full than when it is nearly empty.

8.7.6   Ideal checking conditions

It is not difficult to lay down conditions which checking should fulfil. They are

i)   the shuttle should come to rest in contact with the picker at the same place in the shuttle box after each pick,

ii)   the maximum value of retardation should be kept as small as possible, and

iii)   the impact velocity of the shuttle with the picker should be as small as possible.

These conditions are entirely independent of each other, since (iii) is largely implied by (ii), for impact always results in rapid deceleration. The first condition aims at ensuring that the shuttle speed is uniform in the following pick, the second aims at reducing to a minimum the force tending to displace and disintegrate the weft package and the third condition aims at reducing the wear rate of the picker.

8.7.7   Checking limits

The checking limit is determined by the ability of the checking system to absorb the kinetic energy of the shuttle in the time available between two successive arrivals in any one box. As the shuttle is brought to rest, its kinetic energy is transformed into heat and the checking limit is therefore essentially dependent on the heat-disposal capabilities of the checking system. If the final velocity of the shuttle is too high the shuttle itself, the swell and other parts of the box mechanism, will become hotter and hotter, and this will eventually cause damage that will make weaving impossible. The heat disposal capabilities of the checking system can be expressed by the amount of heat, in units of mechanical energy that can be dissipated per second. As the checking limit is intimately connected with the picking limit, they should be kept in harmony. Where the picking limit, i.e., the maximum initial velocity of the shuttle, is constant, the full benefits are only realised when the heat disposal potential of the checking system is high enough to bring the whole range of weaving machines within the orbit of this limit. With the lowest value of heat disposal capability, the speed potential of the picking motion is entirely wasted because the whole range is governed by the checking limit. Thus, any raising the picking limit would benefit the wider machines. It is desirable that in order to increase the picking limit, the design and material used in the checking system and the shuttle itself, much careful attention must be given. Ishida et al., during their work have measured the temperature in the checking system which are summarised below and are of considerable interest to the designers as well as those engaged with cloth production:

a)   The maximum temperature of the swell, caused by friction with the shuttle was about 65°C and was at the crest of the swell profile.

b)   The temperature of the swell was affected by the kinetic energy of the shuttle, the maximum being at the same point as that of higher retardation of the shuttle,

c)   The maximum temperature of the swell at a machine speed of 300 ppm would be reasonable to maintain the quality and effectiveness of the swell cover used.

They worked on a 135 cm reed space automatic single shuttle machine using a shuttle weighing, with pirn, 435 g. The above findings are based on machine running at 154 ppm. Such raising of temperature causes charring of the shuttle back which could be seen clearly after certain length of time.

8.7.8   Weft behaviour during unwinding from shuttle

The magnitude of weft tension immediately before it is trapped by the crossing of the warp threads is crucial governing the appearance of the fabric weft way to a large extent. Its effect on the selvedge is also very important; any appreciable variation may lead to defective selvedge and a bad cloth otherwise woven nicely. It is therefore very important to know the behaviour of weft during unwinding form shuttle in orthodox weaving methods. Figure 8.41a shows a simple graphical representation of tension variation during unwinding from a pirn during one cycle of weaving. Here X-axis give the crankshaft movement in degrees and Y-axis give the weft tension in CN in both seen in Figures 8.41a and b diagrams.

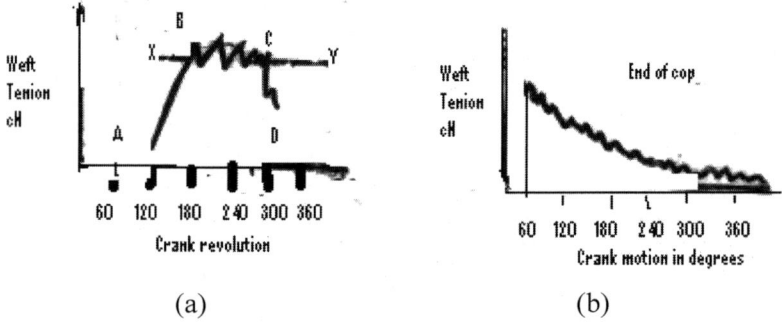

Figure 8.41   (a) Weft tension, (b) weft tension trace

It shows the points where tensions are built-up and how it varies during shuttle flight till the weft is trapped by warp threads. The horizontal scale gives, in degrees, the amount of crankshaft rotation after the front centre position, i.e., 0°. The magnitude of weft tension is measured in the vertical scale. The shuttle moves out of the box at about 80° but the tension in the weft does not develop until all the slack is taken up. At point (A) this process has been completed and weft begins to be withdrawn from the package; the tension then rises until it reaches its value at (B). Therefore it fluctuates about the level (XY) until the point (C) is reached about 250° after front centre. The magnitude of tension level at the level (XY), which is the tension when the shuttle is in free flight across the machine, is determined by the nature of the weft, internal shuttle fittings, the shuttle speed, etc. This period of tension is called the unwinding tension. Between 120° and 250° the shuttle is in free flight, at (C) it makes contact with the swell, and immediately the tension begins to fall. At (D), 270°, the shuttle is at rest in the shuttle box. The closed shed will occur between 260° and 360°, or even later in the cycle, depending on the fabric being woven. Unless the closed shed occurs earlier than 250°, the tension in the weft will be falling when the warp threads cross and arrest any further fall. If the closed shed occurs earlier in the machine cycle, say at 270°, the weft tension at trapping will have value equal to the unwinding tension. If, on the other hand, the closed shed occurs later, the fall in tension will have proceeded further and the tension at trapping will be lower than previously. Thus the actual tension in the weft immediately before it is trapped by the warp threads will depend not only on the unwinding tension but also on time, in the machine cycle, at which the closed shed occurs. The rate at which the tension falls must depend, to some extent, on the efficiency of the shuttle checking arrangement in the box. Obviously, if the shuttle rebounds from the picker the tension will fall to zero with the well-known adverse effect on the fabric. From the above it is clear that the same weft tension at trapping can be obtained by the combination of an early timing of the closed shed and a low value of the unwinding tension or a later shed timing and a higher unwinding tension. Figure 8.42 shows a typical tracing obtained by "Shirely" Unwinding Tension Meter. The weft package for the trace was a cop of 30 tex cotton yarn and the unwinding speed was 13.5 m/s. The vertical scale on the left gives the magnitude of tension while on the horizontal scale the right hand side of the record corresponds to the beginning of the cop and the left hand side to the end. The width of the trace arises from the rapid variation in tension as unwinding proceeds in rapid succession from the nose to the shoulder of the cop. At first, the magnitude of tension is about 250 mN, but as unwinding proceeds the average tension increases and ultimately reaches a value of about 1500 mN. Thus the tension at the end of the cop is about six times its value at the beginning. It is this abrupt tension changes from the end of the cop to beginning of the next that gives rise to, in some fabrics, to the well-known "cop-change" defect. Examinations of the fabrics exhibiting this defect shows that for a short distance corresponding to the end of the cop there are fewer picks than there should be and in the cloth corresponding to the beginning of the next cop there are more. This sharp contrast at the cop changes makes the defect so easily identified. Rise in tension towards the end of the cop has been explained by various workers as that, when unwinding starts from the full cop the yarn balloons away from the tongue but as it proceeds the balloons lengthens and there is some licking of yarn round the shuttle tongue, or in case of pirns round the tip of the pirns, at the end of balloon nearer the cop. As unwinding continues the licking extends over a greater length and it is this that is mainly responsible for the rise in tension on account of frictional resistance to the movement of the yarn. If the licking could be reduced the rise in tension would be so large. It has been found that if the internal shape of the shuttle is a close shaped fit round the nose of the full cop rise in tension could be reduced. Only by shaping of the shuttle wood it cannot be totally minimised as certain allowances have to be made for misalignment of the tongue; it has to be completed by filling the gap with lamb's wool or nylon loops. The cop end should be so fitted that it lies snugly. To this wool or loop surface to get the maximum advantage.

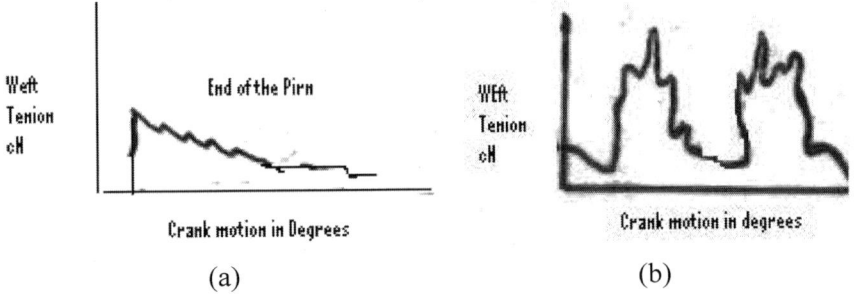

Figure 8.42   (a) Weft tension trace for lamb's wool case, (b) weft tension trace with telemetry system

Figure 8.42a shows the effect of using lamb's wool as lining in the shuttle wall with the same weft and shuttle. The maximum tension that has developed at the end of the cop is only twice compared to the tension at the start. This favours well with other figure where the maximum tension was about six times that of its beginning.

The difficulties associate with the continuous monitoring of weft tension under dynamic conditions on the conventional weaving machines lie due to the fact that the passage of weft yarn had to be diverted to pass a transducer near the selvedge. Selvedge methods suffer from lack of continuity of the tension single and interfere with the weaving cycle itself. It would be better if tensions could be reassured in the region of yarn interlacement at the cloth fell. Radio telemetry technique has been employed where the normal cycle is not describes and the measurements could be done by transmitting signals from a transmitter unit embedded in the shuttle near the weft yarn exit with a receiver unit fitted with aerial antenna a cathode-ray-oscilloscope to record the signal. A typical trace shown in Figure 8.42b illustrates the tension in two cases taken from a Picanol President machine. The figure is labelled according to the timing of the various mechanical motions in the weaving cycle. In the figure is shown tracings of picks, one from right to left across the machine, i.e., 0–360˚ and the second from left in right, i.e., 360–720'. The numbers at top show the stages of weaving cycles, for example – No.1 front centre position, 2 – picking, 3 – weft slack taken and 4 – checking. Weft yarn used was a 20 × 2 tex cotton. There is slight increase in tension as the yarn leaves the signal in two directions, and it is seen from the trace that saturation of tension in two cycles also vary. It may be attributed to the directional flights which has caused this separation, and in the second case the proximity of the yarn position with selvedges. However, though the second stage of measuring gives a better quantitative approach, but basic tracings obtained by two methods are quite the new and phenomenon causing the tension variations of weft and its flight are also seen here. Tension in weft withdrawal tension also persists in the unorthodox weaving, and controls exercised to avoid variations and the magnitude of variations are in the appropriate place flowing the various stages of weft insertion that are in use at the present textile industry.

8.8   Mechanics of picking

The velocity of the shuttle differs from loom to loom from hand loom to power loom. The velocity of the shuttle in a power loom normally depends on the reed space, retardation set up, type of weft yarn, etc. The distance, S, reed space, moved by the shuttle from the time its full depth has entered the shed until the full depth leaves the shed at the other side is approximately equal to = yarn width in reed + length of the shuttle – taped edge length (end of the shuttle is tapered and so does not need so much opening as the rest of the shuttle). The shuttle ought not to enter the shed until the depth of the latter, at the front edge of the shuttle, is equal to the shuttle depth; nor should be shuttle remain in the shed after the time when the shed depth at the shuttle front edge is again equal to the shuttle depth. The time of shuttle stay in the shed depends on the crankshaft movement. The time can be found by the graphical construction (for example Figure 8.43) or it may be obtained by the use of the formula for shed opening at the shuttle front.

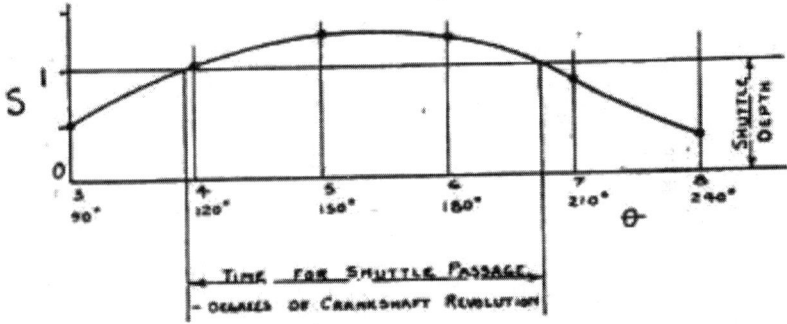

Figure 8.43   Calculation of time of shuttle stay in shed by graphical method
(Source: Hanton)

If formulae is used to calculate the time of shuttle stay in shed, it will be found most convenient to calculate the values of the shed opening for different values of the crank movement θ and by plotting S against θ, and finding the values of θ when S is equal to the depth of the shuttle. The retardation of the shuttle may be due to its contact with the lower line of the shed, shuttle wall with reed, and the top layer of the shed with the shuttle during shed movement, etc. The shuttle is also retarded to some extent by the drag of the weft. The coefficient of friction between the bottom of the shuttle and the warp varies with the nature of the warp from about 0.3–0.5; the average value is about 0.4. This causes a retarding force of 0.4 W, where W is the weight of the shuttle. The backward and forward velocity and acceleration of the shuttle, due to the slay's movement, may be taken to be the same as that of the sword pin, though actually the shuttle moves slightly faster than the sword pin since it is farther away from the rocking shaft. The acceler-

ation of the sword pin varies from $\dfrac{v^2}{r}\left(1+\dfrac{r}{l}\right)$ at the beating up position 0

to $\dfrac{v^2}{r}\left(r-\dfrac{r}{l}\right)$ at the opposite crank pin position 6 and Figure 8.44 can be

taken as a model to understand the concept. The second scale in the diagram also show the force required to accelerate the shuttle. This force is the force pressing the shuttle towards the reed or the sides of the shuttle box. From positions 0 to 3, approximately the shuttle is being accelerated in a backward direction – along with the rest of the slay – and the force needed to accelerate

the shuttle, varying as in the diagram, from $\dfrac{W}{g}\left\{\dfrac{v^2}{r}\left(1+\dfrac{r}{l}\right)\right\}$ where W is the

weight of the shuttle, to zero, is provided by the front of the shuttle box, as the shuttle is in the box during that period. After passing position 3 the parts are retarded and the shuttle presses against the box back and then against the reed when the shuttle leaves the box at about position 4. The force pressing the shuttle towards the reed is greatest at position 6, where its value is

$$\dfrac{W}{g}\left\{\dfrac{v^2}{r}\left(1-\dfrac{r}{l}\right)\right\}.$$

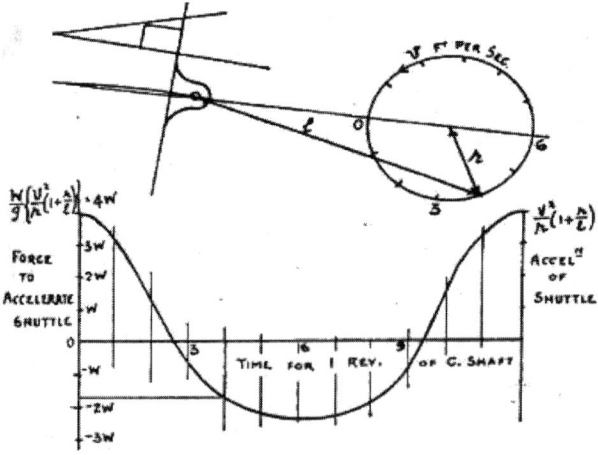

Figure 8.44 Example showing acceleration of the sley in a loom (Source: Hanton)

In an average fast running loom, with crank circle 6 cm diameter, arm 44 cm long, and speed 210 picks per minute, $v = 2\pi \times \dfrac{3.0}{1} \times \dfrac{210}{60} = 0.65 \; m/s$

and the accelerating force on the shuttle at position $6 = \dfrac{W}{9.81} \times \dfrac{0.65^2}{3}$

$(0.932) = 0.0133W$. At position 0, the force is $\dfrac{W}{9.81} \times \dfrac{0.65^2}{3} (1.06) = 0.01533W$.

Thus the force pressing the shuttle towards the reed, during the shuttle passage, when the reed is full back, and then decreases and as the shuttle is leaving the shed. The actual pressures between the shuttle and the reed are somewhat less than these values because of the friction between the bottom of the shuttle and the warp; this friction force provides some of the force needed to change the shuttle's velocity and reduces the pressure between shuttle and reed by the value of the friction force. The coefficient of friction between shuttle and reed is also to be noted; consequently the retarding force due to this friction; the total retarding force due to friction at the bottom and side of the shuttle etc. can be considered for the study. For slow running looms the values of the friction at the reed are much less, the force pressing the shuttle towards the reed varying as the square of the speed, other things being the same. Allowing a little for the drag of the weft, which is very variable, but not very much except for special cases, the total retarding force on the shuttle will be about 0.5 W for looms running about 100 picks/min, and about equal to W for looms running about 200–210 picks/min, W being the weight of the shuttle. The retardation caused by this force is in proportion to the force and inversely proportional to the mass of the shuttle

Force F = Mass × Acceleration (or retardation).

Retardation $= \dfrac{F}{M} = \dfrac{0.5\,W}{W/g} = 0.5g =$ say 5 m/s for slow looms

and $\dfrac{W}{W/g} = g = 10$ m/s for fast running looms.

With the knowledge of the distance travelled by shuttle in loom, the time taken, and the retardation to which the shuttle is subjected, the velocity of the shuttle as it enters the shed is easily found from the law of motion

$$S = Vt - \frac{1}{2}f^2 \text{ and } V = \frac{S + \dfrac{1}{2}f^2}{t}$$

where S = space passed over, in meters;

V = starting velocity, in meter per second;

f = retardation in meter per second per second;

t = time in seconds.

It is found that the shuttle movement resembles SHM giving it its maximum speed V; the shuttle starts up gradually and speeds up as though it were to move harmonically, but it leaves the picker when the maximum speed is attained. Figure 8.45 shows approximate curves of shuttle velocity during acceleration for a number of looms; the loom was turned slowly by hand and the shuttle movement for given crankshaft movements noted, from which the shuttle velocity can be calculated. The two dotted curves are velocity curves for simple harmonic motions, and it will be seen that these agree fairly well with the actual curves.

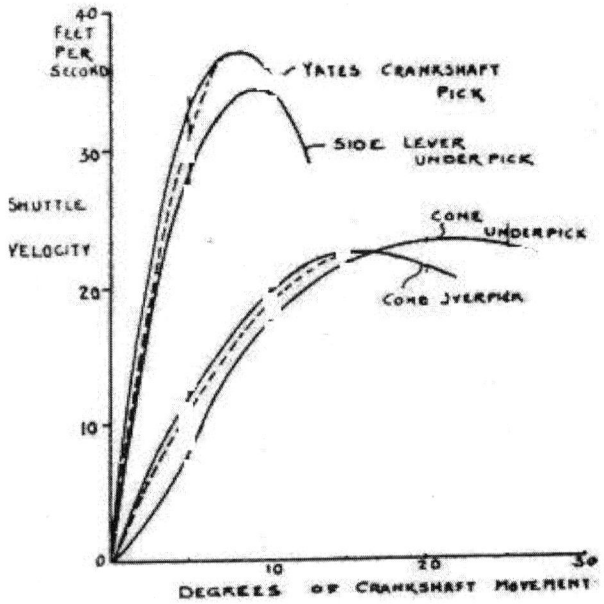

Figure 8.45   Shuttle velocity during its acceleration (Source: Hanton)

8.8.1   Mechanics of cone over pick mechanism

In the cone over pick mechanism the picking stick is moved by the cam thro various elements like cone, vertical shaft, picking band, picker, buffer, etc., through an angle of 30°–40° from the time it starts until the maximum speed is attained. The stick movement is nearly parallel with shuttle path when maximum speed is achieved and this is known as linear motion of picking stick or shuttle. The initial stages of the stick movement is used for tightening up the picking band and it is only after the band is tight that movement of the picker and shuttle starts. Sometimes the shell of the picking tappet (Figure 8.46) is fixed in eccentric position so that tightening of the band about the point A is achieved. But the maximum tightening is achieved by the nose N of the tappet, and often no movement of the stick takes place until the nose begins to act. From the time the nose begins to act until the shuttle reaches its maximum speed the tappet rotates through from 30° to 40°; about two-thirds of this time is generally spent in tightening the strap, leaving only about 10°–20° for accelerating the shuttle.

Analysis of picking mechanism (cone over pick) – Reed space of loom = 45″ or 1.125 m; width of yarn in reed = 41″ or 1.025 m; length of shuttle 13″ or 0.325 m; weight of shuttle 10 oz or 285 g; loom speed 180 picks/min; time for the passage of the shuttle = 96° of the crankshaft revolution; time for development of the pick = 45° of the tappet shaft revolution;

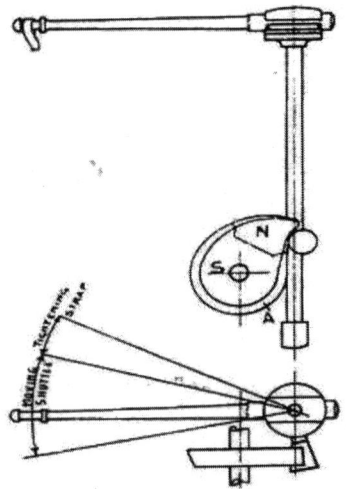

Figure 8.46   Cone over pick mechanism (Source: Hanton)

effective length of picking arm = 21″ or 0.525 m; angle moved by arm and cone during acceleration = 36°. The distance moved by the shuttle during its passage will be approximately = (41″ + 13″ – 6″) = 48″ or 4 ft or 1.2 m = S; the time for this movement

$$= \frac{96}{369} \times \frac{60}{180} = 0.089 \text{ s} = t.$$

∴ Average shuttle speed during passage $= \dfrac{S}{t} = \dfrac{4}{0.089} = 45$ ft or 13.5 m/s.

The retardation of the shuttle at this speed of loom will be about

f =24 ft.7.2 /s see, corresponding to a retarding force of

$$\frac{24}{32}W = 0.75W = 0.75 \times 10 = 7.5 \text{ oz or } 214 \text{ g.}$$

The maximum shuttle velocity

$$V = \frac{S + \frac{1}{2}ft^2}{t} = \frac{4 + \left(\frac{1}{2} \times 24 \times 0.089^2\right)}{0.089} = \frac{4.095}{0.089} = 46 \text{ ft or } 1.15 \text{ m/s}$$

Or the maximum velocity may be found as follows

Kinetic energy in shuttle at average speed $= \dfrac{Wv^2}{2g},$

where v = average velocity = 45 ft/s or 1.125 m $= \dfrac{10}{16} \times \dfrac{45^2}{2 \times 32.2} = 19.8$

ft-lb or 0.1840 N

This may be assumed to be the energy in the shuttle when halfway through its passage, though that is not quite correct. The energy lost by the shuttle in moving from its maximum speed position to its average speed position is equal to the work done in overcoming the resistance to the shuttle through that distance, viz., energy lost $= \dfrac{7.5}{16} \times 2 \; ft. = 0.94$ foot-lb approximately (reader is instructed to convert these units into SI system).

∴ Kinetic energy in shuttle at maximum speed

$$= 19.8 + 0.94 = 20.74 \; ft. - lbs = \frac{WV^2}{2g}$$

and maximum velocity –

$$V = \sqrt{\frac{2gKE}{W}} = \sqrt{\frac{2 \times 32.2 \times 20.74 \times 16}{10}} = 46 \text{ ft or } 1.15 \text{ m/s approx.}$$

The time for developing the pick is 45° of the tappet shaft or 90° of the crankshaft, the time in seconds

$$\sqrt{\frac{2 \times 32.2 \times 20.74 \times 16}{10}} \; s = 0.0835 \text{ s.}$$

Assuming that the shuttle is accelerated in the last third of this time – the rest being occupied in tightening the strap, and that the shuttle has a movement similar to part of a simple harmonic motion, the curve of shuttle velocity during the time of its acceleration can be drawn. The end of the picking arm, to which the picker is attached by means of the picking strap, is generally set so that it moves at the same rate as the picker for the last part of the acceleration. The time occupied by the arm in moving 36° is, as found above, 0.0835 s, and in that time the arm end moves

$$\frac{36}{360} \times 2\pi \times 21" = 13.2" \text{ or } 33 \text{ cm.}$$

Hence the average arm end velocity must be $\dfrac{13.2}{12 \times 0.0835} = 13.2$ ft 4 m (nearly) per second.

8.8.2   Example of a cone under pick mechanism

Consider an example of a side lever picking: Reed space is 36" 90 cm; cloth width, 32" 80 cm; shuttle length 13" 0.325 m; speed 150 picks/min, time for passage of shuttle 90° of crankshaft revolution; time for development of pick, 20° of crankshaft or 10° of bottom shaft revolution, movement of picker during acceleration 4" or 10 cm. Centre of picking bowl vertically below centre of bottom shaft when maximum shuttle speed is reached (Figure 8.47).

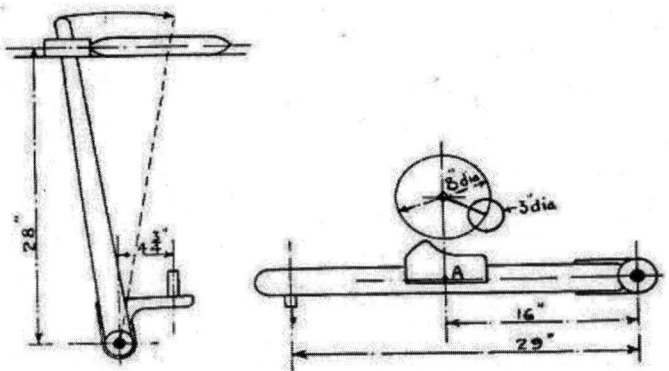

Figure 8.47   Cone under pick (side lever) mechanism (Source: Hanton)

Distance moved by shuttle during passage
$S = 32 + 13 - 6 = 39" = 3.25$ ft = approximately equal to 1 m.

Time for shuttle passage $= \dfrac{90}{360} \times \dfrac{60}{150} = 0.1$ s

Average shuttle velocity during passage $= \dfrac{3.25}{0.1} = 32.5$ ft or 9.75 m/s.

Taking the retardation of the shuttle, $f = 20$ ft 6 m/s/s, then maximum shuttle velocity,

$$V = \frac{S + \frac{1}{2}ft^2}{t} = \frac{3.25 + \left(\frac{1}{2} \times 20 \times 0.1^2\right)}{0.1} = 33.5 \text{ ft } 1.05 \text{ m/s.}$$

Time for development of pick

$$= \frac{20}{360} \times \frac{60}{150} = \frac{1}{45} \text{ s} = 0.0222 \text{ s}$$

Average velocity of picker during acceleration

$$= \frac{4"}{12 \times 0.0222} = 15 \text{ ft or } 4.5 \text{ m/s}$$

Shuttle after a pick at box: We know that the shuttle starts its journey with zero initial velocity and then increases to maximum velocity and when it leaves the shed and enters the shuttle box, its velocity will be brought to zero and this will be achieved by swell spring of the shuttle box. Some of the energy due to this velocity is used in overcoming the resistance of the swell. The velocity of the shuttle when leaving the shed is not much less than the average shuttle velocity; in medium width looms the shuttle loses approximately as much velocity in moving from the centre of the shed to its leaving position as it lost in moving from the position at entry to the midway position or the velocity when leaving may found from the formula $v = V - ft$, where v = leaving velocity, V = maximum velocity, f = retardation, t = time. For example, in the cone over pick motion considered above, the maximum velocity V was found to be 46 ft/s, f = 24 ft/s/s, and t = 0.089 s. Then v = 46 – (24 × 0.089) = 46 – 2.1 = 43.9 ft 13.17 m/s. The shuttle moves about 6", as a rule, after clearing the shed before it reaches the swell; it loses a little in speed during this movement, but its average speed would be about 43 ft 12.9 m/s and the time it would take to move 6" $t = \frac{0.5}{43} = 0.011$ s, or about 12° of the crankshaft's rotation at 180° picks per minute.

Overcoming the resistance of the swell entails a movement of about 3", and the shuttle has to work some energy. The kinetic energy in the shuttle, weighing 10 oz or 285 g and moving at, say, 43.5 ft or 13.20 m/s when it reaches the swell will be reduced. The average velocity during the 8 cm of movement is about 42 ft1.1 m/s, and the time taken about $\frac{3}{12} \times \frac{1}{42} = 0.006$ s, or about 30 degrees of the crankshaft revolution.

Relation between "e" and picking – Fast forward movement and slow backward movement of sley has significant effect on picking. We know that "e" can be varied by changing either "*l*" or "*r*" in loom. For example, if the connecting arm's length is reduced in relation to the crank size, gives longer time for the passage of the shuttle, which enables a weaker pick to be used, or a higher loom speed. Reducing the length of the arm, for example, from 16″ to 6″, with a 2½″ crank, the other particulars of the loom being unchanged which results in about 20% more available for the shuttle passage. Figure 8.48 shows the effect on the shed opening at the shuttle front and also on the displacement, velocity, and acceleration of the slay. The increase of 20%, in the time available for the passage of the shuttle would permit of a 20% increase in the speed of the loom, which is generally the object aimed at when short arms are used. The pick would have to be made earlier so that the shuttle enters the shed at crank position A instead of B. It would also have to be weakened, as if the pick remains of the same strength the shuttle would leave at C, where AC = BD. Generally, in any loom an early pick at A is weaker than one at B, with the same setting. This is mainly due to the different position of the slay, when the shuttle is leaving the picker, but here the slay is as far back at A with the short arm as at B with the long one.

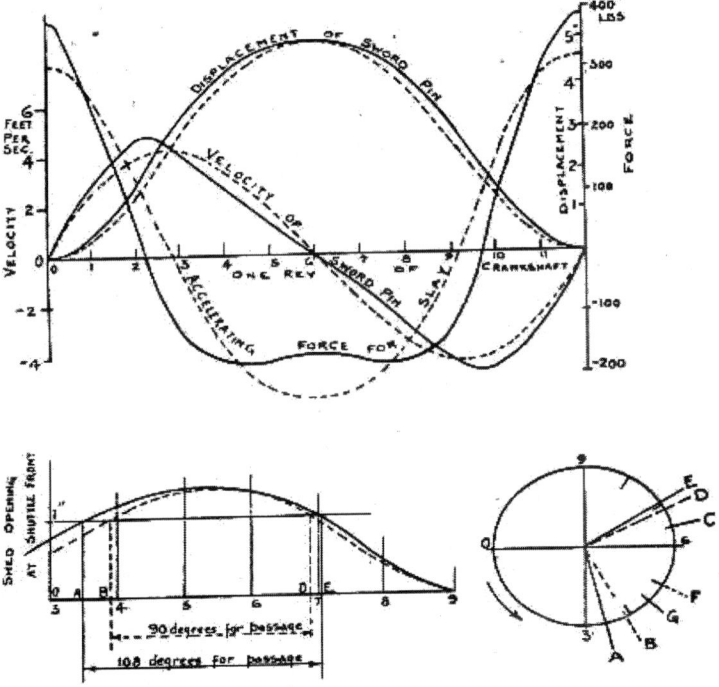

Figure 8.48 The effect of shed opening at the shuttle displacement, velocity and acceleration of sley (Source: Hanton)

If the pick is kept of the same strength and the shuttle passes through the shed from A to C instead of from B to D, then evidently the closing of the shed could be started earlier by reducing the dwell of the tappet; thus instead of the dwell ending at F it could end at G, where GF may be taken equal to AB or CD. Such a change in the dwell will have a slight effect on the shed opening at beating up and on the "cover" of the cloth. Assuming the dwell to start at the same time the healds will be level slightly earlier with the reduced dwell and the shed opening at beating up will be slightly greater. The displacement and velocity diagrams show that the reduced slay velocity about the back centre position, which gives the longer time for shuttle passage, is accompanied by an increased slay velocity near the beating up position. It is often claimed as an advantage of short arms that this change in the sley's velocity is better for beating up; but the actual velocity of the reed when pushing the weft into place – during the last fraction of an inch of the reed's forward movement – is so small that the slight increase in velocity cannot have much effect, especially in positive uptake looms. The greatest disadvantage of short arms is the excessive vibration they cause, due to slay inertia. The effect on the accelerating forces needed to move the slay is seen in Figure 8.48. If the weight of the slay, considered concentrated at the sword pin, is 96 lb, then the accelerating forces at the beating up and back centre position at a speed of 200 picks/min, and a crank pin speed $v = 4.36$ ft/s, would be: long arm, beating up

$$F = \frac{W}{g} \cdot \frac{v^2}{r}\left(1 + \frac{r}{l}\right) = \left(\frac{96}{32} \times \frac{4.36^2}{2.5} \times 12\right)\left(1 + \frac{2.5}{16}\right) = 274 \times 1.156 = 316 \, lb \, or \, 143.31 \, kg$$

And at the back centre,

$$F = \frac{W}{g} \cdot \frac{v^2}{r}\left(1 - \frac{r}{l}\right) = 274 \times 0.844 = 230 \, lb \, 104.30 \, kg$$

With the short arm, at beating up

$$F = 274\left(1 + \frac{2.5}{6}\right) = 274 \times 1.418 = 387 \, lb \, 175.51 \, kg$$

and at the back centre $F = 274 \times 0.583 = 160$ lb 72.56 kg. The greatest force causing vibration, which at the front centre, will thus get increased by shortening the arms, whilst the difference between the forward and backward forces is also increased. And if, as is generally the case, the shortening of the arms is accompanied by an increased loom speed, the maximum accelerating force is much further increased. If, for example, the speed of the loom is increased 20% then, since accelerating forces are proportional to the square of the speed, the force at beating up would he increased to $175.5 \times \dfrac{120^2}{100^2} =$

252.15 kg, which is an increase of 76%, over the 143.31 kg of the long arm loom. For this reason it is not common practice to use very short arms on fast running looms, by keeping the mass of the slay as small as possible and the crank throw a minimum. Short arms are more commonly fitted to very vide looms, which must always run at a low speed. In looms weaving plain and other simple cloths, the factor which generally decides the loom speed is the speed of the shuttle; it is not found possible or desirable to have a higher shuttle speed than about 1.25 m/s. Consequently in a wide loom, the speed of rotation must be low, and this enables short arms to be used, since vibration due to the sley is proportional to the square of the speed and is not the troublesome at the low speed. In such a loom the dwell of the shedding tappets can also be increased to give yet longer time for the shuttle passage and enable a higher loom speed to be used, since at the low loom speed there is ample time for changing the healds even with the increased dwell. If the vibration at high speeds could he reduced, short arms would be of greater advantage to narrow looms than they are to wide ones, since the latter, even at their reduced speed, generally produce a greater number of square yards of cloth in a given time than the former.

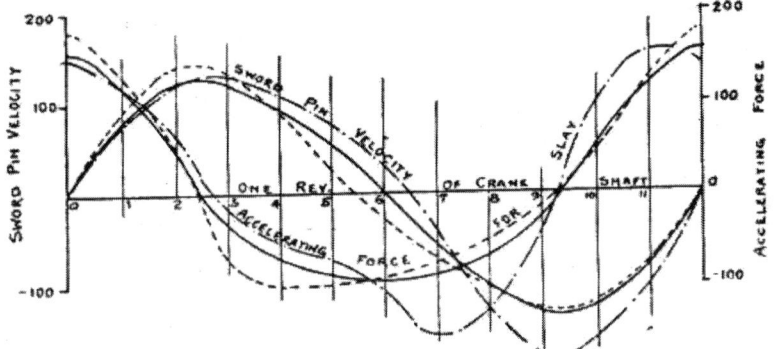

Figure 8.49   Effect of lowering the crank wrt sword pin on sley performance
(Source: Hanton)

This is due partly to the shorter arms used, but also to the longer dwell permissible in the wide loom and to the smaller proportion that the shuttle length bears to the width of the cloth. Shorter arms can also be used with some merits in dobby and Jacquard looms. In these looms the shedding motions give only a very little dwell. Shorter arms would also increase the time for the passage of shuttle and allow a weaker pick to be formed. The loom speed can't be increased as the limiting factor is not the shuttle speed but the shedding motion can also govern the situation. The effects of lowering the loom crank with respect to sword pin is shown in Figures 8.49 and 8.50.

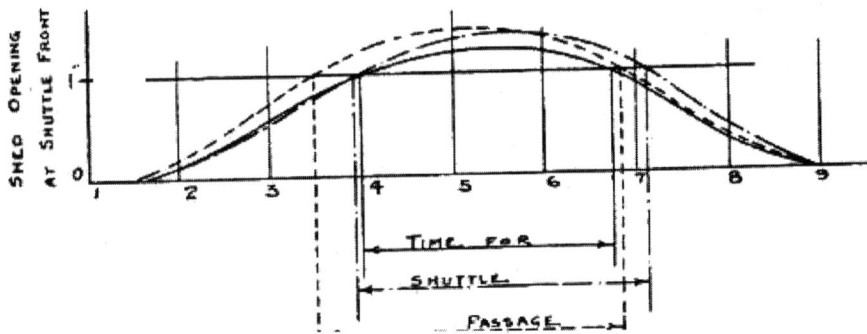

Figure 8.50 Effect of lowering the crank wrt sword pin on sley performance
(Source: Hanton)

8.8.3   Picking mechanism as an elastic mechanism

Picking mechanism can be represented by a simple elastic system. "A" mass "M" rests on a smooth horizontal surface, which can represent the shuttle and the spring represent the elasticity of the picking mechanism. "A" represents the situation when picking starts. B shows the movement of mass (shuttle) through a distance corresponding to the actual movement of the picker at a given instant during picking and mass "M" move through M distance equal to actual movement of the picker (Figure 8.51).

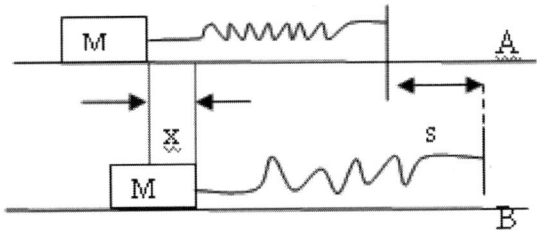

Figure 8.51   Picking as an elastic mechanism

λ = be the rigidity of the mechanism (i.e., force/unit extension)
S = nominal movement
x = actual movement
and x = Actual velocity
x = Actual acceleration
      The force accelerating on mass M is nothing but the tension in the spring which is equal to the product of rigidity and extension.
      Force = $\lambda (S - x)$

But   Force = Mass × Acceleration

$$= M x$$

$$Mx = \lambda (S - x)$$

$$\text{``}x\text{''} = \frac{x}{M} (S - x)$$

For a given mechanism λ and M are constant. Hence λ/m is called as "ALACRITY" of the system which is a constant.

Thomas write $\dfrac{\lambda}{M} = x^2$

$$\therefore \ddot{x} = n^2 (s - x)$$

$$\ddot{x} = n^2 s - n^2 x \qquad\qquad n^2 x + \ddot{x} = n^2 s$$

The picking mechanisms are designed to give a straight line nominal movement. The picker displacement is proportional to angular movement of the crankshaft θ = crank movement i.e., $S = P\,\theta$, where P = constant of proportionality.

But $\theta = Wt$ where "W" is the angular

$$\therefore S = P\,Wt$$

Differentiating $\dfrac{ds}{dt} = PW$

But we know that

$$n^2 s = \ddot{x} + n^2 x$$

$$n^2 PWt = \ddot{x} + n^2 x$$

This is the 2nd order differential equation and solution is given by

$$x = PW \left(t - \frac{\sin nt}{n}\right)$$

Differentiating the expression

$$\dot{x} = PW (1 - \cos nt)$$

The value of $\dot{x}$ will be maximum when $\cos nt = -1$ i.e., $\pi = nt$ or $t = \pi/n$

$$\dot{x}_{max} = 2PW \quad \therefore \quad PW = \frac{\dot{x}\,max}{2}$$

Thus max actual velocity is twice the constant nominal velocity. If "L" is the picking stroke at $t = \pi/n$ for max actual velocity.

$$L = PW \left(\frac{\pi}{3} - \frac{\sin \pi}{n}\right)$$

But $\sin \pi = 0$

$$= \frac{PW\pi}{3}$$

Substituting the value of $PW = \dfrac{x\,\text{max}}{2}$ in

$$L = \frac{x\,\text{max}\times \pi}{2n}$$

or

$$n = \frac{x\,\text{max}\times \pi}{2L}$$

From 7 it can be said that max actual velocity α to effective stroke of picker

Now let us derive the relation x max, x max. for that purpose consider equation no.4a

$$x = PW(1\text{-}\cos nt)$$

Differentiating

$$x = PW \sin \text{nit } (n)$$

The maximum value of x occurs at $\sin t = 1$ i.e., $nt = \pi/2 \quad t = \pi/2n$

$$x\,\text{max} = PWn$$

$\therefore$ The maximum acceleration occurs at half the time required to produce maximum velocity if half way through the effective stroke of the picker.

$$\text{gives } PW = \frac{x\,\text{max}}{2}$$

Substituting for PW in equation

$$x\,\text{max} = \frac{x\,\text{max}\,n}{2}$$

But from equation 8 value of $n = \dfrac{x\,\text{max}\,\pi}{2L}$

$$x\,\text{max} = \frac{x\,\text{max}}{2}\, x\,\frac{x\,\text{max}\,\pi}{2L}$$

Or $x\,\text{max} = \dfrac{(x\,\text{max})^2\,\pi}{4L}$

$\therefore$ Maximum acceleration is inversely proportional to effective stroke for a given maximum velocity.

8.8.4   Power required for picking

The energy usefully expended in accelerating the shuttle is equal to its kinetic energy when it leaves the picker, so that:

$$\text{Energy /pick} = \frac{mv^2}{2} \text{ J.}$$

Where m is the mass of the shuttle in kg and
v is the maximum velocity in m/s.

and:

Power $= t\,W = 1$ J/s.

Thus, if P is the loom speed in picks/min then:

$$\text{Power for picking} = \frac{mv^2}{2} \times \frac{P}{60} \times \frac{1}{1000} \text{ } kW.$$

Let R be the useful reed space in cm (i.e., the width of the warp in the reed when the reed space is being fully utilised), "l" be the length of the shuttle in cm excluding its tapered ends and θ be the number of degrees of crankshaft rotation occupied by the passage of the shuttle through the warp shed.

Then the time for the passage of the shuttle is:

$$t = \frac{\theta}{360} \times \frac{60}{P} = \frac{\theta}{6P} s.$$

And the distance moved by the shuttle is:

$$d = \frac{R+L}{100} m.$$

If v is now the average speed of the shuttle during its passage through the shed then:

$$v = \frac{R+L}{100} \times \frac{6P}{\theta} = \frac{6P(R+L) \times 10^{-2}}{\theta} m/s.$$

We then have

$$\text{Work done / pick} = \frac{mv^2}{2} = \frac{36mP^2(R+L)^2 \times 10^{-4}}{2\theta^2}$$

$$= \frac{18mP^2(R+L)^2 \times 10^{-4}}{\theta^2} J.$$

Hence

$$\text{Power for picking} = \frac{mv^2}{2} \times \frac{P}{60} \times \frac{1}{1000} = \frac{18mP^3(R+L)^2 \times 10^{-4}}{\theta^2 \times 6 \times 10^4} kW.$$

$$= \frac{3mP^{3}(R+L)^{2} \times 10^{-8}}{\theta^{2}} kW.$$

Examples

1. Consider a cotton or rayon loom of 110 cm reed space, running at 216 picks/min with a shuttle of mass 450 g and length 28 cm. Assume the passage of the shuttle to occupy 135°. We then have

$$\text{Work done/pick} = \frac{18 \times 0.45 \times 216^{2} \times 138^{2} \times 10^{-4}}{135^{2}} = 39.51 \text{ J.}$$

And

$$\text{Power for picking} = \frac{39.51 \times 216}{60} \times \frac{1}{1000} = 0.142 \text{ kW.}$$

2. Consider a heavy blanket loom with a reed space of 533 cm running at 65 picks/min with a shuttle 47 cm long and of mass 9 g. Since the loom has a high sley – eccentricity ratio (0.54), assume the passage of the shuttle to occupy 150°. We then have

$$\text{Work done/pick} = \frac{18 \times 0.9 \times 65^{2} \times 580^{2} \times 10^{-4}}{150^{2}} = 100.2 \text{ J.}$$

$$\text{Power for picking} = \frac{100.2 \times 65}{60} \times \frac{1}{1000} = 0.109 \text{ kW.}$$

3. Consider a loom with reed space 1.15 m with average shuttle speed of 13.75 m/s with a shuttle length of 30 cm when crank turns through 135°, then maximum loom speed is

$n = (13.75 \times 135) / 6 (1.15 + 0.3)$

$= 213$ rpm or ppm.

8.8.5   Mechanics of torsion bar picking (Sulzer)

Velocity – Acceleration

- Acceleration is far from uniform and reaches a peak value of 6630 m/s^2 when the projectile has travelled nearly 15 mm. The peak deceleration has a value of 9920 m/s^2 and occurs about 12.5 mm before the picker comes to rest.

Projection / picking

- The picker accelerates the projectile over a distance of about 65 mm in 0.007s.

- The energy utilisation of the projectile system is not much more efficient than that of shuttle projection and all other weaving machines are less efficient.
- The residual energy in the picking system, some 62% of the whole, is absorbed in the hydraulic buffer, the body and plunger.
- The picker and the rest of the lever system are then brought to rest over the next 40 mm of picker movement by the oil brake.
- The speed of the projectile as it leaves the picker is normally about 24.4 m/s.

Velocity – acceleration

- Acceleration of the projectile occupies 0.007 s.
- The velocity of the projectile varies during its acceleration, flight across the loom (the loss in the speed), and retardation to its rest position at the opposite side of the loom.
- Retardation occurs over a substantially longer time than acceleration and is therefore relatively gentle.
- The peak forces acting on the projectile are of the same order as those acting on the shuttle loom because the projectile has a mass of less than 40 g.
- This gives a peak accelerating force of about 262.5 N and a peak retarding force of about 397 N.

8.8.6   The picking mechanism – The torsion bar picking system of the machine

The formula for the axial twisting of circular rods is:

$$T = \frac{CJ\theta}{l}$$

Where T is the torque applied, C is the modules of rigidity of the bar material, J is the polar second movement of area of the rod section, θ is the twisting angle in radians, l is the twisting length, and r is the radius of the rod. The polar second movement for a circular section (J) is $\pi d^4 / 32$. The work done in loading up the torsion bar is the area under the $T\theta$ curve,

$$\text{Work done} \quad \frac{T\theta}{2} = \frac{CJ\theta^2}{2l}$$

Applied torque is 25.34 kg m units and the work done on the rod per pick is 7.62 kg m units. Only 14% of the work done in energizing the torsion bar is effectively transferred to the projectile inserting the weft. To achieve higher initial projection speeds the diameter of the torsion bar was increased from 15 mm to 17 mm, and more recently to 19 mm in subsequent models. As the energy in the system varies as the fourth power of the torsion rod diameter, all else remaining equal, the increase from 17 to 19 mm diameter will provide 56% more strain energy

When a rod of radius, r, of length, l, is subjected to a pure torsion by torque equal to T at one end of the rod, as shown in Figure 8.51, the straight line (AM) parallel to the rod axis takes a position (BM). So the position of (A) relative to (M) is proportional to the length, 1.

So, the arc (AB) œ 1, and $r\theta = AB$ x; So that,

$$r\theta \; \alpha \; 1$$

From figure, we have shear strain = x' / 1' = x/1. When the rod is twisted it will be subjected to a shear stress F and shear strain of x/1.

$$\text{Modulus of rigidity, } C \; \frac{\text{shear stress}}{\text{shear strain}} = (F.1)/x.$$

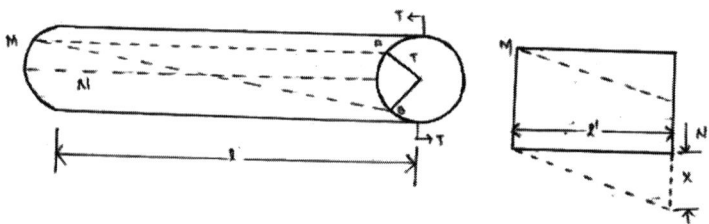

Figure 8.52   Mechanics of torsion bar picking in Gripper projectile loom

Or, $x = (F.1)/C$ (1)

Let θ be the angle of twist resulting out of torque, and we can write

$\theta = x/r = (2x)/d$, where, d = rod diameter.

Combining this with the equation (1), we have

$\theta\, d/2 = F\, 1/C$ (2)

Rearranging equation (2),

$$F = (\theta\, Cd) / 21 \tag{3}$$

Torque, $T = (2F.J)/d$,

where, J = polar second moment of area

$= (\pi\, d^4)/32$

So, $T = \dfrac{2\theta C dJ}{2\, d.1} = \dfrac{\theta C J}{1}$ (4)

after equation (4) gives us the torque required to twist the torsion bar.

Work done in one loading cycle of the torsion bar is equal to the area under the torque-twist $(T - \theta)$ curve, so that

$$= \frac{T\theta}{2} = \frac{\theta^2 CJ}{2\,1} \tag{5}$$

From equation (5) work done on the torsion bar can be evaluated. Gripper energy when it leaves the shoe is given by

$$= \frac{mV^2}{2} \tag{6}$$

With an effective length of the torsion bar of 0.72 m having a diameter of 0.015 m, and the angle of twist being 28° radians, the modulus of rigidity of the material being 8.4×10^9 kg/m².

From equation (4) we can estimate the torque as:

$$T = \frac{\pi d^2\, \theta\, c}{32\,1} = \frac{22 \times 8.4 \times 10^9 \times 28 \times (0.015)^4\ 9.81}{7 \times 57.3 \times 0.72 \times 32}\ Nm,$$

$$= 28.34 \times 9.81 \ Nm = 278 \ Nm.$$

$$1 \ rad = 57.3°$$

And from equation (5), work done in one loading of the torsion bar is estimated to be

$$\frac{278}{2} \times \frac{28}{57.3} = 67.9 \ J$$

Gripper energy when it leaves the shoe, assuming that the weight of the gripper is 40 g and has a velocity of 24.3 m/s (Figure 8.52).

$$= \frac{mV^2}{2} = \frac{0.04 \times 24.3^2}{2} = 11.81 \ J$$

So utilisation of the work done energizing the torsion bar to the gripper is

$$\frac{11.81 \times 100}{67.9} = 17.3\%$$

Now power for picking in a machine running at 300 picks/min with 330 cm reed space and gripper length of 9 cm and gripper taking 190° to clear the shed is found to be

$$= \frac{3 \times 0.04 \times 300^3 \times 3.39^2 \times 10^{-4}}{190^2} = 0.103 kW$$

8.9   Kinetics of beat-up or sley

8.9.1   Significance of reed and its selection

Reed is an important element in weaving irrespective of type of fabric formation technique. Reed is a part of beat-up mechanism in loom and is selected in relation to ends/inch desired in fabric and width of the fabric. Normally reed count is defined (stock port system) as number of dents per 2″. With 2 ends/dent order, reed count is equal to ends/inches in reed. Reeds are selected basically on warp count and number of ends/inch of the finished fabric.

$$\text{Reed count is given by} = \frac{n_1}{1 + C_2} \quad \text{where } n_1 = \text{ends/inch,} \quad C_2 - \text{weft crimp}$$

(%)

while applying the above formulae, sometimes it is practically impossible to use the reed count obtained through calculation. In such situations, it is the trend to select the half the number of reed count with 4 ends/dent. For e.g., warp count = 34^s, ends/inch = 98, picks/inch – 85, weft count – 30^s, weft crimp 8%, reed count as per formulae = $\dfrac{98}{1 + \dfrac{8}{100}} = 90^s$

(approximate).

But 34^s yarn cannot be woven using 90^s reed (finer reed) even though if an attempt is made, end breakages, selvedge breakages may be more. Therefore it is better to recommend 4/44^s reed. Similar is the case of poplin (40^s K × 40^s C), (112 × 78), (8% × 6.5%) give 104^s reed, but in weaving we use 4/52^s reed. Another example is commonly sateen cloth using filament warp and filament weft has (194 × 100) where the weft crimp is minimum, it is better to use 94^s or 98^s reed with 4 ends per dent. Reed count is available with maximum value as 120^s or 128^s, as normally 60% air space is considered, while designing a reed. Reed also control the fabric texture and by using 4 ends/dent, reed marks are observed in grey fabric. These marks can be eliminated if fabric is subjected to chemical processing, Figures 8.53–8.55 show sley in a loom.

It can be observed from Figure 8.56 that with simple harmonic motion (corresponding to e value as zero) and indefinitely long crank arms, the sword pin attains its maximum velocity and exactly half of its maximum displacement at 90° and again at 270° (Table 8.4). With a finite value of "e", with any arrangement possible in practice if the sley is crank driven, the sword pin attains its maximum velocity and exactly half of its path.

Figure 8.52 also show sley as four and six bar linkage mechanism. In most of the looms sley is moved by crank and connecting arm and forming a four bar linkage mechanism. Figure 8.53 show the set of four bar unit. The machine frame forms the fixed bar and the crank arm, crank and sley sword from the moving parts. The crankshaft centre O is connected to the sword pin P through the crank OT and the crank arm PT. The line of movement of P passes through the crankshaft centre O and P may be thought of sliding along a horizontal path.

The double beat-up uses a six bar linkage mechanism. The aim of this mechanism is to make the sley to move forward and backward twice for one rotation of crankshaft. Figure 8.53 show the six bar linkage motion and it is assumed this as two number of four bar linkage mechanism which have two bars in common. The first of these four bar linkages consists of the links numbered 1, 2, 3 in diagram together with the frame of the machine which forms the fixed link. Link 1 gets drive from crankshaft. If the length of links are chosen correctly, the rocking link 3 will rock backwards and forwards for each revolution of link 1. The second four bar linkage uses link 3 as a driving link and consists of links numbered 3, 4 and 5, the frame of the machine again being the fixed link. The rocking link of this four bar linkage is link 5 which is also the sley. Therefore if the lengths of the link are again chosen correctly, the link 5 will rock backwards and forwards twice for each rotation of the link 1 thus producing the required double beat-up.

The double beat-up has sword A and crank B and C as connecting arm and Knuckle joint D and arm E hinged to the machine framing and another arm F attached to the sley by usual sword pin. The crank moves about a point G to the top centre and arms EF are bent and the sley moves backwards. Again when the crank reached the point H the arms are again straightened resulting in second beat-up.

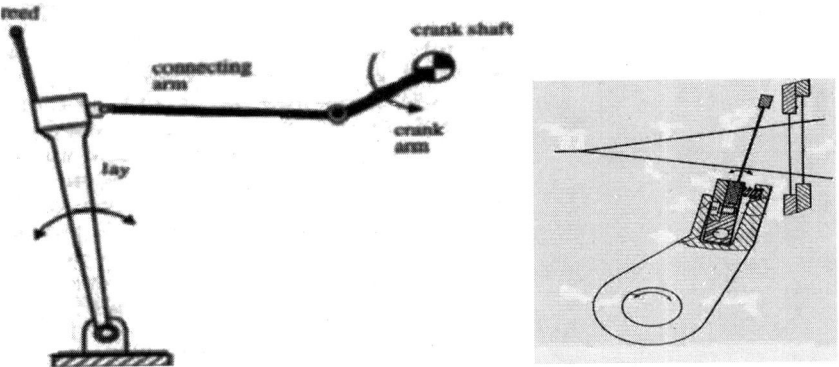

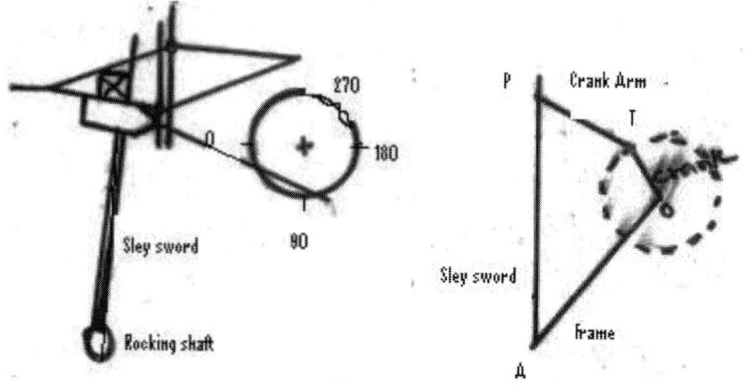

Conventional beat-up   Beat-up as 4 – Bar linkage mechanism

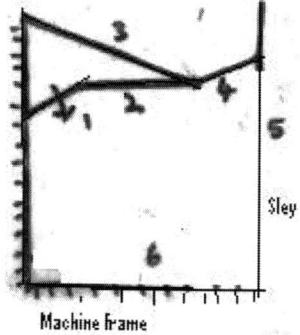

6 – Bar linkage or double beat-up

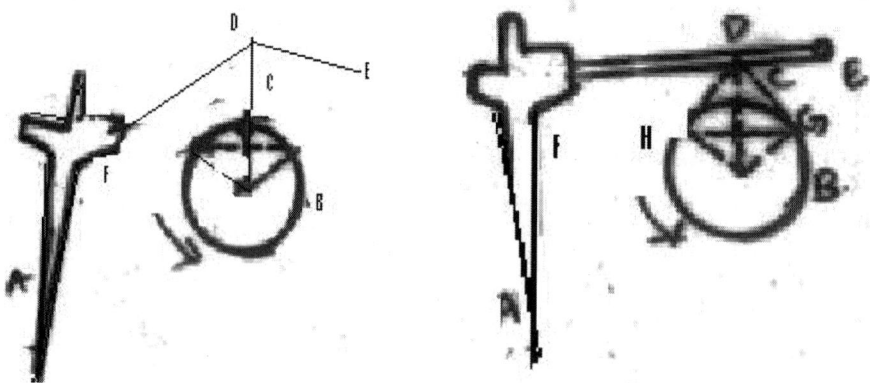

Double beat-up mechanism

Figure 8.53   Sley mechanisms

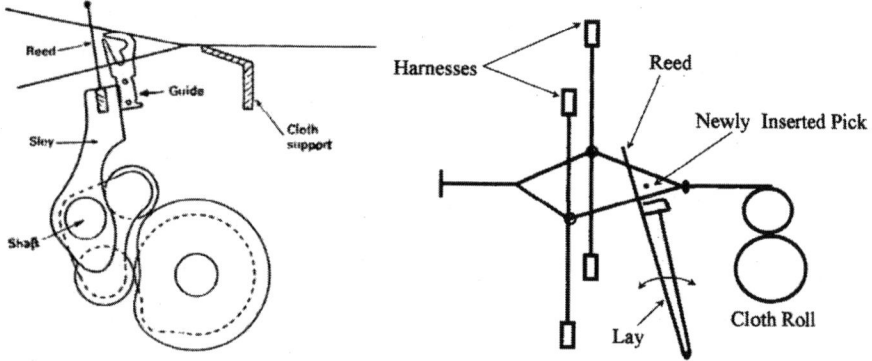

Figure 8.54 Cam beat-up   **Figure 8.55** Sley position in a loom

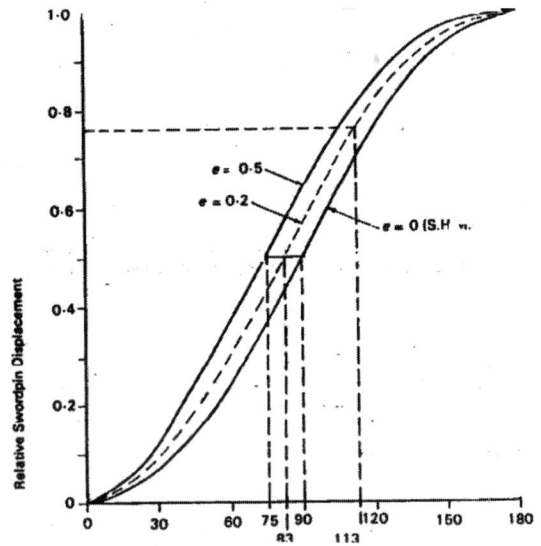

Crankshaft rotation in degrees (Source: Hanton)

Figure 8.56 Eccentricity of sley

Table 8.4 Eccentricity of the Sley

Value of "e"	Crankshaft position at half maximum displacement	Period during which displacement is at least half maximum
0	90° and 270°	180°
0.2	83° and 277°	194°
0.5	75° and 285°	210°

8.9.2   Expression for sley displacement, velocity and acceleration

Let "*l*" be the length of connecting arm and let "*r*" be the crank radius (Figure 8.57).

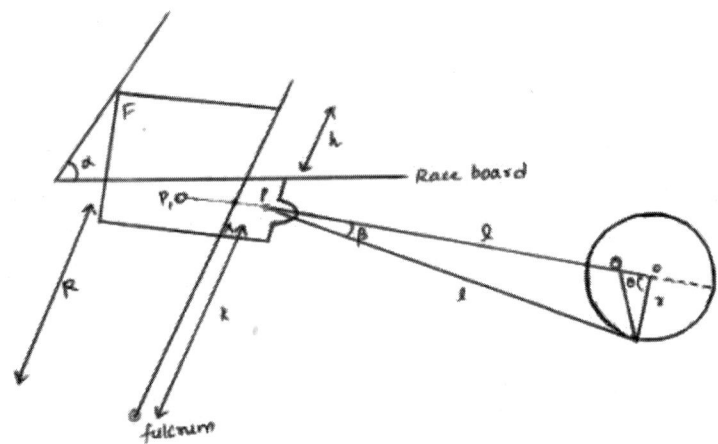

Figure 8.57   Displacement, velocity and acceleration of sley

Inside Δle PQT & POT

QT is a common side. Hence $l \sin \beta = r \sin \theta$

$$\sin \beta = \frac{r}{l} \sin \theta \rightarrow \sin^2 \beta = \frac{r^2}{l^2} \sin^2 \theta \rightarrow (1 - \cos^2 \beta) = \frac{r^2}{l^2} \sin^2 \theta$$

$$\rightarrow \cos^2 \beta = \left[\frac{r^2}{l^2} \sin^2 \theta + 1 \right]$$

$$\rightarrow \cos \beta = \left[1 - \frac{r^2 \sin^2 \theta}{2l^2} \right]$$

Sub the value of $\cos \beta$ in equation.1

$$PP_1 = (r + 1) - \left[l \left(\frac{1 - r^2 \sin^2 \theta}{2l^2} \right) + r \cos \theta \right]$$

$$PP_1 = (r + 1) - \left[l \left(\frac{1 - r^2 l \sin^2 \theta}{2l^2} \right) + r \cos \theta \right]$$

$$PP_1 = r + 1 - 1 + \frac{r^2 \sin^2 \theta}{sl} - r\cos\theta \left(\frac{r}{l} = \frac{1}{n} \right)$$

$$PP_1 = r + \frac{r\sin^2 \theta}{2n} - r\cos\theta = r \left[1 + \frac{\sin^2 \theta}{2n} - \cos\theta \right] = r \left[1 - \cos\theta + \frac{\sin^2 \theta}{2n} \right]$$

Reed displacement $S = PP_1 (R/K) \, PP_1 = r \left[1 - \cos\theta + \frac{\sin^2 \theta}{2n} \right]$

Velocity

$$V = \frac{ds}{dt} = \frac{ds}{d\theta} x \frac{d\theta}{dt} \Rightarrow V = \omega \frac{ds}{d\theta}$$

$$= \omega \left\{ r\sin\theta + \frac{r2\sin\theta\cos\theta}{2n} \right\}$$

$$V = \quad \omega \left\{ r\sin\theta + \frac{r\sin 2\theta}{2n} \right\} = r\omega \left\{ \sin\theta + \frac{\sin 2\theta}{2n} \right\}$$

$WKT \, r\omega = v$

$$V = v \left\{ \sin\theta + \frac{\sin 2\theta}{2n} \right\}$$

When $\theta = 90°$ $V = v$ i.e., max velocity of sley occurs at 90° and 270°.

Acceleration

$$A = \frac{dv}{dt} = \frac{dv}{d\theta} x \frac{d\theta}{dt} \Rightarrow \frac{dv}{d\theta} x\omega = \omega x \left(v\sin\theta + \frac{v\sin 2\theta}{2n} \right)$$

$$A = v \cdot \frac{v}{r} \left[\sin\theta + \frac{\sin 2\theta}{2n} \right]$$

Differentiation $A = \frac{v^2}{r} \left[\cos\theta + \frac{\cos 2\theta}{n} \right]$

Max acceleration occurs when $\theta = 0°$ and 180°

Illustrative example: Examine the velocity and acceleration and the angular position of the crankshaft for a sley having eccentricity of 0.5.

Solution :

Given, $e = 0.5$

We know that velocity for sley is given by

$$v_e = r\omega \left(\sin\theta + \frac{(e \sin 2\theta)}{2} \right)$$

Velocity for SMH is given by $v_{SHM} = r\omega \sin\theta$

$\sin 2\theta = 0$, when $\theta = 90°, 180°, 360°$

So, v_e and VSHM are equal at those θ values.

For, maximum velocity of sley, acceleration will be zero.

$$\theta_{V\,max} = \cos^{-1} \frac{\sqrt{1 + 28\,e^2} - 1}{4\,e} = \cos^{-1} \frac{\sqrt{1+2} - 1}{2}$$

$$= \cos^{-1} 0.366 = 68.5°, 291.5°$$

So, at 68.5° and 291.5° maximum velocity will be attained. Now for $\theta = $ 68.5° and 291.5° in the expression of velocity, the value of maximum velocity can be obtained in terms of $r\omega$.

This will be equal to 1.1 $r\omega$.

When θ is less than 90°, the term $\dfrac{e \sin 2\theta}{2}$ is positive and therefore, the velocity for sley will be more than that of SHM. However, when θ is greater than 90°, the term $\dfrac{e \sin 2\theta}{2}$ is negative and therefore, the velocity for sley will be less than that of SHM.

$$f_2 = r\,\omega^2 (\cos\theta + e \cos 2\theta)$$
$$f_{SHM} = r\,\omega^2 (\cos\theta)$$

When $\theta = 0°$

$$f_e(\theta = 0°) = r\omega^2 (1 + e) = 1.5\, r\omega^2$$

When $\theta = 180°$

$$f_e(\theta = 180°) = r\omega^2 (1 - e) = 0.5\, r\omega^2$$

On differentiating,

$$\frac{df_e}{dt} = r\omega^3 \left(-\sin\theta + 2e \sin 2\theta \right)$$

$$= -r\omega^3 \left(\sin\theta + \sin 2\theta \right)$$

If $\dfrac{da_\theta}{dt} = 0$ then $(\sin\theta + \sin 2\theta) = 0$

Or, $\sin\theta (1 + 2 \cos\theta) = 0$

Solving we get, $\theta = 0°, 180°, 120°$ and $240°$.

Therefore, at the above four values of θ, the slope of the acceleration curve will be zero. Using these values of θ in the expression of acceleration of sley, the values of acceleration can be calculated in terms of rw^2.

When $\theta = 120°$ or $240°$

$F_e (\theta = 120/240°) = 0.75\ r\omega^2$

Now, using the above information, the acceleration curve of the sley can be plotted as the maximum velocity was attained when $\theta = 68.5°$ and $291.5°$, acceleration at these two point will be zero.

Estimation of "e"

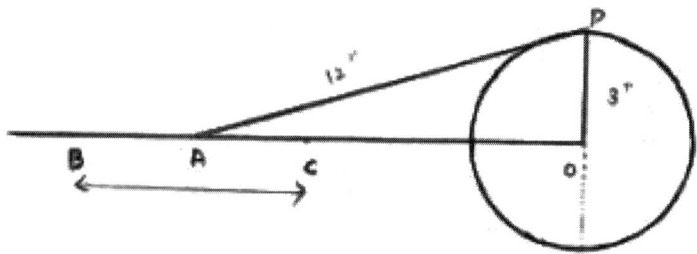

Figure 8.58   Derivation for "e"

Let $12''$ be the (Figure 8.58) connecting arm length let $3''$ be the crank arm length consider the following diagram.

Consider triangle AOD

$= OD^2 + OA^2 = AD^2$

$OA^2 = AD^2 - DO^2$

$= 12^2 - 3^2$

$OA = 11.61''$

$ABO = 15''$

$OB = 15''$

$OA = 11.61''$

$BO = OA + AB = 12'' + 3''$

$AB = 15'' - 11.61'' = 3.39''$

$BC = 6''$

$BC - AB = AC$

$AC = 6 - 3.39'' = 2.61''$

$e = 3.39 - 2.61 = 0.78$ this value is observed in Jute looms

8.9.3   Length of the crank and crank arm

The shed size or shed depth do depends on the shed angle and the sley sweep and in turn both will decide the size of the shuttle. Let us say that the crank sweep is increased, of the circle described by the crank and hence the sweep of the sley will also get increased leading to the increase in shed angle. In other words we need to use a larger shuttle. Chances are also high to notice the high end break or slubs may get buried into cloth due to high sley sweep. Increased sley velocity may result in high wear and tear of loom parts. Therefore it is suggested to use higher crank size for larger shuttle. The length of the crank will also depend on the weft size and the type of the cloth woven. In practice the looms weaving medium to fine fabrics will have crank lengths varying from 6 to 7 cm and those weaving heavy fabrics like suiting the crank length may vary from 7 to 9 cm. The crank size of 12 cam is recommended for extra heavy fabrics weaving.

Few examples as shown will indicate some information about the type of loom, weaving and crank details.

Saurer – cotton – Tappet $r = 6.25$ and $L = 15$ cm giving $e = 0.42$; Ruti – cotton – Dobby, $r = 7.6$ and $L = ?$ cm giving $e = 0.23$; Piconol – cotton – Tappet, $r = ?$ and $L = 32.4$ cm giving $e = 0.225$; Water jet – Viscose – Tappet $r = 3.33$ and $L = 22.9$ cm giving $e = ?$; Northrop – Industrial blanket – Tappet, $r = 10.8$ and $L = 20.3$ cm giving $e = 0.54$.

(**Note**: **The reader is advised to calculate the necessary data in the above information**)

8.10   Eccentricity of sley – Significance

Eccentricity of sley is defined as covering equal distance in unequal intervals of time in other reeds fast forward motion slow back ward motion. Fast forward motion is required for firm beat-up and slow back ward motion for dwell of the sley or crank at back centre for long period to allow the shuttle to pass through. Eccentricity as indicated by "e" is given by eccentric ratio r/l where $r \rightarrow$ crank radius and $l \rightarrow$ connectivity arm length for a smaller value of r "e" will be $0....$ and the arc of a circle described by the crank is small when compared to large sweeps employing larger crank and smaller connecting arm length. It is necessary to consider "e" from the following aspects (Figure 8.59).

1. Count of weft processed
2. High wear and tear of the loom
3. Beat-up force and effective beat-up
4. Mass of the loom in relation to "e"

8.10.1   Count of weft processed

When processing coarser counts like 2s 4s 6s or jute or waste yarn etc., higher beat-up force required due to the space occupied by weft as it is coarse. This calls for a firm beat-up with heavy force. This is achieved by employing larger sweep for the sley or larger crank and smaller connecting arm.

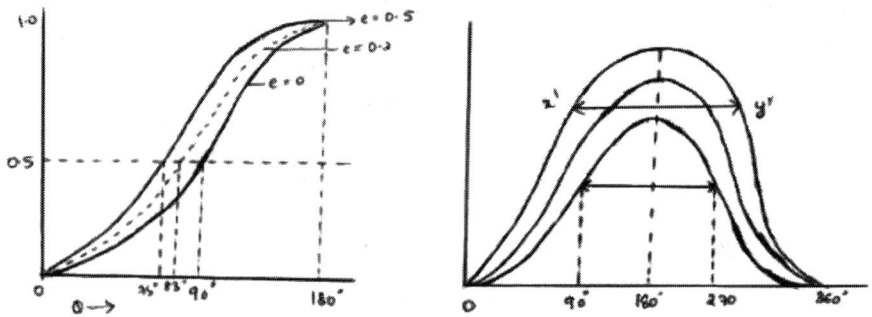

Figure 8.59   Displacement of sley

8.10.2   High wear and tear of the loom

When higher values of e are employed due to larger size of r and smaller l the accelerating force of beat-up increases and this results in accumulated pressure on crank bushes and bearings which cannot sustain for heavy forces. The result is high wear and tear of loom crank. Hence the manufacturers of loom (conventional, semi-auto and auto) do consider the aspects like width of loom, speed of loom with achievable and maximum speed, range of weft's employed, etc., while manufacturing a loom. It is observed that higher values of "e" will give high wear and tear resulting in decreased efficiency of the loom.

8.10.3   Beat-up force and effective beat-up

When warp is divided into 2 layers there exists two tangents T_1, T_2 and the resultant T which will be opposed in magnitude and direction by weaving resistance to have stable fall. This is possible only when firm beat-up is expected. The effectiveness of the beat-up is judged by sizes r and l, by arc of the crank movement and the height or the position of sword pin in relation to crank. In the conventional set-up the sley is rocking forward and backward with respect to rocking shaft which is a fixed point. Hence the distance like sword pin and rocking shaft compared to rocking shaft and race board or sley race will have effect on beat-up.

8.10.4   Mass of the loom in relation to "e"

Consider following example: Loom A is with 1.14 m reed space with 220 rpm and 7 cm crank. Compare to loom B with 5.33 m width and 65 picks/min speed with 108 cm crank length. Let m be the mass of A loom, M be the mass of B loom. We know that force required in accelerating sley is nothing but MV^2

$$\therefore \frac{F_A}{F_B} = \left(\frac{220 \times 7}{65 \times 108} \right)^2 \times \frac{m}{M}$$

$$= 4.8 \, \frac{m}{M}$$

which indicates that a loom B should be at least 4.8 times heavier than loom A to have a crank size of 108 cm. The significance of "e" can be further explained as follows

We know that reed displacement $S = r \left\{ (1 - \cos \theta) + \frac{Sin^2 \theta}{2n} \right\}$

$$\frac{1}{n} = \frac{r}{l}$$

If $n = 0$. $S = r (1 - \cos \theta)$ which resembles SHM
i.e., if $e = 0$ long crank arms are required. Sword pin reaching (reed displacement) its half maximum displacement at 90° for $e = 0$ whereas at 83° and 75° for $e = 0.2$ and 0.5 when sley eccentricity is increased effectiveness of beat-up increases and long-time is available for the shuttle passage. In short it can be concluded that for high values of "e" the machine parts and machine construction must be robust. The significance of e can be further understood by considering effect of change of crank length by loom.

Case (i)
Effect of change of crank length: Let mass of sley be 44 kg, Speed of the sley be 1.33 m/s.
Crank rad 6 cm, crank length 40 cm, Accelerating force F is given by

$$F = \frac{mv^2}{0.06} \left(1 + \frac{0.06}{0.4} \right) N$$

$$= 1.49 \text{ KN.}$$

Case (ii)
If the crank is reduced to 0.15 m than accelerating force, $F = 1.8$ KN, % increase in force = 21%. Thus by changing the crank length from 40 cm to 15 cm the force increases by 21% which cause vibration in the loom ports.

Case (iii)

Effect of change of loom speed
Consider loom speed is increased by 20% than the accelerating force in proportion to square of the speed

i.e. $\dfrac{1.8 \times (120)^2}{(100)^2} = 2.59.$

Percentage increase from 1.49 to 2.59

$\dfrac{2.59 - 1.49}{2.59} \times 100 = 42\%$

It is concluded here that by changing the speed and crank length the force increase by about 42%, if shorter cranks are employed. The above description explains the significance of beat up in terms of eccentricity.

Example 4. You are given connect of cum length 37 cm and radius 6 cm. Calculate "*e*". If "*e*" is to reduced by 20% what should be the crank length. If crank length is reduced by 12.5 cm, what is the value of "*e*" by how much % value changes?

Connecting arm length $l = 37$ cm (14.56″), $d = 6$ cm (4.72″)

$e = 2\left\{ l - \sqrt{l^2 - r^2} \right\}$

$= 2\left\{ 14.56 - \sqrt{212 - 5.56} \right\}$

$= 2\left\{ 14.56 - 14.36 \right\}$

$= 0.4$, to reduce "*e*" by 20%, $e = (0.4 - 0.08) = 0.32$

Now $e = 0.32$, $d = 4.72″$, $l = ?$

$e = 2\left\{ l - \sqrt{l^2 - r^2} \right\}$

$0.32 = 2\left\{ l - \sqrt{l^2 - (2.36)^2} \right\}$

$(0.32)^2 = 4\left[\left\{ l - \sqrt{l^2 - (2.36)^2} \right\} \right]^2$

$= 4\left\{ l^2 + l^2 - (2.36)^2 \right\} - 2xl\sqrt{l^2 - (2.36)^2}$

$= 2x2\left\{ l^2 + l^2 - (2.36)^2 - 2l\sqrt{l^2 - (2.36)^2} \right\}$

$= 4\left\{ 2l^2 - (2.36)^2 - 2l\sqrt{l^2 - (2.36)^2} \right\}$

$$= 4\left\{ -(2.36)^2 + l\left[2\left(l - \sqrt{l^2 - (2.36)^2}\right)\right]\right\}$$

$$= 4\left\{-(2.36)^2 + 2lx0.32\right\}$$

$$= 4\left\{-5.57 + 0.64l\right\}$$

$$0.1024 = -22.28 + 2.56\ l$$

$$22.38 = 2.56\ l,\quad l = 8.74''$$

The slay gets its movement from the crankshaft A (Figure 8.60), and the timing of most of the parts is regulated from the crankshaft. Approximate timings for a Lancashire loom weaving plain calico are shown. Starting at the beating up position, when the slay is in its forward position, the slay is full back approximately half a revolution later. About 30° after beat-up the warp shed is full open and remains so for about 120°, after which the warp threads begin to move, the changing threads being level with each other about 90° before heat up. The shuttle carrying the weft does not enter the warp shed until about 120° after beat-up, long after the shed is full open; the shuttle is in the shed for from 90° to 120°, leaving the shed long after the warp threads begin to move for closing the shed. The shed is opened so long before the shuttle can enter, so that beating up takes place on a crossed. And nearly open shed; also the shed begins to close before the shuttle leaves the shed, so as to give longer time for the changing of the warp threads. For easy movement of the warp threads the dwell should be short and the time for changing as long as possible, while for an easy shuttle movement a long dwell is desirable. The movement of the slay is generally described as an eccentric one, the term being used to denote divergence from a simple harmonic motion. The difference is due partly to the circular movement of the slay and to the relative positions of the parts, but mainly to the relatively short connecting arms used for driving it. Figure 8.60 show the typical sley displacement diagrams for a plain loom.

The shorter the arm – relatively to the crank length – the more rapidly does the slay move away from the beating up position and the earlier does it reach its maximum velocity; on the other band, the rate of movement is less when the slay is near the full back position. The relative positions of the crankshaft A (Figure 8.60), the sword pin B, and the rocking shaft C also affect the movement of the lay.

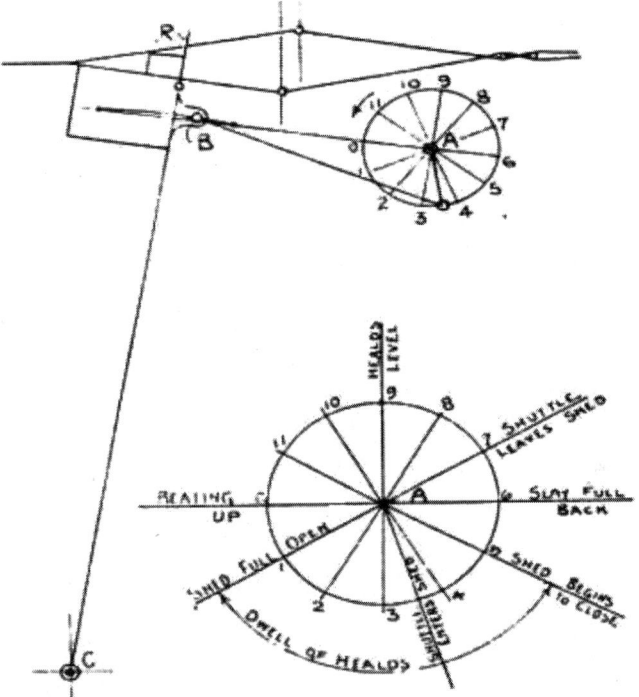

Figure 8.60   Sley drive and loom timing circle (Source: Hanton)

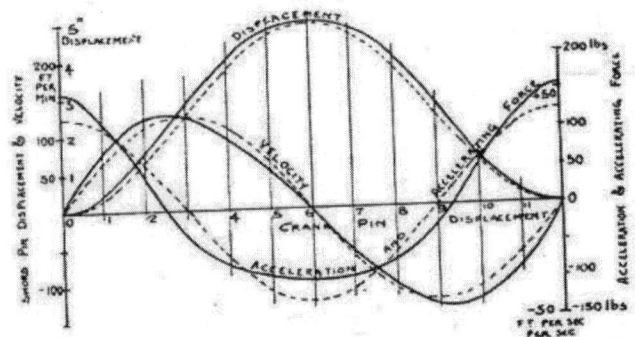

Figure 8.61   Sley displacement diagram in a plain loom (Source: Hanton)

The reed R is not often made to pass the vertical line through the rocking shaft C, so that the force it exerts on the weft when beating up may be approximately horizontal. The crankshaft A should lie on the line produced through the extreme positions reached by the sword pin, although often it is placed higher or lower than this. The position stated gives the most uniform movement to the slay, and the travel at the sword pin is then equal to the diameter of the circle described by the centre of the crank pin. Lifting or lowering the crankshaft increases slightly the travel of the sword pin, as seen in Figure 8.62, which shows the effect of lifting and lowering the crankshaft 4" from the normal position. Lifting the crankshaft makes very little difference during the backward movement of the slay, whilst lowering it makes practically no difference during the forward movement. Figure 8.63 shows the method adopted on a silk loom to enable

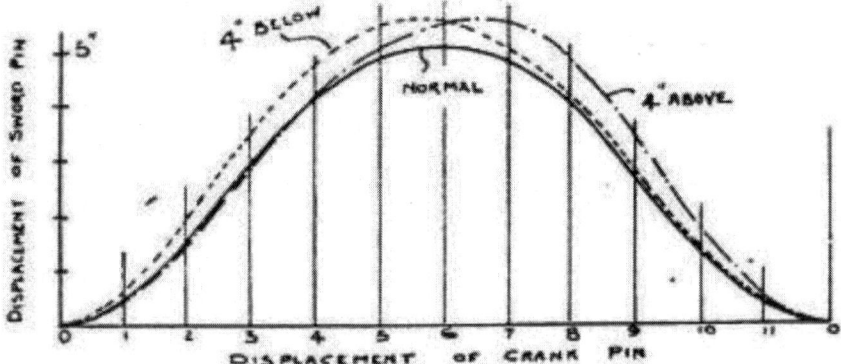

Figure 8.62   Sley eccentricity

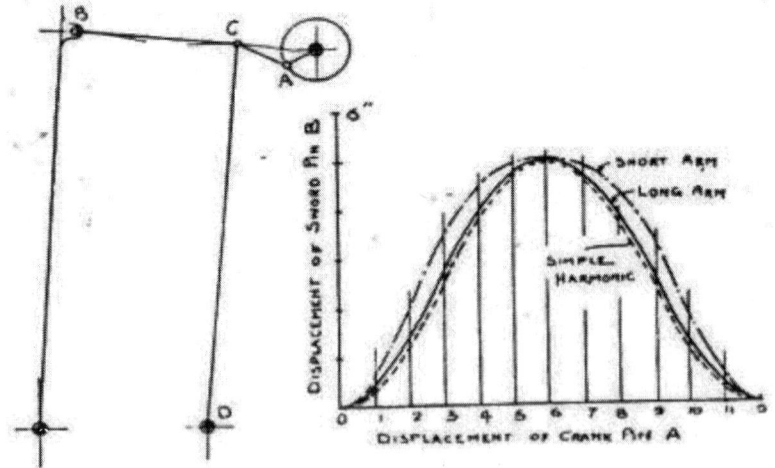

Figure 8.63   Long arm in a silk loom (Source: Hanton)

the arm to be used long or short. As shown the arm length is AC, but by removing the link CD and putting in a clamping pin at C the arm can be lengthened to AB. The effect is shown by the displacement diagrams in Figure 8.63. Another arrangement, where the driving shaft is placed low down in the loom framework, is shown in Figure 8.64, with short arm.

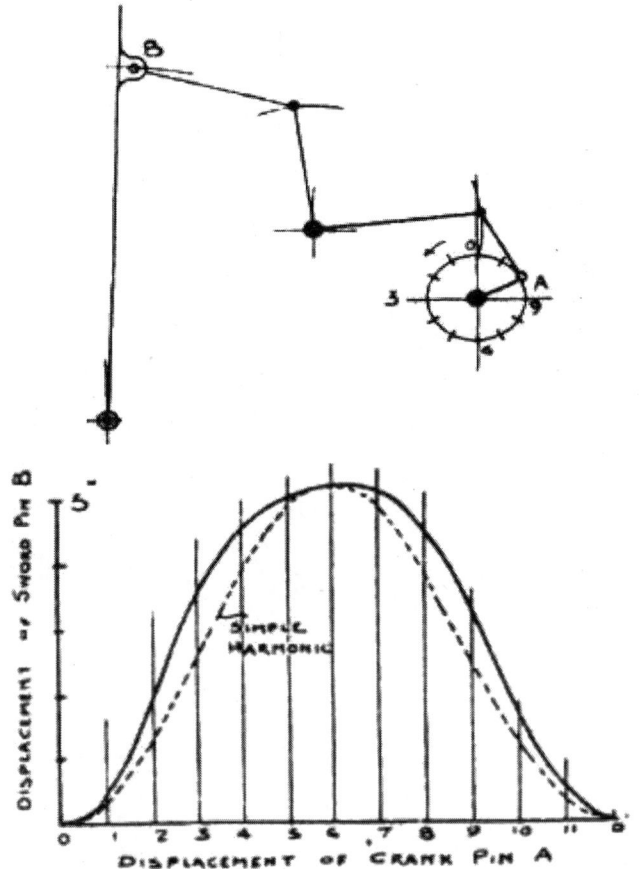

Figure 8.64   Effect of change of sley arrangement (Source: Hanton)

The accelerating force F (Figure 8.65) to be applied at the loom sword pin, tangentially to its path, to accelerate the slay is in direct proportion to the acceleration of the sword pin, and the curve of acceleration represents to some scale the accelerating force F. The value of the force also depends on the moment of inertia of the slay about the rocking shaft, i.e., on the sum of the products of the mass of each particle of the slay and the square of the distance of the particle from the rocking shaft.

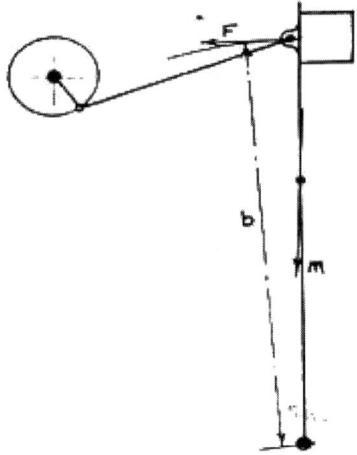

Figure 8.65   Accelerating force of sley (Source: Hanton)

Now the turning moment applied = Moment of inertia × Angular acceleration.

The turning moment applied = $F \times b$

Moment of inertia = mK^2

Where, m = mass of slay and K is radius of gyration about rocking shaft.

$$\text{Angular acceleration of slay} = \frac{= \text{linear acceleration of sword pin}}{b} = \frac{v}{b}$$

Thus $Fb = mK^2 \dfrac{v}{b}$

and $F = \dfrac{mK^2 v}{b^2} = Mv$

where M = equivalent mass at the sword pin, i.e., that mass, which if concentrated at the sword pin and moving with acceleration "v" would require the same accelerating force F as the actual slay requires.

Then $M = m \dfrac{K^2}{b^2}$

Example 5. Weight of slay, m = 124 lb or 56.23 kg; radius of gyration K = 23.8″ or 0.595 m; distance from rocking shaft to sword pin b = 27″ or 67.5 cm; then the equivalent mass $M = 56.23 \times \dfrac{0.595^2}{67.5^2} = 43.53 kg \longrightarrow 96\, lb.$ The radius of gyration of the slay can be calculated, by summing up the weight of each part multiplied by the square of the distance of the part from the rocking shaft; The maximum values of the accelerating force F are reached at the beating up position o, and directly opposite at position 6. These accelerating forces are respectively

$$F_0 = \frac{W}{g} \cdot \frac{v^2}{r} \left(1 + \frac{r}{l} \right)$$

and
$$F_6 = \frac{W}{g} \cdot \frac{v^2}{r} \left(r - \frac{r}{l} \right)$$

where W is the equivalent mass of the slay at the sword pin; g = acceleration due to gravity = 9.81 m/s or 32.2 ft/s;

v = crank pin speed in meter or feet per second;
l = length of connecting arm and
r = length of crank

In addition to the accelerating force F required at the sword pin, force has to be applied to counteract the effect of gravity, but this force is insignificant, in fast and medium speed looms, in comparison with the accelerating forces. In Figure 8.66, G is the position of the centre of gravity of the slay, the weight of the slay, W lb, acting always vertically through G. Then the force F_1 which has to be applied to the sword pin tangentially to its path to overcome gravity, can be found by taking moments about the rocking shaft.

Thus $F_1 \times b = W \times h$ and $F_1 = \dfrac{Wh}{b}$

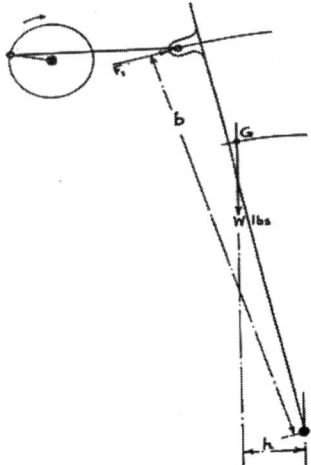

Figure 8.66   Effect of centre of gravity of sley (Source: Hanton)

1. If the angular position of the crank is represented by angle θ when the acceleration of sley is zero (or velocity is maximum) then show that

$$\theta = \cos^{-1} \frac{l\left(\sqrt{1 - 8\left(\frac{r}{l}\right)^2} - 1\right)}{4r}$$

Solution :

Acceleration of sley $= a = r\omega^2 (\cos\theta + e \cos 2\theta)$

$$= r\omega^2 \left(\cos\theta + \frac{r}{l}\left(2\cos^2\theta - 1\right)\right)$$

For acceleration to be zero, $a = 0$

So, $2 r \cos_2 \theta + l \cos\theta - r = 0$

$$\cos\theta = \frac{-l \pm \sqrt{l^2 + 4 \times 2 r \times r}}{4r}$$

Or, $\theta = \cos^{-1} \dfrac{-l \pm l\sqrt{1 + 8\left(\frac{r}{l}\right)^2}}{4r}$

Considering the position sign only,

$$\theta = \cos^{-1} \frac{l\left(\sqrt{1 + 8\left(\frac{r}{l}\right)^2} - 1\right)}{4r}$$

$$= \cos^{-1} \frac{\left(\sqrt{1 + 8 e^2} - 1\right)}{4e}; \quad \left(if \ \frac{r}{l} = e\right)$$

8.10.5   Force, torque and power needed to drive sley

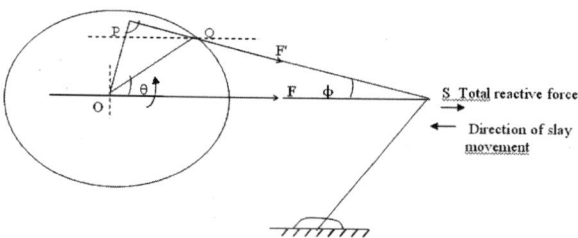

Figure 8.67   Forces acting during sley motion

The various forces acting on the sley is shown in Figure 8.67.
From figure, we have

$$\angle QOS = \theta, \angle QSO = \varphi, \angle OPQ = 90^0, OQ = r,$$

$$\angle OQP = (\theta + \varphi), OP = r\sin(\theta + \varphi)$$

Therefore total reaction force action on sley $= F = Mf$
where $M =$ mass of sley including its accessories and $f =$ sley acceleration.
We know that

$$F = M\omega^2 r\left(\cos\theta + \frac{r}{l}\cos 2\theta\right)$$

Force acting on the crank arm $= F = F\sec\phi$
$\therefore$ Torque at point O is:

$r = F'\ OP$

or, $r = F\sec\theta\, r\sin(\theta + \phi)$

Assuming ϕ as too small in comparison to θ, then

$r = Fr\sin\theta$

$$= M\omega^2 r^2\sin\theta\left(\cos\theta + \frac{r}{l}\cos 2\theta\right)$$

$$= \frac{M\omega^2 r^2}{2}\left\{\sin 2\theta + \frac{r}{l}(2\cos 2\theta\sin\theta)\right\}$$

Or, $r = \dfrac{M\omega^2 r^2}{2}\left\{\sin 2\theta + \dfrac{r}{l}(\sin 3\theta - \sin\theta)\right\}$

Power required drive sley $= P = r\omega$

Or $P = \dfrac{M\omega^2 r^2}{2}\left\{\sin 2\theta + \dfrac{r}{l}(\sin 3\theta - \sin\theta)\right\}$

It is noticed that the power required is proportional to the equivalent
mass of sley, cube of the loom speed and square of the crank radius.

8.10.6   Analysis of motion of various points on the sley

The motion of the points A, B, E and P during sley movement can be
analysed as follows (Figures 8.68 and 8.69).

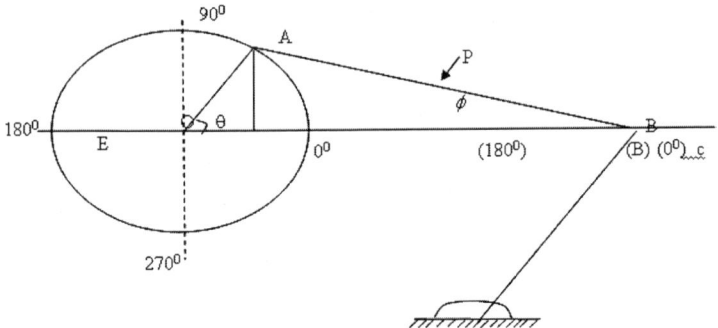

Figure 8.68 Sley geometry

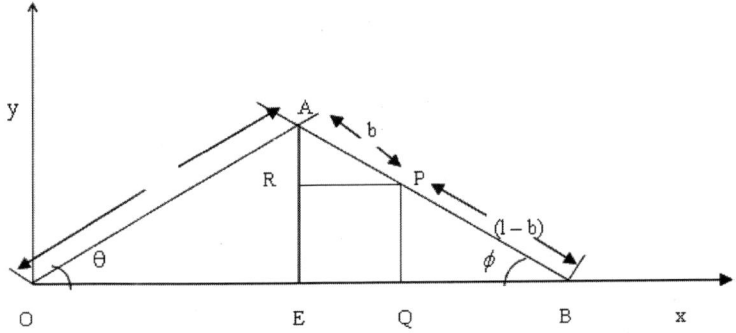

Figure 8.69 Coordinates of crank and crank-arm

Let the crank radius OA be equal to r

Then coordinates of point A is given by $(r \cos \theta, r \sin \theta)$

As locus of the point A describes a circle we can write the x coordinate of point B is

$$x = r \left(1 - \cos \theta\right) + \left(\frac{r^2}{2l}\right) \sin^2 \theta$$

It is to be noted that, projection E of the crankpin on the x axis performs SHM, the motion of the point B is more complicated which deviates from SHM.

Therefore, $AE = r \sin \theta = l \sin \theta$

and $AR = b \sin \phi$

$\therefore RE = AE - AR$

$= (l - b) \sin \phi$

$= (l - b) \dfrac{r}{l} \sin \theta$

At any point P on the axis of the crank arm at the distance b from the crank point A, it can be written as

$$y = PQ = RE = \frac{i-b}{l} r \sin \theta \quad \text{and} \quad x = OQ = OE + EQ$$

or, $x = OE + RP$

or, $x = r \cos \theta + b \cos \theta$

or, $x = r \cos \theta + b \sqrt{1 - \sin^2 \theta}$

or, $x = r \cos \theta + b \sqrt{1 - \left(\frac{r}{1}\right)^2 \sin^2 \theta}$

If, for instance, $r = l$

$$x = (r + b) \cos \theta \quad , \quad y = (r - b) \sin \theta$$

$$\sin^2 \theta + \cos^2 \theta = 1$$

$$or, \left(\frac{x}{r+b}\right)^2 + \left(\frac{y}{r-b}\right)^2 = 1$$

Thus in each point on the axis of the crank arm describes an ellipse with major and minor axis.

8.10.7 A special discussion on sley as four bar linkage mechanism

As already in the beginning of the chapter it is mentioned that sley can be modulated as four bar linkage mechanism and let us consider the following discussion.

Figure 8.70 shows a four-link mechanism, such as, for example, the slay-driving mechanism of the loom. The link AD is fixed, AB rotates about A and CD rocks about D. In the loom the link AD is replaced by the frame work of the machine. If AB rotates at uniform speed, the linear velocity of B can be calculated. The velocity of C can be found most easily by using the instantaneous centre of the link BC. This instantaneous centre is an imaginary point so situated that if the link were fulcrumed there and given a small displacement, its ends would move in the same directions and at the same speeds as in the actual mechanism. Thus if the link AB (Figure 8.71) is moving so that its ends trace out paths as shown, then, when in the position AB, A is moving in the direction AC and B in the direction BD. But if the link were fulcrumed anywhere on the line OAP – the normal to the curve at A – the end A, when displaced, would move in the direction AC. Similarly the end B would move along *BD* if the lever were fulcrumed anywhere on the normal *QBO*. The intersecting point *O* of *PA* and *QB* is the instantaneous centre of link *AB* for the position shown. For if the link were fulcrumed at *O*, as indicated by the dotted arm *OR*, and given a small displacement, the ends *A* and *B* would move along the correct paths. Evidently the position of the instantaneous centre of *AB* will be continually changing; it traces out an imaginary path known as the axode or centrode. Now if the link *AB* were fulcrumed at *O* and displaced, its ends *A* and *B* would move with velocities in proportion to their distances from *O*, i.e., velocity of *A*: velocity of *B* = *AO*: *BO*.

Applying this to the mechanism (Figure 8.72), the end *B* of link *BC* is moving at right angles to *AB*, therefore the instantaneous centre lies somewhere along *AB* or *AB* produced; similarly, since *C* moves at right angles to *CD*, the instantaneous centre must lie along *CD* or *CD* produced. *O* is therefore the instantaneous centre of the link, and the velocity of *C*: velocity of *B* = *OC*: *OB*. But the velocity of *B* is known, therefore the velocity of *C* can be found. Draw *AE* parallel to *CD* and produce *CB* to cut *AE* in *F*. Then the triangle *ABF* is similar to *BCO*, and

$$\frac{AF}{BA} = \frac{OC}{OB} = \frac{velocity\, C}{Velocity\, B}$$

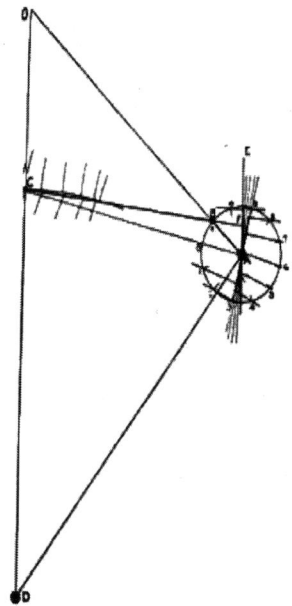

Figure 8.70 Sley as Four bar linkage mechanism

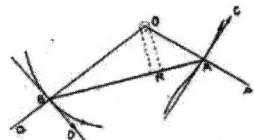

Figure 8.71   Aanalysis of Four bar linkage mechanism

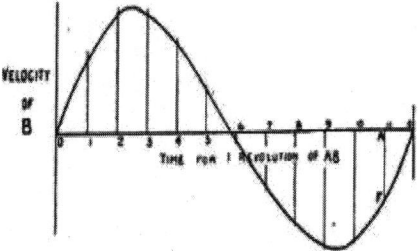

Figure 8.72   Displacement diagram of Four bar linkage mechanism

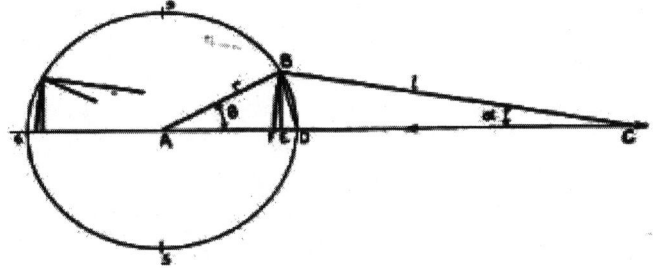

Figure 8.73   Geometry of Four bar linkage mechanism

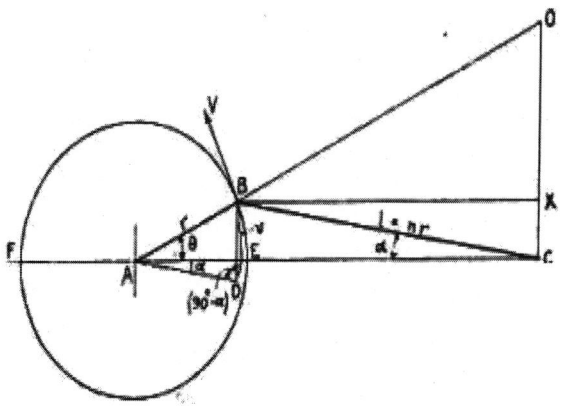

Figure 8.74   Geometry of Four bar linkage mechanism

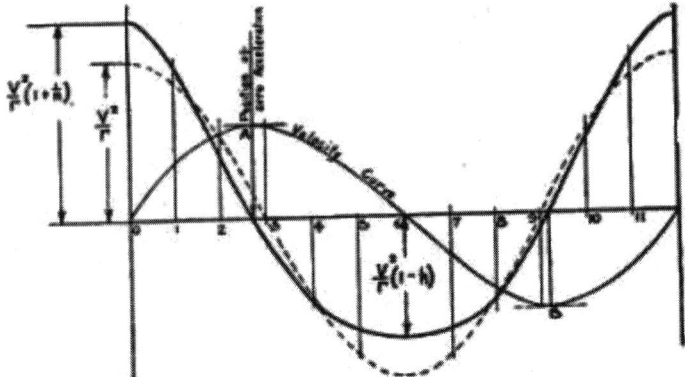

Figure 8.75   Velocity curve in four bar linkage mechanism (All Figures from 8.70 to 8.75)

Thus, if the constant length of the crank is taken to represent the constant velocity of the crank pin B, the velocity of C is represented by the distance from the centre of the crank circle to where the link BC, or BC produced, cuts a line through the centre, parallel to CD. If this construction is repeated for a number of equidistant positions of B and the values of AF plotted as ordinates at the corresponding positions on a time base the velocity diagram for C can be constructed. This is the easiest method of getting the velocity diagram for C and the most accurate graphical method. If the movement of C is small compared to the length of CD, as in the loom mechanism, no great error results if all the intercepts AF are measured on one line, drawn parallel to the mid position of CD. When the pin C moves to and fro in a straight line path passing through the crankshaft centre A, the acceleration of C can readily be calculated for position 0 and 6, and an approximately correct diagram of acceleration can be obtained. Let the crank pin B move through a very small angle θ. With an infinitely long arm C would be displaced a distance DE, but with the short arm the displacement would be DF, where $CF = 1$. The average velocities of C during the movement would be in proportion to these distances DE and DF, since the time is the same in both cases. Since the velocity is zero when the crank pin is at D, these distances DE and DF represent the changes in velocity in equal time intervals, that is, they represent the accelerations with infinite and short arms respectively. Now when θ is very small BD will be approximately at right angles to BA, and the angle DBE will be equal to θ. Also BF will be approximately at right angles to BC and the angle FBE equal to α (Figure 8.73–8.75).

But the angles θ and α are very small, and the cosines of these angles may be taken equal to 1.

Kinetics of take-up and let-off motions

9.1   Role of fabric take-up and warp let-off in loom

The take-up motion has the objective of taking of the cloth as it is woven during the loom working. But it is interesting to note the kinetic aspects like
What force is acting on the warp during unwinding?
What is the force acting on the reed during beat-up?
What is the force offered by the fell during the reed at front centre?
What happens if the teeth on the wheels driving the cloth roller is eccentric?
What happens if the cloth roller diameter is increased?
What happens if all the wheels in take-up motion except change wheel is kept constant etc.
In order to answer all these it is necessary to consider the kinetics of take- and let-off motions. The following subtitles will give the reader the complete picture of the kinetics.

9.2   Mechanics of take-up motion

As mentioned above, the take-up motion controls the rate at which the cloth is drawn forward by the train of wheels as the reed advances. The take-up motion must ensure uniform pick spacing or uniform density of weft depending on the use of positive or negative take-up motion. It is to be noted that both the objectives are achieved only when the warp tension remains constant from D_0 to D_{max}. We know that as the beam weaves down the angle subtended by warp on the back rest will change. Fabric cover will mainly depend on the thread density like the ends per unit area and picks per unit areas in relation to their count. Uniform pick spacing is most desirable in a majority of the cloths produced.

9.3   Beat-up force vs weaving resistance – The cloth fell position

The shed is completely open when crank is at back centre and the beat of the weft inserted by picking takes place when crank or reed is at front centre.

When reed is coming forward i.e., from bottom centre to top centre and intern top centre to front centre, the reed will move fast and during the backward journey it moves slow and we all know this concept as sley eccentricity. In deed the fast forward motion will introduce a force known as beat-up force which is responsible for positioning of the pick inserted. Therefore as the reed moves the newly inserted pick to the fell to position it in the cloth and the reed will experience resistance due to (i) friction between warp and weft (ii) crimp behaviour of warp and weft. Fabric cover factor will play a key role in deciding the pick density and cloth appearance. The reed will encounter lesser resistance when fabrics with low weft covers are produced than with high weft covers. During the weaving of the fabrics with low weft cover, the pick is pushed to its position by reed and cloth fell will not move during beat-up. However during weaving of cloths with normal to high weft cover, the desired pick spacing will not be achieved unless the reed exerts some extra pressure on the fell at beat-up. This pressure is called as beat-up force. This is possible only when the fell offers resistance to displacement and this is known as weaving resistance. Therefore the two factors namely beat-up force and weaving resistance must be equal and opposite for equilibrium conditions or to produce firm cloth. The reason for fell resistance arises from the fact that warp is tensioned by let-off motion.

9.4 Kinetics of take-up motion – warp tension and cloth control

The rate of cloth withdrawal is controlled by the "take-up" motion, and its function is to maintain equal pick spacing throughout the cloth and keep the cloth fell in its correct position. In both conventional and un-conventional weaving uniform pick spacing is most desirable. It is the function of the "let-off" motion to control the warp tension and release equal length of warp during weaving pick by pick. Therefore, both let-off and take-up motions must function in perfect unison to achieve correct pick spacing in all types of fabrics. Fabrics that are made from uniform yarns are liable to be more uniform in density, or the visibility of unevenness in such yarns will be more pronounced. If the yarn is uneven, the variation in pick spacing is not likely to be visible. Fabrics made out of man-made fibres require much care and in case of staple yarns it is a common practice to resort to weft mixing so that any long- or short-term variations in the yarn would be levelled out.

In forming a cloth the warp threads take-up a certain width in the weaving machine, and with shed open a pick of weft is inserted and again the shed closes, the weft is trapped between the wrap threads and beaten up to the fell of the cloth. The length of weft that has gone into the cloth be either equal to the width in reed of the cloth or more. As because, after shed closing both warp and weft threads have to move round each other, the length needed for such configuration will be greater than what has been inserted. The cloth shrinks width wise; at least this phenomenon is visible to the naked eyes. Shrinkage of warp on the machine is not visible, but that does not mean that the warp threads do not shrink – they do shrink. The shrinking in the weft direction is more pronounced due to weft yarn being under low tension than the warp threads. When the weft is woven into the cloth, it will crimp due to closing of the shed. Further, as the warp threads are normally under greater tension, the weft yarns are compelled to take a tortuous path than warp. This crimp will lead to cloth shrinkage, as the length of the weft can only be increased by tension, and this is absent, and the cloth must shrink width wise which in turn results in greater number of ends per unit space in the cloth than what has been in the reed.

During the course of beat-up the reed compels the weft to take the same configuration as the reed width, but under considerable tension. As soon as the reed recedes the weft yarn springs back to its crimped state resulting in cloth shrinkage. If this type of shrinking is continued in the cloth fell region in the subsequent sequences of beat-up the warp threads will be severely abraded against the edge reed wires and weaving cannot proceed due to yarn breaks. Under the action of the temples the weft yarn is under high tension until they clear the region. It has been found that weft crimp in the vicinity of the temples are small. As the cloth moves towards the take-up roller weft crimp gradually increases and the cloth shrinks, and to this the warp crimp gradually reduces, thus a balance is struck between the two sets of threads.

9.4.1   Pick spacing

As the reed moves to push the newly laid weft to the cloth fell a negligible force is required for this purpose when the pick spacing is very large, i.e., weaving low pick cloth. But when normal cloth with greater number of picks per unit space are woven a higher intensity of beat-up will be required for closer pick spacing. The beat-up intensity depends on the cloth fell distance. There is no movement of the cloth fell during beat-up. Unless the reed beats up with greater intensity closer pick spacing cannot be obtained, and in doing so the reed exerts considerable force on the fell and is known as "beat-up force". The reed can exert force on the cloth fell only when it resists displacement and is called "weaving resistance". As the reed approaches the cloth fell, the velocity of the reed decreases. This would mean that the kinetic energy of the sley at the impact of the reed with the fell, and hence the intensity of beat-up, depends on the cloth fell distance. This is termed as the "velocity theory". The intensity depends on the length of period of contact between the reed and the cloth fell. This is termed as "contact theory" and is dependent on the cloth fell distance, and hence the beat-up intensity depends on the cloth fell distance. In the third theory, postulated by Greenwood and Cowhig, we have the "excess tension" theory. It is based on balancing of beat-up force by an excess tension over the fabric tension. This excess tension can happen only when the cloth fell is displaced during beat-up, this displacement will depend on the cloth fell distance and the beat-up intensity (Figures 9.1–9.3).

Let T_1 be the instantaneous warp tension and T_2 be the instantaneous fabric tension and the resultant is the weaving resistance

$$R = T_1 - T_2 \tag{1}$$

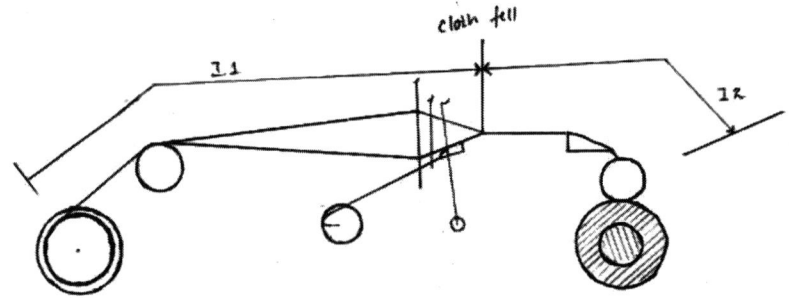

Figure 9.1   The free lengths of warp and woven fabric on roll in the loom

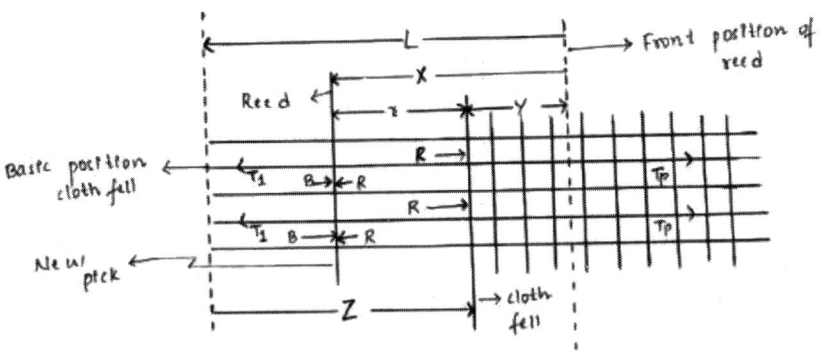

Figure 9.2   The cloth fell during a beat-up cycle

Equation 1 expresses the basis of excess tension theory. R represents both the pressure exerted by the reed and the resistance faced by newly inserted weft by shuttle at any time "t" second. We know that prior to beat-up the weaving resistance is equal to zero or $T_1 = T_2 = T_0$, and T_0 is the basic warp tension determined by the weights of let-off mechanism and the shed height. At this situation the cloth fell position is given by the cloth fell distance L.

As reed beats the new weft the cloth fell moves by distance equal to Z and this will cause the extension in warp by an amount equal to Z and an equal amount of contraction of the fabric. But it is necessary to assume that during beat-up no let-off warp is occurring. The warp tension will raise due to extension of warp by an amount dT_1 and the contraction of the cloth will result in fall of the fabric tension by dT_2.

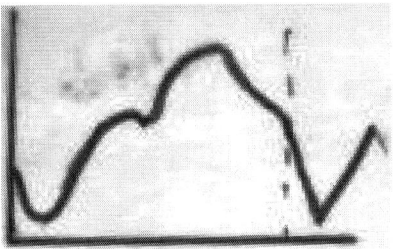

Figure 9.3   Variation in speed of take-up roller

So $dT_1 = E_1 Z/ l_1$ and $dT_2 = E_2 Z /l_2$ (Equation 2 and 3)

where E_1 and E_2 are yarn and fabric moduli and l_1 and l_2 are free lengths of warp and fabric, respectively (refer Figure 9.1). In Figure 9.2 we find the cloth fell during a beat-up along with other forces acting on the cloth fell. The instantaneous warp and cloth tensions during beat-up are then

$$T_1 = T_0 + E_1 Z / l_1 \tag{4}$$
$$T_2 = T_0 - E_2 Z / l_2 \tag{5}$$

Combining above equations with equation 1 we have

$$R = T_0 + E_1 Z / l_1 - T_0 + E_2 Z / l_2$$
$$= Z (E_1 / l_1 + E_2 / l_2) \tag{6}$$

Equation 6 is applicable at any instant during a beat-up and thus it can be said that weaving resistance R is equal and opposite to the beat-up force and proportional to the displacement of the cloth fell from its initial position. The peak beat-up force is therefore proportional to the maximum cloth fell movement which is equal to the cloth fell position. Further it can also be concluded that beat-up force is independent of the basic warp tension as the term T_0 is not seen the final equation i.e., 6. The beat-up force at any instant is equal to the weaving resistance. The latter arises from many factors; the friction between warp and weft, the rigidity and tension of warp and weft, etc. the total weaving resistance. R, can be regarded as the sum of a frictional resistance and an elastic resistance, the difference between the two follows. The energy used in overcoming the frictional resistance is dissipated in heat and, to some extent, in the form of static electricity and, when the new pick has been forced to a point near the cloth fell, the frictional force will tend to keep it there. The energy used in overcoming the elastic resistance is stored in the form of potential energy and when the new pick has been forced to a point near the cloth fell, the elastic resistance will tend to eject it from the fabric. In spite of these differences the elastic and frictional resistances have may common features. Both resistances increase as the new pick approaches the fell and tend to infinity for a finite value of the distance of the new pick from the fell. Both increases with an increase in warp tension. The distinction between elastic and friction a resistance is obviously complex though of great importance and is beyond the scope at the present juncture. It may be assured that the weaving resistance acts as a simple repulsive force between the cloth fell and the new pick. If it is assured that a simple relationship between R and the distance, r, of the new pick exists which could be measured experimentally, we write

$$R = \frac{k}{r - D} \tag{7}$$

This shows that R varies inversely with r, and tends to infinity for a finite value of r, D and k are considered to be constants for a particular fabric. D can be considered the theoretical minimum pick spacing (though the practical minimum is always greater than D), and should be of the order of the yarn diameter, K is termed the co-efficient of weaving resistance and is a measure of the difficulty of weft insertion into a given warp. The value of K will vary with warp tension, speed of the weaving machine, some machine setting or atmospheric condition and on the surface characteristics of the yarns themselves. Equation (7) will be called the "inverse distance equation". It applies at any instant during beat-up and when the reed reaches its front position it becomes

$$R_s = \frac{k}{S - D} \tag{8}$$

This gives the required relationship between beat-up force and pick spacing with S indicating the pick spacing. When the reed is at its front position, $X = 0$ by definition so that $r = S$, and from the geometrical relations at any instant during beat-up, according to Figure 9.4 we find

$$Y = L + Z \tag{9}$$

This is obtained by remembering that in Figure 9.4 the distances L, and Y are measured from the front position of the reed and are therefore negative, so

$Y = L - (+Z)$
or $Y = (L + Z)$, and eventually
$Y = L + Z$

Again, remembering the sign convention

$$r = Y - X \tag{10}$$
$$r = L + Z - X$$

From the equations (9) and (10) it follows that

$$X = L + Z - r \tag{11}$$

An equation represents the relation between the cloth distance and the force of beat-up on the basis of the less tension theory.

$$R_s = (S\,L) \tag{12}$$

Now, the value of R_s in equation (8) is put in equation 12 the following relation is obtained:

$$L = -\frac{k}{S - D} + S \tag{13}$$

where

$$K = k / (E_1/1_1 + E_2/1_2) \tag{14}$$

It is termed as the cloth fell coefficient.

Equation (13) represents the cloth fell equation arrived at on the basis of the combined excess tension theory and the reverse distance equation and implicitly determines the pick spacing in terms of the cloth fell distance. Equation (13) applies irrespective of whether the cloth fell position is correct to give desired pick spacing P or incorrect. When the condition is stable, that is, there is no change in the cloth fell distance from pick to pick so that $S = P$ under stable condition, equation becomes.

$$L_p = \frac{K}{P - D} + P \tag{15}$$

and can be used to determine the correct cloth fell distance for any rate of take-up. It must be remembered that the elastic modules of the fabric E_2 is a function of pick spacing in the majority of the woven fabric. Therefore K cannot be considered as a constant in a rigorous treatment.

9.4.2   Bumping condition

Equation (13) applies only as long as the fabric remains constant at beat-up. When beat-up slackens the fabric, a different treatment is required, because the fabric can offer no resistance to longitudinal compression, with the result that T_2 cannot assume negative value. The fact that the fabric is completely slack at beat-up is easily recognizable by the peculiar noise made when it becomes taut again as the reed recedes. This noise is known to the weavers as "bumping". Though this term is not well-defined in the industry and is often used merely to describe unsatisfactory weaving conditions. But weaving of heavier fabrics may normally lead to such state and is quite normal. Such conditions are reached no sooner a saturated value of weft density appears. This condition can be recognised if one feels the cloth fell with ones finger during running of the machine. The criterion for bumping is

$$T_0 + dT_2 \leq 0 \qquad (16)$$

The value of dT_2 when the reed is at its front position is given by equation (3) and by substituting it in equation (16), we obtain.

$$E_2 Z_S / l_2 \geq T_0 \qquad (17)$$

Substituting Z from equations (6) and (8) combined, we get

$$\frac{k}{(S - D)(E_1 l_2 / E_2 l_1 + 1)} \geq T_0 \qquad (18)$$

Equation (18) applies to the general case where S does not equal P and conditions are unstable. Under such conditions, it is the cloth fell distance which is known rather than pick spacing, and S will have to be derived in terms of L using equation (13). When the conditions are stable equation (18) takes the form

$$\frac{K}{(P-D)(E_1 + l_2 / E_2 l_1 + 1)} \geq T_0 \qquad (19)$$

The above relation shows that bumping conditions can be brought about by decrease in pick spacing, i.e., more picks per unit space. A decrease in pick spacing also leads to a decrease in the value of E_2 and partly cancels this effect. A decrease in the free length of fabric, an increase in free length of warp, a decrease in warp tension, and increase in weft linear density can also lead to bumping. When bumping conditions prevail, equation (5) no longer holds because T_2 always equals zero when the reed is at its front position. Substituting for T_1 and T_2 in equation (1)

$$R_s = T_0 + E_1 Z_S / l_1 \qquad (20)$$

Substituting for Z from equation (11) and remembering that, when the reed is at front position, $X = 0$ and $r = S$.

$$R_S = (S - L) E_1 / l_1 + T_0 \qquad (21)$$

This relation represents the relation between the cloth fell distance and the force of beat-up on the basis of excess tension theory applied to bumping conditions. The relation between the beat-up force and pick spacing is not affected by the onset of bumping and one can substitute for R from the inverse distance equation (8). This gives

$$L = \frac{k/(S-D)-T_0}{E_1/1_1} + S \tag{22}$$

This is the general form of the cloth fell equation under bumping conditions. For stable conditions, when $S = P$, it takes the form

$$L_p = \frac{k/(P-D)-T_0}{E_1/1_1} + P \tag{23}$$

Equations (22) and (23) are the equivalent under bumping conditions of equations (13) and (15). The essential difference between these equations is that in (22) and (23) the fabric term $E_2/1_2$ is eliminated and the warp tension T_0 appears. Thus under bumping conditions, the elastic modulus of the fabric and its free length becomes unimportant, where the warp tension acquires an additional importance. In order to rectify any bumping condition, the basis warp tension is increased so that cloth tension also rises, and under this state the cloth fell can sustain a large displacement without becoming slack Greenwood and Cowhig, in their work, obtained traces of warp and cloth tensions under normal and bumping conditions which are shown in Figure 9.4.

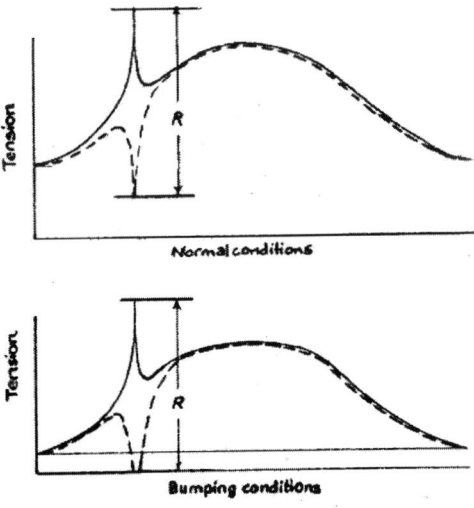

Figure 9.4 Bumping conditions

Both warp and cloth tensions are very substantially equal but for the beat-up period. Traces show that during shed opening the cloth and warp tension rises gradually, and remain constant during the period of shed dwell. Both of them fall gradually with shed closure. During beat-up the warp tension rises sharply and the cloth tension falls almost sharply. The rise in warp tension is due to stretching of warp. The weaving resistance, R, is the difference between the warp and cloth tension, as shown in equation (1) and during this condition the cloth fell does not move and this condition exists just before beat-up. Under normal condition the cloth tension does not fall to zero which is shown in Figure 9.4. When bumping occurs the cloth tension falls below zero tension line and the value of R is reduced showing the effectiveness of beat-up being reduced as depicted in Figure 9.4. To restore the normal weaving state the basic warp tension should be increased so that the cloth tension does not fall below the zero line. Thus for a particular fabric and conditions of weaving there is a value to which the basic warp tension has to be set. This minimum tension will reduce bumping, and it is useless to increase the warp tension unnecessarily which may lead to increased end breakage.

1. The tensile modulus of a woven fabric and corresponding warp sheet is 20 N/cm width and 30 N/cm. If the cloth fell displacement is 5 mm during beat-up and the free length of fabric and warp is 50 cm and 100 cm, respectively, then calculate the weaving resistance. Fabric width is 150 cm. Determine the minimum basic tension to prevent bumping.

Solution :

Here weaving resistance = R
Modulus of fabric and warp sheet = E_f and E_w, respectively.
Length of fabric and warp sheet = L_f and L_w, respectively.
Cloth fell displacement during beat-up = S
Basic warp tension = T

We know that $R = S \left(\dfrac{E_f}{L_f} + \dfrac{E_w}{L_w} \right)$.

$$R \, / \, cm = 0.5 \left(\frac{20}{50} + \frac{30}{100} \right)$$

$$= 0.5 \, (0.4 + 0.3) = 3.5 \text{ N/cm}.$$

So, total weaving resistance = $3.5 \times 150 = 525$ N

Minimum tension to prevent bumping

At limiting condition for bumping the fabric tension will be zero during beat-up.

So, $0 = T - S \dfrac{E_f}{L_f}$

$= T - 0.5 \times \dfrac{20}{50}$

$= T - 0.2$

Or, $T = 0.2$ N/cm.

So, basic tension $= 0.2 \times 150 = 30$ N.

So, weaving resistance is 525 N and basic warp tension to prevent bumping is 30 N.

9.4.3   Disturbed weaving conditions

The positing of the cloth fell does not change from pick to pick during beat-up under normal conditions. Any new pick inserted merely takes up the cloth fell position before take-up motion has actuated. The take-up motion must actuate at the same time and without any variation. Disturbed condition would prevail in case the cloth fell position changes due to some reason. The positive take-up motion being self-correcting, the cloth fell position corrects spontaneously and gradually till the correct fell position is reached. But, in the meantime, faulty cloth will have been produced for the change in the cloth fell position. If the fell moved towards the weaver, the beat-up force will be reduced and weft density is also reduced, producing a thin place in the cloth. On the other hand, if the cloth fell moved away from the weaver, the beat-up force will rise and consequently the pick density will increase thus producing the characteristic thick place in the cloth. The change in pick spacing may be sudden or gradual, and in the former case a visible thin or a thick place will happen which will gradually merge into proper pick spacing. These defects will not be visible in the grey state, but after the fabric is dyed the variation in the shade will be very much apparent. The variations are more apparent in the fabrics woven from uniform and smooth yarns than from yarns with considerable unevenness and hairiness which will conceal these defects. High fabric set will reflect these defects in a more magnified state due to high warp crimp and greater reflecting surfaces. For higher weft density the large change in the cloth fell position takes place and it is just the reverse in case of low weft density fabric. In other words, a slight change in the pick spacing the cloth fell position is affected largely and vice-verse. Thus for producing low weft density fabric any change in the cloth fell position is more visible. This leads to the conclusion that it is difficult to produce evenly distributed weft in low weft density fabrics.

9.4.4   Causes of pick spacing variation

It has been observed that any disturbance in the cloth for position will lead to variation in pick spacing, and if it is severe in nature, it will produce faulty fabric. The variation in pick spacing may be caused by one of the following reasons of combination of two or over all the factor that may contribute to the fault: The take-up motion; The let-off motion; and Machine stoppage (stress relaxation). The variation caused by the first two factors are mainly mechanical though the variation in the warp tension is more pronounced in the second case. The faults caused by the first two factors will be discussed under proper sections.

9.4.5   Stress relaxation in warp and woven fabric

It is a well-known fact that, during a machine stoppage the cloth fell tends to creep away from its correct position, this phenomenon must be regarded as one of the main causes of "setting-on" places. The setting-on place may lead to repping and incorrect pick spacing. But enough is known about the relaxation rates in yarns and fabrics. It would be useful to discuss in a very simple way the picture which will, at least, describe the phenomenon in a qualitative way. It will be assumed that the cloth fell is fixed in its correct position during the stoppage and that, because the difference in the properties of warp and fabric, the tension in the two materials falls from its original value to T_0 a lower value T_1 in the warp and T_2 in the fa where in general $T_1 \neq T_2$. Immediately before starting the machine, the cloth fell is released to find its equilibrium position. Equilibrium again exists when the warp and fell tensions are again equal to a value T_0 and it is necessary to determine the displacement of the cloth fell (dL) which would bring T_1 and T_2 to the same value T_0.

By definition, $dL = dl_1 = -dl_2$ (24)

From Hooke's law, $dT_1 = dl_1 E_1 / l_1$ and $dT_2 = -dl_2 E_2 / l_2$ (25)

It follows that: $dT_1 = dLE_1 / l_1$ and $dT_2 = -dLE_2 / l_2$ (26)

For equilibrium, $T_1 + dT_1 = T_2 + dT_2$ (27)

Substituting from equation (26), it follows that

$$T_1 + dLE_1 / l_1 = T_2 + dLE_2 / l_2$$ (28)

Rearranging, $$dL = \frac{T_2 - T_1}{E_1 / l_1 + E_2 / l_2}$$ (29)

The common warp and fabric tension resulting from this displacement or relaxation is given by

$$T_0 = \frac{T_1 E_2 / l_2 + T_2 E_1 / l_1}{E_1 / l_1 + E_2 / l_2} \qquad (30)$$

Equation (29) describes the effect of differential relaxation of warp and fabric on the cloth fell position. In practice, of course, the tensions in warp and fabric remain equal unless the reed is pressing against the fell during the stoppage. All the quantities on the right hand side of equation (29) can be determined experimentally, the equation enables the drift of the cloth fell to be produced with fair degree of accuracy from tests on the relaxation properties of the warp yarn and fabric. Equations (29) and (30) apply only when a brake type let-off motion is used. In this case it will be seen that, if T_2 is smaller than T_1 i.e., if the fabric relaxes more than the warp, the cloth fell will move away from the weaver. This is the more usual case. Thus the direction of the movement is determined by the relative values of T_1 and T_2 alone. Its magnitude, however, will also depend on the yarn and fabric modulus and, more important still, on the free length of warp and fabric. It will be apparent that the displacement will be greater with longer free lengths of either or both. In comparing the brake type with the dead-weight type of let-off, it is often held in favour of the later that it keeps warp tension constant during a stoppage. This undoubtedly true, but it remains to be seen that how the two types of let-off motion compare with regard to drift of the cloth fell. Assuming that the cloth fell is fixed in its correct position during stoppage, and owing to the action of the dead weight, the warp tension in the present case remains constant at T_0, whereas the fabric tension falls to a value T_2 as before. Immediately before starting the machine, the cloth fell is released and the dead weight brings the fabric tension back to its original value T_0 by stretching the fabric by an amount dL given by:

$$dL = \frac{T_0 - T_2}{E_2 / l_2} \qquad (31)$$

Since the movement of the cloth drift is always away from the weaver, a thick place is always produced by a dead-weight let-off. If we compare equations (31) with (29) we find that, when the fabric relaxes more than the warp, i.e., T_2 is less than T_1, the displacement of the cloth fell will always be grater with a dead-weight than with a brake type let-off motion. Thus it may be said that the constancy of the warp tension in dead-weight let-off is obtained in all practical cases at the expenses of a greater drift of the cloth fell. With a lighter fabric, which will normally be woven under non-bumping condition, the cloth fell position is more important. The drift of the cloth fell will depend on the relaxation behaviour of the warp and the fabric. If the fabric relaxes more than warp the fell will drift away from the weaver and will produce a thick place. On the other hand if the warp relaxation is greater the fell will drift towards the weaver and will consequently produce a thin place in the fabric. It is therefore advisable to leave the reed in contact with the fell and healds level during stops.

9.4.6   Tension control

In positive motions the most common form of tension arrangement consists of back rail which is made to press against the warp sheet as shown in Figure 9.5. X represents the force applied to the warp sheet by the back rail. It will be obvious that if this force remains constant and if the warp tension depends on no other factor then the warp tension must also remain constant; in fact, the conditions are similar to those on the crude machines, the back rail and the means used for applying force X virtually hanging on the warp sheet, as illustrated in Figure 9.5 weight on lever (A) pivoted at (N), applies a downward force (f), to one end of the lever (c) pivoted at (B). At the other end of this lever is mounted the back rail, which is, therefore made to press upwards under the warp sheet with force X. This force can be changed by moving the weight (W) to a different position on the weight lever (A) but for any position X must have a fixed value. In fact, in the simple case illustrated, this force can be calculated

Taking moments about (B) $\qquad X.d = f.c$

$\qquad\qquad\qquad$ Or $\qquad X = (f.c.) / d.$

Again we have $\qquad\qquad\qquad f.b = w.a$

$\qquad\qquad\qquad$ Or $\qquad f = (w.a) / b.$

So that the force, X, exerted by the warp would become

$$w \times \frac{a \times c}{b \times d} + m,$$

where m is the force which would be applied by the weight of the lever and the connecting rod (D). It will be seen later that the swing lever (C) is duplicated at the other end of the machine, the back rail being suspended between the two. These two levers are carried on studs (B) which are bolted to the machine framing. These studs area adjustable in height in order to raise or lower the back rail as required. This adjustment enables the cross sectional shape of the warp sheet to be altered, which is of considerable advantage when weaving under some conditions. But so far as operation of positive let-off is concerned this adjustment is of no importance (Figure 9.6).

9.4.7   Variation in warp tension

It was stated earlier that a constant tension would be obtained with such a system if the tension dependent mainly of force X. This is not really the case in the more common positive motions. There are mainly three sources of variation in the warp tension: Effect of warp beam diameter, Effect at the back rail, and Effects at the pivot of the system.

9.4.8   7 wheel take-up motion

Consider, for example a typical seven-wheel take-up motion, such as is commonly used on cotton looms. The gearing of such a motion is represented in Figure 9.7, and typical sizes for the gears would be as follows

Figure 9.5   7 wheel take-up motion in Rapier

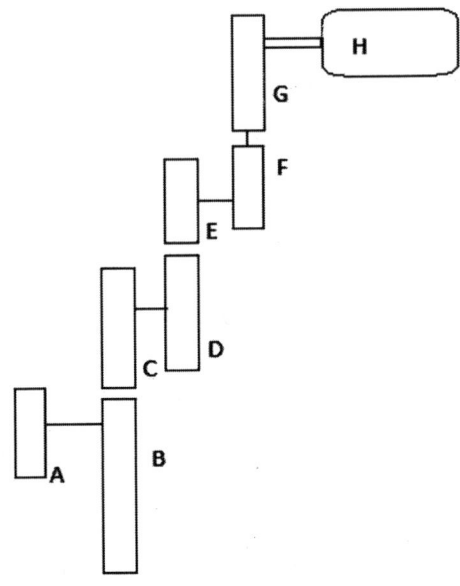

Figure 9.6   Pickle's seven wheel take-up motion

(A)	ratchet wheel, one tooth per pick	24 teeth
(B)	ratchet pinion	36 teeth
(C)	change wheel	
(D)	change pinion	24 teeth
(E)	compound wheel	89 teeth
(F)	compound pinion	15 teeth
(G)	beam wheel	90 teeth
(H)	take-up-roller circumference	15.05 inches.

In this example, the ratchet wheel is turned by one tooth for each pick, and the amount of cloth taken up for each pick may be calculated as follows:

$$\text{Pick-spacing} = \frac{1 \times 36 \times 24 \times 15 \times 15.05\,\text{inches}}{24 \times CW \times 89 \times 90} = \frac{1.015\,\text{inches}}{CW}.$$

where *CW* denotes the number of teeth in the change wheel, and thus,

$$\text{picks/inch} = \frac{\text{teeth in change wheel}}{1.015}$$

Figure 9.7   Fabric take-up motion in a loom

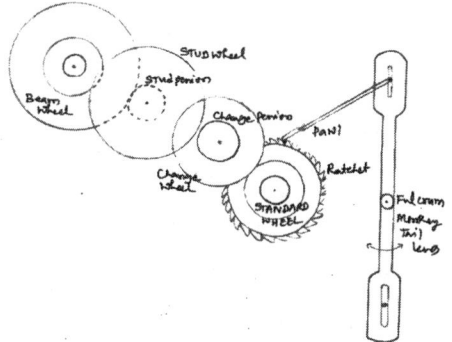

Figure 9.8   Train of wheels of pickles motion

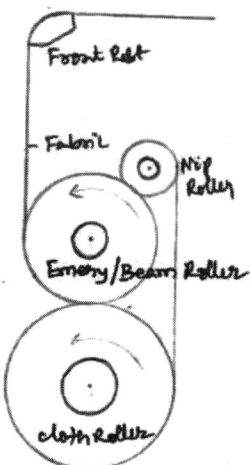

Figure 9.9   The cloth wind up arrangement in take-up motion

Consider now the periodicities that may result from eccentricity. It is obvious that eccentricity of the take-up roller H or the beam wheel G will produce a period having a wavelength equal to the circumference of the take-up roller, which is 15.05″ in this case (Figures 9.8 and 9.9). If either the compound wheel E or the compound pinion F is eccentric, the wavelength of the period will be 15/90 × 15.05 = 2.51″. These and other possible periods are summarised below.

Periods due to eccentricity are:

G or H		15.05″
E or F	15 /90 x 15.05	2.51″
C or D	24 /89 x 2.51	0.68″
A or B	36 /CW x 0.68	24.35 /CW″

The wavelength of the period produced by eccentricity of wheels A or B will be 0.61″ for a cloth with 40 picks /inches and 0.24″ for a cloth with 100 picks /inches. We see that eccentricity of any one of the wheels $A - F$ inclusive will produce a period within the prescribed limits of 1/8 – 10″. Consider now the effect produced if all the teeth on any one of the gear wheels are faulty, owing to faulty design, inaccurate construction, or wear. This will result in a continuous succession of periodicities, each of which will have a wavelength corresponding to one tooth of the wheel. If either of the gears F or G has faulty teeth, the wavelength of the period will be 15.05 /90 = 0.17″. If either of the gears D or E has faulty teeth, the wavelength of the period will be 0.17 × 24 /89 = 0.046″. The various periods due to faulty teeth are summarised as

Periods if all teeth on one wheel are faulty are:

F or G	15.05 /90	0.17″
D or E	15 /89 × 0.17	0.029″
B or C	24 /CW × 0.046	1.104 /CW″
A	36 /24 × 24 /CW × 0.046	1.506 /CW″

Finally, if only one tooth in a gear were faulty, the variation in pick-spacing would extend over a distance corresponding to one tooth of that wheel, and it would recur with a frequency corresponding to one revolution of that wheel. For example, if one tooth in F were faulty, the variation in pick-pacing would extend over a distance of 0.17″ and it would recur every 2.55″. We can see from the above calculations that gears $A - F$, inclusive, are liable to produce dangerous periodicities if eccentric or eccentrically mounted and that faulty teeth in F or G are liable to produce periodicities just within the prescribed limits. The mechanism is clearly not designed to avoid periodicities, but eccentricity is a much more likely cause of faults than faulty teeth.

9.5   Kinetics of let-off

9.5.1   Mechanics of length of yarn delivered during let-off

Derivation of warp beam angular velocity

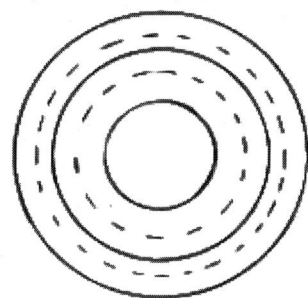

Let "D_o" be the diameter of empty beam "D_n" be the diameter of full beam. Let "d" thickness of warp wound for layers. Let "D" be the diameter of beam at stage in which "l" length of yarn is let-off from the total length "L".

$L_1, L_2, L_3 \ldots\ldots L_n$ be the length of yarn wound along the central line of the respective layers then.

(i) The total length of yarn wound along the central line of 1st layer will

be $l_1 = \pi\left[D_o + \left(\dfrac{d}{2} + \dfrac{d}{2}\right)\right] = \pi[D_o + d]$

(ii) Length of yarn wound along the central line of 2nd layer will be

$l_2 = \pi [D_o + (1\ \frac{1}{2}d + 1\ \frac{1}{2}d)] = \pi [D_o + 3d]$

(iii) Similarly length of yarn at top layer on nth layer is

$l_n = \pi [D_o + (2n - 1)\, d]$

The total length $L = l_1 + l_2 + l_3 + \ldots\ldots\ldots + l_n$

$L = \pi [D_o + d] + \pi [D_o + 3d] + \pi [D_o + 5d] + \ldots\ldots \pi [D_o + 2n = Dd]$

$\quad = \pi D_o n + \pi d [1 + 3 + 5 \ldots..]$

$\quad = \pi D_o n + \pi n^2 d$ 					1

(Because $1 + 3 + 5 \ldots..$ is an odd series and summation of odd series is n^2)

A "D_n" is a dia of full bean with "n" layer we can write

$D_n - D_o = nd + nd = 2nd$

$\quad n = \dfrac{D_n - D_o}{2d}$ 					2

Substitute the value of "n" in equation 1 from equation 2 we have

$$\left[L_n = \pi D_o \left(\frac{D_n - D_o}{2d}\right) + \pi d \left(\frac{D_n - D_o}{2d}\right)^2\right]$$

$$= \pi D_o \left(\frac{D_n - D_o}{2d}\right) + \pi \left(\frac{(D_n - D_o)^2}{4d}\right)$$

$$= \frac{2\pi D_o (D_n - D_o) + \pi (D_n^2 + D_o - 2D_n D_o)}{4d}$$

$$= \frac{2\pi \{D_o D_n - D_o^2\} + \pi D_n^2 + \pi D_o^2 - 2\pi D_n D_o}{4d}$$

$$= \frac{2\pi D_o^2 + \pi D_n^2 + \pi D_o^2 - 2\pi D_o}{4d}$$

$$= \frac{\pi}{4d}(-2D_o^2 + D_n^2 + D_o^2)$$

$$L_n = \frac{\pi}{4d}(D_n^2 - D_o^2)$$

$$L_k = \frac{\pi}{4d}(D_R^2 - D_o^2)$$

Consider the case of a partially filled beam with "k" layers. Length of yarn let-off from kth to nth layer is

$$l_k \Rightarrow L_n - L_K = \frac{\pi}{4d}\left[D_n^2 - D_k^2\right]$$

The fraction of full beam let-off is given by

$$\frac{l_k}{L_n} = \frac{(Dn^2 - D_K^2)}{(D_n^2 - D_o^2)} \Rightarrow L_n(D_n^2 - D_k^2) = l_k(D_n^2 - D_o^2)$$

$$D_n^2 - D_k^2 = \frac{l_k}{L_n}(D_n^2 - D_o^2)$$

$$-D_k^2 = -D_n^2 + \frac{l_k}{L_n}(D_n^2 - D_o^2)$$

$$= D_n^2\left[1 - \frac{lk}{L_n}\left(\frac{D_n^2 - D_o^2}{D_n^2}\right)\right]$$

Replacing K as $\left(\dfrac{D_n^2 - D_o^2}{D_n^2}\right) \Rightarrow D_k^2 = D_n^2\left[1 - K\left(\dfrac{l_k}{L_n}\right)\right]$

In general when a length of sheet "l" has been let-off from total length "L" and diameter of beam is given by

$$D^2 = D_n^2\left[1 - K\left(\dfrac{l}{L}\right)\right] \; orD = D_n\sqrt{\left[1 - K\left(\dfrac{l}{L}\right)\right]}.$$

As the fundamental principle of any positive let-off is that the velocity beam should correspond with diameter of beam at different stages of weaving from full to empty so as to achieve a constant linear velocity. We can write $\omega D = D_n \omega_n$ where ω = angular velocity beam at ω_n = angular velocity of beam at a particular a particular diameter "d", Then

$$\omega = \dfrac{D_n}{D}\omega_n$$

Substitute for $\dfrac{D_n}{D}$, $\omega = \dfrac{\omega_n}{\sqrt{1 - k\left(\dfrac{l}{2}\right)}}$

9.5.2   Effect of beam diameter in let-off

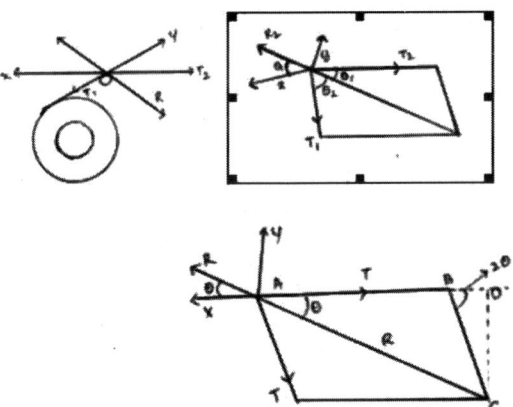

Figure 9.10   Effect of beam diameter in warp let-off

From $\triangle$/e BOC $\dfrac{BO}{BC} = \cos 2\theta$ $BO = T\cos 2\theta$ $AB + BO = 4O$

$$\Delta/e\ AOC\ \frac{AO}{AC} = \cos\theta\ \ AO = R\cos\theta\ \ T + T\cos 2\theta = R\cos\theta\ T(1 + \cos 2\theta) = R\cos\theta$$

$$\frac{X}{R} = \cos\theta \rightarrow R = \frac{X}{\cos\theta}$$

$$T = \frac{X}{2\cos^2\theta}, T_1 = \frac{x}{2\cos^2\theta_1}, T_2 = \frac{x}{2\cos^2\theta_2}$$

$$\frac{T_2}{T_1} = \frac{\cos^2\theta_1}{\cos^2\theta_2},\quad \theta_1 = 52.5,\quad \theta_2 = 37.5$$

$$\frac{T_2}{T_1} \sim 0.6$$

When the beam dia reduces the ten n also reduce (theoretical) (Figure 9.10).

9.5.3   Forces acting at a floating back roller

The case of a freely rotating back roller

Assume that the force F in Figure 9.11 does not vary in direction or magnitude as the beam weaves down. It will not vary in magnitude if the beam-feeler device compensates correctly, but it will vary slightly in direction owing to the gradual fall of the back roller as the beam weaves down. This is neglected in what follows.

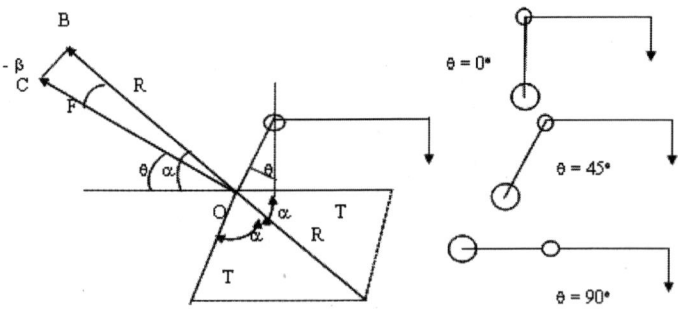

Figure 9.11   A freely rotating backrest

The tension in the warp sheet is given by:

$$T = \frac{R}{2\cos a} \tag{1}$$

We also have

$$A\hat{O}B = \alpha \qquad \text{and} \qquad A\hat{O}C = \theta$$

So that:

$$B\hat{O}C = (a - \theta)$$

and

$$R = \frac{F}{\cos(a - \theta)} \qquad (2)$$

Combining equations (6.3) and (6.4), we have

$$T = \frac{F}{2\cos a \cos(a - \theta)} \qquad (3)$$

Special cases occur when $\theta = 0$ and 90°, when equation (6.5) simplifies as follows:

When $\theta = 0°$

$$T = \frac{F}{2\cos^2 a} \qquad (4)$$

When $\theta = 90°$

$$T = \frac{F}{2\cos a \cos(a - \theta)} = \frac{F}{\sin 2\alpha} \qquad (5)$$

The case of friction at the back roller

This is represented in Figure 9.11, in which $T_2 > T_1$ and $\beta < \alpha$. As in Figure 9.12,

$$R = \frac{F}{\cos(\alpha - \theta)} \qquad (1)$$

$$B\hat{O}C = (\alpha - \theta), \quad A\hat{O}B = \alpha \qquad \text{and} \qquad A\hat{O}C = \theta$$

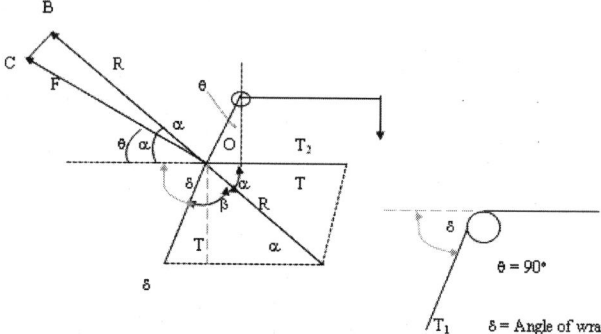

Figure 9.12 Kinetics of back rest

Friction at the back rest

We also have

$$OD = R \sin \alpha = T_1 \sin \delta$$

Hence

$$T_1 = \frac{R \sin \alpha}{\sin \delta} = \frac{F \sin \alpha}{\cos(\alpha - \theta)\sin \delta}$$

But, since $(180 - \delta) = (\alpha + \beta)$ and because $\sin(180 - \delta) = \sin \delta$, we may write equation (6.8) as follows

$$T_1 = \frac{F \sin \alpha}{\cos(\alpha - \theta)\sin(\alpha + \beta)} \tag{3}$$

In any triangle, the length of a side is proportional to the sine of the opposite angle so that:

$$\frac{T_1}{\sin \alpha} = \frac{T_2}{\sin \beta}$$

and

$$T_2 = \frac{T_1 \sin \beta}{\sin \alpha}$$

For which substitute for T_1

$$T_2 = \frac{F \sin \alpha}{\cos(\alpha - \theta)\sin(\alpha + \beta)} \times \frac{\sin \beta}{\sin \alpha}$$

$$T_2 = \frac{F \sin \beta}{\cos(\alpha - \theta)\sin(\alpha + \beta)}$$

i.e.,

9.5.4   Mechanism of negative let-off ($F \alpha X$)

In negative let-off mechanism, the warp tension is felt by operator by his experience. The weight is moved nearly fulcrum point however this procedure is not free from variation in the warp tension. For example assume beam diameter 60 cm and 15 cm of ruffle diameter then long-term variation will be in the 4:1 ratio in addition to it there shall be medium- and short-term variations. The tension variation on warp will have significant effect on geometrical properties of fabric and hence the dimensional stability itself. Consider a negative let-off motion with the following data:

Let the length of the warp be L. Let the distance of the dead weight when at the end of the dead weight lever from the fulcrum be X. Let Y be the distance between the let-off hook in the let-off lever and the fulcrum and is constant for a given loom. Let "T" be the warp tension, "R" be the ruffle radius,

"*F*" be the force (or) tension on tight side, let T_s be the slack side tension, Let "*B*" be the radius of the beam, then taking the moments

$$T \times B = F \times R$$

$$T = \frac{F \times R}{B}$$

The ruffle radius is constant for a given loom, to maintain a constant "*T*". $F \, \alpha \, B$ i.e., when beam 50% in diameter the frictional force must also reduce by 50%,

Further we know that $F = T_t - T_s$

When slip occur $\dfrac{T_t}{T_s} = e^{\mu\theta}$ (1)

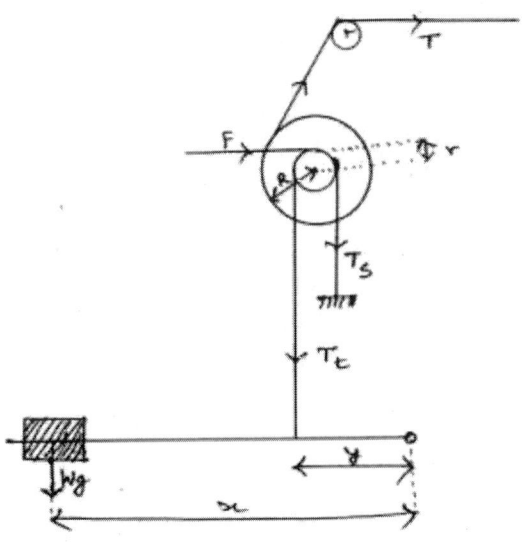

Figure 9.13   Negative let-off

Where "θ" is in radians, the contact angle of warp with back rest. Let "μ" coefficient of friction between chain or rope and ruffle.

Let $e\mu\theta = k_1$ then equation (1) can be written as

$$\frac{T_t}{T_s} = k_1 \qquad (2)$$

By taking the moments about the tight side of the beam (Refer Figure 9.13)

$$T_t = W \times X / Y \qquad (3)$$

$\therefore$ W is constant for and given loom and "Y" is also constant
Let where $k_2 = W / Y$

$$T_t = k_2\ X \qquad (4)$$

We know that differential tension or frictional force $F = T_{t-}\ T_s$

$$F = T_t - \frac{T_t}{k_1} \qquad \text{(from 2)}$$

$$F = T_t\left(1 - \frac{1}{k_1}\right)$$

$$F = k_3 T_t \qquad (5)$$

where $k_3 = 1 - \dfrac{1}{k_1}$

Substituting the value of T_t from 5
$F = k_2 X \times k_3$
$F = k_4\ X$
where $k_4 = k_2 \times k_3$
$F \alpha X$

1. Simple weight lever system at each end of beam is provided with weight of 250 N the leverage of system i.e., x/y ratio is 4 the full beam radius 27 cm. Ruffle $r - 7$ cm rope are given one and half a turnaround ruffle. Note $\mu = 0.15$. Estimate warp tension at the slipping point.

Solution :

$$\frac{T_t}{T_s} = e^{\mu\theta} = e^{0.15 \times 3\pi} = 4.1$$

$$T_t = 250 \times 4$$
$$= 1000 \text{ N}$$
$$T_s = 1000 / 4.1 = 244 \text{ N}$$
$$F = T_t - T_s = 1000 - 244 = 756 \text{ N}$$
Total tension $= 756 \times 2 = 1512 \text{ N}$
W_p tension $= 1512 \times (7/20)$
$$= 529.2.$$

2. Given $W = 30$ lb; $x = 4''$; ruffle diameter $r = 5''$; coefficient of friction μ between chain and ruffle $= 0.15$. When the lap is ½ turn, $\theta = \pi$ and $\dfrac{T_t}{T_s} = 2.718^{0.5\,\pi}$.

Solution :

$$\log \frac{T_t}{T_s} = 0.15\pi \log 2.718$$

$$= 0.471 \times 0.4343 = 0.204$$

and $\quad \dfrac{T_t}{T_s} = 1.6$

Now $\quad T_t = \dfrac{30\,lb.\ x\ y}{4} = 7.5\,y$

Friction $F = T_t - T_s = T_t - \dfrac{T_t}{1.6} = 0.375 T_t = 0.375 \times 7.5\,y = 2.81\,y$

Also the warp tension $\quad P = \dfrac{2Fr}{K} = \dfrac{2 \times 2.81y \times 5''}{K} = \dfrac{28.1y}{K}$

Thus when $y = 12''$ and the beam diameter K is $8''$.

Warp tension $P = 28.1 \times \dfrac{12}{8} = 42$ lb, approximately.

The warp tension P_1 is greater than P because of the friction at the back rest; P_1, by experiment, is about 1.5 P. Also P_2 is greater than P_1 because of friction at lease rods, heald eyes and reed; $P_2 = 2.5\ P$ to 3 P. Cloth tension P_3 at taking up roller $= 3\ P$ to 4 P. The values of the tensions, etc., have been worked out similarly for laps of 1½, 2 ½ and 3 ½ turns, respectively and are tabulated below

Angle of lap		T/T_s	Friction F	Warp tension P	Warp tension when $y = 12$ and $K = 8$	
Turns	θ radians				P lb	$P_3 = 3p$ lb
0.5	π	1.6	2.81 y	28.1 $\dfrac{y}{K}$	42	126
1.5	3π	4.1	5.67 y	56.7 $\dfrac{y}{K}$	85	255
2.5	5π	10.5	6.79 y	67.9 $\dfrac{y}{K}$	102	306
3.5	7π	26.8	7.2 y	72.0 $\dfrac{y}{K}$	108	324

Relation between P and the taking up tension W:

If the cloth roller makes one revolution, then W moves $\pi \times 6''$. The worm S.W. makes 40 revolutions, and P moves downwards $40 \times 2\pi \times 4.5''$. Now, by the principle of work, when friction is ignored, the work done by P = work done on W.

$$P \times 40 \times 2\,\pi \times 4.5 = W \times \pi \times 6$$

$$\therefore W = \frac{p \times 40 \times 2 \times 4.5}{6} = 60\,P$$

Taking the efficiency of the gearing between P and W as 0.25

$$W = 60\,P \times 0.25 = 15\,p.$$

The ruffles of the loom beam are 10" diameter, the chain has 1½ laps and is attached to the lever 4", from its fulcrum. The weight on each lever is 16½ lb, and the weight is 12" from the fulcrum when the beam diameter is 6".

The tight side chain tension $T_t = 16.5 \times \dfrac{12}{4} = 49.5$ say 50 lb.

Taking $\mu = 0.15$, the ratio $\dfrac{T_t}{T_s} = e^{1.5 \times 3\pi} = 4.1$, say 4.

Then $T_s = \dfrac{T_t}{4} = \dfrac{50}{4} = 12.5\ lb.$

Friction at each ruffle = $T_t - T_s = 50 - 12.5 = 37.5$ lb.

$\therefore$ Total friction acting at beam ruffles $= 2 \times 37.5 = 75$ lb and yarn tension at back beam = $75 \times 10/6 = 125$ lb.

The yarn tension at the reed, needed to draw warp-off the beam when the reed is beating up, will be about $2.5 \times 125 = 312$ lb. If $P = 3$ lb then the cloth tension $W = 15\,P = 15 \times 3 = 45$ lb. The tension T between front rest and reed will be less than W, because of the friction at the front rest, T will be about 40 lb. Therefore, pressure between reed and weft = $312 - 40 = 272$ lb, since this force, together with the tension T, must be equal to the 312 lb needed at the reed to overcome the various resistances offered to warp movement between reed and back beam, and the resistance offered by the beam itself.

3. A negative let-off motion uses a total weight of 300 N. The full beam radius is 20 cm, ruffle radius is 8 cm, beam pipe radius is 6 cm, chain takes 1.5 wraps with the ruffle, coefficient of friction between chain and ruffle is 0.14, distance between fulcrum point and chain on tight side is 5 cm. It is required to keep the tension of the warp sheet at 500 N. Determine the position of weight, when (a) the warp beam is full, (b) the warp beam is half and (c) the warp beam is almost empty.

Solution :

From the given data

$W = 300$ N,	$T = 500$N,	$r = 8$ cm,
$x = 5$ cm,	$\mu = 0.015$,	$\theta = 1.5 \times \pi = 9.42$ radian

From equation 4 we have

Distance between fulcrum point and weight

$$y = \frac{T \times R}{W\,r\left(1 - e^{-\mu\theta}\right)} = \frac{500 \times 5 \times R}{300 \times 8 \times \left(1 - e^{0.15 \times 9.42}\right)} = 1.3 \times R$$

When the warp beam is full, $R = 20$ cm,

$\therefore y = 1.3 \times 20 = 26$ cm.

When the warp beam is half, $R = 6 + \dfrac{(20 - 6)}{2} = 13$ cm,

$\therefore y = 1.3 \times 13 = 16.9$ cm

When the warp beam is almost empty, $R = 6$ cm,

$\therefore y = 1.3 \times 6 = 7.8$ cm.

4. The weighting lever for a loom let-off is 50 cm in length, pivoted at one end and supported by a chain placed 10 cm from that end. If the mass of the lever is 2 kg and masses of 8 kg and 12 kg are placed 40 cm and 45 cm, respectively from the fulcrum find the tension in the chain.

Solution :

The mass of the bar may be considered being located at its mid-point i.e., at 25 cm from the fulcrum.

There are thus four forces maintaining equilibrium in the bar, these being:

a) The tension, T, in the chain:

b) The force due to the lever's mass, (2×9.81) N at a distance of 0.25 m from the fulcrum.

c) A force of (8×9.81) N at 0.40 m from the fulcrum, and

d)   A force of (12×9.81) N at 0.45 from the fulcrum.

The tension acts anti-clockwise and the other three forces act clockwise around. The fulcrum.

Taking moments about the fulcrum.

Anti-clockwise moment = $T \times 0.1$

Clock wise moment = $(2 \times 9.81 \times 0.25) + (8 \times 9.81 \times 0.45) + (12 \times 9.81 \times 0.45)$

$$= 4.905 + 31.392 + 52.974$$
$$= 89.271.$$

Then, by principle of moments we can write

$$0.1T = 89.271$$

i.e.,

$$T = 892.71$$

i.e., tension in chain = 892.7 N.

5. A let-off for a loom beam is arranged with a band, fixed to the frame of the loom, is coiled around the weighting ruffle and attached to the weighting lever, which is pulled down by the weight. The ruffle diameter is 15 cm, and the effect of the weight is equivalent to a 100 kg. Load suspended from the band. If the coefficient of friction between the band and the ruffle is 0.18, calculate the work done against friction for one revolution of the beam if identical braking arrangements may be assumed at each of it. In this application, the ruffle is moving against the resistance of the band, so the tension T_2 must be greater than T_1.

Solution :

The tension T_2 is equivalent to a load of 100 kg suspended from the band i.e.,

$$T_2 = 9.81 \times 100 \text{ N}$$
$$= 981 \text{ N}.$$

When slip occurs

$$T_2/T_1 = e^{\mu\theta},$$

i.e.,

$$981 / T_1 = e^{0.18\pi}$$

Since $\theta = \pi$ rad.

Hence

$$T_1 = 557.3 \text{ N}$$

Thus, frictional force at each ruffle = $(981 - 557.3)$ N

and

Total frictional force = 847.4 N.

In one revolution, the ruffle moves $0.15\,\pi$ m, i.e.,

Work done per revolution = $(847.4)\,(0.15\pi)$ J,

i.e.,

Work done = 399.3 J/rev.

6. A simple weight lever based let-off system is attached at each end of a wrapper's beam. It is provided with weights of 400 N at each side. The leverage of the system (y/x ratio) is 5:1. The full beam radius is 50 cm and the ruffle radius is 10 cm. If the ropes are given 1.5 warps around the ruffles and the coefficient of friction is 0.20, determine the warp tension at the slipping point.

Solution :

We know that

$$\text{Warp tension} = T = \frac{r}{R} \times \frac{y}{x} \times W \left(1 - e^{-\mu\theta}\right)$$

Here ruffle radius = r = 10 cm.

Beam radius = R = 50 cm.

$\theta = 2\pi \times 1.5 = 3\pi$

$\mu = 0.2$

$y/x = 5:1$

$W = 400$ N

So, $T = \dfrac{10}{50} \times \dfrac{5}{1} \left(1 - e^{-0.2 \times 3\pi}\right) \times 400$

$= 1 \times (1 - 0.152) \times 400$

$= 399.2$ N

Considering two sides of the loom the warp tension will be $= 399.2 \times 2 = 678.4$ N. So, the warp tension at the slipping point is 678.4 N.

7. A loom is running with negative let of motions. The full and empty diameters of weavers beam is 60 cm and 20 cm, respectively. The weaver does not want tension variation to exceed by 20% during the weaving. How many times the weight has to be shifted during the weaving?
(Source: NPTL)

Solution :

$$\text{Warp tension} = T = \frac{r}{R} \times \frac{y}{x} \times W \left(1 - e^{-\mu\theta}\right)$$

Here full beam diameter = R_1 = 60 cm and empty beam diameter R_2 = 20 cm. Let the final tension in the warp is T_p and starting tension in the warp is T_1. Tension in the warp varies inversely with the weaver's beam diameter.

So, $\dfrac{T_f}{T_1} = \dfrac{R_1}{Rf} = \dfrac{60}{20} = 3$ (if "y" and W are not change during weaving)

So, warp tension will increase by 300%

Permissible increase in tension is 20%.

So, $T_f = 3T_1 = 1.2^n \times T^1$

So, $1.2^n = 3$

Or, $n \times \ln(1.2) = \ln(3) = 0.477$

Or, $n \times 0.079 = 0.477$

Or, $\times = 6$ times.

$*1.2T_1 \times 1.2 \times 1.2$ $\qquad$ nth term $= 3T_1$.

Diameters at which the weight has to be shifted to bring down the tension to T_1 are as follows:

1st change at $60 \times 0.833 = 50$ cm

2nd change at $50 \times 0.833 = 41.67$ cm

3rd change at $41.67 \times 0.833 = 34.72$ cm

4th change at $34.72 \times 0.833 = 28.93$ cm

5th change at $28.93 \times 0.833 - 24.1$ cm

6th change at $24.1 \times 0.833 = 20$ cm

So, the weight has to be shifted six times. The last one may be avoided as the beam has become empty (20 cm).

Construction of cams and tappets

10.1 Introduction

Cams or tappets are the pieces of power transmission which convert rotary reciprocatory motion. In most of the cases of textile machines it is necessary to have reciprocatory motion either for traverse or for any other member of the machine. A number of applications do exits like the traverse for precision winder in double flanged bobbin winding machine, linear traverse for pirn, linear traverse for each section in sectional warping, etc. Cams and Ttappets may be positive or negative in nature. Cam and tappet are indiscriminately used words. Tappets in general are found in looms for shedding and picking mechanisms. On the other hand, cams are found in spinning, weaving, wet processing machines. Cams in certain cases will be the integral part of a mechanism, e.g., cam beat-up, cam driven, heald, cam operated back rest, etc. Cam or tappets are found in followers or anti-friction bowls. Followers may be different types like knife-edged follower or roller follower or plate follower or flat follower or mushroom type follower.

A cam is a mechanical member used to impart desired motion to a follower by direct contact. The cam may be rotating or reciprocating whereas the follower may be rotating, reciprocating or oscillating. Complicated output motions which are otherwise difficult to achieve can easily be produced with the help of cams. Cams are widely used in automatic machines, internal combustion engines, machine tools, printing control mechanisms, and so on. They are manufactured usually by die-casting, milling or by punch-presses. A cam and the follower combination belongs to the category of higher pairs.

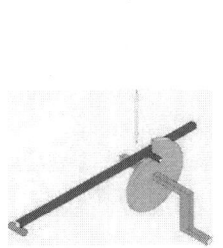
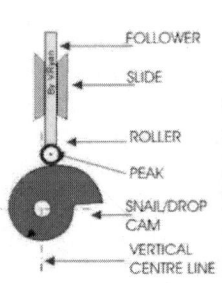
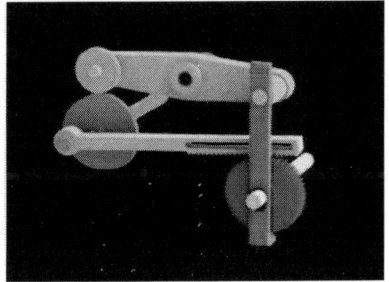

Arrangement of cams in matched cam shedding

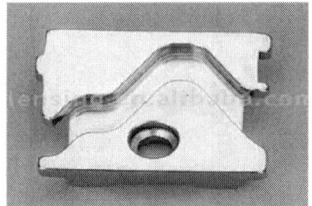

Knit cam

Non-linear cam

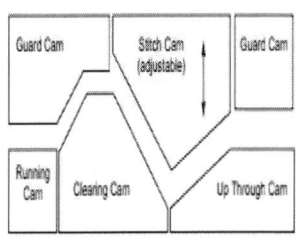

Linear knit cam

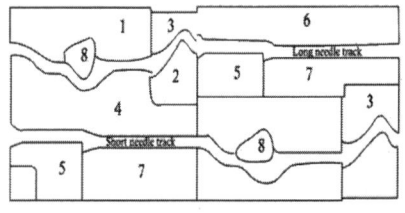

Rib cam

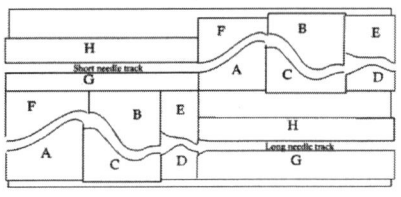

Interlock cam

Cam shedding set

Cam dobby

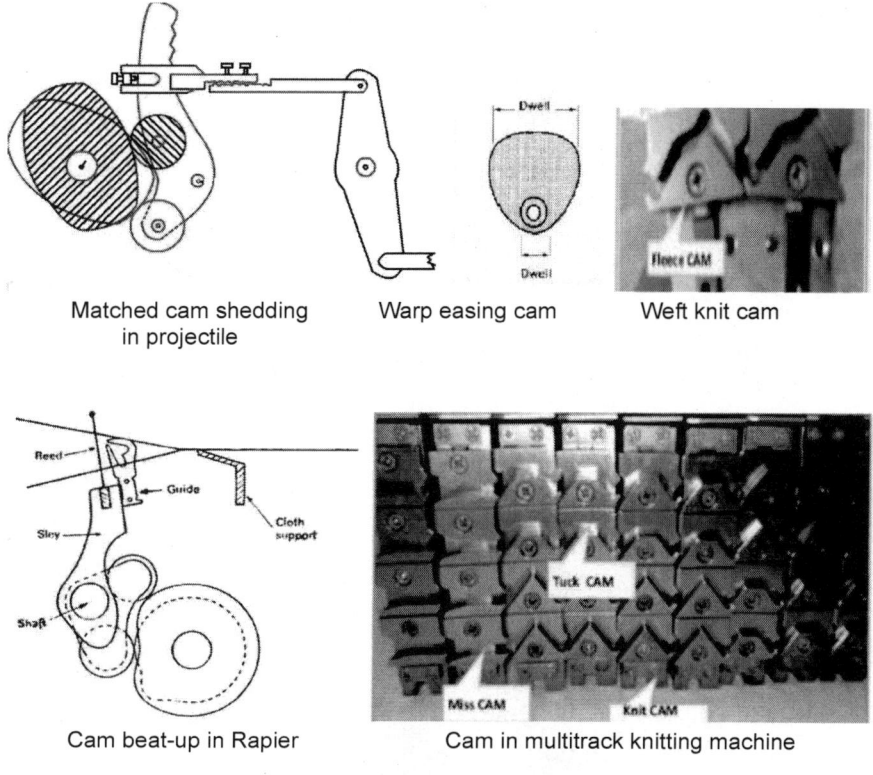

Matched cam shedding
in projectile

Warp easing cam

Weft knit cam

Cam beat-up in Rapier

Cam in multitrack knitting machine

Three-leaved cam for builder motion in ring frame

Figure 10.1   Applications of cams in textile industry

10.2   Applications of cams in textile industry

1. Ring frame – for producing roving build, cop build and combined build
2. Face cam in double flanged bobbin winder for linear traverse
3. Eccentric cam for rocking shaft for lifting the cradle in the case of yarn breakage
4. Sheeting rollers for warping layers before they are wound on to wrapper's beam
5. As eccentric rollers for wavy motion to sized warp prior to their passage through comb in sizing
6. As traversing mechanism for beam press motion
7. As face cam for linear traverse in semi-automatic pirn winder
8. As linear traverse for yarn on pirn in Hacoba auto pirn winder
9. As a set for warp easing motion in loom
10. As matched cam shedding in shuttles looms
11. As cam shedding in modern looms
12. As fabric folding unit in fabric folding machine
13. As a cam for needle motion in weft knitting
14. As a sinker cam for movement of sinkers in circular weft knit machine.
15. As a pair of cams in cam dobby in weaving.
16. As a cam for purl and interlock structures in weft knitting (Figure 10.1).

10.3   Types of cams

Cams are classified according to shape, follower movement, and manner of constraint of the follower.

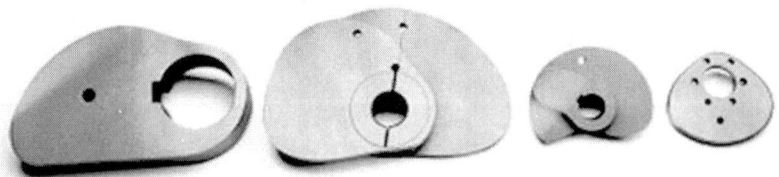

Figure 10.2 Different types of cams

10.3.1   According to shape

Cams can be classified into the following three types based on their shapes. They are explained below.

1. Plate or disk cams: Plate or disk cams are the simplest and most common type of cam. A plate cam is illustrated in Figure 10.2. This type of cam is formed on a disk or plate. The radial distance from the centre of the disk is varied throughout the circumference of the cam. Allowing a follower to ride on this outer edge gives the follower a radial motion.

2. Drum cams: A cylindrical or drum cam is illustrated in Figure 10.2. This type of cam is formed on a cylinder. A groove is cut into the cylinder, with a varying location along the axis of rotation. Attaching a follower that rides in the groove gives the follower motion along the axis of rotation.

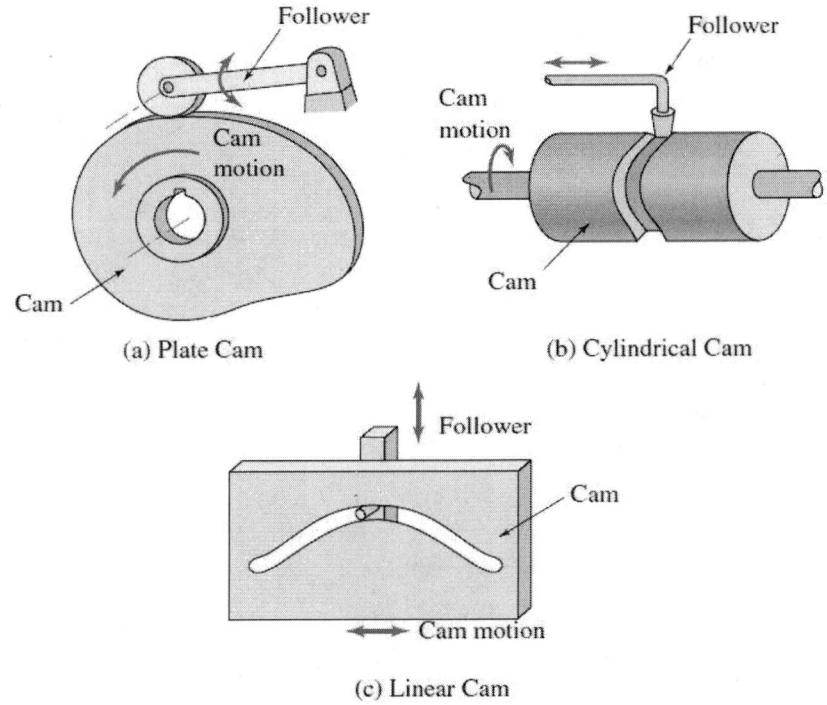

Figure 10.3   Different types of cams based on shape

3. Linear cam: A linear cam is illustrated in Figure 10.a3. This type of cam is formed on a translated block. A groove is cut into the block with a distance that varies from the plane of translation. Attaching a follower that rides in the groove gives the follower motion perpendicular to the plane of translation.

4. **Wedge and flat Ccams:** A wedge cam has a wedge which, in general, has a translational motion. The follower can either translate or oscillate. A spring is, usually, used to maintain the contact between the cam and the follower. In one of the arrangement the cam is stationary and the follower constraint or guide causes the relative motion of the cam and the follower. Instead of using a wedge, a flat plate with a groove can also be used. In the groove the follower is held. Thus a positive drive is achieved without the use of a spring.

5. **Radial or disc cams:** A cam in which the follower moves radially from the centre of rotation of the cam is known as a radial or a disc cam. Radial cams are very popular due to their simplicity and compactness.

6. **Spiral cams:** A spiral cam is a face cam in which a groove is cut in the form of a spiral. The spiral groove consists of teeth which mesh with a pin gear follower. The velocity of the follower is proportional to the radial distance of the groove from the axis of the cam. The use of such a cam is limited as the cam has to reverse the direction to reset the position of the follower. It finds its use in computers.

7. **Cylindrical cams:** In a cylindrical cam, a cylinder which has a circumferential contour cut in the surface, rotates about its axis. The follower motion can be of two types as follows:
   In the first type, a groove is cut on the surface of the cam and a roller follower has a constrained (or positive) oscillating motion. Another type is an end cam in which end of the cylinder is the working surface. A spring-loaded follower translates along or parallel to the axis of the rotating cylinder. Cylindrical cams are also known as barrel or drum cams.

8. **Conjugate cams:** A conjugate cam is a double-disc cam, the two discs being keyed together and are in constant touch with the two rollers of a follower. Thus, the follower has a positive constraint. Such type of cam is preferred when the requirements are low wear, low noise, better control of the follower, high speed, high dynamic loads, etc.

9. **Globoidal cams:** A globoidal cam can have two types of surfaces, convex or concave. A circumferential contour is cut on the surface of rotation of the cam to impart motion to the follower which has an oscillatory motion. The application of such cams is limited to moderate speeds and where the angle of oscillation of the follower is large.

10. **Spherical cams:** In a spherical cam, the follower oscillates about an axis perpendicular to the axis perpendicular to the axis of rotation of the cam. Note that in a disc cam, the follower oscillates about an axis parallel to the axis of rotation of the cam. A spherical cam is in the form of a spherical surface which transmits motion to the follower.

10.3.2   According to follower movement

The motions of the followers are distinguished from each other by the dwells they have. A dwell is the zero displacement or the absence of motion of the follower during the motion of the cam. Cams are classified according to the motions of the followers in the following ways:

1. **Rise-Return-Rise (R-R-R):** In this, there is alternate rise and return of the follower with no periods of dwells (Figure 10.4). It's use is very limited in the industry. The follower has a linear or an angular displacement.

2. **Dwell-Rise-Return-Dwell (D-R-R-D):** In such a type of cam, there is rise and return of the follower after a dwell (Figure 10.5). This type is used more frequently than the R-R-R type of cam.

3. **Dwell-Rise-Dwell-Return-Dwell (D-R-D-R-D):** It is the most widely used type of cam. The dwelling of the cam is followed by rise and dwell and subsequently by return and dwell as shown in Figure 10.6. In case the return of the follower is by a fall (Figure 10.7), the motion may be known as Dwell-Rise-Dwell (D-R-D).

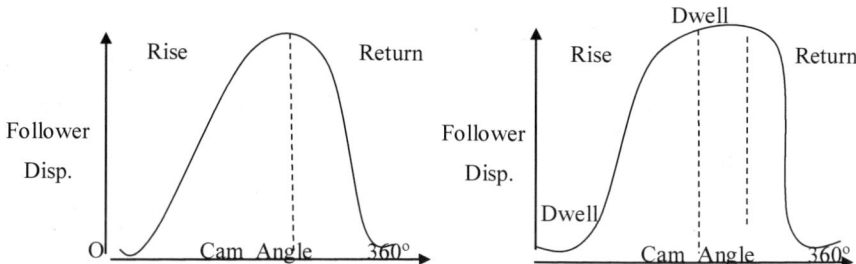

Figure 10.4   Raise Return Raise Cam   **Figure 10.5**   Dwell-Rise-Return-Dwell

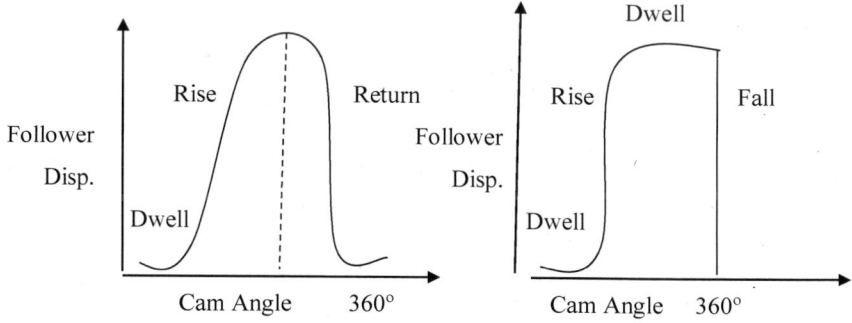

Figure 10.6   Dwell-Rise-Dwell-Return-Dwell   **Figure 10.7**   Dwell-Rise-Dwell

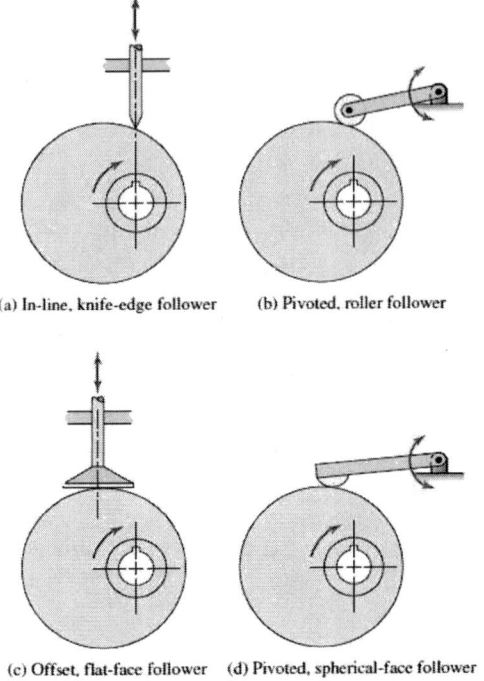

(a) In-line, knife-edge follower   (b) Pivoted, roller follower

(c) Offset, flat-face follower   (d) Pivoted, spherical-face follower

Figure 10.8   Different types of followers

10.3.3   According to manner of constraint of the follower

To reproduce exactly the motion transmitted by the cam to the follower, it is necessary that the two remain in touch at all speeds and at all times. The cams can be classified according to the manner in which this is achieved.

1. **Pre-loaded spring cam:** A pre-loaded compression spring is used for the purpose of keeping the contact between the cam and the follower.

2. **Positive-drive cam:** In this type, constant touch between the cam and the follower is maintained by a roller follower operating in the groove of a cam. The follower cannot go out of the groove under the normal working operations. A constrained or positive drive is also obtained by the use of a conjugate cam.

3. **Gravity cam:** If the rise of the cam is achieved by the rising surface of the cam and the return by the force of gravity or due to the weight of the

cam, the cam is known as a gravity cam. However, these cams are not preferred due to their uncertain behaviour.

10.4   Types of followers

Cam followers are classified according to the shape, movement, and location of line of movement (Figure 10.8).

10.4.1   According to shape

1. **Knife-edge follower:** It is quite simple in construction. However, its use is limited as it produces a great wear of the surface at the point of contact.

2. **Roller follower:** It is a widely used cam follower and has a cylindrical roller free to rotate about a pin point. At low speeds, the follower has a pure rolling action, but at high speeds, some sliding also occurs. In case of steep rise, a roller follower jams the cam and, therefore, is not preferred.

3. **Mushroom follower:** A mushroom follower has the advantage that it does not pose the problem of jamming the cam. However, high surface stresses and wear are quite high due to deflection and misalignment if a flat-faced follower is used. These disadvantages are reduced if a spherical-faced follower is used instead of a flat-faced follower.

10.4.2   According to movement

1. **Reciprocating follower:** In this type, as the cam rotates, the follower reciprocates or translates in the guides.

2. **Oscillating follower:** The follower is pivoted at a suitable point on the frame and oscillates as the cam makes the rotary motion.

10.4.3   According to location of line of movement

1. **Radial follower:** The follower is known as a radial follower if the line of movement of the follower passes through the centre of rotation of the cam.

2. **Offset follower:** If the line of movement of the roller follower is offset from the centre of rotation of the cam, the follower is known as an offset follower.

10.5   Cam nomenclature

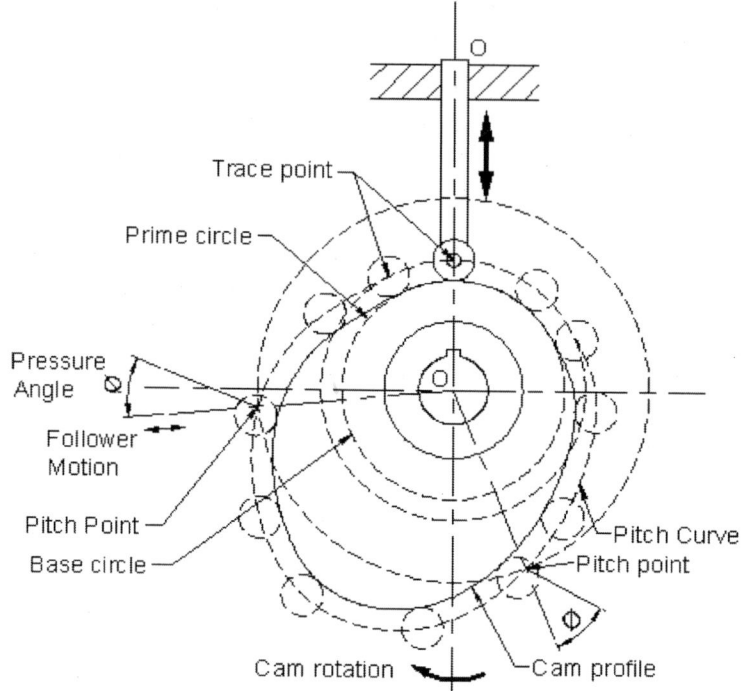

Figure 10.9   Cam nomenclature

Cam profile: Cam profile is outer surface of the disc cam (Figure 10.9).

Base circle: Base circle is the smallest circle, drawn tangential to the cam profile.

Trace point: Trace point is a point on the follower, trace point motion describes the movement of the follower.

Pitch curve: Pitch curve is the path generated by the trace point as the follower is rotated about a stationery cam.

Prime circle: Prime circle is the smallest circle that can be drawn so as to be tangential to the pitch curve, with its centre at the cam centre.

Pressure angle: The pressure angle is the angle between the direction of the follower movement and the normal to the pitch curve.

Pitch point: Pitch point corresponds to the point of maximum pressure angle.

Pitch circle: A circle drawn from the cam centre and passes through the pitch point is called pitch circle.

Stroke: The greatest distance or angle through which the follower moves or rotates.

10.6   Follower displacement programming

As a cam rotates about its axis, it imparts a specific motion to the follower which is repeated with each revolution of the cam. Thus it is enough to know the motion of the follower for only one revolution. The motion of the cam can be represented on a graph, the *x*-axis of which may represent the angular displacements of the cam and the *y*-axis, the angular or the linear displacements of the follower. The follower displacement is measured from its lowest position and is plotted with the same scale as is to be used in the layout of the cam profile. The following terms are used with reference to the angular motion of the cam.

Angle of ascent (φ_a) – It is the angle through which the cam turns during the time of the follower rises.

Angle of dwell (δ) – Angle of dwell is the angle through which the cam turns while the follower remains stationary at the highest or the lowest position.

Angle of descent (φ_d) – It is the angle through which the cam turns during the time the follower returns to the initial position.

Angle of action – Angle of action is the total angle moved by the cam during the time, between the beginning of rise and the end of return of the follower.

To satisfy the given requirements of the follower displacement, a programme can be made keeping in view the following points:

1. In a given specific interval of time, due consideration to the velocity and the acceleration must be given, the effects of which are manifested as inertia loads. The dynamic effects of acceleration, usually, limit the speed of the cams. Moreover, effects of jerks (rate of change of acceleration) in case of high-speed mechanisms produce vibrations of the system, which is undesirable for a follower motion. Though it is very difficult to completely eliminate jerk, efforts are to be made to keep it within tolerable limits.

2. The force exerted by a cam on the follower is always normal to the surface of the cam at the point of contact. The vertical component ($\mathbf{F} \cos \alpha$) lifts the follower whereas the horizontal component ($\mathbf{F} \sin \alpha$) exerts lateral pressure on the bearing. In order to reduce the lateral pressure or $\mathbf{F} \sin \alpha$, α has to be decreased which means making the surface more convex and longer (dotted lines). This results in reduced velocity of the follower and more time for the same rise. This also reduces the dwell period for a fixed angle of action. In internal combustion engines, a shorter dwell period means a smaller period of valve-opening, resulting in less fuel per cycle and lesser power production. Thus the minimum value of α cannot be reduced from a certain value. The size of the base circle controls the pressure angle. Increase in the base circle diameter increases the length of the arc of the circle upon which the wedge (the raised portion) is to be made. A short

wedge for a given rise requires a steep rise or a higher pressure angle, thus increasing the lateral force.

10.7   Motions of the follower

Following are some basic displacement programmes.

10.7.1   Simple harmonic motion (SHM)

This is a popular follower motion and is easy to lay out. Let s = follower displacement (instantaneous), h = maximum follower displacement, v = velocity of the follower = acceleration of the follower, θ = cam rotation angle (instantaneous), φ = cam rotation angle for the maximum follower displacement, β = angle on the harmonic circle.

Construction: The follower rises through a distance h while the cam turns through an angle φ. Construct the follower displacement curve a follows.

i)   Draw a semi-circle with cam rise (or fall) as the diameter. This is, usually, known as the harmonic (semi) circle. Divide this semi-circle into n equal arcs (n even).

ii)   Divide the cam displacement interval into n equal divisions.

iii)   Project the intercepts of the harmonic semicircle to the corresponding divisions of the cam displacement interval.

iv)   Join the points with a smooth curve to obtain the required harmonic curve.

Displacement – At any instant, displacement of the follower is given by,

$$s = \frac{h}{2} - \frac{h}{2}\cos\beta$$

$$= \frac{h}{2}(1 - \cos\beta)$$

For the rise (or fall) h of the follower displacement, the cam is rotated through an angle φ whereas a point on the harmonic semicircle traverses an angle π. Thus the cam rotation is proportional to the angle turned by the point on the harmonic semicircle. That is,

$$\beta = \pi\frac{\theta}{\phi}$$

Thus β can be replaced by θ and φ in equation (i) above,

$$S = \frac{h}{2}\left(1 - \cos\frac{\pi\theta}{\phi}\right)$$

The expression is also valid for β more than 90°. In that case $\cos \beta$ or $\cos \pi\theta/\varphi$ becomes negative so that s is again positive and more than $h/2$.

Let $\varpi = Angular\ velocitty\ of\ the\ cam$

$\therefore \qquad \theta = \omega t$

And $\quad s = \dfrac{h}{2}\left(1 - \cos\dfrac{\pi\omega t}{\phi}\right)$

$$v = \frac{ds}{dt} = \frac{h}{2}\frac{\pi\omega}{\phi}\sin\frac{\pi\omega t}{\phi}$$

$$= \frac{h}{2}\frac{\pi\omega}{\phi}\sin\frac{\pi\theta}{\phi}$$

$$v_{max} = \frac{h}{2}\frac{\pi\omega}{\phi}\ at\ \ \theta = \frac{\phi}{2}$$

$$f = \frac{dv}{dt} = \frac{h}{2}\left(\frac{\pi\omega}{\phi}\right)\cos\frac{\pi\omega\,t}{\phi}$$

$$= -\left(\frac{\pi\omega}{\phi}\right)\cos\frac{\pi\theta}{\phi}$$

$$f_{max} = \frac{h}{2}\left(\frac{\pi\omega}{\phi}\right)^2\ at\ \theta = 0$$

Let ϕ_a = Angle of ascent , ϕ_d = Angle of descent , $\delta_{1.2}$ = Angle of dwells

It can be seen normally from the action of the cam that there is an abrupt change of acceleration from zero to maximum at the beginning of the follower motion and also from maximum (negative) to zero at the end of the follower motion when the follower rises. Similar abruption would also be there at the start and end of the return motion. As these abrupt changes result in infinite jerk, vibration and noise, etc., the programme should be adopted only for low or moderate cam speeds.

10.7.2   Constant acceleration and deceleration

In such a follower programme, there is acceleration in the first half of the follower motion whereas it is deceleration during the latter half. The magnitude of the acceleration and the deceleration is the same and constant in the two halves.

Construction

i)   Divide each half of the cam displacement interval into η equal divisions.

ii) Divide half the follower rise into n^2 equal divisions.

iii) Project 1^2 displacement interval to the first ordinate of the cam displacement, 2^2 to the second ordinate, 3^2 to the third and so on.

iv) The second half of the curve is similar to the first half.

Alternatively, divide half of the follower rise at the central ordinate of the cam displacement into η equal divisions. Joining the zero point with the first division gives 1^2 on the first ordinate, with the second division 2^2 on the second ordinate and so on. The equation for the linear motion with constant acceleration f (during the first half of the follower motion) is found as follows:

$$S = v_0 t + \frac{1}{2} ft^2$$

Where v_0 is is the initial velocity at the start of the motion (rise or fall) and is zero in this case.

$$\therefore S = \frac{1}{2} ft^2$$

Or

$$f = \frac{2s}{t^2} = \text{constant}$$

As f is constant during the accelerating period, considering the follower at the midway,

$$s = \frac{h}{2} \text{ and } \quad t = \frac{\phi/2}{\omega}$$

$$\therefore f = \frac{2h/2}{\kappa^2/4\omega^2} = \frac{4h\omega^2}{\phi^2}$$

The velocity is linear during the period and is given by

$$v = \frac{ds}{dt} = \frac{1}{2} \times 2 ft = ft$$

$$= \frac{4h\omega^2}{\phi^2} \frac{\theta}{\omega} (\theta = \omega t)$$

$$= \frac{4h\omega}{\phi^2}\theta$$

The velocity is maximum when θ is maximum or the follower is at the midway i.e., when $\theta = \varphi / 2$

$$v_{max} = \frac{4h\omega}{\phi^2}\frac{\phi}{2} = \frac{2h\omega}{\phi}.$$

During the second half of the follower motion, the follower is decelerated at constant rate so that the velocity reduces to zero at the end. It is observed from that there are abrupt changes in the acceleration at the beginning, midway and the end of the follower motion. At midway, an infinite jerk is produced. Thus, this programme of the follower is adopted only up to moderate speeds.

10.7.3   Constant velocity

Constant velocity of the follower implies that the displacement of the follower is proportional to the cam displacement and the slope of the displacement curve is constant.

Displacement of the follower for the angular displacement θ of the cam is given by

$$s = h\frac{\theta}{\phi} = h\frac{\omega t}{\phi}$$

$$v = \frac{ds}{dt} = \frac{h\omega}{\phi} = \text{Constant}$$

$$f = \frac{dv}{dt} = 0$$

though acceleration is zero during the rise or the fall of the follower, it is infinite at the beginning and end of the motion as there are abrupt changes in velocity at these points. This results in infinite inertia forces and thus is not suitable from practical point of view. A modified programme for the follower motion can be evolved in which the accelerations are reduced to finite values. This can be done by rounding the sharp corners of the displacement curve so that the velocity changes are gradual at the beginning and end of the follower motion. During these periods, the accelerations may be assumed to be constant and of finite values. A modified constant velocity programme is shown in Figure 10.10.

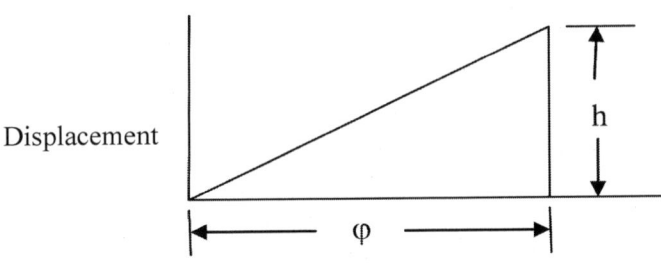

Figure 10.10   Velocity in Cam

10.7.4   Cycloidal

A cycloid is the locus of a point on a circle rolling on a straight line.
Construction

i)   Divide the cam displacement interval into n equal parts (*n* even).

ii)   Draw the diagonal of the diagram and extend it below.

iii)   Draw a circle with the centre anywhere on the lower portion of the diagonal such that its circumference is equal to the follower displacement, i.e., $2\pi r = h$ or $r = h/2\,\pi$.

iv)   Divide the circle into *n* equal arcs and number them as shown in the diagram.

v)   Project the circle points to its vertical diameter and then in a direction parallel to the diagonal of the diagram to the corresponding ordinates.

vi)   Joining the points with a curve gives the required cycloidal. Mathematically, a cycloidal is expressed by

$$s = \frac{h}{\pi}\left(\frac{\pi\theta}{\phi} - \frac{1}{2}\sin\frac{2\pi\theta}{\phi}\right)$$

$$v = \frac{ds}{dt} = \frac{ds}{d\theta}\frac{d\theta}{dt}$$

$$= \left[\frac{h}{\phi} - \frac{h}{2\pi}\frac{2\pi}{\phi}\cos\frac{2\pi\theta}{\phi}\right]\omega$$

$$= \frac{h\omega}{\phi} - \frac{h\omega}{\phi}\cos\frac{2\pi\theta}{\phi}$$

$$= \frac{h\omega}{\phi}\left[1 - \cos\frac{2\pi\theta}{\phi}\right]$$

$$v_{max} = \frac{2h\omega}{\phi} \, at\, \theta = \frac{\phi}{2}$$

$$f = \frac{dv}{dt} = \frac{dv}{d\theta}\frac{d\theta}{dt}$$

$$= \left[\frac{h\omega}{\phi}\frac{2\pi}{\phi} sin\frac{2\pi\theta}{\phi}\right]\omega$$

$$= \frac{2h\pi\omega^2}{\phi^2} sin\frac{2\pi\theta}{\phi}$$

$$f_{max} = \frac{2h\pi\omega^2}{\phi^2} \, at\, \theta = \frac{\phi}{4}$$

It is observed that there are no abrupt changes in the velocity and the acceleration at any stage of the motion. Thus it is the most ideal programme for high speed follower motion.

10.8   Illustrative examples on cam construction

1. Draw the profile of the cam with the following details
Follower is knife edge and its axis passes through cam axis. Outstroke during 60° of cam rotation, dwell for the next 30, return stroke for 60, remaining to follow dwell, the lift of the cam is 40 mm and minimum radius of the cams is 20 mm. Type of motion of follower during outward and return is UV.

Solution :

The displacement diagram is shown in Figure 10.11 and is constructed as follows:

   i)   Draw a horizontal line equal to 14 cm in size.
   ii) Mark first 6 cm as outward rotation and divide this part into six parts.
   iii) Mark the next 1 cm as dwell.
   iv) Mark next 6 cm as return stroke and divide this part also into 6 parts.
   v)   The last 1 cm will be dwell.
   vi) As it is mentioned as UV erect a perpendicular at first zone of 6 cm and divide this also into 6 parts.
   vii) Extend the points as horizontal lines on this lift line so that they cut the vertical lines.
   viii) Mark points where both meet and join to form a smooth curve.
   ix) Procedure for cam profile

x)   Draw the minimum circle first and second circle with radius equals to minimum radius + lift.

xi)   Divide the circle into outward, dwell, return and dwell angles.

xii)   Divide the outward and return portions also in to six parts.

xiii)   Transfer the corresponding points (the height) from displacement diagram and cut the generators (parts) to form the trace points.

xiv)   Join these points to get a pitch curve as the follower is knife- edged.

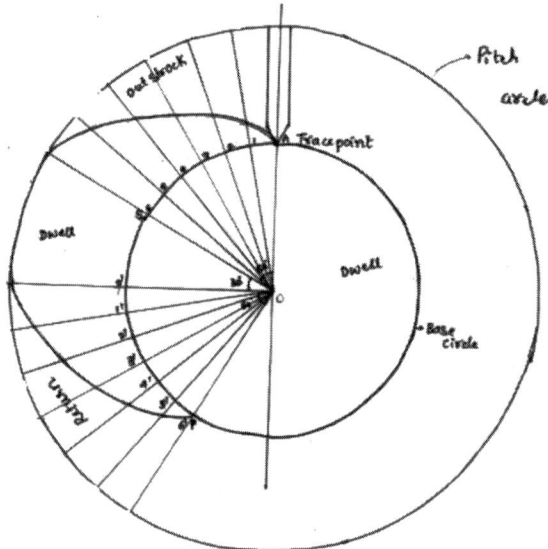

Figure 10.11   Cam profile for the problem number 1

2. Refer to the above problem and draw the profile if the follower is offset by 20 mm.

Solution

The construction of the displacement diagram remains unaltered but the cam profile construction is different and is explained as follows (refer Figure 10.12):

i)   Draw the minimum radius circle.

ii)   Draw the second consecutive circle with 20 mm as radius to indicate the offset circle.

iii)   Draw the last consecutive circle with lift as radius.

iv)   Now mark the position of the follower and divide the circle into outward, dwell, return and dwell potions as explained above.

v)   Divide each portion i.e., outward and return parts into 6 parts and extend the generator lines beyond the lift circle..

vi)   Now draw tangents at these points.

vii) Cut these tangent generator with distances equal to the displacement diagram..

viii) These points are called the trace points and are joined to form a pitch curve.

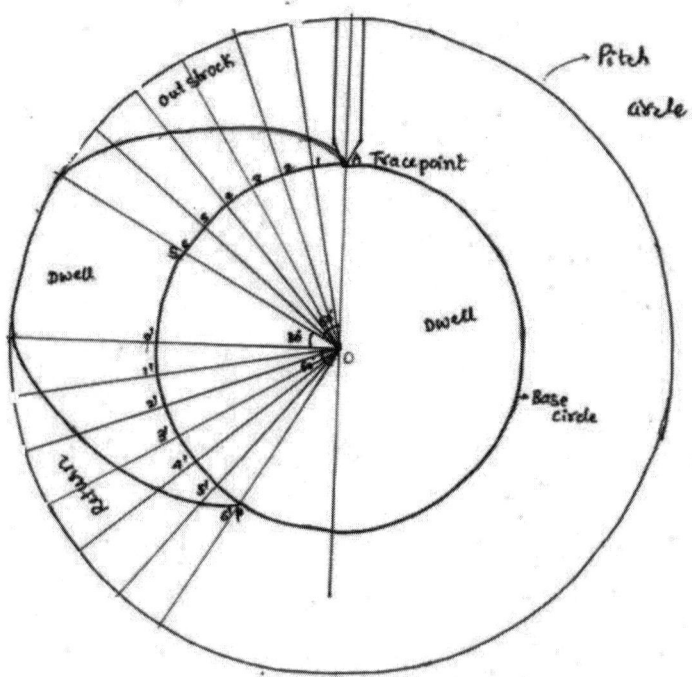

Figure 10.12 Cam profile for the problem number 2

3. Draw the profile of a cam operating a knife-edge follower when the axis of the follower passes through the axis of cam shaft from the following data:

a)   Follower to move outwards through 40 mm during 60° of cam rotation,

b)   Follower to dwell for the next 45°,

c)   Follower to return to its original position during next 90°,

d)   Follower to dwell for the rest of the cam rotation.

The displacement of the follower is to take place with simple harmonic motion during both the outward and the return strokes. The least radius of cam is 50 mm. If the cam rotates at 300 rpm, determine the maximum velocity and acceleration of the follower during the outward stroke and return stroke.

Solution :

Given:

Lift of follower	$S = 40$ mm
Outside angle	$\theta_0 = 60°$
Angle of dwell after outstroke	$\theta_{DO} = 45°$
Angle of return	$\theta_r = 90°$
Angle of dwell after return stroke	$\theta_{DR} = 360 - (60 + 45 + 90) = 165°$
Least radius of cam	$= 50$ mm
Speed of cam	$N = 300$ rpm

$$\therefore \text{Angular velocity of cam} \quad \omega = \frac{2\pi N}{60} = \frac{2\pi \times 300}{60} = 31.42\, rad\,/\,s.$$

First draw a displacement diagram shown in Figure 10.5a13 as per method given below:

i) Choose a suitable scale for displacement diagram. Let horizontal scale is 10 mm = 30° and vertical scale is 10 mm = 10 mm. Now draw horizontal line $AB = 360$. On this line, take $AC = 60°$ to represent outstroke, $CD = 45°$ to represent dwell after outstroke, $DE = 90°$ to represent return stroke and $EF = 165°$ to represent dwell after return stroke.

ii) From A, draw vertical line $AF = 40$ mm to represent the lift (or stroke) of the follower. Complete the angle $AFGB$ as shown in Figure 10.13.

iii) Now draw a semi-circle with AF as diameter. Divide the semi-circle into any equal number of even parts (say in this case six equal parts marked as 1, 2, 3, 4, 5, 6 on semi-circle). Join them with the centre of semi-circle.

iv) Through each division points 1, 2, 3, 4, 5, 6 draw horizontal lines as shown.

v) Divide the distances AC and DE into six equal parts marked as 1, 2, 3, ..., 6 for outstroke and 6', 5', 4', 3', 2', 1' for return stroke. Draw vertical lines through these points as shown in Figure 10.13. These vertical lines intersect the horizontal lines at a, b, c, d, e and f for outstroke and at a', b', c', d', e', f' for return stroke.

vi) Join the points $A, a, b\,c, d, e, f, f', e', ... a', 0', B$ by a smooth curve as shown in Figure 10.5a13. This is the required displacement diagram.

To draw the cam profile

The cam profile is drawn as shown in Figure 10.14 as per method given below: (Assuming the axis of the follower passing through the axis of cam shaft and also assuming that the cam rotates in the clockwise direction).

i) With O as centre and radius equal to 50 mm i.e., minimum radius of the cam, draw a prime circle.

ii) Mark angles of 60°, 45° and 90° on the prime circle to represent the outstroke, dwell and return stroke, respectively such that angle AOC = 60°, angle $COD = 45°$ and angle $DOE = 90°$.

iii) Divide the angles AOC and DOE into the same number of even parts (i.e. six parts) as in displacement diagram.

iv) From O, draw radial lines 0-1, 0-2, 0-3, ... 0-6 for outstroke and 0-6', 0-5',. 0-1' for return stroke. Produce these radial lines as shown in Figure 10.14.

v) Now mark distances 1-a, 2-b, 3-c, ... 6-f and 6'-f', 5'-e', ..., 2'-b', 1'-a' equal to the corresponding displacement as measured from the displacement diagram shown in Figure 10.13.

vi) A smooth curve passing through A, a, b, c, d, e f in Figure 10.14 shows the profile of the cam during the outstroke whereas the curve f', e', d', c', b', a', E shows the profile of the cam during return stroke. For the dwell period, with O as centre and radius equal to of, draw an arc ff' as shown in Figure 10.14 this arc of circle will be parallel to the prime circle.

vii) Since the follower remains at rest for the rest of the revolution, this portion is shown by EA coinciding with the prime circle as shown in Figure 10.14. Now the complete cam profile is shown by thick line in Figure 10.14.

viii) Maximum velocity and maximum acceleration of the follower during outstroke and return stroke.

For simple harmonic motion of the follower, the maximum velocity during outstroke is given by equation and during return stroke is given by equation.

Maximum velocity during outstroke,

$$(V_0)_{max} = \frac{\pi S}{2} \cdot \frac{\omega}{\theta_0}$$

$$= \frac{\pi \times 40}{2} \times \frac{31.42}{\left(60 \times \frac{\pi}{180}\right)} \qquad \left(\because \theta_0 = 60^0 = \frac{60 \times \pi}{180} radian\right)$$

= 1885 mm/s = **1.885 m/s. Ans.**

Maximum velocity during return stroke,

$$\left(V_r\right)_{max} = \frac{\pi S}{2} \times \frac{\omega}{\theta_R}$$

$$= \frac{\pi \times 40}{2} \times \frac{31.42}{\left(90 \times \dfrac{\pi}{180}\right)} \qquad \left(\because \theta_R = 90^0 = \frac{90 \times \pi}{180}\right)$$

= **1256 mm/s = 1.256 m/s.   Ans.**

Maximum acceleration during outstroke from equation,

$$\left(f_0\right)_{max} = \frac{\pi^2 \times S}{2} \times \frac{\omega^2}{\theta_0^2}$$

$$= \frac{\pi^2 \times 40}{2} \times \frac{31.42^2}{\left(60 \, x \, \dfrac{\pi}{180}\right)^2} mm \, / \, s^2$$

= **1,77,470 mm/s² = 177. 47 m/s²   Ans.**

Maximum acceleration during return stroke from equation,

$$\left(f_r\right)_{max} = \frac{\pi^2 \times S}{2} \times \frac{\omega^2}{\theta_R^2}$$

$$= \frac{\pi^2 \times 40}{2} \times \frac{31.42^2}{\left(90 \times \dfrac{\pi}{180}\right)^2} mm \, / \, s^2$$

= **78,876 mm/s² = 78.876 m/s²   Ans.**

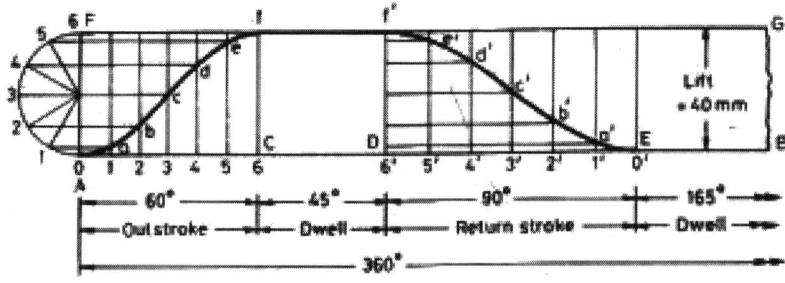

Figure 10.13   Displacement diagram of the cam

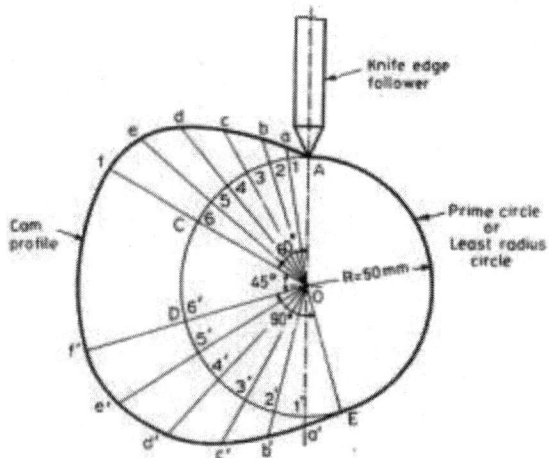

Figure 10.14   Cam profile problem number 3

4. If in above problem the axis of the follower is not passing through the axis of cam shaft but is offset by 20 mm from the axis of the cam shaft, then draw the profile of the cam. All other data remains the same.

Solution :

Given: The data from the above problem is:

$S = 40$ mm, $\theta_0 = 60°$; $\theta_r = 90°$, $\theta_{DO} = 45°$; $\theta_{DR} = 165°$ $R = 50$ mm; $N = 300$ rpm; $\omega = 31.42$ rad/s. First draw the displacement diagram which will be same as shown in Figure 10.15. The curve A-a-b-c-d-e-f-f'-d-'c'-b'-a'-E gives the displacement diagram.

To draw the cam profile when the axis of the follower is offset by 20 mm from the axis of the cam shaft. The cam profile is drawn as shown in Figure 10.15 as per method given below:

i)   With O as centre and radius equal to 50 mm, draw a prime circle.

ii) Draw the axis of the follower at a distance of 20 mm from the axis of the cam, which intersects the prime circle at A.

iii) With O as centre, drawn an offset circle of radius 20 mm.

iv) Join A to O. Make angle $AOC = 60°$ to represent outstroke, angle $COD = 45°$ to represent dwell and angle $DOE = 90°$ to represent return stroke.

v)   Divide the angle AOC and DOE into the same number of even parts (i.e. six parts) as in the displacement diagram.

vi) From points 1, 2, 3, ..., 6 for outstroke and 6',5,'4', ... 1' for return stroke on the prime circle draw tangents to the offset circle and produce these tangents beyond the prime circle as shown in Figure 10.15.

vii) Now mark distance 1-*a*, 2-*b*, 3-*c*, ...6-*f* and 6'-*f*', 5'-*e*', ... 2'-*b*', 1'-*a*' equal to the corresponding displacement as measured from the displacement diagram shown in Figure 10.15.

viii) A smooth curve passing through *A, a, b, c, d, e, f* in Figure 10.15 shows the profile of the cam during the outstroke whereas the *curve f', e', d', c', b', a', E* shows the profile of the cam during return stoke. For the dwell period, with *O* as centre and radius equal to *0-f*', drawn as arc *ff*' as shown in Figure 10.15.

ix) Since the follower remains at rest for the rest of the revolution, this portion is shown by *EA* coinciding with the prime circle as shown in Figure 10.15. Now the complete cam profile is shown by curve *A-a-b-c-d-e-f-f'-e'-d'-c'-b'-a'-E-A*.

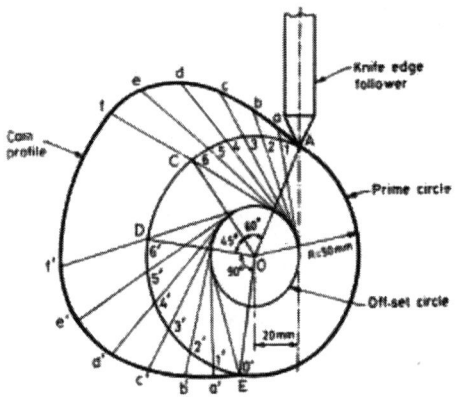

Figure 10.15   Profile of the offset cam

5. Draw the profile of a cam operating a knife-edge follower (when the axis of the follower passes through the axis of the cam shaft) from the following data:

i) Follower to move outward through 30 mm with simple harmonic motion during 120° of cam rotation.

ii) Follower to dwell for the next 60°.

iii) Follower to return to its original position with uniform velocity during 90° of cam rotation.

iv) Follower to dwell for the rest of the cam rotation.

The least radius of the cam is 20 mm and the cam rotates at 240 rpm. Determine:

i) Maximum velocity and maximum acceleration of the follower during outstroke.

ii) Maximum velocity and maximum acceleration of the follower during return stroke.

Solution:

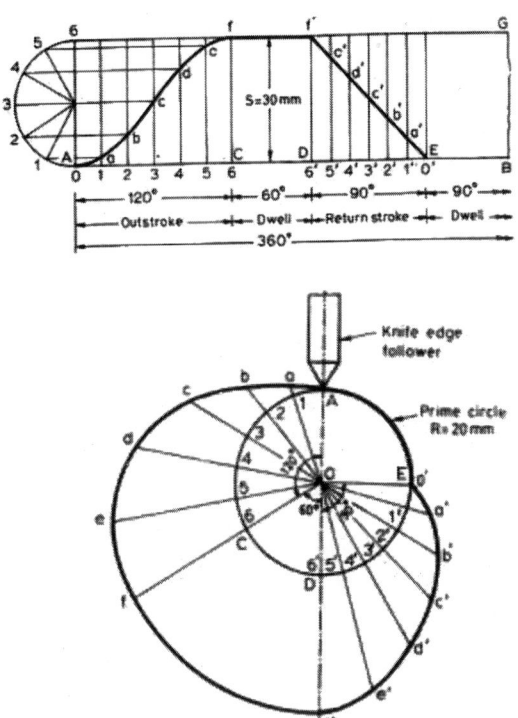

Figure 10.16 Profile of the cam for problem number 5

Solution :

Given:

$S = 30$ mm, $\theta_0 = 120°$; $\theta_{DO} = 60°$; $\theta_r = 90°$; $\theta_{Dr} = 360° - (120 + 60 + 90) = 90°$, least radius of cam, $R = 20$ mm; $N = 240$ rpm or

$$\omega = \frac{2\pi N}{60} = \frac{2\pi \times 240}{60} = 25.14 \; rad / s.$$

First draw the displacement diagram as shown in Figure .10.716. Then the cam profile is drawn as per method given below: (assuming that the cam rotates in the clockwise direction).

i)   With O as centre and radius equal to 20 mm, draw a prime circle.

ii)   Mark angles of 120°, 60° and 90° on the prime circle to represent the outstroke, dwell and return stroke respectively such that angle AOC = 120°, angle $COD = 60°$ and angle $DOE = 90°$.

iii)   Divide the angles AOC and DOE into the same number of even parts (i.e., six parts) as in the displacement diagram.

iv)   From O, draw radial lines 0-1, 0-2, 0-3, ..., 0-6 for outstroke and 0-6', 0-5', ..., 0-1' for return stroke. Produce these radial lines as shown in Figure 10.16.

v)   Now mark distances 1-*a*, 2-*b*, 3-*c*, ..., 6-*f* and 6'-*f*', 5'-*e*', ..., 2'-*b*', 1'-*a*' equal to corresponding displacement as measured from the displacement diagram shown in Figure 10.16.

vi)   A smooth curve passing through *A, a, b c, d, e, f* in Figure 10.16 shows the profile of the cam during outstroke whereas the curve *f*', *e*', *d*', *c*', *b*', *a*', *E* shows the profile of the cam during return stroke. For the dwell period, with O as centre and radius equal to 0-*f* (or equal to 0-*f*') draw an arc *ff*' as shown in Figure 10.16. This arc of circle will be parallel to the prime circle.

vii)   Since the follower remains at rest for the rest of the revolution, this portion is shown by EA coinciding with the prime circle. Now the complete cam profile is shown by thick line in Figure 10.16.

viii) Maximum velocity and maximum acceleration of the follower during outstroke.

During outstroke, the follower moves with simple harmonic motion. The maximum velocity is and maximum acceleration are given by equation

$$(v_0)_{max} = \frac{S}{2} \times \frac{\pi \times \omega}{\theta_0}$$

$$= \frac{30}{2} \times \frac{\pi \times 25.14}{\left(120 \times \dfrac{\pi}{180}\right)}$$

$$\left(\because \quad \theta_0 = 120^0 = \frac{120 \times \pi}{180} \text{ radians } ; \quad \omega = 25.14 rad / s ; \quad S = 30 \text{ mm} \right)$$

$$= \textbf{565.65 mm/s.} \quad \textbf{Ans.}$$

$$(v_0)_{max} = \frac{S}{2} \times \left(\frac{\pi \times \omega}{\theta_0} \right)^2$$

$$= \frac{30}{2} \times \left[\frac{\pi \times 25.14}{\left(120 \times \dfrac{\pi}{180} \right)} \right]^2 = 2130 \text{ mm/s}^2 = \textbf{21.33 mm/s}^2. \text{ \textbf{Ans.}}$$

(ii) Maximum velocity and maximum acceleration of the follower during return stroke

During return stoke, the follower moves with uniform velocity. The velocity is constant during the return stroke.

$$\left(v_r \right)_{max} = v_r = \frac{S}{\theta_r} \times \omega$$

$$= \frac{30}{\left(90 \times \dfrac{\pi}{180} \right)} \times 25.14 \ mm/s \qquad \left(\because \ \theta_r = 90^0 = 90 \times \frac{\pi}{180} \ radius \right)$$

$$= \textbf{480.1 mm/s. \ Ans.}$$

$$\left(f_r \right)_{max} = f_r = 0.$$

6. Draw the profile of a cam operating a knife-edge follower from the following data:

  i) Follower to move outward through a distance of 20 mm during 120° of cam rotation

  ii) Follower to dwell for the next 60° of cam rotation

  iii) Follower to return to its initial position during 90° of cam rotation

  iv) Follower to dwell for the remaining 90° of cam rotation.

   The cam is rotating clockwise at a uniform speed of 500 rpm. The maximum radius of the cam is 40 mm and the line of stroke of the follower is offset 15 mm form the axis of the cam and the displacement of the follower is to take place with uniform and equal acceleration and retardation on both the outward and the return strokes. Determine:

  i) the maximum velocity of the follower during outward and return strokes,

  ii) the maximum acceleration during outward and return strokes.

Solution :

Given:

$S = 20$ mm, $\theta_o = 120°$; $\theta_{DO} = 60°$; $\theta_r = 90°$; $\theta_{Dr} = 360° - (120 + 60 + 90) = 90°$, least radius of cam, $R = 40$ mm; $N = 500$ rpm or; offset of line of stroke of the follower = 15 mm. First draw the displacement diagram as shown in

Figure 10.17. As the displacement of the follower is to take place with uniform acceleration and retardation on both the outstroke and return stroke, hence the displacement curves will be parabolic for both the strokes. (The curve 0-*a*-*b*-*c*-*d*-*e*-*f* is the displacement curve for outstroke whereas curve *f'*-*e'*-*d'*-*c'*-*b'*-*a'*-0' is the displacement curve for return stroke).

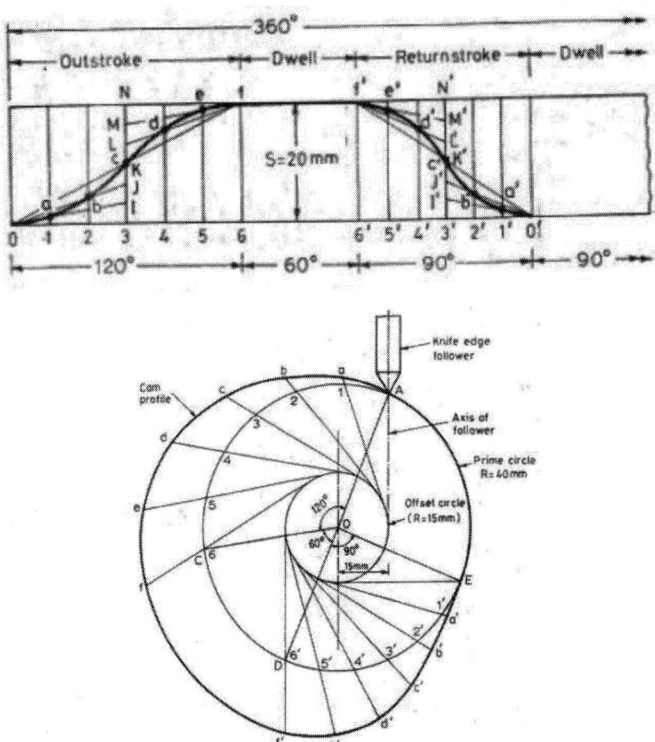

Figure 10.17 Cam profile for the problem number 6

The cam profile is drawn as shown in Figure 10.17 as per method given below:

$$\omega = \frac{2\pi N}{60} = \frac{2\pi \times 500}{60} = 52.36 \; rad \, / \, s.$$

i)   With *O* as centre and radius equal to 40 mm, draw a prime circle.

ii)   Draw the axis of the follower at a distance of 15 mm from the axis of the cam, which intersects the prime circle at *A*.

iii) With *O* as centre, draw an offset circle of radius 15 mm. Join *O* to *A*.

iv)   Make angles *AOC*=120° to represent outstroke, angle *COD* = 60° to represent dwell and angle *DOE* = 90° to represent the return stroke.

v) Divide the angles *AOC* and *DOE* into the same number of even parts (i.e., six parts) as in the displacement diagram.

vi) From   points 1, 2, ..., 5, 6 for outstroke and 6', 5', ..., 2', 1' for return stroke on the prime circle, draw tangents to the offset circle and produce these tangents beyond the prime circle as shown in Figure 10.17.

vii) Now mark distances 1-a, 2-b, 3-c, ..., 6-f and 6'-f', 5'-e', ..., 2'-b', 1'-a' equal to the corresponding displacement as measured from the displacement diagram shown in Figure 10.17.

$$\frac{2 \times \omega \times S}{\theta_r}$$

viii) A smooth curve passing through A, a, b, c, d, e, f in Figure 10.17 shows the profile of the cam during outstroke whereas the curve f', e', d', c', b', a', E shows the profile of the cam during return stroke. For the dwell period, with O as centre and radius equal to 0-f (or equal to 0-f') draw as an arc f-f' as shown in Figure 10.17. This arc of circle will be parallel to the prime circle.

ix) Since the follower remains at rest for the rest of the revolution, this portion is shown by EA coinciding with the prime circle. Now the complete cam profile is shown by thick line in Figure 10.17.

x) Maximum velocity of the follower during the outstroke and return stroke.

xi) The displacement of the follower takes place with uniform acceleration and retardation during outstroke and return stroke.

$\therefore (V_0)_{max}$ = Maximum velocity during outstroke.

$$= \frac{2 \times \omega \times S}{\theta_0}$$

$$= \frac{2 \times 52.36 \times 20}{\left(120 \times \dfrac{\pi}{180}\right)} \qquad \left(\because \ \theta_0 = 120^0 = 120 \times \frac{\pi}{180} \ radius \right)$$

$= 1000$ m/s $= \textbf{1 m/s. Ans.}$

and $(V_r)_{max}$ = Maximum velocity during return stroke

$$= \frac{2S \times \omega}{\theta_r}$$

$$= \frac{2 \times 20 \times 52.36}{\left(90 \times \dfrac{\pi}{180}\right)} \qquad \left(\because \ \theta_r = 90^0 = 90 \times \frac{\pi}{180} \ radius \right)$$

$= 1333.3$ mm/s $= $ **1.333 m/s.   Ans.**

$\therefore (f_0)_{max} = $ Maximum acceleration during outstroke.

$$= \frac{4 \times S \times \varpi^2}{\theta_0^2} = \frac{4 \times 20 \times 52.36^2}{\left(120 \times \dfrac{\pi}{180}\right)^2} = \frac{80 \times 52.36^2}{\left(\dfrac{2\pi}{3}\right)^2}$$

$$= \frac{80 \times 52.36^2}{4\pi^2} = 50,000 \ mm / s^2 = 50 \ m / s^2. \qquad Ans.$$

and (f_r) max $= $ Maximum acceleration during return stroke.

$$= \frac{4 \times S \times \varpi^2}{\theta_r^2} = \frac{4 \times 20 \times 52.36^2}{\left(90 \times \dfrac{\pi}{180}\right)^2}$$

$$= \frac{4 \times 20 \times 52.36^2}{\pi^2} \times 4 = 888889.3 \ \text{mm/s}^2 = 88.8893 \ \text{m/s}^2. \ \text{Ans.}$$

7. A cam with 30 mm as minimum diameter is rotating clock-wise at a uniform speed of 12,000 rpm and has to be given the following motion to a roller-follower 10 mm in diameter:

a)   Follower to complete outward stroke of 25 mm during 120° of cam rotation with equal uniform acceleration and retardation.

b)   Follower to return to its initial position during 90° of cam rotation with equal uniform acceleration and retardation.

c)   Follower to return to its initial position during 90° of cam rotation with equal uniform acceleration and retardation.

d)   Follower to dwell for the remaining 90° of cam rotation.

Draw the cam profile if the axis of the roller-follower passes through the axis of the cam. Determine the maximum velocity of the follower during the outstroke and the return stroke and also the uniform acceleration of the follower on the outstroke and return stroke.

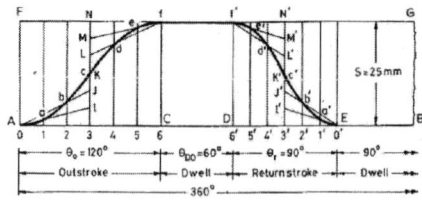

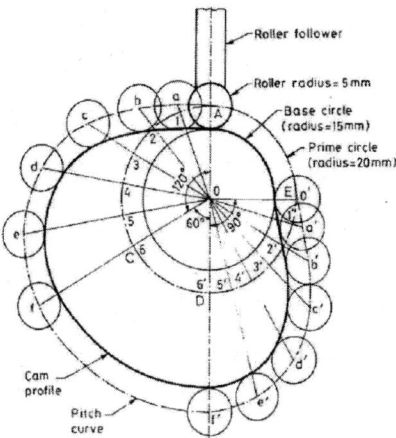

Figure 10.18   Cam profile for problem number 7

Solution :

Minimum dia of cam, D_{min} = 30 mm or
minimum radius of cam, =15 mm;
Speed of cam, N = 1200 rpm.

$$\therefore \text{Angular velocity of cam } \omega = \frac{2 \times \pi \times N}{60} = \frac{2 \times \pi \times 1200}{60} = 125.6\, rad\,/\,s.$$

$$R = \frac{30}{2} \qquad\qquad \frac{10}{2} = 5\, mm.$$

Diameter of roller-follower = 10 mm or radius of roller =
Lift, s = 25 mm
Outstroke angle, θ_0 = 120°
Dwell angle after outstroke, θ_{DO} = 60°
Return stroke angle, θ_r = 90°
Angle of dwell after return stroke, θ_{Dr} = 90°

The outstroke and return stroke has uniform acceleration and retardation.

The displacement diagram is constructed as explained in the previous sections.

Cam profile: The cam profile is drawn as shown in Figure 10.18 as per method given below:

i)   With O as centre, draw a base circle with radius equal to the minimum radius of the cam (i.e., radius = 15 mm).

ii)   With O as centre, draw a prime circle with radius equal to the sum of minimum radius of the cam and radius of roller (i.e., radius = 15 + 5 = 20 mm).

iii) $\therefore$ Radius $OA = 20$ mm.

iv)   Mark angles of 120°, 60° and 90° on the prime circle to represent outstroke, dwell and return stroke, respectively such that angle AOC = 120°, angle COD = 60° and angle DOE = 90°.

v)   Divide the angle AOC and DOE into the same number of even parts (i.e., six parts) as in the displacement diagram.

vi)   From O, draw radial lines 01, 0-2, 0-3, 0-6 for outstroke and 0-6', 0-5', 0-1' for return stroke. Produce these lines beyond prime circle as shown in Figure 10.18.

vii)   Now mark distances 1-a, 2-b, ..., 5-e, 6'-f' and 6'-f', 5'-e', ..., 2'-b', 1'-a' equal to the corresponding displacements as measured form the displacement diagram.

viii) Join the points $A, a, b, c, d, e, f, f', e', d', c', b', a', E$. The curve drawn through these through these points is known as pitch curve.

ix)   From the points $A, a, b, c, d, e, f, f', e', d', c', b', a', E$, draw circle of radius equal to the radius of the roller (i.e., 5 mm).

x)   Join the bottoms of these circles with a smooth curve as shown in Figure 10.18. This is the required profile of the cam.

xi)   Maximum velocity of the follower during outstroke and return stroke.

The displacement of the follower takes place with uniform acceleration and retardation during outstroke and return stroke.

$(V_0)_{max}$ = Maximum velocity during outstroke

$$= \frac{2 \times S \times \omega}{\theta_0} \qquad \left(\because \ \theta_0 = 120^0 = 120 \times \frac{\pi}{180} radius \right)$$

$$= \frac{2 \times 25 \times 125.6}{\left(120 \times \dfrac{\pi}{180} \right)}$$

= 2998.5 mm/s = 2.9985 m/s. = **3 m/s.   Ans.**

and $(V_r)_{max}$ = Maximum velocity during return stroke

$$= \frac{2 \times S \times \omega}{\theta_r} = \frac{2 \times 25 \times 125.6}{\left(90 x \dfrac{\pi}{180} \right)} = 3998 \ mm / s.$$

= 3.998 m/s = **4 m/s.   Ans.**

$(f_0)_{max}$ = Maximum acceleration during outstroke

$$= \frac{4 \times S \times \omega^2}{\theta_0^2} = \frac{4 \times 25 \times 125.6^2}{\left(120 \times \dfrac{\pi}{180}\right)^2} = \frac{100 \times 125.6^2}{\left(\dfrac{2\pi}{3}\right)^2}$$

$$= \frac{100 \times 125.6^2 \times 9}{4\pi^2} = 3,59,635 \; mm/s^2$$

= 359.635 m/s². Ans.

$(f_r)_{max}$ = Maximum acceleration during return stroke

$$= \frac{4 \times S \times \omega^2}{\theta_r^2} = \frac{4 \times 25 \times 125.6^2}{\left(90 \times \dfrac{\pi}{180}\right)^2} = \frac{100 \times 125.6^2}{\left(\dfrac{\pi}{2}\right)^2}$$

$$= \frac{100 \times 125.6^2 \times 4}{\pi^2} = 6,39,351 \; mm/s^2$$

= 639.351 m/s².   Ans.

8. From the following data draw the profile of a cam in which the follower moves with simple harmonic motion (SHM) during ascent while it moves with uniformly accelerated and decelerated motion during descent.

Least radius of cam = 50 mm, Angle of ascent, $\theta_0 = 48°$
Angle of dwell between ascent and descent, $\theta_{DO} = 42°$
Angle of descent, $\theta_r = 60°$, the lift of follower = 40 mm
Diameter of roller = 30 mm
Distance between line of action of the follower and axis of cam = 20 mm
If the cam rotates at 360 rpm and anti-clockwise, find the maximum velocity and acceleration of the follower during descent.

Solution :

Given:

Least radius of cam,	R_{min} = 50 mm
Angle of ascent (outstroke),	$\theta_0 = 48°$
Angle of dwell between ascent and descent,	$\theta_{DO} = 42°$
Angle of descent (or return),	$\theta_r = 60°$
∴ Angle of dwell after return stroke	$= 360 - (48 + 42 + 60) = 210°$
Lift the follower,	$s = 40$ mm
Diameter of roller	$= 30$mm
Distance of offset	$= 20$ mm

Follower moves with SHM during ascent while it moves with uniform acceleration and deceleration during descent.

First draw the displacement diagram as shown in Figure 10.19 as per method given below:

i)   Choose a suitable scale for displacement diagram. Let horizontal scale is 1 mm = 1° and vertical scale is 1 mm= 1 mm. Now draw horizontal line AB = 360°. On this line take AC = 48° to represent ascent (or outstroke), CD = 42° to represent dwell after outstroke, DE = 60° to represent return stroke and EB = 210° to represent dwell after return stroke.

ii) From A, draw vertical line AF = 40 mm to represent the lift of the follower. Complete the rectangle $AFGB$ as shown in Figure 10.19.

iii) Now draw a semi-circle with AF as diameter to represent the SHM of the follower during ascent. Divide the semi-circle into any equal number of even parts (say six). Name these parts as 1, 2, 3, ..., 6. Join them with the centre of semi-circle by radial lines.

iv) Through each points 1, 2, 3, ..., 6 draw horizontal lines as shown.

v)   Now divide the distance AC into same equal number of parts as the number of parts on the semi-circle (i.e., six). Mark these parts as 1, 2, ..., 6. Draw vertical lines through these points. These vertical lines intersect the already drawn horizontal lines at a, b, c, ..., f.

vi) Joint points A, a, b, c, d, e and f. The curve A-a-b- ....-f represent the displacement of diagram for outstroke.

vii) To draw the displacement diagram for return stroke in which case the follower moves with uniform acceleration and deceleration, divide the distance DE into six equal parts. Mark these points as 6', 5', 4,' ....., 1'. Draw the vertical lines through these points.

viii) Divide the vertical lines 3'-H into six equal parts as shown by points I, J, K, L, M and H. The displacement diagram for return stroke will consist of doublt parabola. Join EI, EJ and EK intersecting the vertical lines at a',b', and c', respectively. Join points E, a', b' and c' with a smooth curve.

ix) Join $f'M$,$f'L$ and $f'K$ intersecting vertical lines drawn through 5', 4', and 3' at e', d' and c' respectively. Join points f', e', d' and c' with a smooth curve. The curve A-a-b-c-d-e-f-f' e' ..... a'-E is the required displacement diagram.

$$50 + \frac{30}{2} = 65 \, mm$$

Profile of the cam

First calculate the minimum, pitch circle radius, which is equal to the sum of radii of the minimum radius of the cam and radius of the roller i.e.,

∴ Minimum pitch circle radius = 65 mm.

Now the profile of the cam is drawn as shown in Figure 10.19 as per method given below:

i)   Choose any point O. With O as centre and radius OA equal to minimum pitch circle radius (i.e., 65 mm) draw a prime circle (or min. pitch circle). With O as centre, also draw another circle of radius 20 mm (i.e., equal to eccentricity). This circle is known as offset circle.

ii) Draw the axis of the follower at a distance of 20 mm from point O (i.e. axis of the cam). The axis of the follower meets the prime circle at A. Joint AO.

iii) From OA, draw angle AOC = 48° to represent the ascent (or outstroke), angle COD = 42° to represent the dwell after out stoke and angle DOE = 60° to represent return stoke.

iv) Divide the arc AC and DE into same number of equal parts as in displacement diagram (i.e. six equal parts in each case). The six equal parts on arc AC are marked as 1, 2, 3, 4 5 and 6 whereas on arc DE they are marked as 6', 5', 4', 3', 2', 1'.

v)   Now from points 1, 2 3, ..., 6 and 6', 5', 4', ..., 1' draw tangents to the offset and produce these tangents beyond the prime circle as shown in Figure 10.19.

vi) From points 1, 2, ... 6 and 6', 5', ..., 1' of these tangents, mark distances 1-a, 2-b, b-c, ... 6-f and 6'-f', 5'-e', ..., 1'-a' from displacement diagram. The points a, b, c, ..., f and f', e', ..., a' represents the position of the roller centre.

vii) Now from points A, a, b, ..., f and f', e', ..., a' and E draw circle with radius equal to the radius of the rollower (i.e., = 15 mm).

viii) Join the bottoms of these circles with a smooth curve. This curve represents the cam profile as shown by thick line in Figure 10.19.

ix) Maximum velocity of the follower during descent

i)   The motion of the follower during descent takes place with uniform acceleration and retardation.

Let $(v_r)_{max}$ = Maximum velocity of the follower during ascent.

$$(v_r)_{max} = \frac{2\varpi \times S}{\theta_r} = \frac{2 \times \dfrac{2\pi N}{60} \times 40}{\left(60 \times \dfrac{\pi}{180}\right)} mm/s$$

$$\left(\because \varpi = \frac{2\pi N}{60} \text{ and } \theta_r = 60° = \frac{60 \times \pi}{180} rad\right)$$

$$\left(v_r\right)_{max} = \frac{2\varpi \times S}{\theta_r} = \frac{2 \times \dfrac{2\pi N}{60} \times 40}{\left(60 \times \dfrac{\pi}{180}\right)}\, mm/s$$

$$= 2.88 \text{ m/s}$$

ii) Maximum acceleration of the follower during descent

Let $(f_x)_{max}$ = Maximum acceleration of the following during descent.

It is given by equation (18.8) as,

$$\left(f_r\right)_{max} = \frac{4\varpi^2\, S}{\theta_r^2}$$

$$= \frac{4 \times \left(\dfrac{2\pi \times 360}{60}\right)^2 \times 40}{\left(60 \times \dfrac{\pi}{180}\right)^2}\, mm/s^2$$

$$= 2{,}07{,}360 \text{ mm/s}^2 = 207.36 \text{ m/s}^2$$

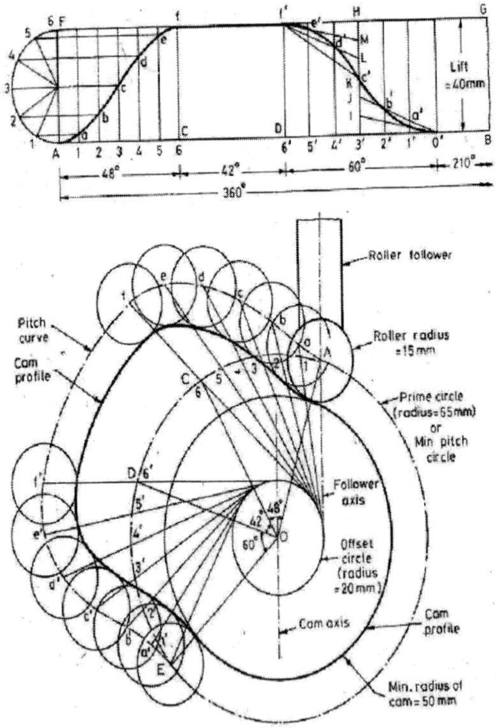

Figure 10.19 Cam profile for problem number 8

9. A cam rotating clockwise with a uniform speed is to give the roller follower of 20 mm diameter the following motion:

i)   Follower to move outwards through a distance of 30 mm during 120° of cam rotation;

ii) Follower to dwell for 60° of cam rotation;

iii) Follower to return to its initial position during 90° of cam rotation;

iv) Follower to dwell for the remaining 90° of cam rotation.

The minimum radius of cam is 45 mm and the line of stroke of the follower is offset 15 mm from the axis of the cam and the displacement of the follower is to take place with simple harmonic motion on both the outward and return strokes. Draw the cam profile.

Solution :

Given:

Diameter of roller follower	=	20 mm
Lift of follower, s	=	30 mm
Outstroke angle, θ_o	=	120°
Angle of dwell after outstroke	= .	60°
Angle of return θ_r	=	90°
Angle of dwell after return stroke		= 90°
Minimum radius of cam,	R_{min}	= 45 mm
Distance of offset		= 15 mm

Displacement of following during outstroke and return stroke follows SHM.

The construction is carried out as per the procedure explained in previous sections

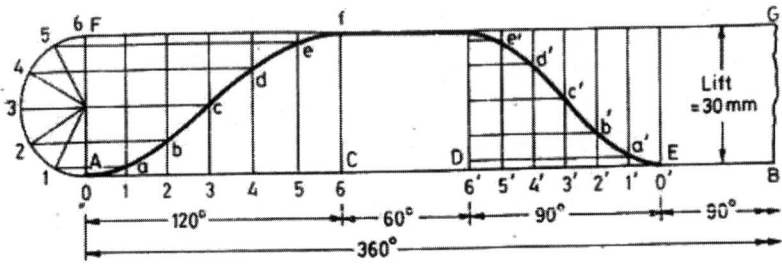

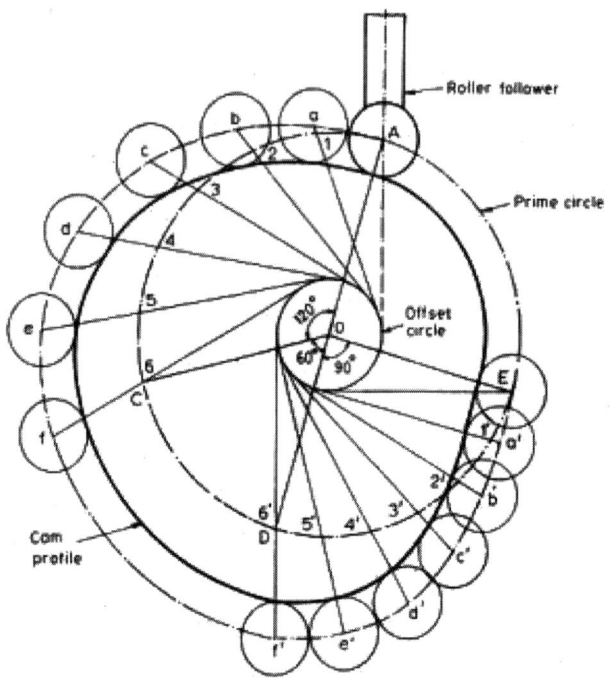

Figure 10.20 Cam profile for the problem number 9

First draw the displacement diagram as shown in Figure 10.20 as per method given below:

i)   Choose a suitable scale for displacement diagram. Let horizontal scale is 10 mm=30° and vertical scale is 10 mm = 10 mm. Now draw horizontal line *AB* = 360°. Take on this line *AC* = 120° to represent outstroke, *CD* =60° to represent dwell after outstroke, *DE* = 90° to represent return stroke and *EF* = 90° to represent dwell after return stroke.

ii)   From *A*, draw vertical line *AF* = 30 mm to represent the lift or stroke of the follower. Complete the rectangle *AFGB* as shown in Figure 10.20.

iii)   Now draw a semi-circle with *AF* as diameter. Divide the semi-circle into any equal number of even parts (say six). Name these parts as 1, 2, 3,.......6. Join them with the centre of semi-circle.

iv)   Through each points 1,2,3,.............6, draw horizontal lines as shown.

v)   Now divide the distances *AC* and *DE* into the same equal number of parts as the number of parts on the semi-circle (i.e. six equal parts).

The six equal parts on distance AC are marked as 1,2,3,.......6, whereas the six equal parts on distance DE are marked as 6', 5',......1'. Now draw the vertical lines through these points. These vertical line intersect the horizontal lines at $a, b, c, ..., f$ for outstroke and at $a', b', c', ...f'$ for return stroke.

vi)   Join points 0, $a, b, c,........,f, f', e', ...a', 0'$ by a smooth curve as shown in Figure 10.20. This is the required displacement diagram.

Cam profile

First circulate the minimum pitch circle radius. This is equal to sum of radii of minimum radius of the cam and radius of the roller i.e. 45 + (20/2) = 55 mm

∴ Minimum pitch circle radius = 55 mm

Now the profit of the cam is drawn as shown in Figure 10.20 as per method given below:

i)   Choose any point O. With O as centre and radius OA equal to minimum pitch circle radius (i.e., 55 mm) draw a prime circle. Also draw another circle of radius equal to eccentricity of 15 mm with O as centre. This circle is known as offset circle.

ii)   Draw the axis of the follower at a distance of 15 mm from point O (i.e. axis of the cam). The axis of follower meets the prime circle at A. Join AO.

iii) From OA, draw angle $AOC = 120°$ to represent outstroke, angle $COD = 60°$ to represent dwell after outstroke and angle $DOE = 90°$ to represent return stroke.

iv)   Divide angle AOC (or Arc AC) and angle DOE (or Arc DE) into same number of equal parts as in displacement diagram (i.e., six equal parts in each case). The six equal parts on arc AC are marked as 1,2,3,......6, whereas on arc DE they are marked as 6', 5', ...1'.

v)   Now from points 1,2,3,............6, and 6', 5',.........1' draw tangents to the offset circle and produce these tangents beyond the prime circle as shown in Figure 10.20.

vi)   Now from points 1,2,3......6 and 6',5',.....1' of these tangents, mark distances 1-a, 2-b, 3-c, ...6-f and 6'-f', 5'-e', ...1'-a' from displacement diagram. The points $a,b,c.....f$ and $f', e', ...a'$ represents the positions of the roller centre.

vii) Now from points $A, a, b, ...f$ and $f', e', ...a'$ and E, draw circles with radius equal to the radius of the roller.

viii) Join the bottoms of these circles with a smooth curve. This smooth curve represents the cam profile as shown by a thick file in Figure 10.20.

10. Draw the profile of a cam to raise a valve with harmonic motion through 50 mm in 1/3 of revolution, keep it fully raised through 1/12 revolution, and to lower it with harmonic motion in 1/6 revolution. The valve remains closed during the rest of the revolution. The diameter of the roller is 20 mm and minimum radius of the cam is to be 25 mm. The diameter of the cam shaft is 25 mm. The axis of the valve rod passes through the axis of the cam shaft. Assume the cam shaft to rotate with a uniform velocity.

Solution :

$$\text{Outstroke angle, } \theta_0 = 1/3\text{rd of revolution} = \frac{1}{3} \times 360^0 = 120^0$$

$$\text{Angle of dwell after outstroke} = 1/12\text{th of the revolution} = \frac{1}{12} \times 360^0 = 30^0$$

$$\text{Return stroke angle} \qquad = \frac{1}{6} \times 360^0 = 60^0$$

Angle of dwell after return stroke = 360 – (120 + 30 + 60) = 150°
Diameter of roller   = 20 mm
Minimum radius of cam, R_{min} = 25 mm
Dia. of cam shaft   = 25 mm
Axis of the valve rod passes through the axis of cam shaft.

Draw the displacement diagram first as shown in Figure 10.21 as per method given below:

i)   Choose a suitable scale for displacement diagram. Let the horizontal scale is 10 mm = 30° and vertical scale is 10 mm = 10 mm. Now draw horizontal line $AB = 360°$ On this line take $AC = 120°$ to represent outstroke, $CD = 30°$ to represent dwell after outstroke, $DE = 60°$ to represent the return stroke and $EB = 150°$ to represent the dwell after return stroke.

ii)   From A, draw vertical line $AF = 50$ mm to represent the lift of the follower. Complete the rectangle $AFGB$ as shown in Figure 10.21.

iii) Now draw a semi-circle with AF as diameter to represent the SHM of the follower. Divide the semi-circle into any equal number of even parts (say eight). Name these parts as 1, 2, 3, ..., 8. Join them with the centre of the semi-circle by radial lines.

iv)   Through each points 1, 2, 3, ..., 8 draw horizontal lines as shown in Figure 10.21.

v)   Now divide the distance AC and DE into the same equal number of parts as the number of parts on the semi-circle (i.e., eight). Mark these parts as 1, 2, 3, ..., 8 for outstroke and 8', 7', ..., 1', 0' for return stroke. Draw vertical lines through these points. These vertical line intersect the already drawn horizontal lines at *a, b, c,* ...., *g, h* for outstroke and at *h', g',* ....., *b', a'* for return stroke.

vi) Join parts *A, a, b,* ......, *g, h, h', g',* ...., *b', a', 0',* B with a smooth curve. This curve represents the displacement diagram for the follower.

Profile of the cam

First calculate the minimum pitch circle radius which is equal to the sum of radii of the minimum radius of the cam and radius of the roller i.e.,

$$25 + \frac{20}{2} = 35 \ mm.$$

∴ Minimum pitch circle radius = 35 mm.

Now the profile of the cam is drawn as shown in Figure 10.21 as per method given below:

i) Choose any point *O*. With *O* as centre and radius equal to *OA* (i.e., minimum pitch circle radius = 35 mm) draw a prime circle.

ii) From *OA*, draw angle *AOC* = 120° to represent the outstroke angle *COD* = 30° to represent the dwell after outstroke and angle *DOE* = 60° to represent the return stroke. The points A, C, D and E line on the prime circle.

iii) Divide the arc *AC* and arc *DE* into the same number of equal parts as in the displacement diagram (i.e., eight equal parts in each case). The eight equal parts on the arc *AC* are marked as 1, 2, 3, 4, 5, 6, 6, 8 whereas on the arc *DE* they are marked as 8', 7', ..., 2', 1'.

iv) Join the above points with *O* and produce these radial lines beyond the prime circle as shown in Figure 10.21.

v) On these radial lines, mark distances 1-*a*, 2-*b*, ..., 7-*g*, 8-*h* and 8'-*h'*, 6-*g*, ..., 2'-*b'*, 1'-*a'* equal to the corresponding displacements as measured from the displacement diagram shown in Figure 10.21. The points *a, b, c,* ...., *g* and *h', g',* ...., *b', a'* represents the position of the roller centre.

vi) Now from points *A, a, b,* ...., *h* and *h', g',* ...., *a* and *E* draw circles with radius equal to the radius of the roller (i.e., = 10 mm).

vii) Join the bottoms of these circles with a smooth curve. This smooth curve represents the cam profile as shown in Figure 10.21 by thick line.

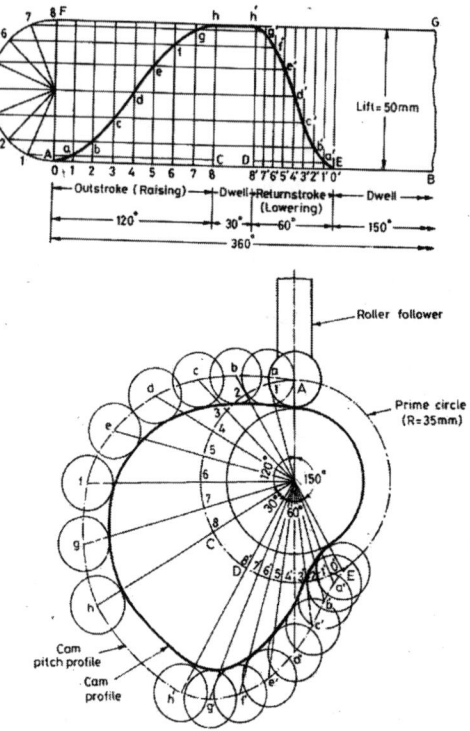

Figure 10.21   Profile of the cam for the problem number 10

11. Draw the cam profile with lift 40 mm, outward 90°, dwell for 30°, return 60° and remaining dwell. Note that outward and return are with SHM. Develop profile without and with offset by 20 mm.

Solution :

Displacement diagram is constructed as explained above (Figure 10.22). The profile shows the two cases like without and with offset.

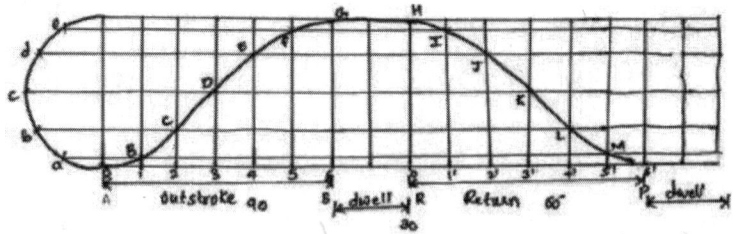

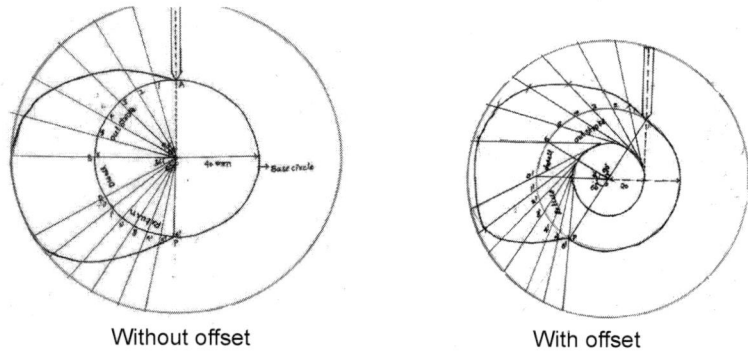

Without offset   With offset

Figure 10.22   Cam Profile for Problem No.11

12. Draw the profile of the cam with the following data:
Minimum radius of cam 10mm, outward and return stroke with SHM for 120 and 150°, after outward dwell for 30 and at the end of return dwell for remaining (Figure 10.23).

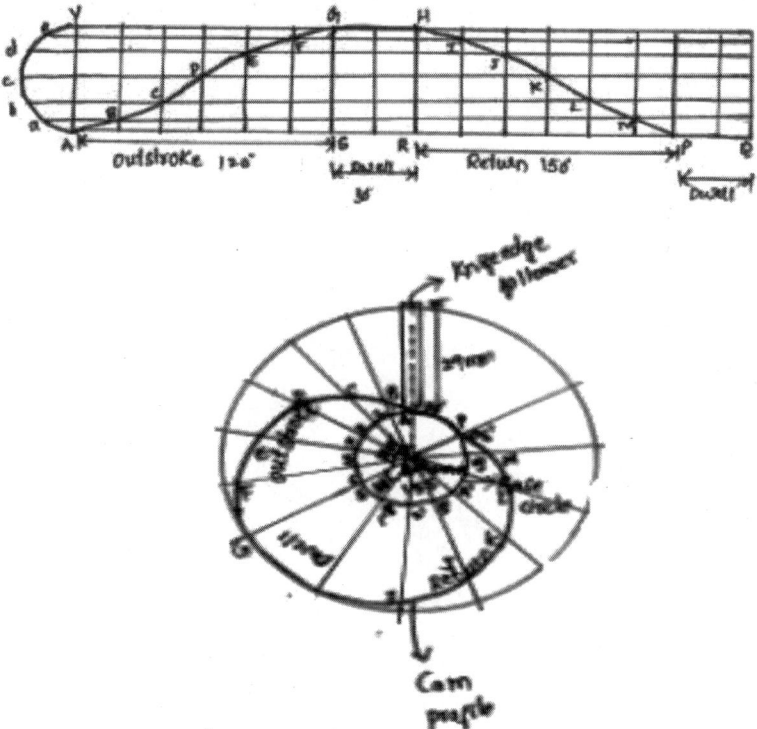

Figure 10.23   Cam Profile for Problem No.12

13. Draw the cam profile for following conditions:
Follower type = Knife-edged, in-line; lift = 50 mm; base circle radius = 50 mm; out stroke with SHM, for 60° cam rotation; dwell for 45° cam rotation; return stroke with SHM, for 90° cam rotation; dwell for the remaining period. Determine max. velocity and acceleration during out stroke and return stroke if the cam rotates at 1000 rpm in clockwise direction.

Solution

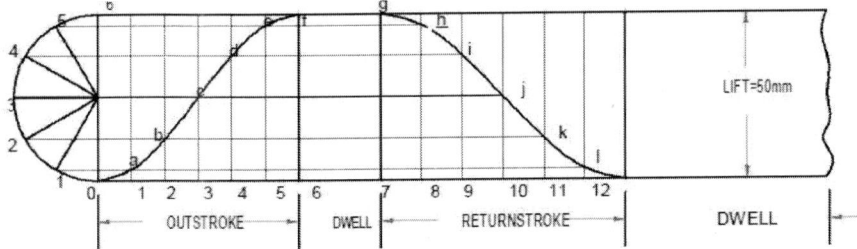

Figure 10.24   Cam Profile for Problem No.13

In the following profile the points "*a, b, c*…. to *i*" are known as trace points or pitch points and the curve by joining these points is known as pitch curve (Figure 10.24 and 10.25).

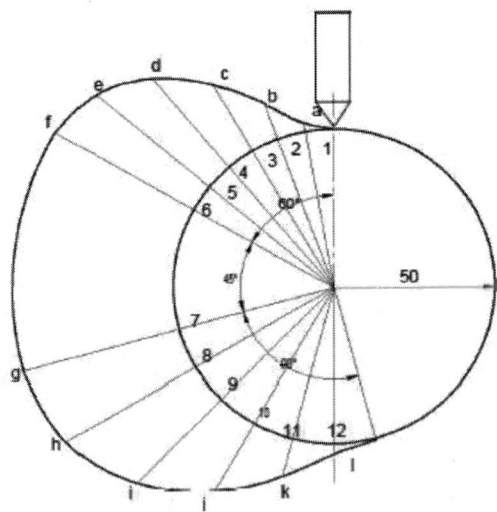

Figure 10.25   Cam Profile for Problem No.14

14. Draw the cam profile for the same operating conditions of above problem, with the follower offset by 10 mm to the left of cam centre (Figure 10.26).

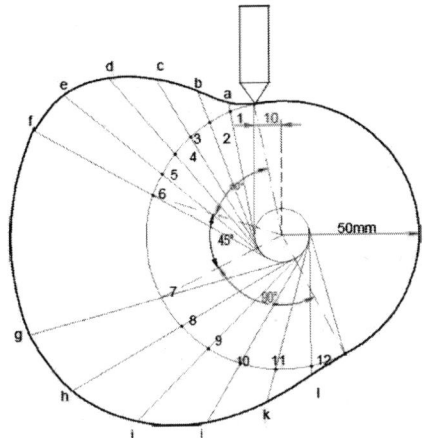

Figure 10.26   Cam Profile for Problem No.15

15. Draw the cam profile with the following data
Lift 40 mm, outstroke 60°, dwell for 40°. And return stroke for 90°, and re-maining dDwell. Both strokes move with SHM. Follower is through the cam axis (Figure 10.27).

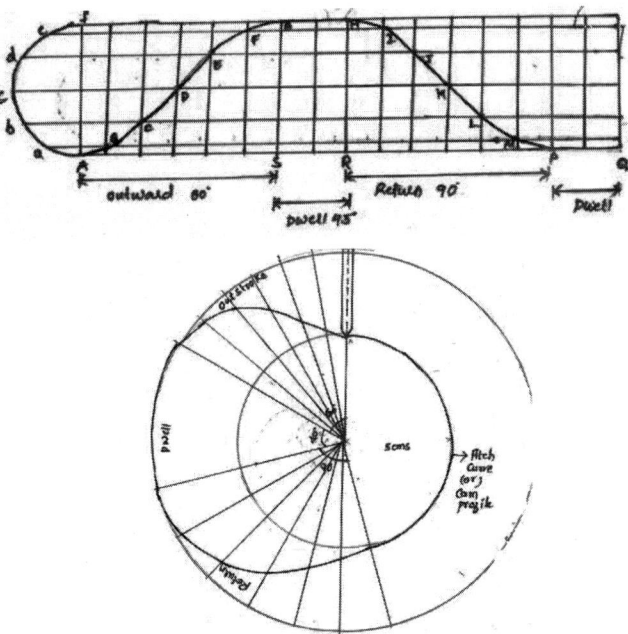

Figure 10.27   Cam Profile for Problem No.16

16. Draw the cam profile for following conditions:
Follower type = roller follower, in-line; lift = 25 mm; base circle radius = 20 mm; roller radius = 5 mm; out stroke with UARM, for 1200 cam rotation; dwell for 600 cam rotation; return stroke with UARM, for 900 cam rotation; dwell for the remaining period. Determine max. velocity and acceleration during out stroke and return stroke if the cam rotates at 1200 rpm in clockwise direction (Figure 10.28).

Solution :

Displacement diagram

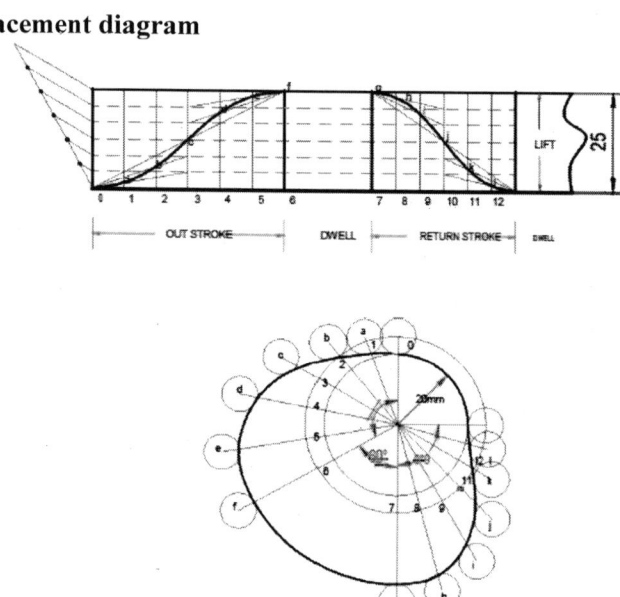

Figure 10.28   Cam Profile for Problem No.17

17. Draw the cam profile for conditions same as above problem, with follower offset to right of cam centre by 5 mm and cam rotating counter clockwise (Figure 10.29).

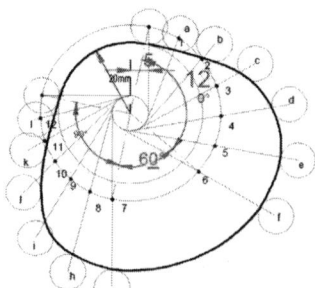

Figure 10.29   Cam Profile for Problem No.18

18. Draw the profile of a cam operating a knife-edged follower having a lift of 30 mm. The cam raises the follower with SHM for 150° of its rotation followed by a period of dwell for 60°. The follower descends for the next 100° rotation of the cam with uniform velocity, again followed by a dwell period. The cam rotates at a uniform velocity of 120 rpm and has a least radius of 20 mm. What will be the maximum velocity and acceleration of the follower during lift and the return?

Solution :

$h = 30$ mm, $\varphi_a = 150°$, $N = 120$ rpm, $\delta_1 = 60°$, $r_c = 20$ mm, $\varphi_d = 100°$, $\delta_2 = (360° - 150° - 100° - 60°) = 50°$

Draw the displacement diagram of the follower as discussed earlier taking a convenient scale. Construct the cam profile as follows:

i) Draw circle with radius r_c.

ii) Divide the angles φ_a and φ_d into same number of parts as is done in the displacement diagram. In this case, each has been divided into 6 equal parts.

iii) Draw radial lines O – 1, O – 2, O – 3, etc. O – 1 represents that after an interval of $\varphi_d/6$ of the cam rotation in the clock wise direction it will take the vertical position of O – O'.

iv) On the radial lines produced, take distances equal to the lift of the follower beyond the circumference of the circle with radius r_c, i.e., 1 – 1', 2 – 2', 3 – 3' etc.

v) Draw a smooth curve passing through O' 1', 2',..........10', 11' and 12'. Draw an arc of radius 0 – 6' for the dwell period δ_1.

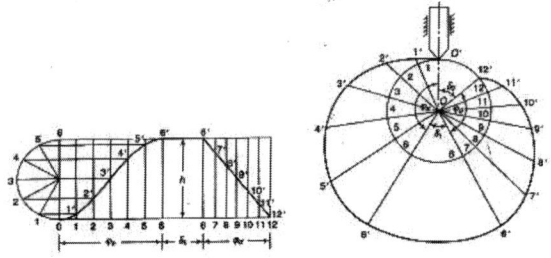

Figure 10.30   Cam Profile for Problem No.18a (Hand Drawn)

During ascent

$$\omega = \frac{2\pi N}{6} = \frac{2\pi \times 120}{60} = 12.57 \, rad/s$$

$$v_{max} = \frac{h}{2} \frac{\pi\omega}{\phi_a}$$

$$f_{max} = \frac{h}{2}\left(\frac{\pi\omega}{\phi_a}\right)^2$$

During descent, $v_{max} = h \dfrac{\omega}{\phi_d}$ $f_{max} = f = 0$

The above cam profile is from the software and the following profile is hand drawn (Figures 10.30 and 10.31).

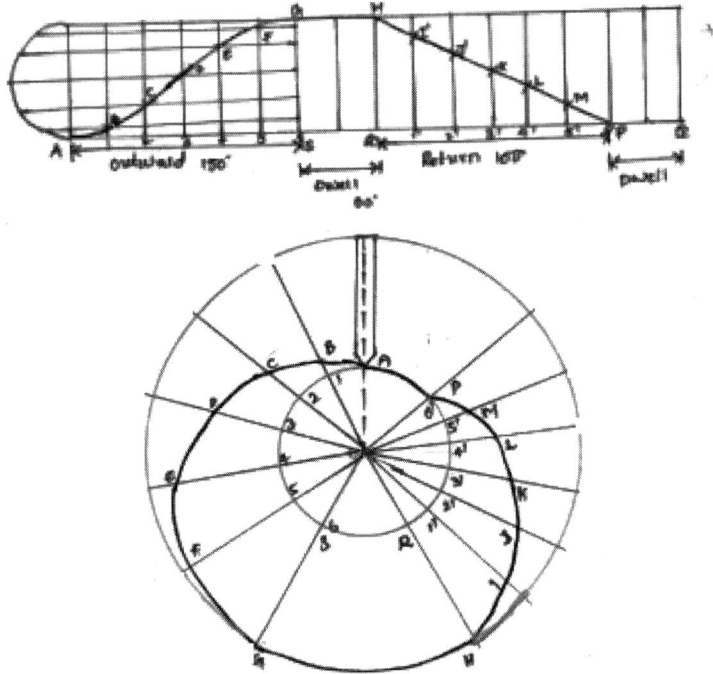

Figure 10.31 Cam Profile for Problem No.19

19. Draw the cam profile with the following data:
Lift 30 mm, outstroke 120°. With SHM, dwell for 30°. And return stroke for 150°. With uniform acceleration and deceleration and remaining Ddwell. Minimum radius of cam 25 mm, roller diameter 15 mm, follower is through the cam axis. Calculate the maximum velocity and acceleration of the follower during the descent period.

Solution :

Lift = 30 mm, Angle of ascent φ_a = 120°, Speed of cam N = 150 rpm, Dwell δ_1 = 30°, Minimum radius of cam r_c = 25 mm, Angle of decent φ_d = 150°, Dwell δ_2 = 60°

Draw the displacement diagram of the follower. Construct the cam profile as described below:

i)   Draw a circle with radius $(r_c + r_r)$.

ii)   From the vertical position, mark angles φ_a, δ_1, φ_d, and δ_2 in the counter – clockwise direction (assuming that the cam is to rotate in the clockwise direction).

iii) Divide the angles φ_a and φ_d into same number of parts as in the displacement diagram. In this case, φ_a has been divided into 6 equal parts whereas φ_d into 8 equal parts.

iv)   On the radial lines produced, mark the distances from the displacement diagram.

v)   Draw a series of arcs of radii equal to r_r as shown in the diagram from the points 1', 2', 3' etc.

vi)   Draw a smooth curve tangential to all the arcs which is the required cam profile.

During the descent period, the acceleration and the deceleration is uniform. Therefore, the maximum velocity is at the end of the acceleration period (Figure 10.32).

$$v_{max} = 2h\,\frac{\omega}{\phi_d}$$

or $$v_{max} = 2 \times 30 \times \frac{\dfrac{2\pi \times 150}{60}}{150 \times \dfrac{\pi}{180}} = \underline{360m\,/\,s}$$

$$f_{max} = f_{uniform} = \frac{4h\omega^2}{\phi_d^2}$$

Or

$$f_{max} = \frac{4 \times 30 \times \left[\dfrac{2\pi \times 150}{60}\right]^2}{\left(150 \times \dfrac{\pi}{180}\right)^2} = 4320mm\,/\,s^2 \quad or \quad 4.32m\,/\,s^2$$

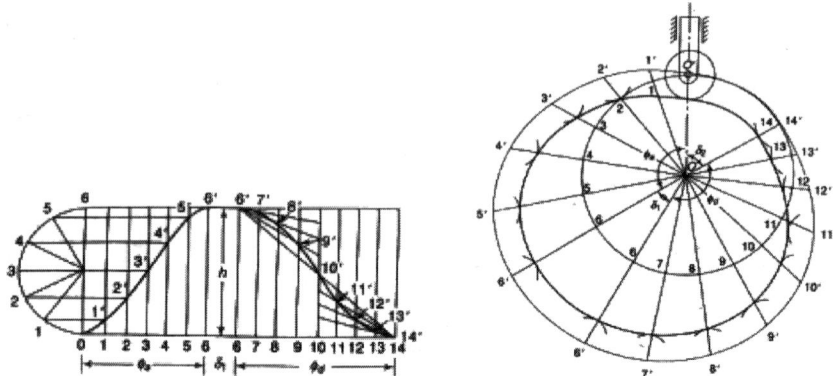

Figure 10.32   Cam Profile for Problem No.20

20. Draw the profile of a cam if the minimum radius is 2.5 cm with lift of 3.5 cm and the cam imparts SHM. The forward is 60° and dwell for 40° and return stroke for 90° with remaining dwell. Note the follower is offset by 1 cm. Work out necessary statistics (Figure 10.33).

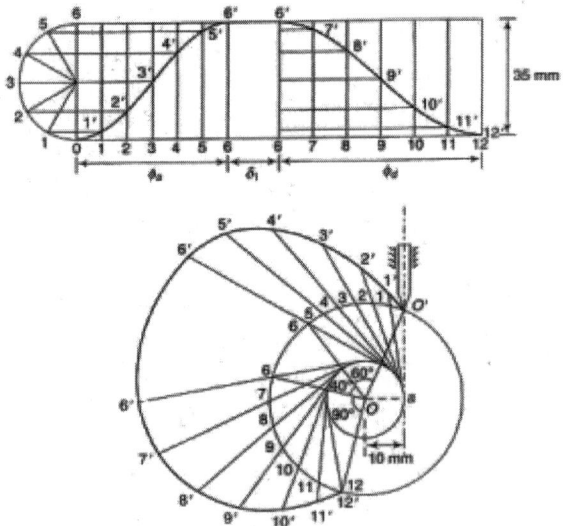

Figure 10.33   Cam Profile for Problem No.21

21. Draw the profile of the cam with the following data:
Minimum radius of the cam 50 mm, Outward and return with uniform acceleration and retardation. The angles of movement of follower is 100°. For outward and dwell for next 80°. And 90° for return and remaining dwell. The cam axis

coincides with follower. The reader is instructed to draw the profile with offset arrangement by 15 mm. Note the stroke of cam is 40 mm (Figure 10.34).

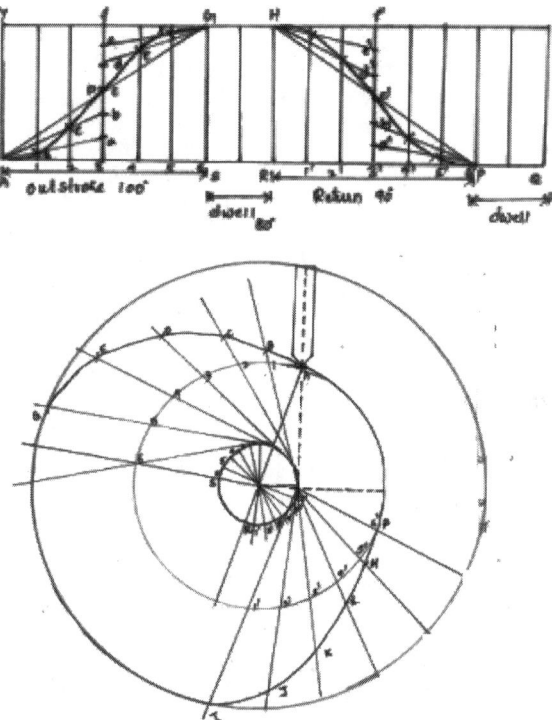

Figure 10.34   Cam Profile for Problem No.22

22. Draw the profile of the cam with roller follower:
Cam shaft diameter 40 mm, Minimum radius of cam 20 mm, Diameter of roller 25 mm, Angle of lift of the follower outwards 120°. Dwell for 45°. And Fall for 150° and remaining is dwell. Lift of follower is with SHM and Fall is with UA (Figure 10.35).

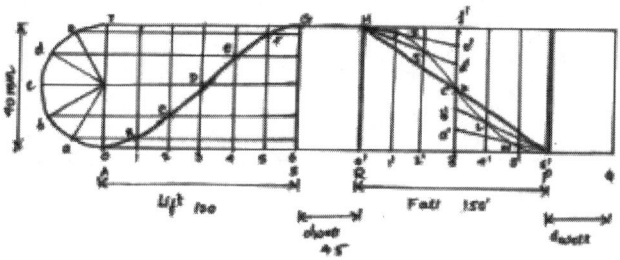

Figure 10.35   Cam Profile for Problem No.23

23. Design a cam for operating exhaust valve of the engine oil. It is required to give equal uniform acceleration and retardation during outward and return motion which corresponds to each 60°. The valve must remain fully open for 20°. Of cam rotation. The lift of the valve is 37.5 mm and the minimum radius of the cam is 40 mm. The follower is provided with a roller of diameter 20 mm and its line passes through the axis of the cam (Figure 10.36).

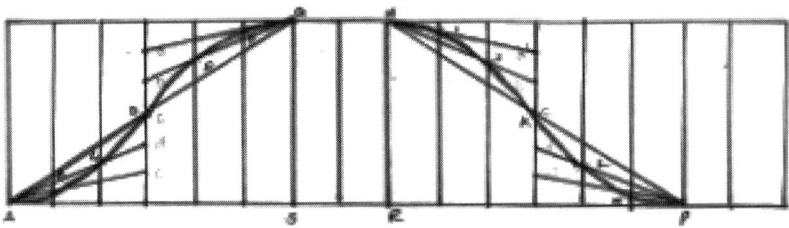

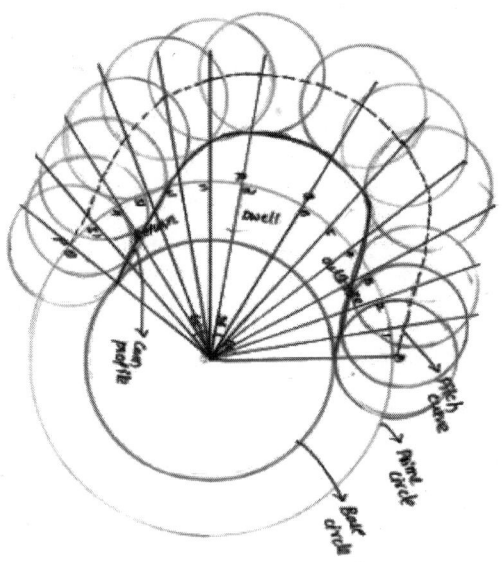

Figure 10.36   Cam Profile for Problem No.24

24. Draw the profile of the cam with data: Minimum radius of cam = 10 mm, Stroke of the follower = 24 mm, 1st 120° outwards with SHM followed by 30° dwell return stroke with SHM for 150° and the dwell for the remaining portion. The cam is provided with knife edged follower and follower axis is passing through the cam axis.

Solution: Reader is directed to work out this profile

25. The following data relate to a cam profile in which the follower moves with uniform acceleration and deceleration during ascent and descent:

Minimum radius of cam 25 mm, Roller radius 2.5 mm, Lift 28 mm, Offset of follower axis 12 mm towards right, Angle of ascent 60°, Angle of descent 90°, Angle of dwell between ascent and descent 45°, Speed of the cam 200 rpm. Draw the profile of the cam and determine the maximum velocity and the uniform acceleration of the follower during the out stroke and the return stroke.

Solution :

Lift = 28 mm, Angle of ascent φ_a = 60°, cam radius r_c = 25 mm, Dwell δ_1 = 45° roller radius r_r = 7.5 mm, Angle of decent φ_d = 90°, Offset distance = 12 mm, Dwell 2 $\delta_2 = (360° - 60° - 45° - 90°)$

From the given data, construct the displacement diagram as usual for the cam profile the procedure is as follows:

i)   Draw a circle with radius $(r_c + r_r)$.

ii)   Draw another circle concentric with the previous circle with radius x. If the cam is assumed stationary, the follower will be tangential to this circle in all the positions. Let initial position be $a - 0'$.

iii) Join $0 - 0'$. Divide the circle of radius $(r_c + r_r)$ into four parts as usual with angles $\varphi_a,$ $\delta_1,$ φ_d and δ_2 starting from $0 - 0'$.

iv)   Divide the angles φ_a and φ_d into same number of parts as is done in the displacement diagram and obtain the points 1, 2, 3, etc., on the circumference of circle with radius $(r_c + r_r)$.

v)   Draw tangents to the circle with radius x from the points 1, 2, 3, etc.

vi)   On the extension of the tangent lines, mark the distances from the displacement diagram.

vii) Draw a smooth curve through 0', 1', 2' etc. This is the pitch curve.

viii) With 1', 2', 3', etc., as centres, draw a series of arcs of radii equal to r_r.

ix)   Draw a smooth curve tangential to all the arcs and obtain the required cam profile (Figure 10.37).

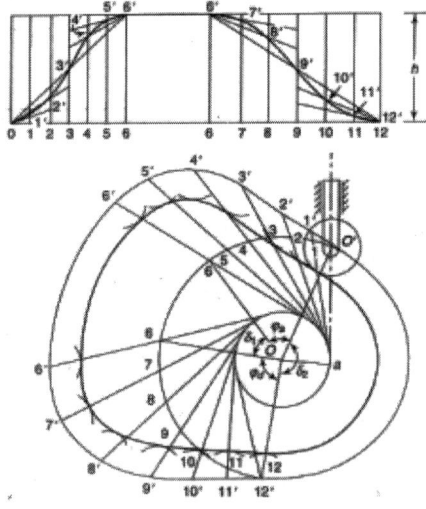

Figure 10.37   Cam Profile for Problem No.25

During outstroke

$$\omega = \frac{2\pi \times 200}{60} = 20.94\, rad\,/\,s$$

$$v_{max} = 2\,h\frac{\omega}{\phi_a} = 2 \times 28 \times \frac{20.94}{60 \times \pi / 180} = 1120\,mm\,/\,s \quad or \quad 1.12\,m\,/\,s$$

$$f_{uniform} = \frac{4h\omega^2}{\phi^2{}_a} = \frac{4 \times 28 \times (20.94)^2}{\left(60 \times \dfrac{\pi}{180}\right)^2} = 44{,}800\,mm\,/\,s^2 \text{ or } 44.8\ m/\,s^2$$

During return stroke

$$v_{max} = \frac{2h\omega}{\phi_d} = \frac{2 \times 28 \times 20.94}{90 \times \pi / 180} = 747\,mm\,/\,s \quad or \quad 0.747\,m\,/\,s$$

$$f_{uniform} = \frac{4 \times 28 \times (20.94)^2}{\left(90 \times \dfrac{\pi}{180}\right)^2} = 19{,}900\,mm\,/\,s^2 \text{ or } 19.9m\,/\,s^2$$

26. A flat-faced mushroom follower is operated by a uniformly rotating cam. The follower is raised through a distance of 25 mm in 120° rotation of the cam, remains at rest for the next 30° and is lowered during further 120° rotation of the cam. The raising of the follower takes place with cycloidal motion and the lowering with uniform acceleration and deceleration. However, the uniform acceleration is 2/3 of the uniform deceleration. The least radius of the cam is 25 mm which rotates at 300 rpm. Draw the cam profile and determine the values of the maximum velocity and maximum acceleration during rising, and maximum velocity and uniform acceleration and deceleration during lowering of the follower.

Solution :

$h = 25$ mm, $\varphi_a = 120°$, $r_c = 25$ mm, $\delta_1 = 30°$, $N = 300$ rpm, $\varphi_d = 120°$, $\delta_2 = 90°$
For uniform acceleration, $v = ft$, during the return stroke of the follower, initial velocity is zero. Then with a uniform acceleration the maximum velocity is attained, followed by a uniform deceleration that again makes the velocity to be zero. Thus, if the uniform acceleration is to be 2/3 of the uniform deceleration, the time for acceleration must be 3/2 times the time for deceleration for the same maximum velocity.

i)   Draw a circle with radius r_c.

ii)   Take angles φ_a, δ_1, φ_d and δ_2 as before (in the counter-clockwise direction if the cam rotation is assumed clockwise).

iii)   Divide φ_a and φ_d into same number of parts as in the displacement diagram.

iv)   Draw radial lines and on them mark the distances 1 – 1′, 2 – 2′, 3 – 3′, etc.

v)   Draw the follower in all positions by drawing perpendiculars to the radial lines at 1′, 2′, 3′, etc. In all the positions, the axis of the follower passes through the centre O.

vi)   Draw a curve tangential to the flat-faces of the follower representing the cam profile (Figure 10.38).

Let f' = uniform deceleration (magnitude), t' = period of deceleration, Then uniform acceleration $f = 2/3\, f'$, Pperiod of acceleration.

$$t = \frac{3}{2}t$$

Displacement, $s = ut + \frac{1}{2}ft^2$

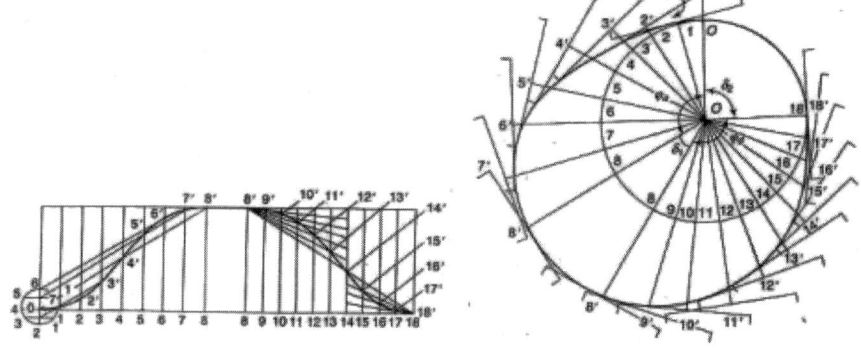

Figure 10.38   Cam Profile for Problem No.26

But initial velocity $u = 0$
Distance moved during acceleration period

$$= \frac{1}{2}\left(\frac{2}{3}f'\right)\left(\frac{3}{2}t'\right)^2$$

$$= \frac{3}{2}\left(\frac{1}{2}f't'^2\right)$$

During ascent

$$V_{max} = \frac{2h\omega}{\phi_a}$$

where $\omega = \dfrac{2\pi \times 300}{60} = 31.4 rad/s$

$$V_{max} = \frac{2 \times 25 \times 31.4}{120 \times \pi / 180} = 750 mm / s \quad or \quad 0.75 m / s$$

$$f_{max} = \frac{2 \times 25 \times \pi \times (31.4)^2}{(120 \times \pi / 180)^2} = 35,310 mm / s^2 \quad or \quad \underline{35.31 m / s^2}.$$

During descent

$$v = ft$$

or

$$v = \left(\frac{2s}{t^2}\right) t = \frac{2s}{t}$$

v will be maximum at the end of the acceleration period. At the end of the acceleration period,

$$s = \frac{3}{5} \times 25 = 15 mm$$

and the time taken to travel this distance is found as under,

Time for 300 rev = 60 s

Time for 1 rev (= 360°) = $\frac{60}{300}$ = 0.2 s

Time for $\left(\frac{3}{5} \times 120^o\right) = 0.2 \times \frac{72}{360} = 0.04 s$

$$V_{max} = \frac{2 \times 15}{0.04} = \underline{750 \text{ mm } /s^2} \quad or \quad \underline{0.075 \text{ m/s}^2}$$

Uniform acceleration = $\frac{V_{max}}{time} = \frac{0.75}{0.04} = 18.75 m / s^2$

Uniform acceleration = $18.75 \times \frac{3}{2} = \underline{28.13 \text{ m } / s^2}.$

27. The following data relate to a cam operating an oscillating roller follower: Minimum diameter of the cam = 44 mm, Diameter of roller =14 mm, Length of the follower arm = 40 mm, Distance of fulcrum centre from cam centre = 50 mm, Angle of ascent 75°, Angle of descent 105°, Angle of dwell for the follower in the highest position 60°, Angle of oscillation of follower 28°. Draw the profile of the cam if the ascent and descent both take place with SHM.

Solution :

r_c = 22 mm, θ = 28°, r_r = 7 mm, φ_a = 75°, Follower arm length = 40 mm, δ_1 = 60°, δ_2 = 120°, φ_d = 105°, $h = \theta \times$ Arm length $= \left(28^o \times \frac{\pi}{180^o}\right) \times 40$

= 19.5 mm

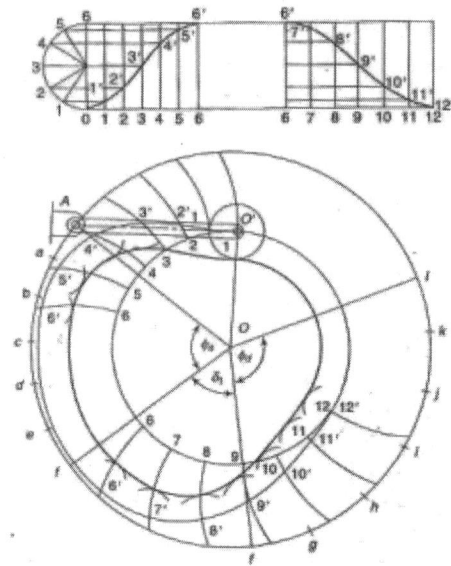

Figure 10.39   Cam Profile for Problem No.27

To draw the cam profile, proceed as follows:

i) Draw a circle with radius $(r_c + r_r)$.

ii) Assuming the initial position of the roller centre vertically above the cam centre O, locate the fulcrum centre as its distances from the cam centre and the roller centre (equal to length of follower arm) are known.

iii) Draw a circle with radius OA and centre at O.

iv) On the circle through A, starting from OA, take angles φ_a, δ_1 and φ_d as usual.

v) Divide the angles φ_a and φ_d into same number of parts as is done in the displacement diagram and obtain the points a, b, c, d, etc., on the circle through A.

vi) With centres A, a, b, etc., draw arcs with radii equal to length of the arm.

vii) Mark distances $1 - 1'$, $2 - 2'$, $3 - 3'$, etc., on these arcs as shown in the diagram (Figure 10.39). It is on the assumption that for small angular displacements, the linear displacements on the arcs and on the straight lines are the same.

viii) With $1'$, $2'$, $3'$, etc., draw a series of arcs of radii equal to r_r.

ix) Draw a smooth curve tangential to all the arcs and obtain the required cam profile.

28. Construct the profile of a disk cam that follows the displacement diagram shown below. The follower is a radial roller and has a diameter of 10 mm. The base circle diameter of the cam is to be 40 mm and the cam rotates clockwise. The reader is directed to analyse the construction and proceed (Figure 10.40).

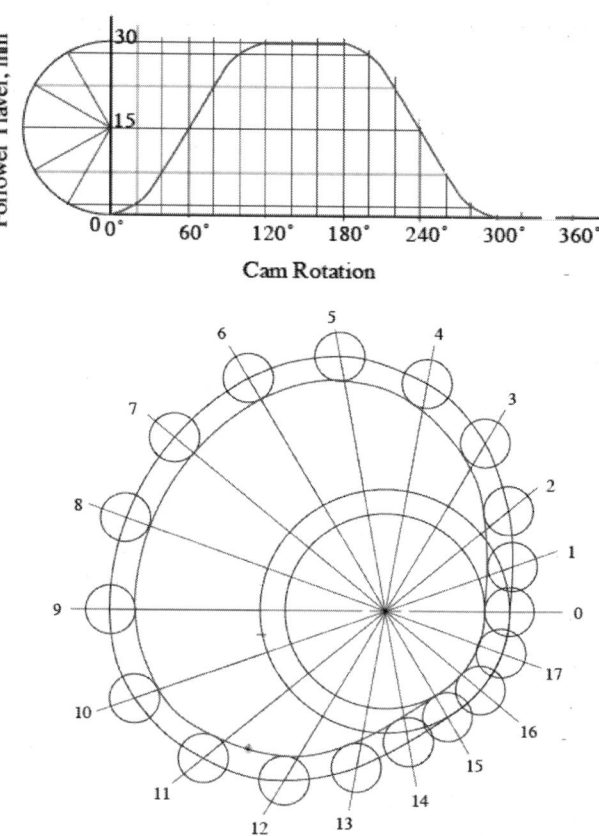

Figure 10.40   Cam Profile for Problem No.28

Note: **Reader should note the different way of asking the question of cam construction as indicated below**

29. Lay out a cam profile using a harmonic follower displacement (both rise and return). Assume that the cam is to dwell at zero lift for the first 100° of the motion cycle and to dwell at a 1 in lift for cam angles from 160° to 210°. The cam is to have a translating, radial, roller follower with a 1-in. roller diameter, and the base circle radius is to be 1.5″. The cam will rotate clockwise. Lay out the cam profile using 20° plotting intervals (Figure 10.41).

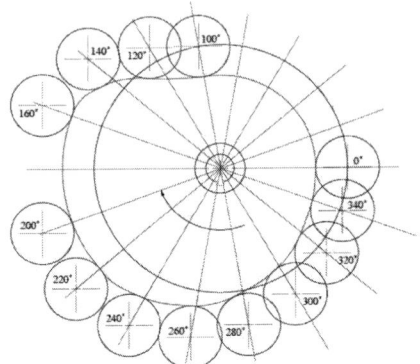

Figure 10.41   Cam Profile for Problem No.29

30. Lay out a cam profile using a cycloidal follower displacement (both rise and return) if the cam is to dwell at zero lift for the first 80° of the motion cycle and to dwell at 2-in. lift for cam angles from 120° to 190°. The cam is to have a translating, radial, roller follower with a roller diameter of 0.8″. The cam will rotate counter clockwise, and the base circle diameter is 2″. Lay out the cam profile using 20° plotting intervals (Figure 10.42).

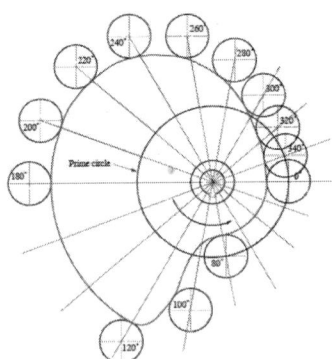

Figure 10.42   Cam Profile for Problem No.30

31. Lay out a cam profile assuming that an oscillating, roller follower starts from a dwell for 0° to 140° of cam rotation, and the cam rotates clockwise. The rise occurs with parabolic motion during the cam rotation from 140° to 220°. The follower then dwells for 40° of cam rotation, and the return occurs with parabolic motion for the cam rotation from 260° to 360°. The amplitude of the follower rotation is 35°, and the follower radius is 1″. The base circle radius is 2″, and the distance between the cam axis and follower rotation axis is 4″. Lay out the cam profile using 20° plotting intervals such that the pressure angle is 0 when the follower is in the bottom dwell position (Figure 10.43).

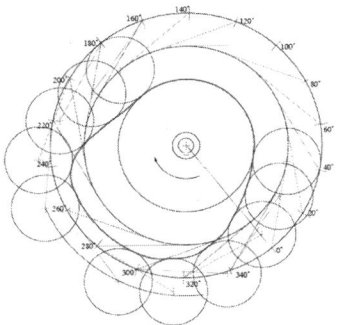

Figure 10.43   Cam Profile for Problem No 31

32. Lay out the rise portion of the cam profile if a flat-faced, translating, radial follower's motion is uniform. The total rise is 1.5″, and the rise occurs over 100° of can rotation. The follower dwells for 90° of cam rotation prior to the beginning of the rise, and dwells for 80° of cam rotation at the end of the rise. The cam will rotate counter clockwise, and the base circle radius is 3″ (Figure 10.44).

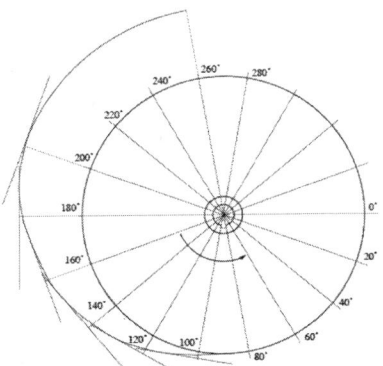

Figure 10.44   Cam Profile for Problem No.32

Note: **Alternate method of constructing the cam profile**
Note: **The reader is informed that the displacement diagram can be constructed inside the cam profile only rather than a separate diagram.**

33. Draw the profile of the cam with the following data:

Lift 40 mm, outward and return stroke with UV. Outward for first 60°. Dwell for the next 30° and return with 60°. Base circle is 50 mm radius.

i) Draw a horizontal line $AX = 360°$ to some suitable scale. On this line, mark $AS = 60°$ to represent outstroke of the follower, $ST = 30°$ to represent dwell, $TP = 60°$ to represent return stroke and $PX = 210°$ to represent dwell.

ii) Draw vertical line AY equal to the stroke of the follower (i.e., 40 mm) and complete the rectangle as shown in Figure 10.45.

iii) Divide the angular displacement during outstroke and return stroke into any equal number of even parts (say six) and draw vertical lines through each point.

iv) Since the follower moves with uniform velocity during outstroke and return stroke, therefore the displacement diagram consists of straight lines. Join AG and HP.

v) The complete displacement diagram is shown by $AGHPX$ in Figure 10.45.

Case (i) Collinear: Profile of the cam when the axis of follower passes through the axis of cam shaft.

The profile of the cam when the axis of the follower passes through the axis of the cam shaft, as shown in Figure 10.45 is drawn as discussed in the following steps:

i) Draw a base circle with radius equal to the minimum radius of the cam (i.e., 50 mm) with O as centre.

ii) Since the axis of the follower passes through the axis of the cam shaft, therefore mark trace point A, as shown in Figure 10.45.

iii) From OA, mark angle $AOS = 60°$ to represent outstroke, angle $SOT = 30°$ to represent dwell and angle $TOP = 60°$ to represent return stroke.

iv) Divide the angular displacements during outstroke and return stroke (i.e., angle AOS and angle TOP) into the same number of equal even part as in displacement diagram.

v) Join the points 1, 2, 3 etc., and o′, 1′, 2′, 3′....., etc., with centre O and produce beyond the base circle as shown in Figure 10.45.

vi) Now set-off 1B, 2C, 3D...etc. and 0′ H, 1′ J ... etc., from displacement diagram.

vii) Join the points A, B, C ...M, N, P with a smooth curve. The curve $AGHPA$ is the complete profile of the cam.

Note to reader: The points *B, C, D ... L, M, N* may also be obtained as follows:

i)   Mark *AY* = 40 mm on the axis of the follower, and set-off, *Ab, Ac, Ad,...*, etc., equal to the distance 1B, 2C, 3D, ... etc., as in displacement diagram.

ii)   From the centre of the cam *O*, draw arcs with radii *Ob, Oc, Od,* etc. The arcs intersect the produced lines *O*1, *O*2 ... etc., at *B, C, D, ... L, M, N.*

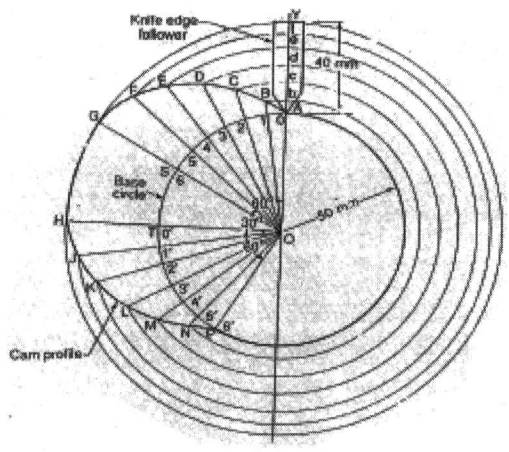

Figure 10.45   Cam Profile for Problem No.33 Case(i)

Case (ii): Non-collinear case: Cam axis and follower axis are not coinciding or the follower is OFF set by a distance

The profile of the cam when the axis of the follower is offset from the axis of the cam shaft, as shown in Figure 10.46 is drawn as discussed in the following steps:

i)   Draw a base circle with radius equal to the minimum radius of the cam (i.e., 50 mm) with *O* as centre.

ii)   Draw the axis of the follower at a distance of 20 mm from the axis of the cam, which intersects the base circle at *A*.

iii) Join *AO* and draw an offset circle of radius 20 mm with centre *O*.

iv)   From *OA*, mark angle *AOS* = 60° to represent outstroke, angle *SOT* = 30° to represent dwell and angle *TOP* = 60° to represent return stroke.

v)   Divide the angular displacement during outstroke and return stroke (i.e., angle *AOS* and angle *TOP*) into the same number of equal even parts as in displacement diagram.

vi)   Now from the points 1, 2, 3…etc., and 0', 1', 2', 3'… etc., on the base circle, draw tangents to the offset circle and produce these tangents beyond the base circle as shown in Figure 10.46.

vii) Now set off 1B, 2C, 3D…etc., and 0' H, 1' J … etc., from the displacement diagram.

viii) Join the points A, B, C …M, N, P with a smooth curve. The curve AGHPA is the complete profile of the cam.

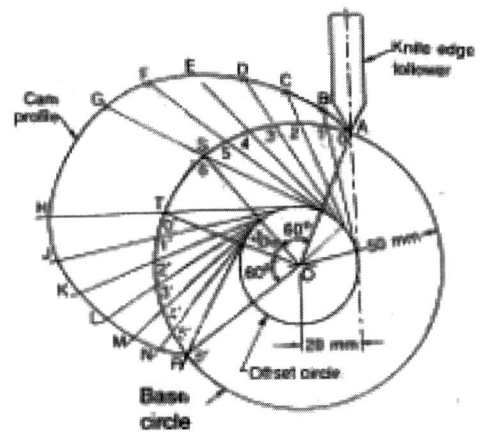

Figure 10.46   Cam Profile for Problem No.33 Case(ii)

34. A cam is to be designed for a knife edge follower with the data: cam lift = 40 mm, during 90° of cam rotation with simple harmonic motion. Dwell for the next 30°. During the next 60° of cam rotation, the follower returns to its original position with simple harmonic motion. Dwell during the remaining 180°. Draw the profile of the cam when (a) the line of stroke of the follower passes through the axis of the cam shaft, and (b) the line of stroke is offset 20 mm from the axis of the cam shaft. The radius of the base circle of the cam is 40 mm (Figure 10.47).

Determine the maximum velocity and acceleration of the follower during its ascent and descent, if the cam rotates at 240 rpm.

Solution:

The procedure of construction is similar to the previous example and the reader is advised to follow the same.

Maximum velocity of the follower during its ascent and descent.

We know that angular velocity of the cam,

$$\omega = \frac{2\pi N}{60} = \frac{2\pi \times 240}{60} = 25.14 \, rad/s$$

We also know that the maximum velocity of the follower during its ascent,

$$v_o = \frac{\pi \omega S}{2\theta_o} = \frac{\pi \times 25.14 \times 0.04}{2 \times 1.571} = 1m/s \quad Ans.$$

and maximum velocity of the follower during its descent,

$$v_R = \frac{\pi \omega S}{2\theta_R} = \frac{\pi \times 25.14 \times 0.04}{2 \times 1.047} = 1.51m/s \quad Ans.$$

Maximum acceleration of the follower during its ascent and descent.

We know that the maximum acceleration of the follower during its ascent,

$$a_o = \frac{\pi^2 \omega^2 \cdot S}{2(\theta_o)^2} = \frac{\pi^2 (25.14)^2 0.04}{2(1.571)^2} = 50.6m/s^2 \quad Ans.$$

And maximum acceleration of the follower during its descent,

$$aR = \frac{\pi^2 \omega^2 \cdot S}{2(\theta_R)^2} = \frac{\pi^2 (25.14)^2 0.04}{2(1.047)^2} = 113.8m/s^2 \quad Ans.$$

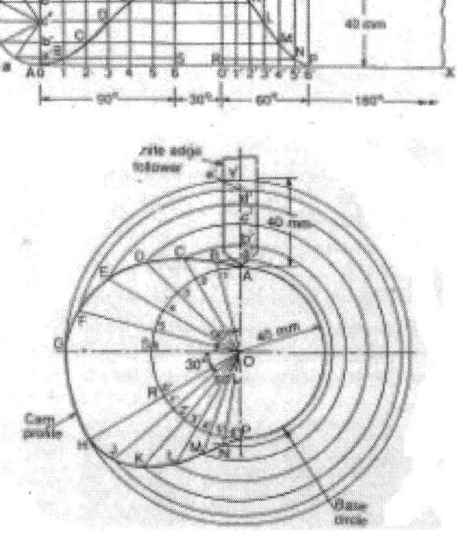

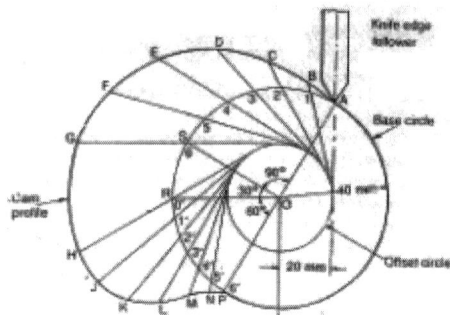

Figure 10.47   Cam Profile for Problem No.34

35. A cam, with a minimum radius of 25 mm, rotating clockwise at a uniform speed is to be designed to give a roller follower, at the end of a valve rod, motion described as: To raise the valve through 50 mm during 120° rotation of the cam; To keep the valve fully raised through next 30°; To lower the valve during next 60°; and To keep the valve closed during rest of the revolution i.e., 150°. The diameter of the roller is 20 mm and the diameter of the cam shaft is 25 mm. Draw the profile of the cam when (a) the line of stroke of the valve rod passes through the axis of the cam shaft, and (b) the line of the stroke is offset 15mm from the axis of the cam shaft. The displacement of the valve, while being raised and lowered, is to take place with simple harmonic motion.

Determine the maximum acceleration of the valve rod when the cam shaft rotates at 100 rpm. Draw the displacement, the velocity and the acceleration diagrams for one complete revolution of the cam.

Solution

Given: $S = 50$ mm $= 0.05$ m; $\theta_o = 120° = 2\pi/3$ rad $= 2.1$ rad; $\theta_R = 60° = \pi/3$ rad $= 1.047$ rad; $N = 100$ rpm.

Since the valve is being raised and lowered with simple harmonic motion, therefore, the displacement diagram, is drawn in the similar manner as discussed in the previous example (Figure 10.48).

(a) Profile of the cam when the line of stroke of the valve rod passes through the axis of the cam shaft.

The profile of the cam, is drawn as discussed in the following steps:

i)   Draw a base circle with centre O and radius equal to the minimum radius of the cam (i.e., 25 mm.)

ii)   Draw a prime circle with centre O and radius.

iii) $OA = \text{Min. radius of cam} + \dfrac{1}{2} \text{ Dia. of roller} = 25 + \dfrac{1}{2} \times 20 = 35mm.$

iv) Draw angle *AOS* = 120° to represent raising or out stroke of the valve, angle *SOT* =30° to represent dwell and angle *TOP* = 60° to represent lowering or return stroke of the valve.

v) Divide the angular displacements of the cam during raising and lowering of the valve (i.e., angle *AOS* and *TOP*) into the same number of equal even parts as in displacement diagram.

vi) Join the points 1, 2, 3, etc. with the centre *O* and produce the lines beyond prime circle.

vii) Set off 1B, 2C, 3D etc. equal to the displacements from displacement diagram.

viii) Join the points *A, B, C...N, P, A*. The curve drawn through these points is known as *pitch curve.*

ix) From the points *A, B, C ... N, P*, draw circles of radius equal to the radius of the roller.

x) Join the bottoms of the circles with a smooth curve. This is the required profile of the cam.

(b) Profile of the cam when the line of stroke is offset 15 mm from the axis of the cam shaft.

The profile of the cam when the line of stroke is offset from the axis of the cam shaft, may be drawn as discussed in the following steps:

i) Draw a base circle with centre *O* and radius equal to 25 mm.

ii) Draw a prime circle with centre *O* and radius *OA* =35 mm.

iii) Draw an offset circle with centre *O* and radius equal to 15 mm.

iv) Join *OA*. From *OA* draw the angular displacements of cam i.e., draw angle, *AOS* = 120° angle *SOT* = 30° and angle *TOP* = 60°.

v) Divide the angular displacements of the cam during raising and lowering of the valve into the same number of equal even parts (i.e., six parts) as in displacement diagram.

vi) From points 1, 2, 3...etc. and 0′, 1′, 2′, 3′... etc. on the prime circle, draw tangents to the offset circle.

vii) Set off 1B, 2C, 3D...etc. equal to displacement as measured from displacement diagram.

viii) By joining the points *A, B, C... M, N, P*, with a smooth curve, we get a pitch curve.

ix) Now *A, B, C ...* etc., as centre, draw circles with radius equal to the radius of roller.

x) Join the bottoms of the circles with a smooth curve. This is the required profile of the cam.

Maximum acceleration of the valve rod.

We know that angular velocity of the cam shaft,

$$\omega = \frac{2\pi N}{60} = \frac{2\pi \times 100}{60} = 10.47 rad / s$$

We also know that maximum velocity of the valve rod to raise valve,

$$v_o \frac{\pi\omega.S}{2\theta_o} = \frac{\pi \, x \, 10.47 \times 0.05}{2 \times 2.1} = 0.39 m / s.$$

And maximum velocity of the valve rod to lower the valve,

$$v_R = \frac{\pi\omega.S}{2\theta_R} = \frac{\pi \times 10.47 \times 0.05}{2 \times 1.047} = 0.785 m / s.$$

The maximum acceleration of the valve rod to raise the valve,

$$a_o = \frac{\pi^2 \omega^2.S}{2(\theta_o)^2} = \frac{\pi^2 (10.47)^2 \, 0.05}{2.(2.1)^2} = 6.13 m / s^2 \qquad Ans.$$

And the maximum acceleration of the valve rod to lower the valve,

$$a_R = \frac{\pi^2 \omega^2.S}{2(\theta_R)^2} = \frac{\pi^2 (10.47)^2 \, 0.05}{2.(1.047)^2} = 24.67 \ m / s^2 \qquad Ans.$$

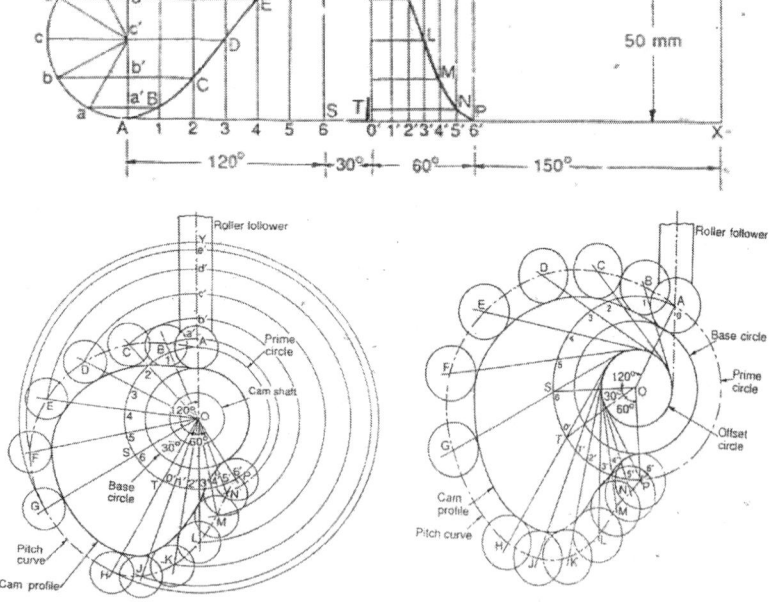

Figure 10.48   Cam Profile for Problem No.35

36. A cam drives a flat reciprocating follower in the following manner: During first 120° rotation of the cam, follower moves outwards through a distance of 20 mm with simple harmonic motion. The follower dwells during next 30° of cam rotation. During next 120° of cam rotation, the follower moves inwards with simple harmonic motion. The follower dwells for the next 90° of cam rotation. The minimum radius of the cam is 25 mm. Draw the profile of the cam.

Solution:

The follower moves outwards and inwards with simple harmonic motion. The profile of the cam driving a flat reciprocating follower is drawn as mentioned below:

i) Draw a base circle with centre O and radius OA equal to the minimum radius of the cam (i.e., 25 mm)

ii) Draw angle $AOS = 120°$ to represent the outward stroke, angle $SOT = 30°$ to represent dwell and angle $TOP = 120°$ to represent inward stroke.

iii) Divide the angular displacement during outwards stroke and inwards stroke (i.e., angle AOS and TOP) into the same number of equal even parts as in the displacement diagram.

iv) Join the points 1, 2, 3 ... etc., with centre O and produce beyond the base circle.

v) From points 1, 2, 3 ... etc., set-off 1B, 2C, 3D ... etc., equal to the distances measured from the displacement diagram.

vi) Now at points $B, C, D, ... M, N, P$, draw the position of the flat-faced follower. The axis of the follower at all these positions passes through the cam centre.

vii) The curve drawn tangentially to the flat side of the follower is the required profile of the cam, as shown in Figure 10.49.

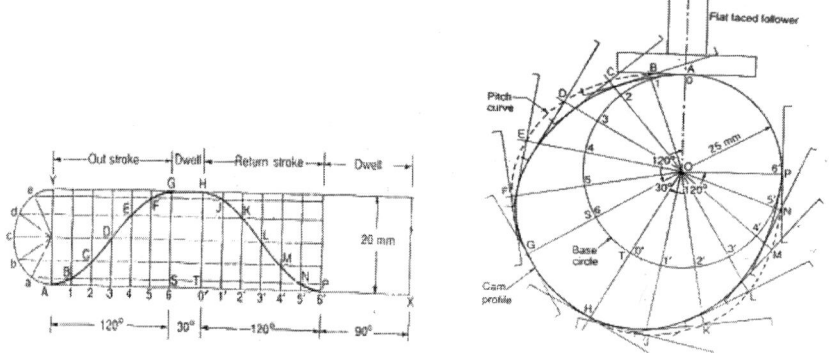

Figure 10.49 Mashroom Cam Profile for Problem No.36

37. A cam that is designed for cycloidal motion drives a flat-faced follower. During the rise, the follower displaces 1 in for 180° of cam rotation. If the cam angular velocity is constant at 100 rpm, determine the displacement, velocity, and acceleration of the follower at a cam angle of 60°.

Solution :

The equation for cycloidal motion is:

$$y = L\left(\frac{\theta}{\beta} - \frac{1}{2\pi}\sin\frac{2\pi\theta}{\beta}\right)$$

For $L = 1$, and $\beta = 180° = \pi$, then

$$y = L\left(\frac{\theta}{\beta} - \frac{1}{2\pi}\sin\frac{2\pi\theta}{\beta}\right) = 1\left(\frac{\theta}{\pi} - \frac{1}{2\pi}\sin\frac{2\pi\theta}{\pi}\right) = \left(\frac{\theta}{\pi} - \frac{1}{2\pi}\sin 2\theta\right)$$

$$\overset{.}{y} = \frac{L\omega}{\beta}\left(1 - \cos\frac{2\pi\theta}{\beta}\right) = \frac{\omega}{\pi}(1 - \cos 2\theta)$$

$$\overset{..}{y} = 2L\pi\left(\frac{\omega}{\beta}\right)^2\sin\frac{2\pi\theta}{\beta} = 2\pi\left(\frac{\omega}{\pi}\right)^2\sin 2\theta$$

The angular velocity is $\theta = 100\ rpm = 100\frac{2\pi}{60} = 10.472\ rad\ /\ s$

When $\theta = 60° = \frac{\pi}{3}$,

$$y = \left(\frac{\pi/3}{\pi} - \frac{1}{2\pi}\sin 2(\pi/3)\right) = \left(\frac{1}{3} - \frac{1}{2\pi}\sin(2\pi/3)\right) = 0.195\ in.$$

$$\overset{.}{y} = \frac{\omega}{\pi}(1 - \cos 2\theta) = \frac{10.472}{\pi}(1 - \cos(2\pi/3)) = 5.000\ \frac{in.}{sec.}$$

$$\overset{..}{y} = 2\pi\left(\frac{\omega}{\pi}\right)^2\sin 2\theta = 2\pi\left(\frac{10.472}{\pi}\right)^2\sin(2\pi/3) = 60.46\ \frac{in.}{sec^2.}$$

38. A constant-velocity cam is designed for simple harmonic motion. If the flat-faced follower displaces 2″ for 180° of cam rotation and the cam angular velocity is 100 rpm, determine the displacement, velocity, and acceleration when the cam angle is 45°.

Solution :

The equation for simple harmonic motion is:

$$y = \frac{L}{2}\left(1 - \cos\frac{\pi\,\theta}{\beta}\right)$$

For $L = 2$, and $\beta = 180° = \pi$, then,

$$y = \frac{L}{2}\left(1 - \cos\frac{\pi\,\theta}{\beta}\right) = \frac{2}{2}\left(1 - \cos\frac{\pi\theta}{\pi}\right) = (1 - \cos\theta)$$

$$\dot{y} = \frac{d\,y}{d\,t} = \frac{d}{d\,t}(1 - \cos\theta) = \dot{\theta}\,\sin\theta$$

$$\ddot{y} = \frac{d^2 y}{d\,t^2} = \frac{d}{d\,t}\left(\dot{\theta}\,\sin\theta\right) = \dot{\theta}^2\,\cos\theta$$

The angular velocity is $\dot{\theta} = 100\ rpm = 100\,\dfrac{2\,\pi}{60} = 10.472\ rad\,/\,s$

When $\theta = 45°$,

$$y = (1 - \cos\theta) = (1 - \cos 45°) = 1 - 0.707 = 0.292\ in.$$

$$\dot{y} = \dot{\theta}\,\sin\theta = 10.472\,\sin 45° = 10.472\,(0.707) = 7.405\,\frac{in.}{s}$$

$$\ddot{y} = \dot{\theta}^2\,\cos\theta = 10.472^2\,(0.707) = 77.531\,\frac{in.}{s^2}$$

39. A cam drives a radial, knife-edged follower through a 1.5″ rise in 180° of cycloidal motion. Give the displacement at 60° and 100°. If this cam is rotating at 200 rpm, what are the velocity (ds/dt) and the acceleration (d^2s/dt^2) at $\theta = 60°$?

Solution :

The equation for cycloidal motion is:

$$y = L\left(\frac{\theta}{\beta} - \frac{1}{2\pi}\sin\frac{2\pi\theta}{\beta}\right)$$

For $L = 1.5$, and $\beta = 180° = \pi$, then

$$y = L\left(\frac{\theta}{\beta} - \frac{1}{2\pi}\sin\frac{2\pi\theta}{\beta}\right) = 1.5\left(\frac{\theta}{\pi} - \frac{1}{2\pi}\sin\frac{2\pi\theta}{\pi}\right) = 1.5\left(\frac{\theta}{\pi} - \frac{1}{2\pi}\sin 2\theta\right)$$

$$y = \frac{L\omega}{\beta}\left(1 - \cos\frac{2\pi\theta}{\beta}\right) = \frac{1.5\omega}{\pi}\left(1 - \cos 2\theta\right)$$

$$y = 2L\pi\left(\frac{\omega}{\beta}\right)^2 \sin\frac{2\pi\theta}{\beta} = 2(1.5)\pi\left(\frac{\omega}{\pi}\right)^2 \sin 2\theta = 3\pi\left(\frac{\omega}{\pi}\right)^2 \sin 2\theta$$

The angular velocity is $\dot{\theta} = 200$ rpm $= 200\frac{2\pi}{60} = 20.944\ rad\ /\ s$

When $\theta = 60° = \frac{\pi}{3}$,

$$y = 1.5\left(\frac{\theta}{\pi} - \frac{1}{2\pi}\sin 2\theta\right) = 1.5\left(\frac{\pi/3}{\pi} - \frac{1}{2\pi}\sin\frac{2\pi}{3}\right) = 0.293\ in.$$

$$\dot{y} = \frac{1.5\ \omega}{\pi}\left(1 - \cos 2\theta\right) = \frac{1.5(20.944)}{\pi}\left(1 - \cos\frac{2\pi}{3}\right) = 15.00\frac{in.}{s}$$

$$\ddot{y} = 3\pi\left(\frac{\omega}{\pi}\right)^2 \sin 2\theta = 3\pi\left(\frac{20.944}{\pi}\right)^2 \sin\frac{2\pi}{3} = 362.76\frac{in.}{s^2}$$

When $\theta = 100° = \frac{100\pi}{180} = \frac{5\pi}{9}$,

$$y = 1.5\left(\frac{\theta}{\pi} - \frac{1}{2\pi}\sin 2\theta\right) = 1.5\left(\frac{5\pi/9}{2\pi} - \frac{1}{2\pi}\sin\frac{10\pi}{9}\right) = 0.915\ in.$$

40. Draw the displacement schedule for a follower that rises through a total displacement of 1.5″ with constant acceleration for 1/4th revolution, constant velocity for 1/8th revolution, and constant deceleration for 1/4th revolution of the cam. The cam then dwells for 1/8th revolution, and returns with simple harmonic motion in 1/4th revolution of the cam.

Solution :

The displacement profile can be easily computed using the equations using Matlab. The curves are matched at the endpoints of each segment. The profile equations are:

For $0 \le \theta \le \frac{\pi}{2}$

$$y_1 = a_0 + a_1\theta + a_2\theta^2$$

The boundary conditions at $\theta = 0$ are $y_1 = 0$ and $y_2 = 0$. Therefore,

$a_0 = a_1 0$

So,

$y_1 = a_2 \theta^2$

and

$y'_1 = 2 a_2 \theta$

where a_2 is yet to be determined.

For $\dfrac{\pi}{2} \leq \theta \leq \dfrac{3\pi}{4}$

$y_2 = b_0 + b_1 \theta$

The boundary conditions at $\theta = \dfrac{\pi}{2}$ are $y_1 = a_2 \left(\dfrac{\pi}{2} \right)^2$

and $y'_1 = 2a_2 \dfrac{\pi}{2} = a_2 \pi.$

Then $\quad b_0 + b_1 \dfrac{\pi}{2} = a_2 \left(\dfrac{\pi}{2} \right)^2$

and

$0 = - a_2 \left(\dfrac{\pi}{2} \right)^2 + b_0 + b_1 \dfrac{\pi}{2} .$

Also

$y'_2 = b_1 = a_2 \pi$

or

$0 = a_2 \pi - b_1$

For $\dfrac{3\pi}{4} \leq \theta \leq \dfrac{5\pi}{4}$

$y_3 = c_0 + c_1 \theta + c_2 \theta^2$

$y'3 = c_1 + 2c_2 \theta$

$y''_3 = 2c_2.$

The boundary conditions at $\theta = \dfrac{3\pi}{4}$ are

$y_2 = a_2 \pi \left(-\dfrac{\pi}{4} + \theta \right) = a_2 \pi \left(-\dfrac{\pi}{4} + \dfrac{3\pi}{4} \right) = a_2 \dfrac{\pi^2}{2}$

and $y'_2 = a_2 \pi.$ Also, at $\theta = \dfrac{5\pi}{4}, y_3 = 1.5$ and $y'_3 = 0.$

Then, matching the conditions,

$$Y_3 = a_2 \frac{\pi^2}{2} = c_0 + c_1 \frac{3\pi}{4} + c_2 \left(\frac{3\pi}{4}\right)^2$$

$$y'3 = c_1 + 2c_2 \frac{3\pi}{4} = a_2 \pi$$

$$1.5 = c_0 + c_1 \frac{5\pi}{4} + c_2 \left(\frac{5\pi}{4}\right)^2$$

$$y'_3 = c_1 + 2c_2 \frac{5\pi}{4} = 0$$

The boundary condition equations can be written as:

$$0 = -a_2 \left(\frac{\pi}{2}\right)^2 + b_0 + b_1 \frac{\pi}{2}$$

$$0 = a_2 \pi - b_1$$

$$0 - a_2 \frac{\pi^2}{2} + c_0 + c_1 \frac{3\pi}{4} + c_2 \left(\frac{3\pi}{4}\right)^2$$

$$0 = -a_2 \pi + c_1 + 2c_2 \frac{3\pi}{4}$$

$$1.5 = c_0 + c_1 \frac{5\pi}{4} + c_2 \left(\frac{5\pi}{4}\right)^2$$

$$0 = c_1 + 2c_2 \frac{5\pi}{4}$$

In matrix form,

$$
\begin{bmatrix} 0 \\ 0 \\ 0 \\ 0 \\ 0 \\ 1.5 \end{bmatrix} =
\begin{bmatrix}
-\left(\frac{\pi}{2}\right)^2 & 1 & \frac{\pi}{2} & 0 & 0 & 0 \\
\pi & 0 & -1 & 0 & 0 & 0 \\
-\frac{\pi^2}{2} & 0 & 0 & 1 & \frac{3\pi}{4} & \left(\frac{3\pi}{4}\right)^2 \\
-\pi & 0 & 0 & 0 & 1 & \frac{3\pi}{2} \\
0 & 0 & 0 & 0 & 1 & \frac{5\pi}{2} \\
0 & 0 & 0 & 1 & \frac{5\pi}{2} & \left(\frac{5\pi}{4}\right)^2
\end{bmatrix}
\begin{bmatrix} a_2 \\ b_0 \\ b_1 \\ c_0 \\ c_1 \\ c_2 \end{bmatrix}
$$

Solving for the constraints using Matlab,

$$\begin{bmatrix} a_2 \\ b_0 \\ b_1 \\ c_0 \\ c_1 \\ c_2 \end{bmatrix} = \begin{bmatrix} 0.2026 \\ -0.5000 \\ 0.6366 \\ -1.6250 \\ 1.5915 \\ -0.2026 \end{bmatrix}$$

The equations are then given in the following:

For $0 \le \theta \le \dfrac{\pi}{2}$

$y_1 = a_2 \theta^2 = 0.2026 \theta^2$

For $\dfrac{\pi}{2} \le \theta \le \dfrac{3\pi}{4}$

$y_2 = b_0 + b_1 \theta = -0.5000 + 0.6366\theta$

For $\dfrac{3\pi}{4} \le \theta \le \dfrac{5\pi}{4}$

$y_3 = c_0 + c_1\theta + c_2 \theta^2 = -1.6250 + 1.5915\theta - 0.2026\theta^2$

For $\dfrac{5\pi}{4} \le \theta \le \dfrac{3\pi}{2}$

$y_4 = 1.5.$

For the return, $\dfrac{3\pi}{2} \le \theta \le 2\pi$, and

$y_5 = \dfrac{L}{2}\left(1 + \cos\dfrac{\pi\theta}{\beta}\right) = \dfrac{3}{4}(1 + \cos 2\theta))$

10.9   Construction of textile cams and tappets

Textile machines like ring frame, precision winding machine, loom crank shaft, shedding mechanism, picking mechanism, beat-up motion, let-off motion, etc., are also using cams and tappets for various purposes (Figure 10.50).

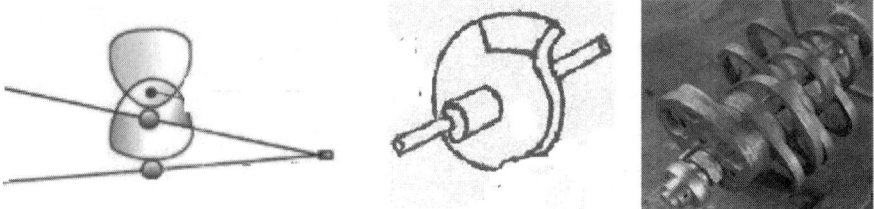

Figure 10.50 Shedding Tappet for Problem No.37

10.9.1 Generation of tappet profile through CAD software

In these days the mechanical engineering has produced a CADD (Computer Aided Design and Drafting) and is very useful in developing the profile of the tappets. Following example is one such instance of designing the tappet through software. (Source: Internet) (Figure 10.51).

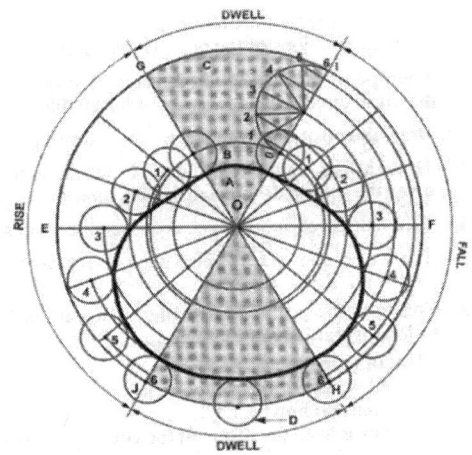

Figure 10.51 Shedding Tappet Profile for Problem No.38

Tappets are the shedding devises used for shedding process. The minimum number is 2 and the maximum number may be 5 or 7.

10.9.2 Construction of shedding tappet (Hhand drawn)

General procedure for all plain tappets

(i) *Minimum data required for tappet construction*

1. Least distance of contact or LDC, 2. Bowl or Roller diameter, 3. Lift, 4. Weave, 5. Dwell.

Generally the dwell is one third of a pick as the healds in shedding require 120° of stationary period or dwell which is one third of 360° and for every one revolution of crank shaft one pick is inserted.

(ii) *Procedure for constructing the profile of shedding plain tappet*

1. First draw a circle equal to radius of shaft (if given).
2. Draw next concentric circle with radius equals to shaft radius + LDC.
3. Draw next concentric circle with radius equals to shaft radius + LDC + Lift.
4. Divide the circles into number of picks per repeat given for example in plain it is 2 picks per repeat and therefore divide them into two equal parts.
5. Name one part as up and the other as down.
6. Now divide each part into three parts out of which one is dwell and the remaining two will be change.
7. Now mark the centre point on the lift line.
8. Now construct the semicircle and divide the semi-circle into 6 parts and extend these points on the lift line and name them as 1, 2, ... 6.
9. Now consider the both change parts and divide them into 6 parts.
10. From the centre take the radius equal to the divided points on the lift line, cut the corresponding generator in each change sections.
11. With these as centres describe a semi or circle with radius equal to bowl.
12. Join the tangents of these circles in change area.
13. In dwell sections, describe the part of the circle with end point of change and centre of main circle as radius to touch the other portion of dwell and repeat the same procedure for the second dwell also.

10.9.3   General procedure for all twill tappets

The procedure for constructing the twill tappet remains the same as that of the plain tappet but the circles drawn will be divided into number of picks as mentioned in the weave. For example, if it is 2/1 weave then divide the circles into 3 parts and if it is given as 2/2 twill, divide the circles into 4 parts etc., Then other point to be noted is that if once the construction is started, the next part of the pick to be seen. In other words consider 2/1 weave, in which we will have UP, UP and DOWN. Now if the construction is started in first UP, the last point in the 1st UP is continued till the 2nd UP is closing and then the construction is reversed in DOWN. Similarly if it is given as 1/2 weave, the construction in UP potion is outwards and in first DOWN is reverse and the same is continued for the second DOWN also. The reader is instructed to observe carefully the twill tappets.

1. Construct the profile of a shedding tappet if the data is LDC – 2 cm, Type of motion to follower – SHM, Follower – roller, Dwell 1/3rd of a pick, Weave 1/2 Twill, Lift 4 – cm (Figure 10.52).
2. Develop the profile of a plain shedding tappet if LDC – 2 cm, Type of motion to follower – SHM, Follower – roller, Dwell 1/3rd of a pick, Weave 1/1 Plain, Lift – 4 cm (Figure 10.53).
3. Develop the profile of a plain shedding tappet if LDC – 2 cm, type of motion to follower – SHM, Follower – roller, bowl diameter – 1 cm, Dwell 1/3rd of a Ppick, Weave 1/1 Plain, lift 4 cm (Figure 10.54).
4. Construct the profile of a shedding tappet if the data is LDC – 2 cm, type of motion to follower – SHM, Follower – roller, Dwell 1/3rd of a pick, Weave 2/1 Twill, Lift 4 cm (Figure 10.55).
5. Construct the profile of a shedding tappet if the data is LDC – 2 cm, type of motion to follower – SHM, Follower – roller, Dwell 1/3rd of a pick, Weave 2/2 Twill, Lift 4 cm (Figure 10.56).
6. Construct the profile of a shedding tappet if the data is LDC – 4 cm, type of motion to follower – SHM, Follower – roller, Dwell 1/3rd of a pick, Weave 2/2 Twill, Lift 4 cm (Figure 10.57).
7. Construct the profile of a shedding tappet if the data is LDC – 4 cm, type of motion to follower – SHM, Follower – roller, Dwell 1/3rd of a pick, Weave 3/1 Twill, Lift 4 cm (Figure 10.58).
8. Construct the profile of a shedding tappet if the data is LDC – 4 cm, type of motion to follower – SHM, Follower – roller, Dwell 1/3rd of a pick, Weave 3/2 Twill, Lift 4 cm (Figure 10.59).

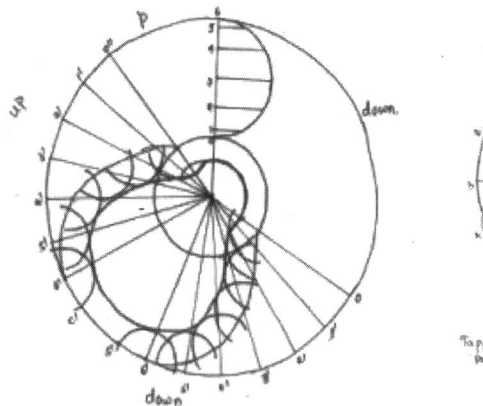

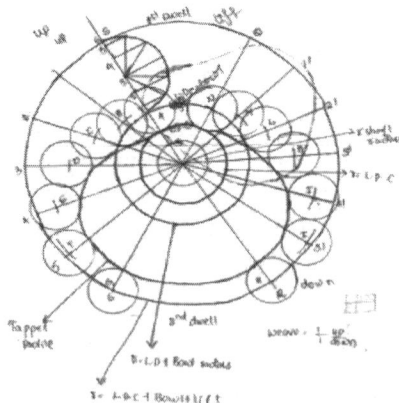

Figure 10.52 1 up 2 down twill tappet   **Figure 10.53** Plain shedding tappet

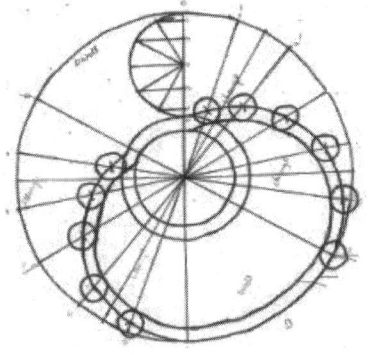

Figure 10.54 Plain shedding tappet

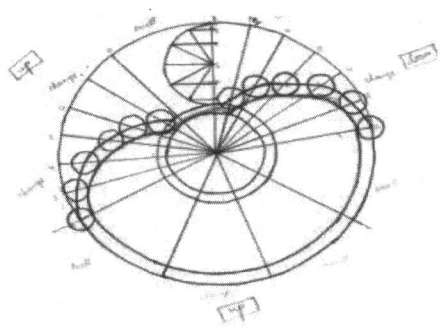

Figure 10.55 Twill 2 up 1 down tappet

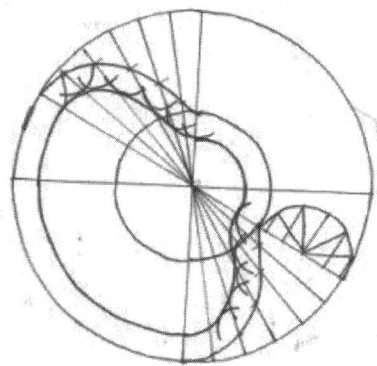

Figure 10.56 Twill 2 up 2 down tappet

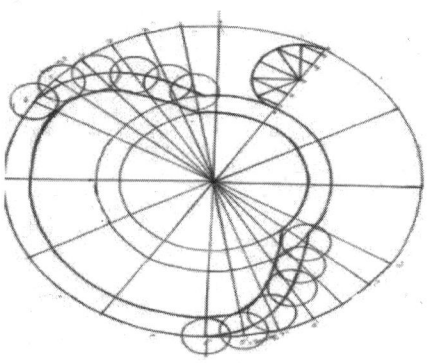

Figure 10.57 Twill 2 up 2 down tappet

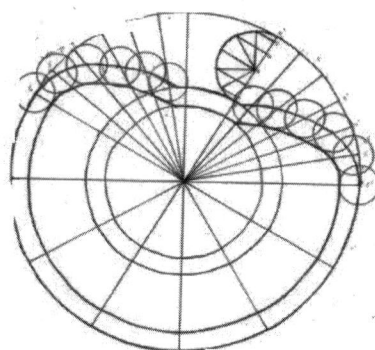

Figure 10.58 Twill 2 up 2 down tappet

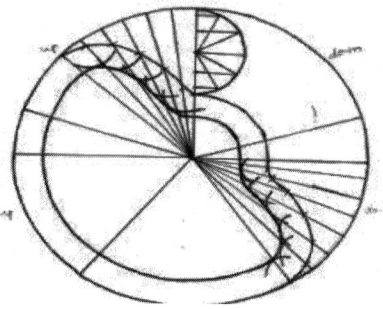

Figure 10.59 Twill 2 up 2 down tappet

10.9.4   Construction of three-leaved cam for cop build in ring spinning

A three-leaved cam as shown in Figure 10.60 used for the production of cop build bobbins. The cam will have two portions names winding and binding. The binding will help in laying the binding coils to support the winding coils. The ring rails moves faster during these binding coils as they will be less in number when compared to winding coils (Figure 10.61).

Figure 10.60   Three-leaved cam for ring frame

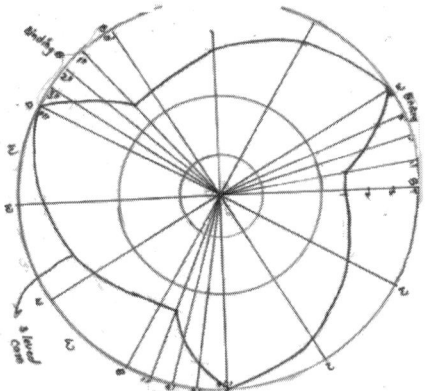

Figure 10.61   Construction of three-leaved cam

Power required to drive spinning and weaving machines

11.1   Introduction

The power required to drive a machine is spent in overcoming various re-sistances, which include (1) frictional resistances, such as bearing friction, air resistance, etc.; (2) resistance due to inertia or to accelerating parts of the machine; (3) resistances offered by the material being treated by the machine. In most textile machines the work spent in the last manner, i.e., the useful work done, is very small compared with the total, and hence the efficiencies of the machines, reckoned in the usual way, are low. Table 11.1 shows the general level of power spent in textile mill.

Work spent in overcoming bearing friction depends largely on the na-ture of the bearings and the system of lubrication; generally the bearings are plain, with intermittent lubrication; the coefficient of friction i.e., the ratio $\frac{\text{Friction}}{\text{Resultant load on bearing}}$, varies very much with the quality and quantity of the lu-brication; for the majority of bearings the value is probably about 0.1 though under good conditions it will be as low as 0.01. Its value depends on the na-ture of the bearing surfaces, on the lubrication, and also on the surface speed. No definite rule can be given for the latter, but generally the coefficient de-creases with increase of speed up to about 200 ft/min, after which the coeffi-cient rises.

Table 11.1   General level of power used in textile mills

Machine	Power	Number of units	Production
Blow room			
Bale plucker	11.9 kw		800–1000 kg/h
Material transport fan	7.5 kw		
Vario cleaner	3		
Contamination scanner with ventilator	3		
Unimix	7.5		
Flexi clean	3		
Condenser	5.5		
Total blow room	41.4		

Carding			
Card 3 lick	16.33 kw	10 cards	
Card single lick	14.5 kw		
Waste recovery system	90 kw		
Total cards and waste recovery	253.3		
Draw frame			
Breaker draw frame	5.5	2	11 kw
Finisher draw frame	7.5	4	30 kw
Speed frame	18	4	72 kw
Ring frame 1440 spindles	75	15	1125 kw
1632 spindles	90		
Warp winding			
Autoconer	35	6	210 kw

For small changes in speed, the coefficient can be assumed to be constant. When all bearings are used the value of the coefficient of friction may be as slow as 0.002, and the power absorbed by such bearings maybe taken to be only about 10% of that required when ordinary bearings are used.

If the coefficient of friction = μ; the total load carried by the bearings = R lb; the speed W rpm and the diameter of the shaft D, then the friction force acting at the shaft surface $F = \mu R$; the work spent in overcoming the friction per minute will be $\mu R \pi \dfrac{D}{12} N$ foot-lb; and the horse-power (HP) needed to keep the shaft running will be $HP = \dfrac{\mu R \pi DN}{33,000 \times 12}$ or $HP = \dfrac{\mu R \pi DN}{4500}$ watts.

If the speed changes and the coefficient of friction remains unchanged, then the work done per minute and the horse-power will be in direct proportion to the speed. Air resistance, such as has to be overcome in driving a fan, or whenever a surface is moved in the air, is subject to different laws. The resistance depends on the nature of the surface, and varies approximately directly the area of the surface and as the square of the velocity. Thus the power varies approximately as the cube of the speed, the resistance varying as the square of the speed, while the displacement, on which also the power depends, varies directly as the speed. Thus, if 1 HP is needed to over-come air resistance on fan blades at 1000 rpm, then at 1500 rpm the horse power needed will be $1 \times \dfrac{1500_3}{1000_3} = 3.35$ approximately.

Work used in accelerating machine parts is frequently given up again when the parts are retarded and helps to drive other parts of the machine. For example, during the early part of the run in a mule carriage, work has to be expended on the carriage to speed it up; most of this work is given up during the latter part of the run in, while the carriage is slowing down, as it is then helping to drive the spindles and to other parts that are working. Under such circumstances, the net work done in accelerating and retarding is zero, but during the acceleration extra power is needed to drive the machine, whilst during the retardation less power is required. The work done during an acceleration is the product of the resistance offered by the body to having its velocity changed, multiplied by the displacement of the resistance. If the acceleration is a linear one the resistance in lbs is equal to mass (Engineer's units) × Acceleration (ft/s/s) if the acceleration is angular the resisting moment is *Moment of inertia × Angular acceleration*, and the work done is the product of the resting moment and the angular displacement (in radians) of the body during the acceleration. An easy way of getting the work spent during an acceleration, or given up during a retardation, is to find the change in kinetic energy due to the change in velocity. This change in kinetic energy is equal to the work done. The resistance offered by the material depends naturally on its treatment by the machine. Thus in carding, the cotton offers resistance to being detached by the licker in, resistance, due to carding, at the cylinder wires, etc. In the flyer frames the resistance is due to drawing and twisting the cotton, both of which resistances are negligible in comparison with the frictional and other resistances to be overcome in driving the machine parts.

Power required to drive Scutcher – The machine was driven by electric motor, and observations taken first with only the counter shaft of the machine running; next the beater was added, and so on until the machine was running normal. The figures given for the various parts are only approximately correct when the machine is running normally; most of the parts will take more power than that give, on account of the greater loads on the bearings, etc., due to cotton passing through the machine. The figures given are the net owners taken, the necessary allowances having been made for the power taken by the motor itself.

The power required by the countershaft and by the beater is all needed to overcome frictional resistances, mainly at the bearings. Power for the fan is spent in overcoming the air resistance on the blades and, to a small extent, frictional resistance at the bearings. The feeding parts include feed lattice and pedal roller; these parts move very slowly and are only loaded to a moderate extent, hence they absorb very little power. The delivery parts include the cages, calendar rollers and lap rollers these parts do not run very fast, but the calendar and lap rollers are heavily loaded and consequently consume considerable power. The power for the fan, as stated above, increases very rapidly with increase of speed.

11.2   Power for a carding engine

The cylinder of the carding carding engine uses about half the power needed to drive the machine and this power is mainly spent in overcoming bearing friction, although some is needed for air existence on the cylinder. Slow moving parts much such as the doffer, the coiler, and the side shaft, take very little power; the flats, although the speed is very low, have a good deal of friction and so account for rather a high proportion of the power. Fast running parts, like the doffer comb and the licker in, require a proportionately large share of the power. The power for carding the cotton naturally depends on the quantity of cotton passing through the machine in a given time, and tests show that the power required is nearly in direct proportion to the quantity carded in unit time. The power for carding is low, the doffer speed and production being low when the test was made. The power for carding varies form about the value given in the table to about 3 times as s much with a doffer speed of 16 rpm the lap weighting 10 oz/yd and the draft being 120 in both cases. Since the power required, other than for carding the cotton, is increased very slightly with increased production obtained by increasing the offer speed or reducing the draft, the proportion that the power for carding bears to the total will be much higher for a big production than the 11% – at high productions the proportion may be 25 or 30%. Important points in connection with driving of the carding engine are the heavy starting load and the long time machine takes to come to rest after the belt is put on the loose pulley. The high starting load is mainly due to the inertia of the heavy cylinder, mainly concentrated in the rim, that the radius of gyration is nearly equal to the full radius of the cylinder, say 2 ft. The kinetic energy in such a cylinder when rotating say, at

$$160 \text{ rpm} = \frac{W \, \omega^2 K^2}{2g}$$

where W = weight of cylinder I lb or kg
ω = angular velocity in radius per second,
K = radius of gyration in feet and g = 32.2 ft/s/s per 9.81 kg/s²

hence kinetic energy (KE) $= 10 \times 112 \text{ lb} \times \left(2\pi \times \frac{160}{60}\right)^2 \times \frac{2^2}{2 \times 32.2}$

$= 19{,}500$ feet-lb, or $26{,}438.45$ joules.

This energy has to be stored up in the cylinder during starting up, and has to be used up in overcoming bearing and other friction while the machine is being stopped. Compared with that of the cylinder, the energy in the other part of the machine is small, due to the light weight of slow speed of these parts. Assuming the belt to be 3″, wide, capable of applying, at the slipping point, an effective driving force of 200 lb. To the rim of a 15″ diameter pulley, the work done per revolution of the cylinder during starting up is 200 × π × 15/12 = 785 ft-lb or 15.7 joules. Of this a certain about is required to overcome frictional resistances; assuming these to remain constant during starting up, and assuming that the card requires 0.8 hp to keep it running steadily at 160 rpm, the work spent in friction per revolution is $\dfrac{0.8 x 33000}{160}$ =165 ft-lb or 3.3 joules. This leaves (785-165) = 620 foot-lb per revolution or 12.4 joules, available for accelerating the machine. The number of revolutions required to start up

$$= 20,000/620 = 32 \text{ approximately.}$$

The average speed during starting up, assuming uniform acceleration = 160/2 = 80 rpm. Time taken to start up = 32/40 × 60 =24 s. Also, during stopping, 20,000 foot-lb or 400 joules of energy have to be used up, and each revolution takes 165 foot-lb or 3.3 joules number of revolutions required to bring machine to rest = 20,000/165 =120 or 400 / 3.3 = 121.21 approximately. The time taken to stop will be approximately 120/80 rpm = 1½ min. The assumption that the frictional resistances to be overcome during starting and stopping are constant is only approximately true. Actually the resistance increase as the card speeds up, due to increasing air resistance on the cylinder. The effect of this will be that the average work done per revolution will be somewhat less than 165 foot-lb, which will increase the time for stopping and reduce slightly the time for starting up.

Power required to drive a flyer frame – Practically all the power is absorbed in friction, only a negligible amount being spent in drawing and twisting the cotton. Most of the power is absorbed in rotating the spindles, which are fairly heavy and not very well lubricated. The power for rotating the bobbins naturally varies since the speed is not constant.

Power required to drive a ring frame – The power required to drive ring frames varies very much in the conditions of working, the power required varying largely with tension of the spindle bands, which affects the power taken by tin roller and the spindles, and with the counts spun which affects the power needed to overcome traveller resistance. The number of spindles driven per horse power varies from about 60–100 or 110, the average being about 70–75. Thus a frame of, say, 380 spindles will require about $\dfrac{380}{70}$ =5.45 hp. The proportion of the power taken by various parts of the machine also varies; a test on a frame spinning 43s counts spindle speed 9300 rpm, ring diameter 1½", resulted as follows:

Power taken by tin rollers and bare spindles……….	61%
Power taken by rollers, traverse motion and gearing ….	11%
Power taken by weight of bobbins and yarn ……..	11%
Power taken by the pull of the traveller ………	<u>17%</u>
	<u>100%</u>

The power absorbed by the tin rollers and spindles is entirely due to bearing friction, and that is mainly caused by the pull of the driving bands. The resistance of the travellers' accounts for a considerable proportion of the power, and this resistance varies approximately inversely as the counts spun and directly as the diameter of the ring. Assuming the values for 20s yarn, the resistance to traveller movement round the ring averages about 240 grains. The radius of the path moved in by the traveller is 0.905", and the traveller speed about 8000 rpm. Then the horse power expended in moving 380 travellers will be

$$\left(\frac{240}{7000}\right)lb. \times 2\pi \times \frac{0.905}{12} \times \frac{8000}{33,000} \times 380 = 1.5H.P.$$

This would be 25%, or more the total power for the frame, which is a considerably higher proportion than that given above. This is accounted for by the fact that the traveller weight for the 43s yarn would be much less than that needed for the 20s. The resistance to traveller movement is nearly proportional to the centrifugal force of the traveller, which is in direct proportion to traveller weight and also to the ring diameter and the square of the speed. For a given frame, if the spindle speed is kept constant, the power for moving the travellers will be approximately inversely proportional to the counts spun, since traveller weight varies nearly in that proportion. Traveller resistance also depends on air resistance, which varies inversely as $\sqrt{counts}$, but this resistance is much smaller than the frictional resistance due to centrifugal force.

11.3   Mechanics at flyer frame and ring spinning

11.3.1   Flyer frame winding

In the flyer frames the material is wound in layers of closely pitched coils on parallel flangeless bobbins (Figure 11.1). Each successive layer is given a slightly shorter traverse than the preceding one so that conical ends are formed, which make the bobbins easier to handle without damage to the ends.

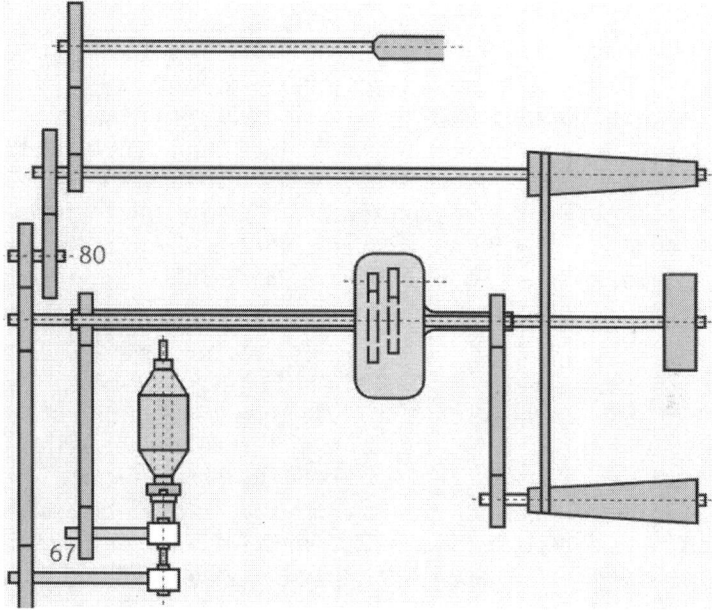

Figure 11.1 Driving arrangements of flyer frame

The material – Roving or roving – is a soft, practically untwisted threat or strand, varying in diameter from about 1/4–1/32″. It is very delicate and easily broken or stretched, and consequently the winding speed has to be regulated very carefully; under proper conditions, the winding speed is kept quite constant from the start to the finish of the bobbin. The bobbin is mounted loosely on a spindle S and is driven positively by gearing. The spindle also is driven by gearing and rotates at constant speed. The roving is led as shown down the hollow leg L of a flyer, fixed on the spindle, and is guided on to the bobbin by a presser P, mounted freely on L and kept pressing against the bobbin by centrifugal force acting on a balance weight W. Each revolution of the flyer puts one twist into the material, and since the delivery of material to the flyer is constant the spindles must rotate at constant speed to keep the number of twists per inch constant. The winding of the roving on the bobbin is effected by running the bobbin at a slightly higher or lower speed than the spindle. The former method, known as bobbin lead, is almost universally used in cotton spinning frames; it has the advantage that the change in speed needed at the end of each traverse is a reduction in speed and not an increase as in flyer lead. There is also reduced friction in the differential motion through which the winding speed regulated; this will be more easily seen when the motions have been considered. Flyer lead is very often used in jute spinning; it has the advantage that the bobbin speeds used are lower than those needed for bobbin lead.

In bobbin lead the bobbin draws the roving on to itself, through the presser eye; in flyer lead the presser lays the roving on the bobbin. In both it is the difference between the speeds of bobbin and flyer that gives the winding revolutions; for every revolution that the bobbin gains over the flyer, or lags behind it, one coil of roving will be put on the bobbin.

As the bobbin increases in diameter the number of winding revolutions per minute must be reduced in inverse proportion to the increase in bobbin diameter, the product of bobbin diameter and winding revolutions per minute remaining constant, this change in winding speed, or the difference in speed between the bobbin and the spindle, is regulated by means of tapered cones, or their equivalent, driving through an epicyclic wheel train, known as a differential motion. At the end of each layer being regulated to suit the count of the material. The use of the differential motion enables the speed variation obtained from the cones to apply only to the difference in speed between bobbin and spindle, i.e., to the winding speed, and not to the total bobbin sped; it also reduces, to a large extent, the power that has to be transmitted through the cone drive, reducing belt slip and making the speed regulation more exact.

The spacing of the coils is done by raising and lowering the bobbin whilst the spindle and flyer rotate in fixed positions. The rate of movement of the lifter rail, which lifts and lowers the bobbins, has to be altered for each successive layer; like the winding speed of the bobbins the rate of the lifter rail's movement must vary in inverse proportion to the bobbin diameter, and so the same cone drive which varies the speed of the bobbins is used to vary the speed of the lifter. The arrangement of the mechanism for a cotton intermediate frame is shown in Figure 11.1. from the driving shaft of the machine a wheel train S gives motion to the spindles. The bobbins are driven by a train B from a bobbin wheel BW which is part of the differential motion. The motion shown in the figure is known as Tweedales. The arm of the differential motion is fixed to the main shaft, and carries loosely the spindle of a compound bevel carrier. The larger of the bevel wheels gears with one which is mounted loosely on the main shaft and which receives is motion from the bottom or driven cone drum. The smaller bevel wheel mounted on the arm gears with a wheel compounded with the bobbin wheel BW, these wheels also being loose on the main shaft. The top or driving cone drum is driven direct from the main shaft; a train of wheels from the top cone drives the drawing rollers, the front pair of rollers delivering the material to the flyer.

Calling the speed of the main shaft "*a*", that of the bobbin wheel "*n*", and that of the wheel driven from the cones "*m*" revolutions per minute, respectively, the velocity ratio of the motion,

$$e = \frac{n-a}{m-a} = +\frac{18}{30} \, x \, \frac{16}{48} = +\frac{1}{5}$$

Where $5n - 5a = m - a$

$$n = \frac{4}{5}a + \frac{1}{5}m$$

i.e., the speed of the bobbin wheel is equal to a constant multiplied by the speed of the main shaft, plus another constant multiplied by the speed of the wheel driven from the cones.

All differential motions give equations of this type; sometimes the cone drums vary the speed not of the first wheel of the train, but of the arm of the motion, but in all of the motions the speed of the bobbin wheel is made up of two parts; one part, derived from the main shaft, is constant, whilst the other, derived from the driven cone, is variable. The constant part ($\frac{4}{5}a$ in the example) gives to the bobbins a speed equal to the spindle speed; the variable part ($\frac{1}{5}m$) gives the bobbins their winding speed. The gearing is always arranged in such a way that when the part driven from the cones is kept stationary, the bobbins rotate at the same speed as the spindles. The part driven from the cones then simply adds, for a bobbin lead frame, or subtracts, for a flyer lead frame, a sufficient number of revolutions to give the necessary winding speed.

Now the winding speed of the bobbin must vary inversely as the bobbin diameter; the speed of the variable part of the differential – that is, the part driven from the cones – must vary in the same way, since the winding speed depends directly on this speed. The speed of the variable part of the differential is in direct proportion to the speed of the bottom cone, and therefore the speed of the bottom cone varies inversely as the bobbin diameter.

$$\therefore \text{Speed of top cone } x \, \frac{D}{d} \propto \frac{1}{bobbin \ diameter}$$

and, since the speed of the top cone is constant,

$$\frac{D}{d} = r \propto \frac{1}{bobbin \ diameter} \qquad (1)$$

Also $D + d = $ constant (approximately)   (2)

At the end of each layer, the reversal of the lifter rail is accompanied by the movement of the cone belt, and the amount of movement given is uniform throughout the building of the bobbin, although it can be altered to suit different counts of material.

It is generally assumed that each layer put on the bobbin increases its diameter by an equal amount, and that the cone diameters, after each movement of the belt, must correspond to this increased bobbin diameter. This is not strictly correct; the reasons for the difference and the effect on the shape of the cone drums will be considered later.

The following example is worked out in detail to show the application of the above to the speed calculations and the design of the cone drums.

$$\text{Speed of spindles} = 350 \, x \, \frac{40}{42} \, x \, \frac{55}{50} = 611 \text{ revolutions per minute.}$$

$$\text{Front roller delivery} = 350 \, x \, \frac{48}{48} \, x \, \frac{48}{130} \, x \, 1'' \, x \, \pi = 406 \quad in. \text{ per minute.}$$

Differential equation: Bobbin wheel speed $n = \dfrac{4}{5} a + \dfrac{1}{5} = m$

When $m = 0$, $n = \dfrac{4}{5} a$

$$\text{And bobbin speed} = \frac{4}{5} \, x \, 350 \, x \, \frac{50}{42} \, x \, \frac{55}{30} = 611 r.p.m. = \text{spindle speed.}$$

$$\text{Bobbin winding speed} = \frac{1}{5} \, m \, x \, \frac{50}{42} \, x \, \frac{55}{30} = 0.437 \, m.$$

$$\text{Also m} = m = 350 \, x \, \frac{48}{48} \, x \, \frac{D}{d} \, x \, \frac{18}{68} \, x \, \frac{40}{34} = 109 \, x \, r$$

where $r = \dfrac{D}{d}$

And bobbin winding speed $= 0.437 \times 109 \times r = 47.6 \, r$

Cone drums – When the bobbin is $1\dfrac{5}{8} \, in$. diameter,

$$\frac{D}{d} = r = \frac{6}{3.5} = 1.714$$

At any other bobbin diameter x,

$$r = 1.714 \, x \, \frac{1.625''}{x} = \frac{2.783}{x}$$

Also $D = d = 6 + 3.5 = 9.5$"
$Rd + d = 9.5$

Whence $d(1 + r) = 9.5$ and $d = \dfrac{9.5}{1 + r} = \dfrac{9.5}{1 + \dfrac{2.783}{x}}$

Then $\hspace{3cm} D = 9.5 - d.$

The surface speed of the bobbin – 416 inch per minute, is calculated on the assumption that the diameter at the centre of the coils laid on the bare bobbin is $1\frac{5}{8}$". Nor has any allowance been made for the fact that the coils are wound on spirally, which slightly increases the length of a coil. When these two factors have been taken into account, and also the contraction which takes place in the material when it is twisted, after leaving the rollers, it will be evident that the actual rate of winding at the bobbin is somewhat higher than the roller delivery – 406 inch per minute – keeping the material taut during winding. Assuming that the bobbin diameter increases equally for each layer, and that the total belt movement between bobbin diameters $1\frac{5}{8}$ and 5" is 27".

11.3.2   Ring frame winding

The ring bobbin is pressed firmly on the spindle, and bobbin and spindle rotate at the same speed, the spindle being driven by band from a tin roller. The yarn, after leaving the drawing rollers, passes through a fixed wire guide eye E (Figure 11.2). Then it passes to the traveller T, a small 'c' shaped piece of steel sprung into position on the flange of the ring R, on which the traveller is free to move. The yarn is passed through the traveller and thence on to the bobbin, which draws the yarn on to itself, through the traveller.

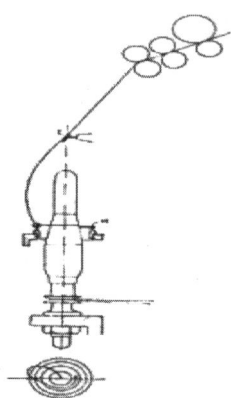

Figure 11.2   Yarn passage at ring frame (Source: Hanton)

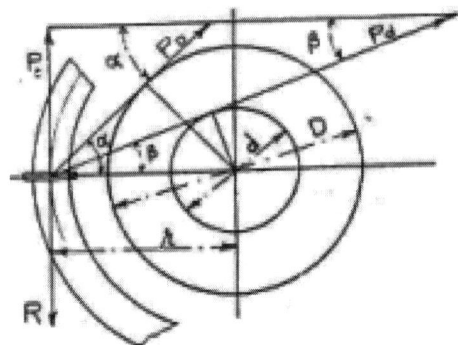

Figure 11.3   Forces acting at ring and traveller (Source: Hanton)

The spindle rotates at a high speed about 10,000 revolutions per minute in order to get the necessary twist into the yarn, every revolution of the traveller about the centre of the ring inserting one turn into the yarn. The surface speed of the bobbin is far higher than the rate of delivery from the rollers and the yarn drags the traveller round the ring at a speed nearly as high as the spindle speed, the traveller aromatically lagging behind a sufficient number of revolutions to wind on the roller delivery. There is no positive regulation of the winding speed, as in the flyer frame. The tension in the yarn during winding depends on the resistance offered by the traveller to movement round the ring, the tension automatically increasing until it exerts sufficient force on the traveller to keep it moving. By varying the weight of the traveller its resistance to movement is changed, and so the yarn tension can be regulated to suit different counts (Figure 11.3).

The yarn is wound on in cop form, the ring rail rising and falling continuously through the depth of the chase, the starting and finishing points for each layer being made slightly higher than for the previous layer. The yarn between traveller and guide eye bellies out into a shape known as the "balloon" caused by centrifugal force acting on the mass of yarn revolving at high speed.

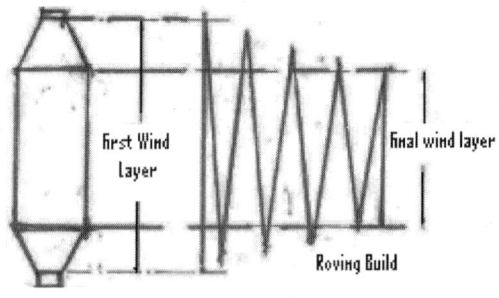

Roving build

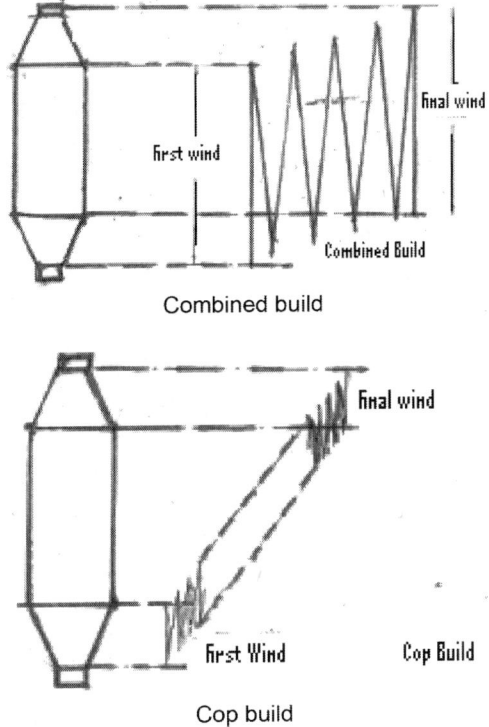

Combined build

Cop build

Figure 11.4   Different types of builds at ring frame and the stages of formation

Figure 11.4 show the difference in winding of layers for roving, combined and cop build in ring frame and accordingly the cams used also differ. However the majority of the cases in industry include the production of cop build only.

Speed of the traveller – The traveller speed is the difference between the spindle speed and the winding speed needed to wind on the roller delivery; since spindle speed is constant and the winding speed varies as the winding on diameter changes, it follows that the traveller speed varies. For example, roller delivery, 394 inch per minute; spindle speed, 10,000 rpm; diameter of full bobbin, $D = 1\frac{1}{4}''$; diameter of bare bobbin $d = 5/8''$. Then winding rpm, or lag of traveller when winding on D, will be

$$\frac{394}{\pi \times 1.25''} = 100 \text{ r.p.m. and when winding on d, speed } \frac{394}{\pi \times 1.25''} = 100 \text{ r.p.m}$$

Thus traveller speed at D, will be $10,000 - 100 = 9900$ rpm, and at d traveller speed will be $10,000 - 200 = 9800$ rpm. The variation therefore is only about 1%, and may generally be ignored.

Variations in yarn tension – Much more important than the variation in traveller speed are the variations in yarn tension which take place when the wining on diameter changes in frames with constant spindle speed. These variations in yarn tension have been very fully investigated a research study conducted at London. The resistance of the traveller to movement round the ring is mainly frictional resistance due to the pressure of the traveller on the ring; there is also a slight backward pull of the yarn due to resistance on the ballooning yarn, but the frictional resistance is by far the greater.

The pressure between the traveller and ring is due to several forces acting on the traveller, which will be considered more fully later. But by far the greatest force acting on the traveller is centrifugal force, which depends on the weight of the traveller, its speed and the radius of its path. Now the traveller speed is nearly constant, and therefore its centrifugal force is nearly constant, and the resistance of the traveller, which depends mainly on this centrifugal force, can roughly be taken as constant. This resistance R (Figure 11.3) acts tangentially to the path of the traveller. It has to be overcome by the yarn pull between the bobbin and the traveller, acting in the directions P_D and P_d when winding on D and d, respectively. Only the tangential component P_t of these yarn tensions is effective for moving the traveller, and since Pt must be constant if R, which P_t overcomes, is constant, it can be seen that the tension P_d when winding on the small diameter, must be much greater than P_D when winding on the largest diameter.

$$Now \quad P_t = p_D \sin \alpha = P_d \sin \beta$$

$$\therefore \frac{P_D}{P_d} = \frac{\sin \beta}{\sin \alpha}$$

$$Also \quad \frac{D}{2} = r \sin \alpha \quad and \quad \frac{d}{2} = r \sin \alpha, \quad i.e. \frac{d}{D} = \frac{\sin \beta}{\sin \alpha} = \frac{P_D}{P_d}$$

That is, with constant speed spindles, if the resistance to movement of the traveller is constant, the winding on tension varies inversely as the winding on diameter. This statement is only approximately correct, because the resistance to the movement of the traveller is not quite constant, as will be evident when all the factors affecting it are considered later, but the variation in tension is always great with constant spindle speed, and the production is low for the speed has to be kept down so that the tensions when winding on the bare bobbin are not so great as to cause excessive breakages.

The actual variations in the tensions from the rollers to the bobbin have been studied in industry, by the following method. Figure 11.5 shows in front, side and plan view the various yarn tensions and the forces acting on the traveller. The yarn has a certain tension, x, when leaving the rollers and the same tension x acts upwards at the guide eye. The tension U at the other side of the eye is greater than x by the amount of the friction at the eye. At the other end of the balloon the yarn is pulling up at the traveller with tension Q. The tensions U and Q do not act in the radial plane joining the traveller with spindle centre, but are deflected backwards, as seen best in the side view, by air resistance acting on the ballooning yarn. The outward bulge of the ballooning yarn as seen in the front view is caused by centrifugal force C_B. This force acts as shown at the meeting point of the lines of action of the radial components U_r and Q_r of U and Q since the three forces C_B, U_r and Q_r keep the balloon in equilibrium. The yarn tension P between the traveller and the bobbin is greater than the tension Q at the other side of the traveller by the friction between yarn and traveller. P is shown in plan view resolved into its tangential and radial components P_t and P_r P_t pulls the traveller round the ring, and must be equal to the sum of Q_t (the backward pull of the yarn due to air resistance) and F, the frictional resistance between the traveller and the ring. The radial component, P_r tends to pull the traveller in towards the centre of the ring. The forces which act on the traveller and cause it to press on the ring are the forces acting in the radial plane. The tangential forces P_t, Q_t, and F do not affect the pressure between traveller and ring. The radial forces acting on the traveller are the centrifugal force C_r of the traveller acting radially outwards; the weight W of the traveller acting downwards; the radial component P_r acting radially inwards; and the radial component Q_r acting upwards.

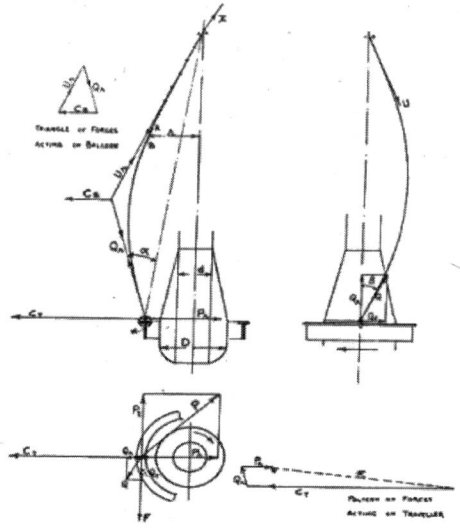

Figure 11.5 Detailed analysis of forces acting on traveller (Source: Hanton)

The resultant R of these forces gives the force pressing the traveller against the ring and causing the friction F. The coefficient of friction between the traveller and the ring is F/R. The starting point for the estimation of the various tensions is the balloon. The exact shape and size of this can be obtain by photographing it, and knowing the counts being spun and speed, the centrifugal force of the ballooning yarn can be found. Subdividing the length of ballooning yarn into equal short lengths such as AB, the radius r of each short length can be measured from the photograph. If w is the weight of AB, its centrifugal force is 0.00034 wrN^2, where N is the traveller speed in revolutions per minute. Treating each short length of yarn in this way the total centrifugal force CB is obtained by summing the centrifugal forces of the various short lengths. U_r and Q_r act tangentially to the line of the yarn at the guide eye and traveller, respectively, and CB passes through the point where U_r and Q_r meet. The values of U_r and Q_r can be found by drawing the triangle of forces acting on the balloon. Experiment with large-size balloons have shown that the backward bulge due to air resistance can be taken to be the same as the bulge caused by centrifugal force, the angle β being equal to the angle α. This enables the actual yarn tensions Q and U to be found and the tangential component Q_t.

Effect of the position of the ring rail – Tensions very not only with the winding on diameter, but according to the position of the ring rail.

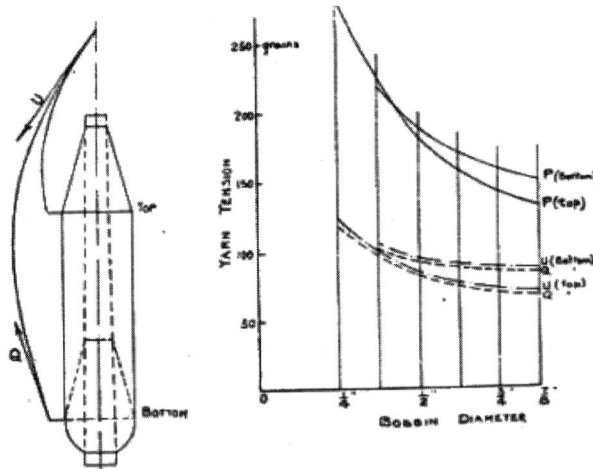

Figure 11.6   Effect of position of ring rail (Source: Hanton)

A study conducted by Mrs. Brown, Boverie and Co., London to find the effect of the ring rail position resulted as follows:

Counts spun: 40s, Diameter of ring 1 1/8"; Traveller No. 7/0; spindle speed 10,000 rpm.

Bobbin diameter	U		Q		P		P / Q	
	Bottom	Top	Bottom	Top	Bottom	Top	Bottom	Top
In.								
7/8	85	69	83	67	148	130	1.78	1.94
3/4	86	71	84	69	155	139	1.84	2.0
5/8	89	76	87	73	168	157	1.93	2.10
1/2	93	82	91	79	186	183	2.04	2.2
3/8	103	98	100	95	215	223	2.15	2.35
1/4	---	122	---	118	---	280	---	2.45

The tabulated results and the curves (Figure 11.6) show that the variation in the tensions is much greater for the top position than for the bottom position, but this is due, not to the position of the rail but to the smaller bare bobbin diameter at the top position; when winding on equal diameters the tensions are generally less at the top than at the bottom position.

Effect of changing the speed of the spindle – If the spindle speed is increased, all the tensions are increased but no change takes place in the shape or size of the balloon. The centrifugal forces of the balloon and traveller and the air resistance on the balloon all vary as the square of the speed, and therefore the yarn tensions, which are in equilibrium with these forces, will vary in the same way. For example, in the frame previously considered, spinning 20s at 8160 rpm spindle speed, if the spindle speed were increased to 9000 rpm all tensions would be increase in the proportion of $\frac{9000^2}{8160^2} = \frac{810}{666}$, an increase of 22%. Thus the winding on tension P at the full diameter was 389 against at 8160 rpm. At a speed of 9000 rpm, the value of P would be $389 \, x \frac{810}{660} = 477$ grains; and similarly for the other tensions. As the speed increases the centrifugal force of the balloon increases and tends to make a larger balloon. But the resistance to movement of the traveller increases the same proportion, so that the yarn tension increases and prevents the balloon from changing.

Effect of changing the weight of the traveller – If the weight of traveller is changed, when spinning a given count, the yarn tensions change and the size and shape of the balloon change. Thus a lighter traveller has less centrifugal force and consequently less frictional resistance, so that a small winding on tension P is needed to pull the traveller round, the tension Q (Figure 11.7) decreases. This allows the centrifugal force of the balloon CB to overcome U and Q and the balloon increases slightly in size until balance again is reached. The increased size of the balloon results in a slight increase in CB, and finally the new balloon comes to a state of equilibrium as shown dotted. Although CB1 is greater than CB the yarn tensions Q_1 and U_1 are less than Q and U, because of the change in inclination of the lines of action of the forces. Similarly a heavier traveller results in a smaller balloon and increased yarn tensions. The exact values of the tensions cannot be worked out readily, since to do so it would be necessary to work out the change in shape and size of the balloon.

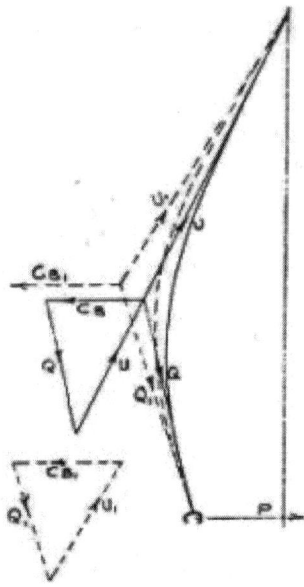

Figure 11.7   Effect of changing the weight of traveller (Source: Hanton)

Effect of the diameter of the ring – If the same counts are spun on rings of different diameters with bobbin sizes in proportion, and if the same weight of traveller is used, then the centrifugal force of the traveller and the yarn tension varies approximately as the diameter of the ring. Thus, if a certain traveller is correct for 1½″, ring, the tension will be too great for a 1¾″ ring, resulting in a smaller balloon than is desirable, and probably too many break-ages. To give approximately the same tension with the 1¾″ ring, as with the 1½″ ring, the traveller weight for the larger ring would require to be reduced in the proportion 1½:1¾.

For example, when spinning 16s on a 1½″ ring, the traveller used is gen-erally about No. 5, weighting 1.4 grain. For a 1¾″ ring the traveller weight should be about $1.4 \times \dfrac{1.5}{1.75} = 1.2$, which is the weight of a No. 3 traveller. This is the traveller used in practice with the 1¾″ ring when a No. 5 is used with the 1½″ ring.

Relationship between traveller weight and counts spun – The weight of the traveller has to be changed when the counts spun are changed. The only safe guide for the change of weight necessary is practical experience, but ring frame makers give the traveller suitable for different counts under normal conditions. The traveller are so chosen that the size and shape of the balloon is nearly the same for all counts. Assuming this to be the case, the centrifugal force of the balloon C_B and the yarn tension Q_r, which depend on the weight of the yarn, will vary inversely as the counts spun. Air resistance, Q_r, depends on the area of the surface of yarn exposed to the resistance, and this area is proportional to the yarn diameter, which varies inversely as $\sqrt{count}$. Taking the yarn tensions for 20s yarn, the traveller weight can be worked out. This has been done, and the results summarised in the table below.

The ring diameter is assumed constant at 1 5/8″ and the spindle speed at 8160 rpm. The spindle speed would actually be lower at the fine and coarse counts, but whilst that would change the yarn tensions it should not affect the weight of the traveller. Referring to the table, the headings of the various columns show how the results are calculated from the know values for 20s count. The values of the centrifugal force of the traveller C_r are obtained from the polygon of forces acting on the traveller. The traveller weight must be in proportion to its centrifugal force. For comparison the actual travellers used are given in the last two columns, and it will be seen that whilst the calculated weights agree fairly well with the weights used in practice between counts 10s and 40s there is a great discrepancy at the lowest and highest counts, the calculated traveller being much too heavy. To a small extent this difference may be due to the lower spindle and traveller speeds used for very fine and coarse counts, this resulting probably in a higher coefficient of friction between traveller and ring. But this does not nearly account for all the difference.

Variable speed spinning – The yarn is subjected to the maximum tension only during a short period with constant speed of spindles, and during the greater part of the spinning process the speed could be increased without increasing the risk of breakage, but with a greatly increased production. It will be clear from the foregoing treatment that the tensions in the yarn depend on the speed, and by suitably varying the speed the tensions can be altered as desired and can be kept constant. The speed may be regulated so as to keep the tension P between traveller and bobbin constant, which improves the winding, or the tension U can be kept constant, with improved spinning. In either case the production is increased, since the speed is increased when winding the larger diameters, to keep the tension nearly up to the maximum. For example, it was found, on a frame, spinning 40s, with traveller 8/o and ring 1 1/8″ diameter, that the tension U could be kept constant by varying the spindle speed form

10,500 rpm to 11,600 at the beginning of the bobbin and from 9700 to 13,100 at the end of the bobbin. Tension U is constant, and about 6% less than the maximum with constant speed of 10,000 rpm; tension P between traveller and bobbin varies about 24%, instead of 23%, whilst the maximum value of P is 7%, less than the maximum with constant speed. Production is increased about 15%, or 13%. If the base is wound at reduced speed. The difficulty with variable speed spinning lies in the mechanism needed to provide the speed variation. So far the most satisfactory method is to use a variable speed motor for driving the frame, the speed being varied automatically according to the rise and fall of the ring rail. The motors are comparatively high in cost and not as efficient as constant speed motors.

11.4   Winding mechanics at modern ring frame

11.4.1   Formation of cop

The cop (Figure 11.8) consists of three visually distinct parts – the barrel-like base **A**, the cylindrical middle part **W**, and the conically convergent tip **K**. It is built-up from bottom to top from many conical layers (Figure 11.8), but constant iconicity is achieved only after the formation of the base. In the base portion itself, winding begins with an almost cylindrical layer on the similarly almost cylindrical tube. With the deposition of one layer on another, the iconicity gradually increases. Each layer consists of a main layer and a cross-layer (Figure 11.9a, b). The main layer is formed during slow raising of the ring rail, the individual wraps being laid close to each other or on each other. The main layers are the effective cop-filling layers. The cross-layers are made up of widely separated, steeply downward inclined wraps of yarn and are formed during rapid lowering of the ring rail. They form the separating layers between the main layers and prevent the pulling down of several layers simultaneously when yarn is drawn-off at high speed in winding machines. In the absence of such separating layers, individual yarn layers would inevitably be pressed into each other, and layer-wise draw-off of yarn would be impossible. Raising and lowering of the ring rail are caused by the heart shaped cam and are transmitted by chains, belts, rollers, etc., to the ring rail (Figure 11.10). The long, flatter part of the cam surface forces the ring rail upwards, slowly but with increasing speed. The short, steep portion causes downward movement that is rapid but occurs with decreasing speed.

Cts.	$C_B\alpha\dfrac{t}{ct.}$	$Q_r\alpha\dfrac{r}{ct.}$	$Q_t\alpha\dfrac{r}{\sqrt{ct}}$	$Q=\sqrt{Q_r^2+Q_r^2}$	$P_t\alpha Q$	$P_r\alpha Q$	$F=P_t-Q_t$	$R=\dfrac{F}{\mu}$	C_T	Traveller weight a C_T	Actual traveller used Wt.	No.
4	560	975	150	987	1270	1330	1120	9350	10,400	6.0	3.3	13s
10	224	390	95	400	517	538	422	3520	3950	2.3	2.1	7s
20	112	195	67	208	269	280	202	1685	1900	1.10	1.10	2s
30	74	130	55	141	182	190	127	1060	1210	0.7	0.6	3/0
40	56	97	48	109	141	147	93	775	890	0.52	0.375	9/0
60	37	65	39	76	98	102	60	500	580	0.33	0.125	19/0

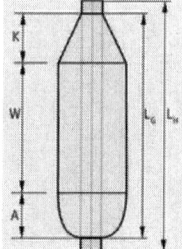

Figure 11.8   The cop as a yarn package

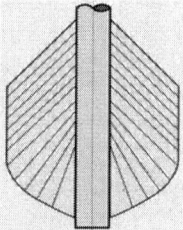

Figure 11.9 (a)   Building up the cop in layers

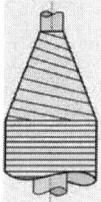

Figure 11.9(b)   Main layers and cross layers

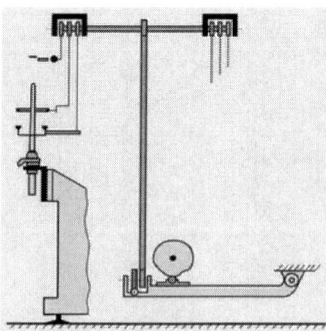

Figure 11.10   The winding mechanism

11.4.2   Formation of the base

The creation of the typical cop form is explained as follows by Johannsen and Walz. The heart-shaped cam and the delivery cylinder are coupled together by the drive gearing. Thus the quantity delivered for each revolution of the cam, and hence per yarn double layer, is always the same. The volumes of the individual double layers are therefore also equal. Deposition of double layers on the tube begins with a small average layer diameter, d_1 (Figure 11.11). The average diameter increases gradually with each newly deposited layer. With constant layer volume, this can have only one result, namely a continual reduction of the layer width from b_1 to b_2 to b_3, and so on. Since the ring rail is also raised by a constant amount h after each deposited layer, it follows that a curve, rather than a straight line, arises automatically in the base portion, at the bottom.

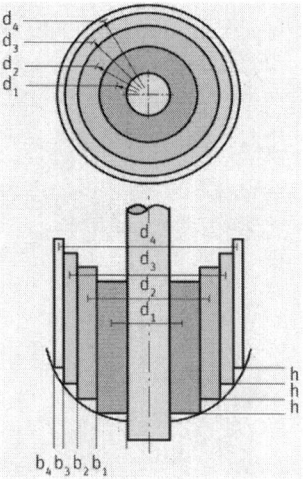

Figure 11.11   Formation of base

11.4.3   The formation of the conical layers

We know that the ring rail is not moved uniformly. Its speed increases during upward movement and falls during downward movement. At the tip of each layer the speed is higher than at the base of the layer, i.e., the ring rail does not dwell as long at the tip as it does at the base – less material is wound, and the layer is thinner at the tip. If it is assumed by way of example that the ring rail is moving twice as fast at the top of its stroke as at the bottom of the stroke, the first layer would be half as thick at the top as at the bottom, i.e., b_1 ½ instead of b_1, (Figure 11.12). The first layer would correspond to a trapezium with the side b1 at the bottom and the side $\mathbf{b}_1$ ½ at the top. This is followed by the deposition of the second layer. Owing to the constant, short-term lifting of the ring rail, the upper portion of the new layer would again be deposited on the bare tube. The average diameter at the top would be the same as that of the first layer, and the volume, and hence the thickness, would also be the same, that is $\mathbf{b}_1$ 1/2. Each newly deposited layer will have this thickness of $\mathbf{b}_1$ 1/2 at the top. At the bottom, however, the winding diameter is increasing continually so that the layer thickness is declining from $\mathbf{b}_1$ to $\mathbf{b}_2$ to $\mathbf{b}_3$ to $\mathbf{b}_4$. Accordingly, continually narrowing trapezia are produced. At some stage, the trapezium will become a parallelogram, i.e., the lower side will be the same size as the upper side – both will be $\mathbf{b}_1$ 1/2. Since all other winding conditions now remain the same, no further variation can now arise in the layering. One conical layer will be laid upon the other until the cop is full, i.e., when the cylindrical portion of the cop is formed.

The gearing change wheel has little influence on this sequence of events. If too many teeth are inserted, the final condition of constant conical layers will be reached too soon, and the cop will be too thin. It will be too thick if the ring rail is lifted too slowly.

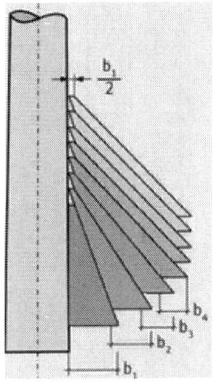

Figure 11.12   The formation of the conical layers

11.5   The winding principle

As in the case of the roving frame, two assemblies with different speeds must be used in order to enable winding to occur. One assembly is the spindle, the other is the traveller representing the remnant of the flyer. Furthermore, the speed difference must be equal over time to the delivery length at the front cylinder. In the roving frame, each assembly has its own regulated drive.

In the ring spinning frame, this is true only for the spindle. The traveller is dragged by the spindle acting through the yarn. The speed of the traveller required to give a pre-determined speed difference arises through more or less strong braking of the traveller on the running surface of the ring. Influence can be exerted on this process by way of the mass of the traveller. For winding with a leading spindle following relationships apply. The delivery is given by:

$$L = V_{spi} - V_T$$

Where V_T is the traveller speed. Thus we have:

$$L = d \times \pi \times n_{spi} - d \times \pi \times n_T$$

and

$$L - d \times \pi (n_{spi} - n_T)$$

The required traveller speed is then:

$$n_T = n_{spi} - \frac{L}{d \times \pi}$$

As in the case of the roving frame, the diameter d is the diameter at the winding point.

11.5.1   Variation in the speed of the traveller

In contrast to the roving frame, the winding diameter in the ring spinning frame changes continually with raising and lowering of the ring rail, since the winding layers are formed conically (Figure 11.13). The traveller must have different speeds at the base and the tip. Assuming, for example, a spindle speed of 13,500 rpm, the layer diameters given (as in Figure 11.13), and a delivery speed of 15 m/min, the traveller speed at the base will be:

$$n_{TB} = 13500 - \frac{15000}{46\pi}$$

$$= 13500 - 104 = 13396 min^{-1}$$

and at the tip it will be

$$n_{TS} = 13500 - \frac{15000}{25\pi}$$

$$= 13500 - 191 = 13309 min^{-1}$$

In comparison with the constant speed of the spindle, the traveller has a changing speed difference of 0.77 to 1.41%.

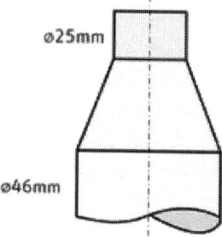

ø25mm

ø46mm

Figure 11.13 Variation in the speed of traveller

11.5.2   Force and tension relationships during winding by using travellers

In the following explanations, certain inaccuracies have been deliber-
ately accepted; for example, representation exclusively in two dimensions when the actual process is three-dimensional. The intention is not to
present either exact scientific theory or a detailed basis for calculations.
Rather, the aim here is to provide the textile specialist involved in every-
day practice with an understanding of the interrelations and in particular
to bring out the interplay of forces. For this purpose, simplified models
have been used; there is much literature available on scientifically exact
usage. The whole treatment is based on the parallelogram of forces, the
normal "school" presentation of which is repeated here briefly for com-
pleteness (see Figure 11.14). If a carriage is to be moved forward on rails,
it can be pulled directly in the direction of the rails (as F_T). In this case
the whole of the force contributes to the forward movement. This is no
longer true if the force is directed with a sideways inclination (pulling in
direction F_F). Now only a part of the total force exerted (F_F) will contribute to the forward movement (F_T). Part of the force F_F (i.e., the force F_R)
will press the carriage against the rails at an angle of 90° to the direction
of movement. This component is lost as far as forward motion is concerned. The pulling force F_F can therefore be resolved into two components, the tangential force F_T, which draws the carriage forward, and the
radial force F_R. Accordingly, if the carriage is to be moved forward with
the required force F_T and the pulling force is effective at an angle α, then
the pulling force must have the magnitude F_F (friction forces being neglected here). These forces can be represented graphically and measured
or calculated in accordance with the formula:

$$F_F = \frac{F_T}{\sin \alpha}$$

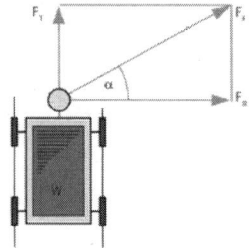

Figure 11.14 Vector analysis

11.5.3   Conditions at the traveller in the plane of the ring

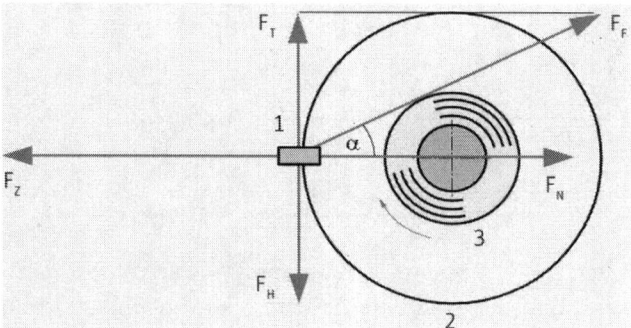

Figure 11.15 Conditions at the traveller in the plane of the ring

The following forces act on the traveller (1) in the plane of the ring (2):

A tensile force F_F, which arises from the winding tension of the yarn and always acts at a tangent to the circumference of the cop (3). A frictional force F_H between the ring and the traveller. In the stationary state, i.e., with constant traveller speed, this braking force F_H is in equilibrium with the forward component F_T of the yarn tension FF. Hence we have (Figure 11.15):

$F_H = F_T$ **or** $F_H \times sin\alpha$

A force F_N normal to the surface of the ring (pulling the traveller in the direction of the cop, diminishing the friction of the traveller at the ring created by the centrifugal force F_Z). The frictional force F_H arises from this normal force in accordance with the relation:

$F_H = \mu \times F_N$, where μ is centrifugal force F_Z, which is the largest force acting on the traveller. This force can be calculated in accordance with the relations:

$$F_Z = m_L \times \omega^2_L \times d_R / 2$$

$$\omega_L = n_{spindle} \times \pi / 30$$

where m_L is the mass of the traveller, ω_L is the angular velocity of the traveller, and d_R is the diameter of the ring.

Professor Krause (ETH, Zurich) identifies the following relationships between these forces, solved for the tensile force:

$$F_F = \frac{\mu \times F_Z}{sin\,\alpha + \mu \times cos\,\alpha}$$

$$F_F = \frac{\mu \times m_L \times \omega^2_L \times d_R}{2 \times (sin\,\alpha + \mu \times cos\,\alpha)}$$

For a rough estimate, the term $\mu \times cos\,\alpha$ can be ignored. Approximately, therefore, we have:

$$F_F = \frac{\mu \times m_L \times \omega^2_L \times d_R}{2 \times sin\,\alpha}$$

11.6   Changes in the force conditions

Continuous variation of the operating conditions arises during winding of a cop. This variation is especially large with regard to changes in the winding diameter, i.e., when wraps have to be formed on the bare tube (small diameter), and then on the full cop circumferences (large diameter). This occurs not only at the start of cop winding (formation of the base); such changes arise at very short intervals in each ring rail stroke as demonstrated by the example illustrated in Figure 11.16. It has already been mentioned that tensile force F_F must be assumed tangential to the cop circumference because it arises from the winding point. Frictional force F_H undergoes only small variations; it can be assumed to be the same in both cases. The components F_T of the yarn tension are then also equal. However, owing to the difference in the angle the tensile forces F_F are different. The same dependence of the tensile force F_F on the angle a can be seen from the formulas given above. The result is that the tensile force exerted on the yarn is much higher during winding on the bare tube than during winding on the full cop diameter because of the difference in the angle of attack of the yarn on the traveller. When the ring rail is at the upper end of its stroke, in spinning onto the tube, yarn tension is substantially higher than when the ring rail is at its lowest position. This can be observed easily in the balloon on any ring spinning machine. If the yarn tension is measured over time, then the picture in Figure 11.17 is obtained. The tube and ring diameters must have a minimum ratio, between approximately 1:2 and 1:2.2, in order to ensure that the yarn tension oscillations do not become too great.

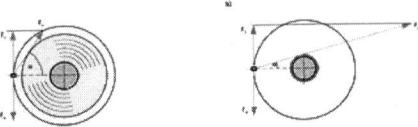

Figure 11.16   The tensile force (F_F) on the yarn; (a) with a large cop diameter; (b) with a small cop diameter (bare tube)

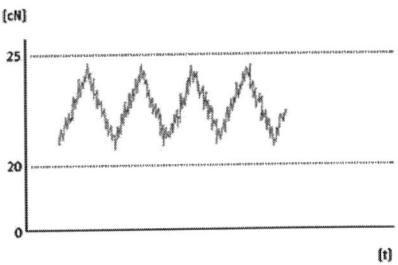

Figure 11.17   Continual changes in yarn tension due to winding on larger and smaller diameters
http://www.rieter.com/en/rikipedia/articles/technology-ofshort-staple-spinning/
handling-material/winding-on-flyer-bobbins/the-winding-principle/

11.7   Arrangements at modern ring frame

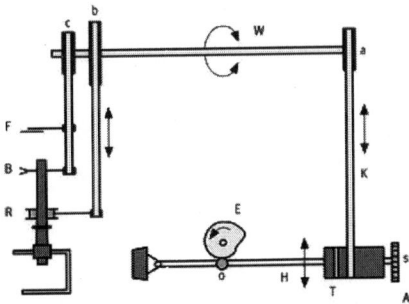

Figure 11.18   Cop building at modern ring frame

Ring rail (**R**) hangs with its entire weight via belts on disc (**b**) mounted on shaft (**W**). At the other end of the shaft is a further disc (**a**), which presses the entire lever (**H**) with roller (**o**) against heart cam (**E**) via chain (**K**) and chain drum (**T**) as a result of the traction of the ring rail. The lever is continuously raised and lowered with the chain drum due to the rotation of the cam. This movement is transmitted to the ring rail via discs (**a** + **b**), the chain and the belt, thus producing the layering traverse. Each time the lever moves down, it presses ratchet wheel (**s**) against a catch, which results in a small turn of drum (**T**) connected to the ratchet wheel. Chain (**K**) is thus wound a small amount onto the drum. This results in a turn of disc (**a**), shaft (**W**) and disc (**b**), and finally a slight raising of ring rail (**R**) (switching traverse). However, disc (**c**) is also mounted on shaft (**W**) with balloon checking rings (**B**) and thread guide eyelets (**F**) suspended on it on belts. These are also raised and lowered accordingly. However, since disc (**c**) is rather smaller than (**b**), the traverse motion is also smaller (Figure 11.18).

11.8   Power required for the loom

Although the cost of power for weaving cloth is only a very small proportion of the total cost, the power ought to be kept as low as possible, especially since excessive vibration and noise generally accompany a high-power consumption. Also a study of the power required, and particularly of the cyclical variation in power, throws light on the working of the loom parts.

The results of a study conducted at a research laboratory gave the following information (Table 11.2).

Table 11.2   Power required for the loom

Part of loom	Power required
	Percentage of Total
Crank and bottom shafts	20.4
Sley	24.5
Shedding	8.2
Vibrating back rest	0.7
Driving end picking motion	7.8
Gearing end motion	7.8
Uptake motion	0.0
Shuttle	30.6
Total	100.0

The study included a loom of reed space of the loom was 26″, the speed 192 picks/min, the cloth woven plain calico with 52 ends per inch of 40s warp and 48 picks per inch of 30s weft.

The time for the shuttle passage was observed to be 85° of the crankshaft revolution, the shuttle weight was 280 g and its length was 36 cm.

Crank and bottom shafts – The power spent in driving these is all required for overcoming frictional resistances at the bearings, and some friction at the gearing connecting the two shafts. The load on the crankshaft bearings is due to the weight of the shaft and pulleys, but also, and to a greater degree, to the belt pull. It is found that the work spent per minute will be about half the observed total for the crank and bottom shafts, and is probably too small (due to μ being taken too small) since the bottom shaft will not take quite so much as the crankshaft. The weight of the bottom shaft is more than that of the crankshaft, due to the picking and shedding tappets, but there is no belt pull; also the speed is only half as great.

Power for sley – The work spent on the sley is entirely spent in overcoming friction at the bearings – rocking shaft, sword pins, and crank pins; with frictionless bearings the work spent in driving the sley would be zero.

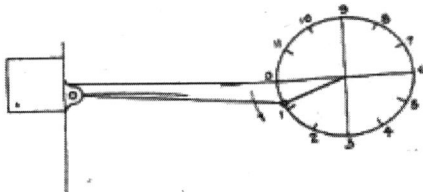

Figure 11.19   Loom sley drive

At the beating up position, 0, Figure 11.19 the sley is at rest; as the crank pin moves round to nearly No. 3 position, where the sley speed is a maximum, the crank shaft does work on the sley, through the medium of the connecting arms, the work done being stored up as kinetic energy in the sley. From about positions 3 to 6 the sley is retarded and brought to rest, the connecting arms applying a driving effort to the crankshaft and returning to it all the energy stored in the sley at position 3. From 3 to 6 the sley helps to drive the other parts of the loom and reduces the pull on the driving belt. The same cycle is repeated during the other revolution, the crankshaft doing work in accelerating the sley from 6 to 9 and the sley returning this to the crankshaft from 9 to 0. The effect of gravity on the sley has been ignored, hut this is small as compared with inertia; moreover, the net result of the work done against gravity is also zero, work being done from 6 to 0 in lifting the sley and returned, from 0 to 6, while the sley is falling. Similarly the net effect of lifting, lowering, and accelerating the connecting arms is zero.

Most of the work required for friction is spent at the crank pins. The load on the rocking shaft is considerable and the coefficient of friction often high, but the movement is small; similarly at the sword pin bearings, the load, which is mainly due to the accelerating forces needed for the sley, is practically the same as the load on the crank pins, hut the displacement of the rubbing surfaces is much less.

Power for shedding – As for the sley, so with the shedding, when friction is ignored the net work done is zero. Work is done in forming the shed, but not only does the shed close of its own accord but during the closing the treadle bowl applies to the tappet a force which helps to turn the bottom shaft. The effects of gravity and acceleration on the treadle levers, heald shafts, etc., are similar, and again the only net work done is in overcoming friction, which is mainly found at the tappet bowls, the treadle lever fulcra, and the top roller mountings.

Power for picking – The work done on the shuttle is the work done in pushing the shuttle from the box against swell friction, together with the work done in accelerating it and work done per minute on shuttle is considerably less than the observed power and the difference is probably mainly due to the extra friction at the bearings of the picking motions, bottom shaft, and crankshaft, due to the greater load on these bearings when the shuttle is working. The work done on the picking motion per pick is the work done in overcoming friction at the bearings together with the work done in accelerating the motion. The work done in accelerating the picker is the kinetic energy in the picker at the maximum shuttle speed.

It is found that none of the work spent in accelerating the shuttle, picker, and picking motion is returned to the crankshaft. Some of the energy in the shuttle as it leaves the picker is spent in carrying the shuttle across the warp against warp friction, but most of the energy still remains in the shuttle as it enters the opposite shuttle box, where it is used to move the swell and to lift the protector blade in a fast reel loom. Finally, it moves the check strap, and any energy then remaining is wasted on impact with the picker. The energy in the picker, after the shuttle leaves it, is wasted on impact with the buffer; much of the energy in the picking motion is also wasted at the buffer, the remainder stretching the returning spring. The spring gives back the energy to the picking motion, which swings back until the cone strikes against the tappet, when any energy in the picking motion is destroyed by the impact. It must be remembered that the cone generally flies clear of the tappet when the maximum shuttle speed is reached.

Work done in taking up cloth – The work spent in drawing the cloth forward is always small, and in the loom tested it was so small that it made no appreciable difference to the power used. If T lb is the cloth tension at the taking up roller, and S ft of cloth are drawn forward per minute, then the work done on the cloth in drawing it forward is $T \times S$ foot-lb per minute. If E is the efficiency of the uptake motion, then work done = $\dfrac{TS}{E}$ foot-lb per minute. If the uptake is driven by ordinary gearing E will be about 75%; if worm gearing is used E can be taken at 25%.

Power for various shedding and picking motions – The power required for shedding naturally depends on the nature of the shedding mechanism, and sometimes much more power is needed than is given in the above table for the plain loom, which had the usual tappet shedding. At the same time the power required is all used, in practically every case, for overcoming frictional resistances. If, for example, heavy parts have to be lifted by the shedding mechanism, as in the Jacquard loom, extra power is needed, but that is clue to the friction caused at the various rubbing surfaces by the heavy loads. Again, a heavily weighted warp increases the work done in shedding, but again this is due to the extra friction caused by the greater loads on the various rubbing surfaces.

The amount of work to be done in picking likewise depends on the type of picking motion. Much depends on whether the pick is developed by sudden impact, as in most under pick motions, or by a gradual acceleration, as in the cone over pick. It makes very little difference to the power required by the shuttle, but the picking motions, apart from the shuttle do require considerably more power when started by sudden impact, due to the loss of energy by the shock of the impact. It makes practically no difference to the power required whether the pick is developed from the crankshaft or the bottom shaft, so long as the mechanism is properly designed to give the necessary shuttle speed.

Power variation with speed – The power required for the crank and bottom shafts being mainly needed to overcome friction due to the weight of the shafts, will not change much with increasing speed. The coefficient of friction for bearings of the kind used generally falls somewhat with increasing speed, which counteracts the greater displacement of the rubbing surfaces at the higher speed. If the coefficient of friction remains constant, power required varies directly as speed.

The power needed for the sley, although entirely spent in over-coming friction at the bearings, will hot follow the same law. The friction at the rocking shaft will remain approximately constant at different speeds, since it is almost entirely due to the weight of the sley; the power required to overcome this friction will vary directly as the speed. But the pressures at the crank pin and sword pin bearings are to a large extent due to the accelerating force required by the sley. This force varies as the square of the speed and the power required for the bearings will vary as the cube of the speed. The total power for the sley will thus vary as the speed raised to some power between 1 and 3.

The power for shedding depends on friction, mainly due to the weight of the parts of the shedding motion and to the warp tension; hence the friction will not change much with change of speed, and power will be approximately proportional to speed.

The work spent in picking is nearly all used in accelerating the shuttle and picking motions, and the work done per pick will therefore vary as the square of the speed, whilst the power spent in picking will vary as the cube of the speed.

It depends on the relative values of these different factors as to how the total power for the loom will vary. It will lie between speed to the power of one and the cube of the speed, and experiment shows that it does not greatly differ, as a rule, from the square of the speed.

Variation in power or driving effort – Most of the motions in the loom act intermittently and in consequence the driving effort needed at the crank-shaft is continually varying.

The work done by a force P acting at radius r on a shaft, when the latter turns through α radians is

work done = $Pd = P\alpha r$, since $\alpha r = d$

Pr is the driving effort applied to the shaft,

$$\therefore \text{driving effort } Pr = \frac{Pd}{a} = \frac{\text{work done}}{\text{angle turned through}}.$$

Effort for friction – It may be assumed, without much error, that the effort for overcoming friction in the loom is constant. This is only true for the friction of some of the parts, the friction of others, such as the crank pin bearings, varying throughout the cycle; but any such variation is small compared with that of the effort needed for moving the sley or the shuttle and no great error is introduced by ignoring it.

Effort for sley – The effort required to accelerate the sley varies continually throughout the revolution of the crankshaft. The accelerating force, which must be applied by the connecting arms to the sley, can be easily found and the diagram of accelerating force are drawn. The accelerating force, F, so obtained, is the force which must be applied at the sword pin, in the direction in which the pin is moving, i.e., at right angles to the sword line AG (Figure 11.20). The force F_1 actually applied by connecting arms at A in the line AB must be greater than F and can be bound from the triangle of forces ANK. Through C draw CD parallel to the sword line AG and produce AB to D. Draw CE at right angles to AD. Then the triangle CED is similar to the triangle ANK.

The moment of F1 about the crankshaft centre C is $F_1 \times CD$ and since

$$\frac{F_1}{F} = \frac{CD}{CE} \text{ or } F_1 \times CE = F \times CD$$

$\therefore$ Moment of F_1 about $C = F \times CD$

The driving effort E, which must be applied to the crankshaft to over-come this resisting moment of F_1 must be equal and opposite to that resisting moment, i.e., $E_s = F \times CD$.

Thus the driving effort required to accelerate the sley at any instant is the product of the accelerating force (as shown by the accelerating force diagram), and the distance from the crankshaft centre to where the arm, or arm produced, meets a line through the centre, parallel to the sword line.

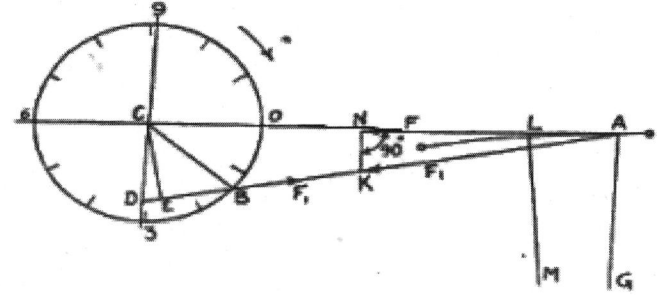

Figure 11.20   Connecting arm and Crank

The driving effort E, which must be applied to the crankshaft to over-come this resisting moment of F_1 must be equal and opposite to that resisting moment, i.e., $E_s = F \times CD$.

Thus the driving effort required to accelerate the sley at any instant is the product of the accelerating force (as shown by the accelerating force diagram), and the distance from the crankshaft centre to where the arm, or arm produced, meets a line through the centre, parallel to the sword line. If all the intercepts are measured on the average position of the line CD, i.e., when CD is parallel to the average sword position LM, the error will only be small.

The driving effort is zero at the beating up position 0, and approximately at the back centre 6, and again when the acceleration is zero, a little before position 3 and after position 9. The effort is positive from 0 to 3 when the sley is storing up energy equal to the work spent in accelerating it, negative from 3 to 6 when the sley returns its energy to the crankshaft, positive from 6 to 9 and negative from 9 to 0.

For better understanding consider a plain loom sley has an equivalent weight, at the sword pin, of 37.18 kg, the crank length r is 2½", the arm length l 10", and the speed 192 rpm:

$$\text{Crank pin speed } v = 2\pi \times \frac{2.5}{12} \times \frac{192}{60} = 4.2 \ ft \ /s \text{ or } 1.29 \text{ m/s.}$$

The accelerating force F at the beating up position

$$= \frac{W}{g}\left[\frac{v^2}{r}\left(1 + \frac{r}{l}\right)\right] = \frac{82}{32.2} \times \frac{4.2^2}{2.5} \times 12 \times 1.25 = 270 \; lb \text{ or } 122.45 \text{ kg}$$

At the back centre

$$F = \frac{W}{g}\left[\frac{v^2}{r}\left(1 - \frac{r}{l}\right)\right] = \frac{82}{32.2} \times \frac{4.2^2}{2.5} \times 12 \times 0.75 = 162 \; lb. \; 73.46 \text{ kg}$$

Another small varying effort is required to overcome the effect of gravity on the cranks and connecting arms, and to accelerate the latter. The weight of the unbalanced mass of each revolving elements should also to be considered for ideal situation.

Effort for picking – The effort for picking is required for accelerating the picking motion and the shuttle and for overcoming swell resistance, the effort for the latter being small compared to the others. The resistance offered by the inertia of the shuttle at any time depends on the shuttle mass and its acceleration, the latter being obtainable from the shuttle velocity curve. The driving effort at the crankshaft, needed for overcoming the shuttle resistance, depends on the velocity ratio between crankshaft and shuttle, which is readily found when the crankshaft speed and the shuttle velocity are known. Similarly, the resistance due to the inertia of the piking motion – the mass of which may be considered as concentrated at the radius of the gyration – may be found, and the driving effort calculated.

The distance moved by the shuttle is equal to the throw of a crank which, if driving the shuttle by an infinitely long connecting rod, would when moving at a speed of 39 ft/s make ¼ revolution in the time for 30° of the loom crankshaft's movement, i.e., $\dfrac{30}{360} \times \dfrac{60}{192} = 0.026$ s. The imaginary crank would move a distance $\dfrac{\pi}{2}r$ in 0.026 s, at a speed 39 ft/s, hence

$$\frac{\pi}{2}\cdot\frac{r}{0.026} = 39 \text{ or } r = \frac{39 \times 2 \times 0.026}{\pi} = 0.645 \; ft \; or \; 7.75 \; in \text{ or } 19.68 \text{ cm.}$$

Or, since average shuttle velocity during acceleration is equal to maximum velocity $\div \dfrac{\pi}{2}$, and distance moved is average velocity × time, hence, distance moved

$$= \frac{\text{maximum velocity}}{\dfrac{\pi}{2}} \times \text{time} = 39 \times \frac{2}{\pi} \times 0.026 = 0.645 \; ft \text{ as above.}$$

The maximum acceleration of the shuttle – as it begins to move is $\dfrac{v^2}{r}$.

And maximum accelerating force = maximum shuttle and picker resistance

$$= \frac{W}{g}\cdot\frac{v^2}{r} = \frac{13}{16 \times 32.2} \times \frac{39^2}{0.645} = 59\ lb.$$

The driving effort, in lb-ft, needed at the crankshaft, is equal numerically to the force in lb needed at a radius of 1 ft.

The linear velocity of the crankshaft, at a radius of 1 ft is

$$2\pi \times \frac{192}{60} = 20\ ft\,/s.$$

The velocity ratio between the driving force and the shuttle is

$$\frac{20\ ft\,/\,s}{\text{Shuttle velocity}} = V.R.$$

And driving effort = driving force = $\dfrac{\text{shuttle resistance}}{V.R.}$.

Effort for shedding – This will evidently depend very much on the nature of the shedding. In tappet looms, where generally the same number of warp threads are lifted for each pick, the effort for shedding will be the same for each pick. The effort will vary since a positive effort is needed during the opening of the shed, whereas the closing of the shed applies an effort to the tapped shaft, through the action of the bowls on the tappets. The effort required is also affected by the acceleration of the moving parts, treadle levers, healds, etc.

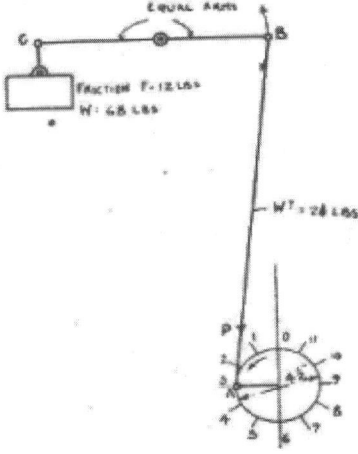

Figure 11.21 Bea Up motion

But for looms weaving medium weight cloths these variations in effort for shedding will be very small compared with the variations in effort for the sley and for picking, and may be ignored. Thus for the plain loom under consideration the total effort E required will vary approximately as shown in Figure 11.21, the value of the effort at any instant being the sum of the efforts for friction, sley, and picking. For dobby and especially for Jacquard shedding the effort required will vary very much form pick to pick, because of the varying number of warp ends lifted.

Index